TECHNICAL WRITING

PROCESS AND PRODUCT

Fifth Edition

Sharon J. Gerson
DeVry University

Steven M. Gerson
Johnson County Community College

PEARSON

Prentice Hall

Upper Saddle River, New Jersey
Columbus, Ohio

Library of Congress Cataloging in Publication Data

Gerson, Sharon J.
   Technical writing : process and product / Sharon J. Gerson, Steven M. Gerson.—5th ed.
     p. cm.
   Includes bibliographical references and index.
   ISBN 0-13-119664-2 (paperback)
    1. English language—Technical English. 2. English language—Rhetoric. 3. Technical
writing. I. Gerson, Steven M., 1948– II. Title.

PE1475.G47 2006
808'.0666—dc22

                         2004058675

Senior Acquisitions Editor: Gary Bauer
Editorial Assistant: Jacqueline Knapke
Development Editor: Monica Ohlinger, Ohlinger Publishing Services
Production Editor: Louise N. Sette
Production Supervision: Kelly Mulligan, Carlisle Publishers Services
Design Coordinator: Diane Ernsberger
Text Designer: Kristina Holmes
Cover Designer: Kristina Holmes
Production Manager: Pat Tonneman
Marketing Manager: Tim Peyton

This book was set in Sabon by Carlisle Communications, Ltd. It was printed and bound by
Courier Kendallville, Inc. The cover was printed by Phoenix Color Corp.

Pearson Education Ltd.   Pearson Education Australia Pty. Limited
Pearson Education Singapore Pte. Ltd.   Pearson Education North Asia Ltd.
Pearson Education Canada, Ltd.   Pearson Educación de Mexico, S.A. de C.V.
Pearson Education—Japan   Pearson Education Malaysia Pte. Ltd.

10 9 8 7 6 5 4 3 2 1
ISBN 0–13–119664–2

For our daughters, Stacy and Stefani

ABOUT THE AUTHORS

Sharon and Steve Gerson are dedicated career professionals who have a combined total of over 65 years teaching experience at the college and university level. They have taught technical writing to thousands of students, attended and presented at dozens of conferences, written numerous articles, and published several textbooks, including *Technical Writing: Process and Product* (fifth edition), *The Red Bridge Reader* (third edition, co-authored by Kin Norman), and *Writing That Works: A Teacher's Guide to Technical Writing* (second edition). They are currently preparing a textbook on business communication to be published in 2006 for Prentice Hall.

In addition to their academic work, Sharon and Steve are involved in business and industry through their business, Steve Gerson Consulting. In this business, they have worked for companies such as Sprint, AlliedSignal–Honeywell, General Electric, JCPenney, Avon, and the Missouri Department of Transportation. Their work for these businesses includes writing, editing, and proofreading many different types of technical documents, such as proposals, marketing collateral, reports, and instructions.

Steve also has presented hundreds of hands-on workshops on technical/business writing and business grammar in the workplace. Over 10,000 business and governmental employees have benefited from these workshops. For the past decade, Steve has worked closely with K–12 teachers. He has presented many well-attended, interactive workshops to give teachers useful tips about technical writing in the classroom.

Both Steve and Sharon have been awarded for teaching excellence and are listed in *Who's Who Among America's Teachers*. In 2003, Steve was named Kansas Professor of the Year by the Carnegie Foundation for the Advancement of Education.

Their wealth of experience and knowledge has been gathered for you in this fifth edition of *Technical Writing: Process and Product*.

Welcome to the fifth edition of *Technical Writing: Process and Product*. In this edition, we have worked to create an even more reader-friendly textbook, coupling technological information with easy-to-follow instructions and real-life examples. You'll also find a substantially rewritten Chapter 1 introducing technical writing and Chapter 18 on oral communication.

DESIGN AND INSTRUCTIONAL AIDS

To enhance readability and enjoyment of this textbook, we have created a colorful new layout and design featuring a larger page size. We have also added several new pedagogical elements to improve learning. On this page and succeeding pages are examples of instructional aids included in this edition:

> *NEW!* **Larger trim size** allows for the use of **margin callouts** to clarify and highlight **key points** in sample documents.

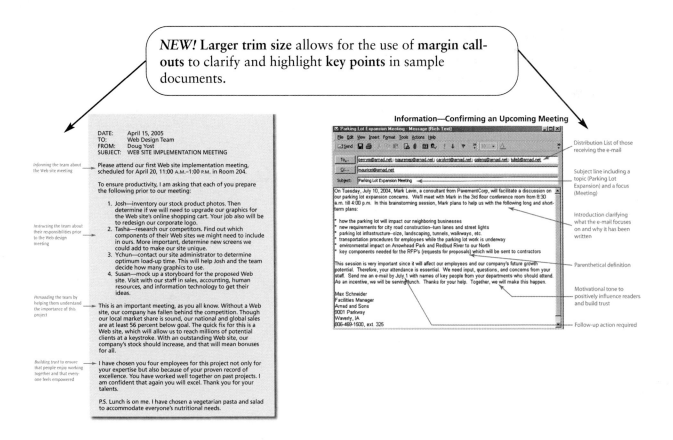

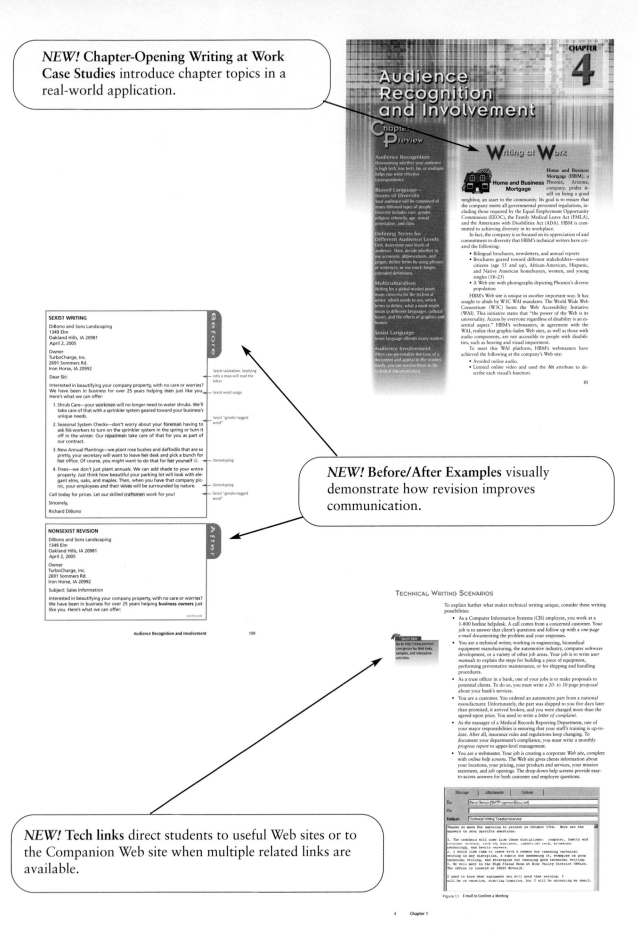

NEW! Chapter-Opening Writing at Work Case Studies introduce chapter topics in a real-world application.

NEW! Before/After Examples visually demonstrate how revision improves communication.

NEW! Tech links direct students to useful Web sites or to the Companion Web site when multiple related links are available.

NEW! **Spotlights** give examples of ways in which corporations use aspects of technical writing in their day-to-day business.

NEW! **Technology Tips** provide helpful instruction on using common computer programs to communicate.

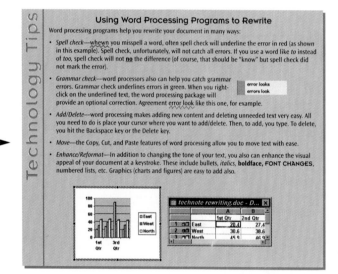

Using Word Processing Programs to Rewrite

Word processing programs help you rewrite your document in many ways:

- *Spell check*—whewn you misspell a word, often spell check will underline the error in red (as shown in this example). Spell check, unfortunately, will not catch all errors. If you use a word like *to* instead of *too*, spell check will not **no** the difference (of course, that should be "know" but spell check did not mark the error).

- *Grammar check*—word processors also can help you catch grammar errors. Grammar check underlines errors in green. When you right-click on the underlined text, the word processing package will provide an optional correction. Agreement error look like this one, for example.

- *Add/Delete*—word processing makes adding new content and deleting unneeded text very easy. All you need to do is place your cursor where you want to add/delete. Then, to add, you type. To delete, you hit the Backspace key or the Delete key.

- *Move*—the Copy, Cut, and Paste features of word processing allow you to move text with ease.

- *Enhance/Reformat*—In addition to changing the tone of your text, you also can enhance the visual appeal of your document at a keystroke. These include bullets, *italics*, **boldface**, FONT CHANGES, numbered lists, etc. Graphics (charts and figures) are easy to add also.

Checklists provide students with guidelines to follow in producing different documents.

Job Search Checklist

Job Openings
- 1. Did you visit your college or university job placement center?
- 2. Did you talk to your professors about job openings?
- 3. Have you networked with friends or past employers?
- 4. Have you checked with your professional affiliations or looked for job openings in trade journals?
- 5. Did you read the want ads in the newspapers?
- 6. Did you search the Internet for job openings?

Resume
- 1. Are your name, address, and phone number correct?
- 2. Is your job objective specific?
- 3. Are the dates within your education, work experience, and military experience sections accurate?
- 4. Have you included your degree, school, city, and state within the education section?
- 5. Have you included your job title, company name, city, and state within the work experience section?
- 6. Have you used verbs to introduce each of your professional skills?
- 7. Have you quantified each of your achievements?
- 8. Have you avoided using sentences and the word *I*?
- 9. Does your resume use highlighting techniques to make it reader-friendly?
- 10. Have you proofread your resume to find grammatical and mechanical errors?
- 11. Have you decided whether you should write a reverse chronological resume or a functional resume?
- 12. Have you decided to create an online resume? If so, have you considered the ways in which it differs from the traditional paper resume?

Letter of Application
- 1. Have you included all of the letter essentials?
- 2. Does your introductory paragraph state where you learned of the job, which job you are applying for, and your interest in the position?
- 3. Does your letter's discussion unit pinpoint the ways in which you will benefit the company?
- 4. Does your letter's concluding paragraph end cordially and explain what you will do next or what you hope your reader will do next?
- 5. Is your letter free of all errors?

Interview
- 1. Will you dress appropriately?
- 2. Will you arrive ahead of time?
- 3. Will you avoid gum, cigarettes, and caffeine?
- 4. Have you practiced answering potential questions?
- 5. Have you researched the company so you can ask informed questions?
- 6. Will you bring to the interview additional examples of your work or copies of your resume?

Follow-up Letter
- 1. Have you included all the letter essentials?
- 2. Does your introductory paragraph remind the readers when you interviewed and what position you interviewed for?
- 3. Does the discussion unit highlight additional ways in which you might benefit the company?
- 4. Does the concluding paragraph thank the readers for their time and consideration?
- 5. Does your letter avoid all errors?

Student Document Samples provide models for each document type.

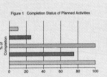
ACTIVITIES

CASE STUDIES
Gulfview Architectural and Engineering Services

You are the team leader of a work project at Gulfview Architectural and Engineering Services. The team has been involved in this project for a year. During the year, the team has met weekly, every Wednesday at 8:00 A.M. It is now time to assess the team's successes and areas needing improvement.

Your goal will be to recommend changes as needed before the team begins its second year on this project. You have encountered these problems:

- One team member, Caroline Jensen, misses meetings regularly. In fact, she has missed at least one meeting a month during the past year. Occasionally, she missed two or three in a row. You have met with Caroline to discuss the problem. She says she has had child care issues that have forced her to use the company's flextime option, allowing her to come to work later than usual, at 9:00 A.M.
- Another team member, Guy Stapleton, tends to talk a lot during the meetings. He has good things to say, but he speaks his mind very loudly and interrupts others as they are speaking. He also elaborates on his points in great detail, even when the point has been made.
- A third team member, Sharon Mitchell, almost never provides her input during the meetings. She will e-mail comments later or talk to people during breaks. Her comments are valid and on-topic, but not everyone gets to hear what she says.
- A fourth team member, Craig Mabrito, is very impatient during the meetings. This is evident from his verbal and nonverbal communication. He grunts, slouches, drums on the table, and gets up to walk around while others are speaking.
- A fifth employee, Julie Jones, is overly aggressive. She is confrontational, both verbally and physically. Julie points her finger at people when she speaks, raises her voice to drown out others as they speak, and uses sarcasm as a weapon. Julie also crowds people, standing very close to them when speaking.

Assignment
How will you handle these challenges? Either individually or as a team, decide on a course of action. To do so, consider the TQM Plan of Action (plan, do, check, act), Six Sigma's DMAIC (define, measure, analyze, improve, control), or a combination of the two approaches to solve the problem.
For example, try this approach:

1. *Analyze* the problem(s). To do so, brainstorm. What gaps might exist causing these problems?
2. *Invent* or envision solutions. How would you solve the problems? Consider Human Performance Improvement issues, as discussed in this chapter.
3. *Plan* your approach. To do so, establish verifiable measures of success (including timeframes and quantifiable actions).

Write an e-mail to your instructor sharing your findings.

End-of Chapter Activities: Each chapter concludes with a variety of **Case Studies, Exercises, and Quiz Questions** providing students with ample opportunities to apply chapter concepts to real-world situations.

ACTIVITIES

your instructor explaining your answer. Or, make a brief oral presentation. Be sure to give reasons for and against each option.
- You are an independent consultant, providing a unique service to companies (assessment of their international business opportunities). You just met with a team of eight managers within a company to discuss ways you could help them. Due to your expertise, they hired you on the spot. When you got back to your office, you thought of several more ideas to benefit their company. Now you need to share these new visions.
How should you communicate to the eight managers? Should you write a memo, an e-mail, or telephone them? Write an e-mail or memo to your instructor explaining your answer. Or, make a brief oral presentation. Be sure to give reasons for and against each option.

WEB WORKSHOP

1. Access an Internet search engine (Google, Yahoo, Excite, Lycos, etc.) and type in a phrase like "government e-mail policies." You will find local, state, and national governmental sites specifying their e-mail policies. In addition, you will find private companies marketing their e-mail management and monitoring systems ("e-policy" hardware/software and consulting), as well as sites denying the validity of such e-mail management.
Review a sampling of these sites and either make an oral presentation on your findings or write a memo documenting the diverse aspects of governmental e-mail policies.

2. Access an Internet search engine (Google, Yahoo, Excite, Lycos, etc.) and type in a phrase like "corporate e-mail policies." You will find the e-mail policies of individual corporations. In addition, you also will find the reasons why corporations have created e-policies (including lawsuits resulting from inappropriate e-mail communications as well as corporate e-mail systems crashing due to inappropriate employee usage). Furthermore, you can find online articles regarding statutes used by companies to legitimize their policies.
Review a sampling of these sites and either make an oral presentation on your findings or write a memo documenting the diverse aspects of corporate e-mail policies.
3. Many universities and colleges provide faculty, staff, and student e-mail opportunities. Thus, the schools usually require adherence to standards and provide specific guidelines.
Access an Internet search engine (Google, Yahoo, Excite, Lycos, etc.) and type in a phrase like "university e-mail policies" or "college e-mail

NEW! **Web Workshops** provide opportunities for students to investigate real-world applications on the Web.

ORGANIZATION

Unit 1: Defining Technical Writing

In this section of the book, we lay the foundation for good technical writing.

NEW—Chapter 1 ("An Introduction to Technical Writing") has been significantly revised. In this chapter, you will find the following: A *Writing at Work* scenario provides a context for technical writing. We then define technical writing and discuss its importance. This chapter emphasizes teamwork, twenty-first-century business management philosophies, conflict resolution in team meetings, and strategies for successful collaboration.

Chapter 2 ("Producing the Product") explains why process—prewriting, writing, and rewriting—will improve your correspondence. We discuss the rationale for the writing process and supply an example of this process in practice.

NEW to this second chapter, we emphasize the recursive nature of the writing process. The writing process is dynamic, with the three parts (prewriting, writing, and rewriting) often occurring simultaneously. The following visual aid is used throughout the book to highlight this fact.

NEW to Chapter 2, we also provide a visual overview of the different types of prewriting techniques discussed throughout the textbook. In addition, this chapter explains the benefits and challenges of communication channels, including e-mail, handheld computers, cell phones, Microsoft PowerPoint, and more.

Chapter 3 ("Objectives in Technical Writing") provides a detailed discussion of five major objectives in successful technical writing: clarity, conciseness, accuracy, organization, and ethics.

NEW to this chapter, we focus on how technology affects conciseness. In addition, we update "Ethical Principles for Technical Communicators."

Chapter 4 ("Audience Recognition and Involvement") teaches how to communicate with different levels of audience and focuses on the importance of definition, multiculturalism, biased language, and personalization.

NEW to Chapter 4, we discuss the need to understand diversity—age, gender, culture, and religion—and avoid biased language.

NEW to this chapter, we emphasize the challenges to multicultural and cross-cultural communication.

Unit 2: Correspondence

How do you get a job? Once you have found employment, what types of writing will you perform on the job? This unit helps you answer these questions.

NEW to Chapter 5 ("Memos and E-mail"), we combine our discussions of memos and e-mail. This chapter provides criteria for successful memos and e-mail messages, samples, and a process log.

NEW to this chapter, we update the importance of e-mail and help you overcome challenges presented by electronic communication.

Chapter 6 ("Letters") provides criteria and examples for many different types of letters.

Chapter 7 ("The Job Search") helps you find employment, write a resume, send a resume, write a letter of application, and interview effectively.

NEW to this chapter, we discuss new ways to look for jobs.

NEW—Chapter 7 also discusses new methods of resume delivery (mail versions, Web resumes, e-mail resumes, and ASCII resumes).

NEW to this chapter, you will learn how to write an e-mail cover message to preface your resume attachment.

Unit 3: Visual Appeal

Effective technical communication is visual as well as verbal. We clarify the relationship between words and graphics in this unit.

Chapter 8 ("Document Design") explains various ways to create an effective page layout.

Chapter 9 ("Graphics") focuses on tables and figures.

These two chapters provide you numerous examples to help you improve the visual appeal of your correspondence.

Unit 4: Technical Applications

Unit 4 illustrates how to effectively write technical documents that combine text and visual aids. These include fliers, brochures, newsletters, technical descriptions, and user manuals.

Chapter 10 ("Fliers, Brochures, and Newsletters") emphasizes the uses of these types of correspondence, as well as organizational techniques, content, design elements, and usability testing.

Chapters 11 and 12 present different types of technical writing, the types of communication that usually accompany manufactured products and corporate services.

Chapter 11 ("Technical Descriptions") discusses how to write specifications for mechanisms, tools, and pieces of equipment.

Chapter 12 ("Instructions and User Manuals") explains how to write the instructions and manuals that show customers how to use or operate the equipment they have purchased.

Unit 5: Electronic Communication

The focus in this unit shifts to online help and Web sites.

NEW to Chapter 13 ("Online Help and Web Sites"), we define characteristics of the online e-reader and ways in which online communication differs from the printed page. We discuss effective online highlighting techniques and Web site criteria. You will find examples of successful Web pages with call-outs.

Unit 6: Report Strategies

As the name of this unit implies, we discuss a wide variety of report options, including research, summaries, short reports, proposals, and oral communication.

Chapter 14 ("Research") teaches you how to find and document information.

NEW to this chapter, we provide sample Works Cited and Reference pages to clarify the differences between MLA and APA formats. We also update documentation formats.

Chapter 15 ("The Summary") shows you how to abstract information for meetings and briefings.

Chapter 16 ("Reports") discusses the similarities and differences among various short reports, including trip reports, progress reports, lab reports, feasibility/recommendation reports, incident reports, investigative reports, and meeting minutes.

NEW to this chapter, we add investigative reports and meeting minutes. We also give multiple examples of how these reports will be used in various professions.

Chapter 17 ("Proposals") focuses on the criteria for writing effective internal and external proposals.

NEW to this chapter, we provide a sample internal proposal.

NEW—Chapter 18 ("Oral Communication") has been completely revised. It now contains instruction on the following:

- Conducting Everyday Oral Communication
- Tips for Using the Telephone and Voice Mail
- Making Informal Oral Presentations
- Tips for Teleconferences and Videoconferences
- Making Formal Oral Presentations
- Creating PowerPoint Presentations
- Using the Writing Process for Oral Communication
- Using Effective Visual Aids in Oral Presentations

Unit 7: Handbook

Chapter 19 ("Grammar, Punctuation, Mechanics, and Spelling") provides you rules and regulations for correct writing.

COMPANION WEB SITE: A WEALTH OF NEW ONLINE MATERIALS

We are especially excited about the wealth of new cases, exercises, activities, and documents that have been developed for each chapter and are available at our Companion Web site located at *www.prenhall.com/gerson*. Online materials for each chapter in the text include the following:

- **Chapter Learning Objectives**—Overview of major chapter concepts.
- **Writing Process Exercises**—Prewriting/Writing/Rewriting assignments.
- *NEW* **Interactive Editing and Revision Exercises**—Interactive documents allow students to see poorly done and corrected versions of documents with additional assignable document revision exercises.
- *NEW* **Communication Cases**—Students encounter real-world situations with links to outside content and a student response box for students to send answers to the professor.
- *NEW* **Activities and Exercises**—Activities specific to a variety of technical and career fields allow students to practice producing communication relevant to their interests.
- **Collaboration Exercise**—Assignments designed to provide practice writing and communicating in teams.
- **Web Resources**—Links to helpful online resources related to chapter content.
- **Document Library**—Additional documents and forms.
- **Chapter Quizzes**—Self-grading multiple-choice quizzes help students master chapter concepts and prepare for tests.

OneKey Distance Learning Solutions

New to this edition are ready-made Blackboard, WebCT, and CourseCompass online courses. If you adopt the text with a OneKey course, student access cards will be packaged with the textbook at no additional charge to the student.

Instructor's Resources

- **Instructor's Manual (ISBN: 0-13-119665-0)**
  New to this edition is an expanded Instructor's Manual loaded with helpful teaching notes for your classroom. Included in the manual are answers to the chapter quiz questions, a test bank, and instructor notes for assignments and activities located on the Companion Web site.
- *NEW* **Instructor's Resource CD (ISBN: 0-13-119669-3)**
  The IRCD includes the following components:
  - *NEW* **Test Generator**
  - *NEW* **PowerPoint Lecture Presentation Package**
  - **Instructor's Manual** (in Microsoft Word)

Acknowledgments

We would like to thank the following reviewers of the fourth edition for helping with the revision of the textbook:

- Myra G. Day, North Carolina State University
- Dr. Cynthia Gillispie-Johnson, North Carolina A&T State University
- Sharon Mouss, Oklahoma State University—Okmulgee
- Esther J. Winter, Central Community College
- Nancy Roberts, Griffin Technical College
- Brian Still, Texas Tech University

We also would like to thank the following people who contributed activities, exercises, and documents to the Companion Web site:

- May Beth Van Ness, Terra Community College, for helping to check and revise our existing online material
- Linda Gray, Oral Roberts University, for creating the innovative Interactive Editing and Revision Exercises
- Catharine Schauer, Visiting Professor, Embry-Riddle Aeronautical University, for contributing a variety of interesting activities
- Malanie Rosen Brown, St. Johns River Community College, for contributing interesting activities and case studies
- Connie Cerniglia, Guilford Technical Community College, for contributing a variety of career-related assignments and documents

We would especially like to thank our Editor, Gary Bauer, and our Developmental Editor, Monica Ohlinger of Ohlinger Publishing, for their efforts, patience, and creativity in helping us bring out this fifth edition.

Sharon J. Gerson
Steven M. Gerson

BRIEF CONTENTS

Contents

UNIT 2   CORRESPONDENCE   129

UNIT 5   ELECTRONIC COMMUNICATION 407

CHAPTER 13 Online Help and Web Sites   **408**

UNIT 6   REPORT STRATEGIES 446

UNIT 7   HANDBOOK 632

DEFINING TECHNICAL WRITING

An Introduction to Technical Writing

Chapter Preview

Technical Writing: A Definition
Technical writing is composed in and for the work environment. At work, you write to supervisors, colleagues, subordinates, vendors, and customers.

Technical Writing Scenarios
Documents you will create on the job include e-mail, memos, letters, reports, proposals, user manuals, Web sites, brochures, and newsletters.

Importance of Technical Writing
Technical writing enables you to conduct business, it requires time, and it costs money.

Teamwork
Every day you interact with others to prepare reports, user manuals, proposals, and Web sites.

21st-Century Business Management Philosophies
Many companies stress a variety of business methodologies geared toward meeting corporate goals. These methodologies focus on the importance of teamwork.

Conflict Resolution in Team Meetings
Coworkers need to use various conflict resolution techniques.

Strategies for Successful Collaboration
Successful team members follow collaboration strategies.

Writing at Work

Gulfview Architectural and Engineering Services is home-based in Gulfview, Texas, with office sites in 10 U.S. cities and 5 locations throughout the world. Gulfview hopes to build a power plant in Saudi Arabia. To accomplish this task, a team of employees is working on two continents.

Through a *series of technical writing activities*, they have had to communicate with each other constantly.

Proposal. First, one team, consisting of engineers, architects, marketing specialists, accountants, lawyers, and technical writers, put together a proposal. In this proposal, they focused on the services they could offer, the expertise of their workforce, the price they would charge for the construction, and a timeline for their work. Despite many competitors, Gulfview won the account.

E-mail. The construction would take Gulfview approximately two years. During that time, Gulfview personnel had to communicate with their Saudi contractors on a daily basis. E-mail answered this need. The team members communicated with each other by writing approximately 50 e-mail messages a day. In these transmittals, the team members focused on construction permits, negotiated costs with vendors, changed construction plans, and asked questions and received answers. They also used these e-mail messages to build rapport with coworkers.

Intranet Web Site. To help all parties involved (those in Saudi Arabia as well as Gulfview employees throughout the United States), Gulfview's Information Technology Department built an intranet site geared specifically toward the power plant project. This firewall protected site, open to Gulfview employees and external vendors associated with the project, helped all construction personnel submit online forms, get corporate updates, and access answers to frequently asked questions. Many of these FAQs were managed through online help screens with pull-down menus.

Letters. To secure and revise construction permits, Gulfview personnel had to write formal letters to government officials in Saudi Arabia. In addition, Gulfview employees had to write letters to vendors, asking for quotes.

Reports. Finally, all of the employees involved in the power plant project had to report on their activities. These included

- progress reports providing updates on the project's status
- incident reports when job-related accidents and injuries occurred
- feasibility reports to recommend changes to the project's plan or scope
- meeting minutes following the many team meetings

Like all companies engaged in job-related projects, Gulfview Architectural and Engineering Services spent much of its time using technical writing to communicate with a diverse audience. The challenges they faced involved teamwork, multicultural and multilingual concerns, a vast array of communication technologies, and a variety of different types of technical writing.

TECHNICAL WRITING: A DEFINITION

Technical writing is communication written for and about business and industry. Technical writing focuses on products and services—how to manufacture them, market them, manage them, deliver them, and use them.

Technical writing is composed primarily in the work environment for supervisors, colleagues, subordinates, vendors, and customers. As either a professional technical writer, an employee at a company, or a consumer, you can expect to write the following types of correspondence for the following reasons (and many more):

- **Memos and electronic mail (e-mail)**—to set meeting agendas and to ask and answer questions
- **Letters**—to sell, complain, hire, fire, ask and answer questions, and explain the contents of attachments
- **Reports**—to report on job-related travel or incidents, to study options and recommend action, to report on the progress of ongoing projects, and to document meeting minutes
- **Proposals**—to highlight problems, to suggest solutions, and to recommend action
- **Brochures**—to sell and inform, using six-to eight-panel (back and front) foldouts
- **Newsletters**—to report on corporate activities to both employees and stakeholders
- **Fliers**—to sell and inform, using brief, single-sided documents
- **Resumes**—to help you find a job
- **Web sites**—to sell and inform, using multiscreened, Internet-based, hypertext-linked communication
- **Online help screens**—to explain, inform, and define, using drop-down menus and pop-ups
- **User manuals**—to explain the steps in a procedure
- **Technical descriptions**—to explain the parts of a mechanism, tool, piece of equipment, or product

In each case, the technical writer's goal is to create text that is clear, concise, easy to understand, and easy to navigate.

To explain further what makes technical writing unique, consider these writing possibilities:

tech link

Go to *http://www.prenhall. com/gerson* for Web links, samples, and interactive activities.

- As a Computer Information Systems (CIS) employee, you work at a 1-800 hotline helpdesk. A call comes from a concerned customer. Your job is to answer that client's questions and follow up with a *one-page e-mail* documenting the problem and your responses.

- You are a technical writer, working in engineering, biomedical equipment manufacturing, the automotive industry, computer software development, or a variety of other job areas. Your job is to write *user manuals* to explain the steps for building a piece of equipment, performing preventative maintenance, or for shipping and handling procedures.

- As a trust officer in a bank, one of your jobs is to make proposals to potential clients. To do so, you must write a *20- to 30-page proposal* about your bank's services.

- You are a customer. You ordered an automotive part from a national manufacturer. Unfortunately, the part was shipped to you five days later than promised, it arrived broken, and you were charged more than the agreed-upon price. You need to write a *letter of complaint*.

- As the manager of a Medical Records Reporting Department, one of your major responsibilities is ensuring that your staff's training is up-to-date. After all, insurance rules and regulations keep changing. To document your department's compliance, you must write a monthly *progress report* to upper-level management.

- You are a webmaster. Your job is creating a corporate *Web site*, complete with *online help screens*. The Web site gives clients information about your locations, your pricing, your products and services, your mission statement, and job openings. The drop-down help screens provide easy-to-access answers for both customer and employee questions.

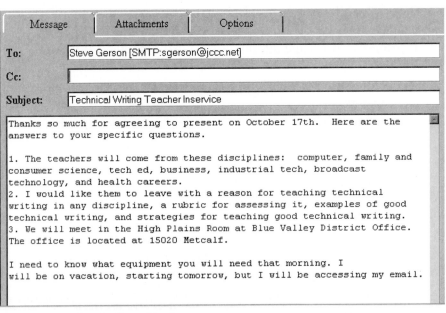

Figure 1.1   E-mail to Confirm a Meeting

- As an entrepreneur, you are opening your own computer-maintenance service (or services for HVAC repair, deck rebuilding, home construction, lawn care, or automotive maintenance). To market your company, you will need to write *fliers, brochures,* or *sales letters.*

- You have just graduated from college (or, you have just been laid off). It's time to get a job. You need to write a *resume* and a *letter of application* to show corporations what assets you will bring to their company.

In other words, you will need to produce technical writing for a host of reasons. See Figures 1.1 through 1.4 for just a few examples of different types of technical writing.

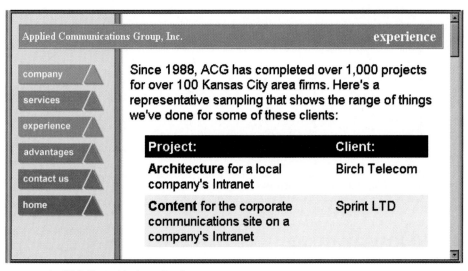

Figure 1.2   Web Site to Market a Service

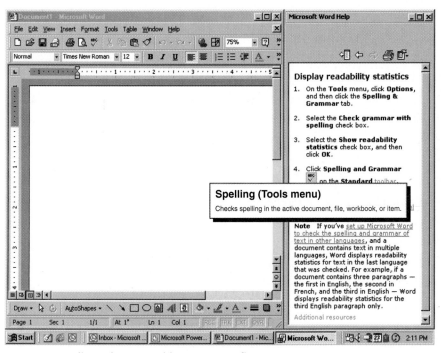

Figure 1.3   Online Help Screen with Pop-up to Define a Term

An Introduction to Technical Writing   5

Attaching Files Using Microsoft Outlook

Many times you will want to send attachments when you e-mail a client, vendor, or colleague. To do so, follow these steps:

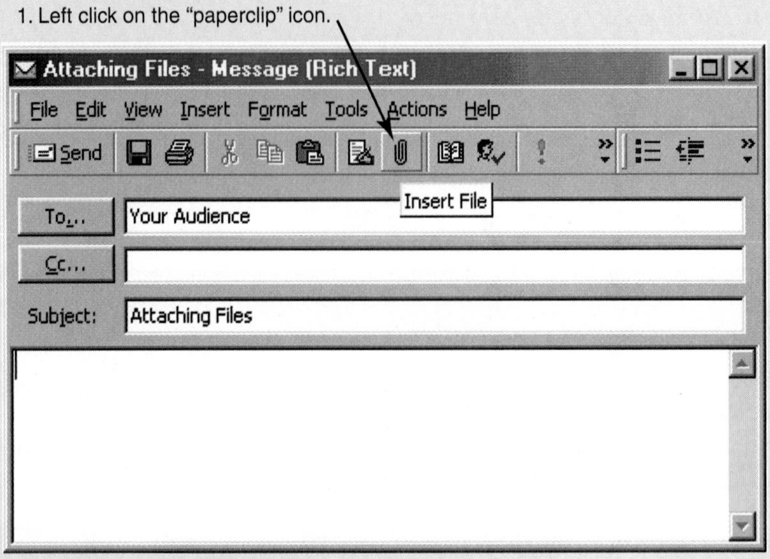

1. Left click on the "paperclip" icon.

Figure 1.4   User Manual Steps with Screen Capture to Instruct

IMPORTANCE OF TECHNICAL WRITING

Technical writing is a significant factor in your work experience for several reasons.

Business

Technical writing is not a frill or an occasional endeavor. It is a major component of the work environment. Through technical correspondence, employees

- Maintain good customer-client relations (follow-up letters)
- Ensure that work is accomplished on time (directive memos or e-mail)
- Provide documentation that work has been completed (progress reports)
- Generate income (sales letters, brochures, and fliers)
- Keep machinery working (user manuals)
- Ensure that correct equipment is purchased (technical descriptions)
- Participate in teleconferences or videoconferences (oral communication)
- Get a job (resumes)
- Define terminology (online help screens)
- Inform the world about your company's products and services (Internet Web sites)

Time

In addition to serving valuable purposes in the workplace, technical writing is important because it requires your time.

TABLE 1.1 NUMBER OF DAILY MESSAGES	
Communication Tools	Number of Messages Sent/Received Daily
Voice-mail options	**Total = 75**
Telephone	48
Voice mail	21
Cellular phone	6
Electronic text tools	**Total = 68**
E-mail	50
Fax	10
Pager	8
Hard-copy tools	**Total = 33**
Interoffice mail	18
Postal mail	15
Miscellaneous tools	**Total = 30**
Sticky notes	12
Telephone message slips	18
Total	**206**

A Pitney Bowes study states that "U.S. workers send and receive an average of 206 messages per day via various text, voice and hardcopy tools" ("Pitney Bowes Study" 2000). Table 1.1 lists the various communication tools used and the total number of messages sent and received daily.

Consider the time you will spend on written communication. As Table 1.1 shows, business employees read and write 50 e-mail messages, 18 memos, and 15 letters a day.

Across all professions, workers spend nearly one-third of their time writing (31 percent). Figure 1.5 shows that 43 percent of the respondents to a survey spend between 11 and 30 percent of their time writing. Another 26 percent of respondents spend between 31 and 50 percent of their time writing (Miller et al. 1996,10).

Note: On the basis of a 40-hour workweek, 31 percent would equal 12 1/2 hours of writing each week.

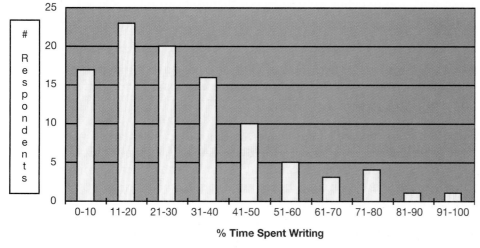

Figure 1.5 Time Spent Writing

To read the article by
Carolyn Miller et al., go to
http://www.chass.ncsu.edu/
ccstm/pubs/no2/index.html.

That 31 percent of time spent writing on the job is just a base number—an average across the board. Generally speaking, new hires might spend less time writing on the job. As a supervisor, you will spend more time directing your subordinates through written correspondence.

Money

You have heard it before—time is money. Here are three simple ways of looking at the cost of your business communication.

- **Cost of correspondence**—A recent study by Dartnell's Institute of Business Research says that the "average cost of producing and mailing a letter is $19.92" ("Business Identity"). That factors in the time it takes a worker to write the letter as well as the cost of the paper, printing, and stamp. If one letter costs almost $20, imagine how much an entire company's correspondence might cost annually, including every employee's e-mail, letters, memos, and reports.

- **Percentage of salary**—Consider how much of your salary is being paid for your communication skills. Let's say you make $35,000 a year. If you are spending 31 percent of your time writing, then your company is paying you approximately $10,850 just to write. That does not include the additional time you spend on oral communication.

    If you are not communicating effectively on the job, then you are asking your bosses to pay you a lot of money for substandard work. Your time spent communicating, both in writing and orally, is part of your salary—and part of your company's expenditures.

- **Generating income**—Your communication skills do more than just cost the company money; these talents can earn money for both you and the company. A well-written sales letter, flier, brochure, proposal, or Web site can generate corporate income. Effectively written newsletters to clients and stakeholders can keep customers happy and bring in new clients. Good written communication is not just part of your salary—it helps pay your wages.

Interpersonal Communication

When you write a memo, letter, or report, you're not just conveying technical information. You're revealing something about yourself to your readers. If you write well, you're telling your audience that you can think logically and communicate your thoughts clearly. If, on the other hand, you write poorly, you give your readers a completely different picture of yourself as a worker. You reveal that you can neither think clearly nor communicate your thoughts effectively. Technical writing is an extension of your interpersonal communication skills at work, and coworkers will judge your competence based on the effectiveness of your correspondence.

For example, a man with a master's degree in criminal justice enrolled in a technical writing class to improve his writing skills. When asked why, he said, "My supervisors are dissatisfied with my writing. In fact, they've given me ninety days to improve my communication abilities. If I can't, then I'm out of a job." On the other hand, a former student got a job in aviation as a technician. One day she wrote a problem/solution report. The report was well received and was passed around the company as an example of good writing. Management was so impressed with her writing skills that she was promoted to a supervisory level, where her communication skills could benefit the company.

Good technical writing can accomplish more than just getting the job done. A well-constructed memo, letter, or report reveals to your readers not only that you know your technical field of expertise but also that you know how to communicate your knowledge thoroughly, accurately, clearly, and concisely. Through good technical writing, you reveal to your audience that you can tell people what to do and can motivate them to do it.

TEAMWORK

Companies have found that teamwork enhances productivity. Teammates help and learn from each other. They provide checks and balances. Through teamwork, employees can develop open lines of communication to ensure that projects are completed successfully.

Collaboration

In business and industry, many user manuals, reports, proposals, PowerPoint presentations, and Web sites are team written. Teams consist of engineers, graphic artists, marketing specialists, and corporate employees in legal, delivery, production, sales, accounting, and management. These collaborative team projects extend beyond the company. A corporate team also will work with subcontractors from other corporations. The collaborative efforts include communicating with companies in other cities and countries through teleconferences, faxes, and e-mail.

Making matters more challenging, these teams will be diverse, consisting of people from different areas of expertise, as well as different ages, genders, cultures, languages, and races (see Chapter 3 for more discussion on the topic of diversity). Teamwork is perceived as so essential that many undergraduate and graduate program directors have mandated team projects in their curriculum.

One director confirmed the importance of teamwork, after having surveyed a cross section of business and industry: "'the results revealed that more than technical skills, employees need skills that allow them to interact and relate effectively with others'" (Bacon 1993). Another program director went further, asserting that "'people don't lose jobs because they don't know how to do things. It's because they fail at relationships'" (Bacon 1993).

The National Association of Colleges and Employers lists the "Top Ten Skills Employers Want" (see Figure 1.6). Notice how interpersonal, teamwork, oral communication, and written communication skills take precedence over work experience.

The Problems with "Silo Building"

Working well with others requires collaboration versus "silo building." The *silo* has become a metaphor for departments and employees that behave as if they have no responsibilities outside their areas. They build bunkers around themselves, failing to collaborate with others. In addition, they act as if no other department's concerns or opinions are valuable.

Such "stand-alone" departments or people isolate themselves from the company as a whole and become inaccessible to other departments. They "focus narrowly" (Hughes 2003, 9). This creates problems. Poor accessibility and poor communication "can cause duplicate efforts, discourage cooperation, and stifle cross-pollination of ideas" (Hughes 2003, 9).

The Skills Employers Want

1. Interpersonal
2. Teamwork
3. Analytical
4. Oral communication
5. Flexibility
6. Computer
7. Written communication
8. Leadership
9. Work experience
10. Internship/co-op experience

Source: Adapted from the National Association of Colleges and Employers 2004.

Figure 1.6   Skills Employers Want

To be effective, companies need "open lines of communication within and between departments" (Hughes 2003, 9). The successful employee must be able to work collaboratively with others to share ideas. In the workplace, teamwork is essential.

Why Teamwork Is Important

Teamwork benefits employees, corporations, and consumers. By allowing all constituents a voice in project development, teamwork helps to create effective work places and ensures product integrity.

Diversity of Opinion

When you look at problems individually, you tend to see issues from limited perspectives—*yours*. In contrast, teams offer many points of view. For instance, if a team has members from accounting, public relations, customer service, engineering, and information technology, then that diverse group can offer diverse opinions. You should always look at a problem from various angles.

Checks and Balances

Diversity of opinion also provides the added benefit of checks and balances. Rarely should one individual or one department determine outcomes. When a team consists of members from different disciplines, those members can say, "Wait a minute. Your idea will negatively impact my department. We had better stop and reconsider."

Broad-Based Understanding

If decisions are made in a silo, by a small group of like-minded individuals, then these conclusions might surprise others in the company. Surprises are rarely good. You always want buy-in from the majority of your stakeholders. An excellent way to achieve this is through team projects. When multiple points of view are shared, a company benefits from broad-based knowledge. Improved communication allows people to see the bigger picture.

Teamwork and Company Culture

The list of companies that have incorporated teamwork into their culture is long and distinguished:

Motorola—Motorola's Connector Team uses collaboration to set performance measures, decrease cycle time, and limit costs.

Miller Brewing Company—Miller's Trenton brewery's goal is to "empower people as partners." To do so, the brewery created "self-directed work teams" to collectively manage the entire plant. The teams are responsible for all plant operations, including personnel, decision making, scheduling, and shift assignments.

FedEx—FedEx uses Quality Action Teams to ensure that "quality initiatives are implemented correctly." The teams accomplish this by working collaboratively to analyze problems and develop solutions. They use the "FADE" approach: Focus-Analyze-Develop-Execute.

Honeywell—Honeywell has developed three levels of teams: Basic, Intermediate, and Advanced. Employees develop skills in a sequence, moving from simple problems to more complex ones. Thus, the teams acquire "responsibility as their experience and skill-levels" grow. Honeywell achieves effective team communication by scheduling weekly team meetings. Through joint discussions, the team members raise issues and solve problems.

Source: Adapted from "Best Practice Database" 2003.

tech link

For further research on best practice strategies, go to *http://www.bestpracticedatabase.com.*

Empowerment

Collaboration gives people from varied disciplines an opportunity to provide their input. When groups are involved in the decision-making process, they have a stake in the project. This allows for better morale and productivity.

Team Building

Everyone in a company should have the same goals—corporate success, customer satisfaction, and quality production. Team projects encourage shared visions, a better work environment, a greater sense of collegiality, and improved performance. Employees can say, "We are all in this together, working toward a common goal."

21st-Century Business Management Philosophies

The way to stay ahead of the competition and to maintain customer satisfaction is to produce a quality product. This is best achieved by anticipating problems and by avoiding them before they occur. A key way to ensure that these goals are met is through collaborative work efforts. Companies have concluded that employees from different areas of expertise, working together in teams, communicating their multidimensional desires for success, can best achieve corporate quality.

Many companies stress a variety of business methodologies geared to ensure quality. The strategies include

- Total Quality Management (TQM)
- Six Sigma
- Human Performance Improvement (HPI)

Total Quality Management

Total Quality Management (TQM) is "an organizational approach to customer satisfaction involving customers, people, and the continuous improvement process" ("SAE's Approach to Total Quality Management" 2003). TQM operates on the basic principle of "involving and respecting people—everyone associated with the organization" ("SAE's Approach" 2003).

The cornerstone to successful implementation of the TQM management model is teamwork. This is best illustrated by looking at the TQM Small Group Activity (SGA) Circle ("SGA" 2003).

The Small Group Activity Circle (Figure 1.7) seeks to achieve these goals:

- Team building
- Improved communication
- Higher involvement
- Analysis and solution of problems

SGA depends upon multifunctional teams composed of members from many different departments. These teams solve work-related problems by searching for root causes. Then, they work to eliminate these problems. To succeed, the teams learn communication skills, which allow them to work together effectively, use each other's unique skills, communicate their opinions and findings, and make decisions based on consensus.

Six Sigma

Six Sigma (6 Sigma) "is a business concept that answers customers' demands for high quality" services ("Six Sigma" 2003).

tech link

Go to *http://www.sae.org/ about/quality.htm* for additional information about Total Quality Management.

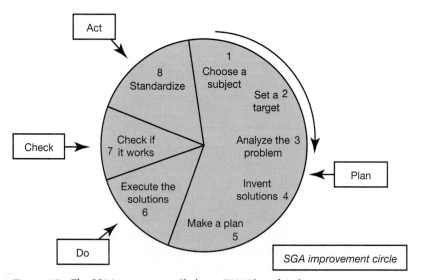

Figure 1.7   The SGA Improvement Circle—a TQM Plan of Action

As with TQM, a key component of Six Sigma is teamwork. Successful companies realize that usually "no one person and no one area" can solve work-related problems. "The challenge to leadership is to harness the ideas and energy of many people across functions, sites, and even business groups. . . . A quantum leap in quality needs people to make it happen" ("Empowering Your Employees" 2003).

tech link

Go to *http://www.sixsigmaco. com/casestudies.shtml* for specific case studies in which Six Sigma was used for product and service improvements.

DMAIC—the Six Sigma Methodology

The primary Six Sigma methodology is abbreviated *DMAIC*:

1. *Define*—What's the problem? Before quality control can be achieved, the team must first define areas needing improvement. In addition, the team needs to determine what its goals will be during a project's lifecycle.

2. *Measure*—Once goals have been set, the next step is to establish valid and verifiable measures of success. This ensures that the team stays on track.

3. *Analyze*—Goal setting and verifiable measurement require an analysis of the gap between "what is" and "what should be." This is the essence of Six Sigma. The Greek Symbol Σ represents the mathematical measurement of variation. The goal of Six Sigma is only 3.4 defects per million. Analyzing tells team members how far off from this goal they are.

4. *Improve*—Once the team knows the extent of the variation from the goal, the team's job is to solve the problem. Its task is to create new processes for manufacturing, engineering, or business applications. These could include brainstorming, simulations, statistical models, or project plans. Whatever method is devised, cross-functional communication—throughout the company—is the key to success.

5. *Control*—Finally, to achieve continuous quality improvement, the company must institutionalize the changes. That means controlling the entire work environment. To do so, a company will modify systems. This might entail changing compensation packages, incentive plans, budgets, documentation, management systems, and communication approaches.

Source: "Six Sigma" 2003.

See Figure 1.8 for a visual representation of DMAIC.

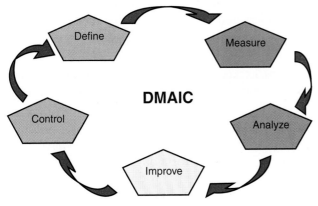

Figure 1.8   Six Sigma Methodology Illustrated

Challenges to Effective Teamwork

Any collaborative activity is challenging to manage: team members do not show up for class or work; one student or employee monopolizes the activity while another individual snoozes; people exert varying amounts of enthusiasm and ability; personalities clash. Some people fight over everything. Occasionally, when a boss participates on a team, employees fear speaking openly. Some team members will not stay on the subject. One team member will not complete her assignments. ("Individual's and Teams' Roles and Responsibilities" 2003).

The goal of programs like TQM and Six Sigma is to encourage employees to work effectively with each other to achieve corporate quality. However, challenges occur. Group dynamics are difficult and can lead to performance gaps.

Human Performance Improvement

tech link

Go to *http://www.ispi.org/* (International Society for Performance Improvement) for additional information about Human Performance Improvement.

Human Performance Improvement (HPI) focuses on "root cause analysis" to assess and overcome the barriers inherent in teamwork. To close performance gaps, HPI analyzes the following possible causes for collaborative breakdowns:

1. *Knowledge*—Perhaps employees do not know how to perform a task. They have never acquired the correct skills or do not understand which skills are needed to complete the specific job. Varying skills of team members can impede the group's progress.

2. *Resources*—Think of these possibilities: tools are broken or missing; the department is out of funds; you do not have enough personnel to do the job; the raw material needed for the job is below par; you ordered one piece of machinery but were shipped something different; you needed 100 items but have only 50 in stock. To complete a project, you often have to solve problems with resources.

3. *Processes*—For teams to succeed in collaborative projects, everyone must know what their responsibilities are. Who reports to whom? How will these reports be handled (orally, in writing)? Who does what job? Are responsibilities shared equally? Structure, of some sort, is needed to avoid chaos, lost time, inefficiency, hurt feelings, and many other challenges to teamwork. To achieve successful collaboration, the team should set and maintain effective procedures.

4. *Information*—A team needs up-to-date and accurate information to function well. If required database information is late or incorrect, then the team will falter. If the information is too high tech for some of the team members, a lack of understanding may undermine the team effort.

5. *Support*—To succeed in any project, a team needs support. This could be financial, attitudinal, or management. When managers from different departments are fighting "turf wars" over ownership of a project, teams cannot succeed. Teams need enough money for staffing, personnel, or equipment.

6. *Wellness*—A final consideration involves the team's health and well-being. People get sick or miss work for health reasons. People have car accidents. If a teammate must miss work for a day or an extended period, this will negatively impact the team's productivity. Stress and absences can lead to arguments, missed deadlines, erratic work schedules, and poor quality.

After assessing root causes that challenge a team's success, HPI creates intervention options. These might include

- Improved compensation packages
- Employee recognition programs
- Revised performance appraisals
- Improved employee training
- Simulations
- Mentoring or coaching
- Restructured work environments to enhance ergonomics
- Safety implementations
- Strategic planning changes
- Improved communication channels
- Health and wellness options—lectures, on-site fitness consultants, incentives to weight loss, and therapist and social worker interventions

People need help in order to work more effectively with each other. A progressive company recognizes these challenges and steps in to help.

CONFLICT RESOLUTION IN TEAM MEETINGS

Members of a team will not always get along. If you are a team leader, you must ensure that people get along with each other. To accomplish this goal, consider these approaches to conflict resolution:

1. *Set guidelines*—One reason that conflicts occur is because people do not know what to expect or what is expected of them. On the other hand, if expectations are clear, then several major sources of conflict can be resolved.

   For example, one simple conflict might be related to time. A team member could be unaware of when the meeting will end and schedules another meeting. If that team member then has to leave the first meeting early, disrupting the team's progress, this can cause a conflict.

   To solve this problem, set guidelines. Tell your team members (before the meeting via e-mail or early in a project) how long the project will last. Also, clarify the team's goals, the chain of command (if one exists), and each team member's responsibilities.

2. *Encourage equal discussion and involvement*—A team's success demands that everyone participate. As team leader, encourage involvement and discussion. Be sure that everyone is allowed a chance to give input.

   Conflicts also arise when one person monopolizes the work. If one person speaks excessively, others will feel left out and disregarded. Your job is to ensure equal participation. Call on others for their opinions and ask for additional input from the team.

   In addition, you could limit an overly aggressive team member's participation by saying, "Thanks, John, for your comments. Now, let's see what others have to say." Or, "Wait one second, John. I'll come back to you after we've heard from a few others."

3. *Discourage taking sides*—Discussion is necessary, but conflict will arise if team members take sides. An "us against them" mentality will harm the team effort. You can avoid this pitfall by seeking consensus, tabling issues, creating subcommittees (each of these points is discussed below), or asking for help from an outside source (boss or teacher, for example).

4. *Seek consensus*—Not every member of the team needs to agree on a course of action. However, a team cannot go forward without majority approval.

   To achieve consensus, your job as team leader is to listen to everyone's opinion and seek compromise. Conflict can be resolved by allowing everyone a chance to speak. Once everyone has spoken, then take a vote.

5. *Table topics when necessary*—If an issue is so controversial that it cannot be agreed upon, take a time out. Tell the team, "Let's break for a few minutes. Then we can reconvene with fresh perspectives." Maybe you need to table the topic for the next meeting. Sometimes, conflicts need a cooling-off period.

6. *Create subcommittees*—If a topic cannot be resolved, teammates are at odds, or sides are being taken, create a subcommittee to resolve the conflict. Let a smaller group tackle the issue and report back to the larger team.

7. *Find the good in the bad*—Occasionally, one team member comes to a meeting with an agenda. This person does not agree with the way things have been handled in the past or the way things are being handled presently. You do not agree, nor do other team members. However, you cannot resolve this issue simply by saying, "That's not how we do things." A disgruntled team member will not accept such a limited viewpoint.

   As team leader, seek compromise. Let the challenging team member speak. Discuss each of the points of dissension. Allow for input from the team. Some of the ideas might have more merit than you originally assumed.

8. *Deal with individuals individually*—From time to time, a team member will cause problems for the group. The teammate might speak out of turn or say inappropriate things. These could include off-color or off-topic comments. A team member might cause problems for the group by habitually showing up late, missing meetings, or monopolizing discussions.

   To handle these conflicts, avoid pointing a finger of blame at this person during the meeting. Do not react aggressively or impatiently. Doing so will lead to several problems:

   • Your reaction might call more attention to this person. Sometimes people come to meetings late or speak out in a group *just* to get attention. If you react, you might give the individual exactly what he wants.

   • Your reaction might embarrass this person.

   • Your reaction might make you look unprofessional.

   • Your reaction might deter others from speaking out. You want an open environment, allowing for a free exchange of ideas.

Speak to any offending team members individually. This could be accomplished at a later date, in your office, or during a coffee break. Speaking to the person later and individually might defuse the conflict.

9. *Stay calm*—Act professionally when dealing with conflict. To resolve conflicts, speak slowly, keep your voice steady and quiet, and stay seated (rising will look too aggressive). You also might want to take notes. This will provide you with a record of the discussion.

10. *Remove, reassign, or replace if necessary*—Finally, if a team member cannot be calmed, cannot agree with the majority, or has too many other conflicts, your best course of action might be to remove, reassign, or replace this individual.

STRATEGIES FOR SUCCESSFUL COLLABORATION

To successfully collaborate with your employees, follow the strategies outlined below.

Develop Your Team

To develop a team, you have two choices:
- Select team members by discipline (engineers with engineers, accountants with accountants, marketing experts with other people in marketing, or computer programmers with computer programmers).
- Choose diverse team members with different skills (teams composed of an engineer, an accountant, a computer programmer, a graphic artist, and a marketing specialist).

The first team profile draws strength from individuals with like skills; the second team benefits from multiple perspectives.

Choose a Team Leader

Either your teacher or your boss selects a team leader, or your group might choose the person best suited to this role. Sometimes, the team leader emerges from the group by consensus. The person most able, the person most involved, or the person most enthusiastic might surface as your team leader. Once this person is chosen or emerges, then the delegation of authority begins. Your team leader might do the following:

1. *Assign duties*—Who will research the material, who will write the various parts of the text, who will interview people, who will proofread and revise, and who will work on graphics?

2. *Create schedules (project milestones)*—When will the group meet, when will research and writing be due, when are revisions required, and when is the finished project due? Creating a Gantt chart before a team project begins is a great way to plot a strategy and to stay on schedule.

3. *Encourage group participation*—How will conflicts among group members be resolved, and how will equal participation be achieved? Consider asking team members to send an e-mail to all participants stating their expectations and individual strengths.

Determine Your Team Goal

Once the teams are formed, brainstorm an area needing improvement, a corporate goal. Decide what project to work on. For example, maybe the group decides to improve a company's work environment, to increase sales, or to improve a company's benefits package.

Identify the Problem

If the group decides that some area needs improvement, then there must be a problem. The key to improvement is identifying the problems impeding the group goal. Perhaps corporate profits are sagging because machines do not work well. Maybe people are uncomfortable, dissatisfied, or overworked. To improve a company's work environment, the group first must identify the problems that diminish the environment.

Analyze the Problem's Causes

Once the group has identified problems impeding a goal, next discover the causes of these problems. For example, imagine that the group goal is to improve a company's work environment. The team has identified two problems: machine malfunctions and employee dissatisfaction. What is causing these problems? Machines might malfunction due to static electricity in the air, excessively hot and humid room temperatures, or outdated equipment.

Maybe employees have difficulties with childcare, distant parking, benefits packages, or ergonomics like heating, cooling, and office space. The team members cannot propose solutions to problems until they analyze the problems' causes.

Determine Potential Improvements

Now is the time to propose the solutions to the problems. Focus on how these potential improvements will be achieved (strategies, timetables, milestones, personnel involved, costs incurred, facilities impacted).

Verify the Suggested Solutions

The team must verify whether the suggested solutions will work. This might require research, interviews, test cases, scenarios, trial runs, simulations, or computer modeling.

Breach the Gaps

For your team to succeed, challenges must be faced. These could include problems caused by verbal or nonverbal communication, poor listening skills, and conflicts within the group. Other issues that create gaps between what your team hopes to achieve and its ability to do so include the following:

1. *Ergonomics*—a room that is too cold or too hot; badly arranged meeting space (too small, too large, not enough seats, etc.).
2. *Teammates with personal problems*—drugs, alcohol, health, family conflicts, etc.
3. *Unskilled teammates*—coworkers who are not aware of the team's rules and goals, or teammates who do not have the correct computer skills or any other technology prerequisites.
4. *Insufficient resources*—this could include finances, numbers of teammates, and management support.

Complete the Team Project

Once the above challenges are overcome, it is possible for the team to collaboratively accomplish its goal. This might entail revising and printing the finished copy (of a proposal or Web site) or developing PowerPoint slides.

Team projects allow students and employees to help and to learn from each other. The team projects also require that people plan; delegate responsibilities; communicate their attitudes toward word choice, tone, organization, development, and grammar; and learn to work with other people.

Often students and employees say that it is easier just to do it themselves rather than confer and compromise. They are right; it is easier. However, the workplace requires team skills. Employers want workers who can work together successfully. Teamwork requires collaborative communication to solve problems. These are valuable skills that will benefit both you and your company.

CHAPTER HIGHLIGHTS

1. Technical writing is communication written for and about business and industry. Technical writing focuses on products and services.

2. Technical writing is an important part of your everyday work life. It can consume as much as 31 percent of a typical workweek.

3. Technical writing costs a company both time and money, so employees must strive to write effectively.

4. A letter can cost almost $20 to produce and mail.

5. The top five skills employers want include interpersonal skills, teamwork, analytical skills, oral communication, and flexibility.

6. Avoid "silo building," isolating yourself on the job.

7. Employees often work in teams.

8. Working in teams allows you to see issues from several points of view.

9. Following a specific Total Quality Management (TQM) strategy will help your team succeed.

10. Like TQM, Six Sigma encourages teamwork.

11. The Six Sigma method helps you define, measure, analyze, improve, and control a project.

12. Human Performance Improvement (HPI) solves problems—"gaps"—inherent in teamwork.

13. Problems teams face include varied knowledge levels, differing motives, and insufficient resources.

14. Conflict resolution strategies are essential to a team's success.

15. To resolve conflicts in a team, you should set guidelines, encourage all to participate, and avoid taking sides.

CASE STUDIES

Gulfview Architectural and Engineering Services

You are the team leader of a work project at Gulfview Architectural and Engineering Services. The team has been involved in this project for a year. During the year, the team has met weekly, every Wednesday at 8:00 A.M. It is now time to assess the team's successes and areas needing improvement.

Your goal will be to recommend changes as needed before the team begins its second year on this project. You have encountered these problems:

- One team member, Caroline Jensen, misses meetings regularly. In fact, she has missed at least one meeting a month during the past year. Occasionally, she missed two or three in a row. You have met with Caroline to discuss the problem. She says she has had child care issues that have forced her to use the company's flextime option, allowing her to come to work later than usual, at 9:00 A.M.

- Another team member, Guy Stapleton, tends to talk a lot during the meetings. He has good things to say, but he speaks his mind very loudly and interrupts others as they are speaking. He also elaborates on his points in great detail, even when the point has been made.

- A third team member, Sharon Mitchell, almost never provides her input during the meetings. She will e-mail comments later or talk to people during breaks. Her comments are valid and on-topic, but not everyone gets to hear what she says.

- A fourth team member, Craig Mabrito, is very impatient during the meetings. This is evident from his verbal and nonverbal communication. He grunts, slouches, drums on the table, and gets up to walk around while others are speaking.

- A fifth employee, Julie Jones, is overly aggressive. She is confrontational, both verbally and physically. Julie points her finger at people when she speaks, raises her voice to drown out others as they speak, and uses sarcasm as a weapon. Julie also crowds people, standing very close to them when speaking.

Assignment

How will you handle these challenges? Either individually or as a team, decide on a course of action. To do so, consider the TQM Plan of Action (plan, do, check, act), Six Sigma's DMAIC (define, measure, analyze, improve, control), or a combination of the two approaches to solve the problem.

For example, try this approach:

1. *Analyze* the problem(s). To do so, brainstorm. What gaps might exist causing these problems?

2. *Invent* or envision solutions. How would you solve the problems? Consider Human Performance Improvement issues, as discussed in this chapter.

3. *Plan* your approach. To do so, establish verifiable measures of success (including timeframes and quantifiable actions).

Write an e-mail to your instructor sharing your findings.

Quick and Sure Delivery

Quick and Sure Delivery (QSD) has not been either quick or sure lately. Customer complaints are up 23 percent this quarter (QSD receives an average of only three complaints a month). Clients are telling customer service representatives that deliveries are arriving up to 10 hours later than promised. In addition, delivered goods are being left unattended outside homes and businesses. This has led to damages due to rain, and on at least five instances, delivered packages have been stolen.

Because QSD guarantees that packages will be handed directly to a home or business owner and never left unattended, these occurrences are actionable under the law. More important, QSD's reputation is being harmed. Already, word is getting around, and customers are taking their business elsewhere. Business is down 12 percent this month, when compared to last year at the same time (QSD made 1,578 deliveries during the month last year, charging an average of $27 per delivery). Something must be done.

Assignment

Form a team to study this problem. Your team's goal will be to improve QSD's quality and performance. First, quantify the damage done (based on the numbers above). Then, use either the TQM Plan of Action (plan, do, check, act), Six Sigma's DMAIC, or a combination of the two approaches to solve the problem.

For example, try this approach:

1. *Analyze* the problem(s). To do so, in the group brainstorm possible causes. What gaps might exist (personnel, checks and balances, management assessments, missing incentives, vehicle problems, weather, etc.)?

2. *Invent* or envision solutions. How would you solve the problems? Consider Human Performance Improvement issues, as discussed earlier.

3. *Plan* your approach. To do so, establish verifiable measures of success (including timeframes and quantifiable actions).

Once you have made your decisions, report your team's findings as follows:

- Write an e-mail or memo to your instructor explaining
  - why you are writing
  - what you are writing about
  - what exactly your team's plan of action is—what your recommendations are for solving QSD's problem
  - when you plan to complete the project (provide tentative due dates and ways in which you can measure success)

EXERCISES

Teamwork—Business and Industry Expectations

Individually or in small groups, visit local banks, hospitals, police or fire stations, city offices, service organizations, manufacturing companies, engineering companies, or architectural firms. Once you and your teammates have visited these sites, have asked your questions (see the following assignments), and have completed your research, share your findings using one of the following methods:

- *Oral*—as a team, give a three- to five-minute briefing to share with your colleagues the results of your research.

- *Oral*—invite employee representatives from other work environments to share with your class their responses to your questions.
- *Written*—write a team memo, letter, or report about your findings.

1. Ask employees at the sites you visit if, how, and how often they are involved in team projects. In your team, assess your findings and report your discoveries.

2. Ask employees at the sites you visit about the challenges they face with conflict resolution. In your team, assess your findings and report your discoveries.

3. Ask employees at the sites you visit what gaps they see between what the company or agency experiences and what the company or agency hopes to achieve. What intervention methods do they think could help close these gaps? In your team, assess your findings and report your discoveries.

4. Use the Internet and/or your library to research companies that rely on team-work. Focus on which industries these companies represent and the goals of their team projects. You could also consider the challenges they encounter, their means of resolving conflicts, the numbers of individuals on each team, and whether the teams are cross-functional. Then report these findings to your professor or classmates, either orally or in writing.

5. Visit one or more companies in your career field and interview employees to see how much time they spend writing, what types of documents they write, and how often they work with others in project teams.

6. Invite guest speakers from corporations to discuss their writing activities at work.

7. Research the Internet for technical writing job opportunities to determine what skills are required, what types of writing are performed, and what types of industries employ technical writers.

8. Visit the Society for Technical Communication (STC) Web site to learn about its membership. See which industries employ technical writers and determine these writers' job responsibilities. Also, learn which colleges and universities have programs in technical writing and what the programs entail. What else can you learn about technical writing from the STC Web site?

9. Research major publications of technical communication, such as *Intercom, Technical Communication,* and *The Journal of Scientific and Technical Communication.* What topics do the articles in these journals focus on?

Conflict Resolution

To understand and practice conflict resolution, complete the following assignments.

1. *Attend a meeting.* This could be at your church, synagogue, or mosque; a city council meeting; your school, college, or university's board of trustees meeting; or a meeting at your place of employment. Was the meeting successful? Did it have room for improvement? To help answer these questions, use the following Conflict Resolution in Team Meetings Matrix. Then report your findings to your professor or classmates, either orally or in writing. Write an e-mail message, memo, or report, for example.

CONFLICT RESOLUTION IN TEAM MEETINGS MATRIX			
Goals	Yes	No	Comments
1. Were meeting guidelines clear?			
2. Did the meeting facilitator encourage equal discussion and involvement?			
3. Were the meeting's attendees discouraged from taking sides?			
4. Did the meeting facilitator seek consensus?			
5. Were topics tabled if necessary?			
6. Were subcommittees created if necessary?			
7. Did the meeting facilitator find the good in the bad?			
8. Did the meeting facilitator deal with individuals *individually?*			
9. Did the meeting's facilitator stay calm?			

2. Have you been involved in a team project at work or at school? Perhaps you and your classmates grouped to write a proposal, research Web sites, create a Web site, or perform mock job interviews. Maybe you were involved in a team project for another class. Did the team work well together? If so, analyze how and why the team succeeded. If the team did not function effectively, why not? Analyze the gaps between what should have been and what was. To help you with this analysis, use the following Human Performance Index Matrix. Then, report your findings to your instructor or classmates either orally or in writing. Write an e-mail message, memo, or report, for example.

HUMAN PERFORMANCE INDEX MATRIX			
Potential Gaps	Yes	No	Comments
1. Did teammates have equal and appropriate levels of knowledge to complete the task?			
2. Did teammates have equal and appropriate levels of motivation to complete the task?			
3. Did the team have sufficient resources to complete the task?			

Potential Gaps	Yes	No	Comments
4. Did teammates understand their roles in the process needed to complete the task?			
5. Did the team have sufficient and up-to-date information to complete the task?			
6. Did the team have sufficient support to complete the task?			
7. Did wellness issues affect the team's success?			

QUIZ QUESTIONS

1. Define technical writing.

2. What are five types of technical writing?

3. List three reasons why technical writing is important in business.

4. Based on Table 1.1, how many e-mail messages, memos, and letters do employees read and write each day?

5. How much does it cost to write and mail a letter today?

6. What is the percentage of time employees spend writing?

7. What are the top five skills employers want?

8. Define *silo building*.

9. Explain why business depends on teamwork to ensure quality.

10. What are three business methodologies geared toward teamwork?

11. What is the Small Group Activity Circle?

12. What is Six Sigma?

13. List five causes for collaborative breakdown, according to Human Performance Improvement (HPI).

14. List four HPI intervention options to help a team solve its problems.

15. List three things a team leader can do to ensure successful teamwork.

Producing the Product

The Writing Process: An Overview
The writing process—prewriting, writing, and rewriting—will help you write well. The three parts of the process are dynamic and frequently overlap.

Prewriting
Before writing your document, you must examine your purposes, determine your goals, consider your audience, gather your data, and determine how the content will be provided.

Writing
After prewriting, the next step is to draft your text by organizing and formatting.

Rewriting
This stage of the writing process allows you to revise, polishing your document to make it perfect.

The Process in Practice
Follow a sample letter from rough draft to final copy to see the writing process at work.

Technical writing is a major part of your daily work experience. It takes time to construct the correspondence, and your writing has an impact on those around you. A well-written memo, letter, report, or e-mail message gets the job done and makes you look good. Poorly written correspondence wastes time and creates a negative image of you and your company.

But recognizing the importance of technical writing does not ensure that your correspondence will be well written. How do you effectively write the memo, letter, or report? How do you successfully produce the finished product?

To produce successful technical writing, you need to approach writing as a process.

THE WRITING PROCESS: AN OVERVIEW

The process approach to writing requires the following sequence:

1. *Prewrite*—Before you can write your technical document, you must have something to say. Prewriting allows you to spend quality time, *prior* to writing the correspondence, generating information. In prewriting, you

- Examine your purposes
- Determine your goals
- Consider your audience
- Gather your data
- Determine how the content will be provided

2. *Write*—Once you have gathered your data and determined your objectives, the next step is to state them. You need to draft your document. To do so, you should (a) *organize* the draft according to some logical sequence that your readers can follow easily and (b) *format* the content to allow for ease of access.

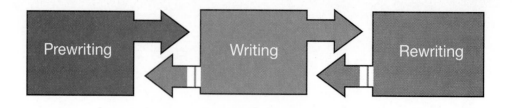

The Writing Process		
Prewriting	Writing	Rewriting
• Examine your purposes • Determine your goals • Consider your audience • Gather your data • Determine how the content will be provided	• Organize the draft according to some logical sequence that your readers can follow easily • Format the content to allow for ease of access	• Revise 　° Add missing details 　° Delete wordiness 　° Simplify word usage 　° Enhance the tone of your communication 　° Reformat your text for ease of access 　° Practice the speech or review the text • Proofread 　° Correct errors

Figure 2.1   The Writing Process

3. *Rewrite*—The final step, and one that is essential to successful writing, is to rewrite your draft. This step requires that you revise the rough draft. Revision allows you to perfect your memo, letter, or report so you can be proud of your final product.

The writing process is dynamic, with the three parts—prewriting, writing, and rewriting—often occurring simultaneously. You may revisit any of these parts of the process at various times as you draft your document. The writing process is shown in Figure 2.1.

PREWRITING

tech link

Go to *http://www.prenhall. com/gerson* for Web links, samples, and interactive activities.

Prewriting, the first stage of the process, allows you to plan your communication. If you do not know where you are going in the correspondence, you will never get there, and your audience will not get there with you. Through prewriting, you accomplish many objectives, including

- Examining your purposes
- Determining your goals
- Considering your audience
- Gathering your data
- Determining how the content will be provided

Examine Your Purposes

Before you write the document, you need to know why you are communicating. Are you planning to write because you have chosen to do so of your own accord or because you have been asked to do so by someone else? In other words, is your motivation *external* or *internal*?

External Motivation

If someone else has requested the correspondence, then your motivation is external. Your boss, for example, expects you to write a monthly status report, a performance appraisal of your subordinate, or a memo suggesting solutions to a current problem. Perhaps a vendor has requested that you write a letter documenting due dates, or a customer asks that you respond to a letter of complaint. In all of these instances, someone else has asked you to communicate.

Internal Motivation

If you have decided to write on your own accord, then your motivation is internal. You need information to perform your job more effectively, so you write a letter of inquiry. You need to meet with colleagues to plan a job, so you write an e-mail message calling a meeting and setting an agenda. Perhaps you recognize a problem in your work environment, so you create a questionnaire and transmit it via the company intranet. Then, analyzing your findings, you call a meeting to report your findings. In all of these instances, you initiate the communication.

Determine Your Goals

Once you have examined why you are planning to communicate, the next step is to determine your goals in the correspondence or presentation. You might be communicating to

- Persuade an audience to accept your point of view
- Instruct an audience by directing actions
- Inform an audience of facts, concerns, or questions you might have
- Build trust and rapport by managing work relationships

These goals can overlap, of course. You might want to inform by providing an instruction. You might want to persuade by informing. You might want to build trust by persuading. Still, it is worthwhile looking at each of these goals individually to clarify their distinctions.

Figures 2.2 and 2.3 depict the interrelationship of these four communication goals.

Communicating to Persuade

If your goal in writing is to change others' opinions or a company's policies, you need to be persuasive. For example, you might want to write a proposal, a brochure, or a flier to sell a product or a service. Maybe you will write your annual progress report to justify a raise or a promotion. As a customer, you might want to write a letter of complaint about poor service. Your goal in each of these cases is to persuade an audience to accept your point of view.

Communicating to Instruct

Instructions will play a large role in your technical communication activities. As a manager, for example, you often will need to direct action. Your job demands

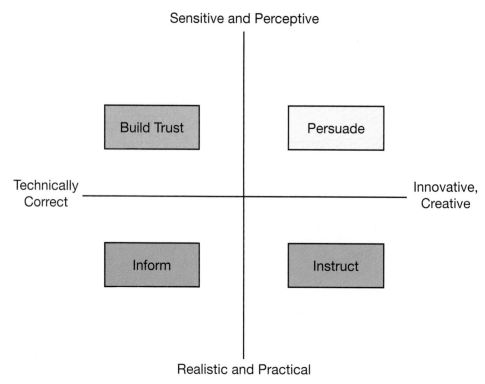

Source: Adapted from Quinn et al. 1991, and Campbell et al. 2003.

Figure 2.2   Communicating Goals

that you tell employees under your supervision what to do. You might need to write an e-mail providing instructions for correctly following procedures. These could include steps for filling out employee forms, researching documents in your company's intranet data bank, using new software, or writing reports according to the company's new standards.

As an employee, you also will provide instructions. As a computer information specialist, maybe you work the 1-800 hotline for customer concerns. When a customer calls about his computer's crisis, your job would be to give instructions for correcting the problem. You either will provide a written instruction in a follow-up e-mail or a verbal instruction while on the phone.

Communicating to Inform

Often, you will write letters, reports, and e-mails merely to inform. In an e-mail message, for instance, you may invite your staff to an upcoming meeting. A trip report will inform your supervisor what conference presentations you attended or what your prospective client's needs are. A letter of inquiry will inform a vendor about questions you might have regarding her services. Maybe you will be asked to write a newsletter informing your coworkers about the corporate picnic, personnel birthdays, or new stock options available to employees. In these situations, your goal is not to instruct or persuade. Instead, you will share information objectively.

Communicating to Build Trust

Building rapport (empathy, understanding, connection, and confidence) is a very important component of your communication challenge. As a manager or

DATE: April 15, 2005
TO: Web Design Team
FROM: Doug Yost
SUBJECT: WEB SITE IMPLEMENTATION MEETING

Please attend our first Web site implementation meeting, scheduled for April 20, 11:00 A.M.–1:00 P.M. in Room 204.

Informing the team about the Web site meeting

To ensure productivity, I am asking that each of you prepare the following prior to our meeting:

1. Josh—inventory our stock product photos. Then determine if we will need to upgrade our graphics for the Web site's online shopping cart. Your job also will be to redesign our corporate logo.
2. Tasha—research our competitors. Find out which components of their Web sites we might need to include in ours. More important, determine new screens we could add to make our site unique.
3. Ychun—contact our site administrator to determine optimum load-up time. This will help Josh and the team decide how many graphics to use.
4. Susan—mock up a storyboard for the proposed Web site. Visit with our staff in sales, accounting, human resources, and information technology to get their ideas.

Instructing the team about their responsibilities prior to the Web design meeting

This is an important meeting, as you all know. Without a Web site, our company has fallen behind the competition. Though our local market share is sound, our national and global sales are at least 56 percent below goal. The quick fix for this is a Web site, which will allow us to reach millions of potential clients at a keystroke. With an outstanding Web site, our company's stock should increase, and that will mean bonuses for all.

Persuading the team by helping them understand the importance of this project

I have chosen you four employees for this project not only for your expertise but also because of your proven record of excellence. You have worked well together on past projects. I am confident that again you will excel. Thank you for your talents.

Building trust to ensure that people enjoy working together and that everyone feels empowered

P.S. Lunch is on me. I have chosen a vegetarian pasta and salad to accommodate everyone's nutritional needs.

Figure 2.3   Communication Goals in Practice

employee, your job is not merely to "dump data" in your written communication. You also need to realize that you are communicating with coworkers, people with whom you will work every day. To maintain a successful work environment, you want to achieve the correct, positive tone in your writing.

This might require nothing more than saying "Thanks for the information," or "You've done a great job reporting your findings." A positive tone shows approval for work accomplished and recognition of the audience's time.

For more detail on audience recognition and involvement, read Chapter 4.

Recognizing the goals for your correspondence makes a difference. Determining your goals allows you to provide the appropriate tone and scope of detail in your communication. In contrast, failure to assess your goals can cause communication breakdowns.

Consider Your Audience

What you say and how you say it is greatly determined by your audience. Are you writing up to management, down to subordinates, or laterally to coworkers? Are you speaking to a high-tech audience (experts in your field), a low-tech audience (people with some knowledge about your field), or a lay audience (people outside your work environment)? Face it—you will not write the same way to your boss as you would to your subordinates. You will not speak the same way to a customer as you would to a team member. You must provide different information to a multicultural audience than you would to individuals with the same language and cultural expectations. You must consider issues of diversity when you communicate. Each of these points is discussed in detail in Chapter 4.

Gather Your Data

Once you know why you are writing and who your audience is, the next step is deciding what to say. You have to gather data. The page remains blank until you fill it with content. Your communication, therefore, will consider personnel, dates, actions required, locations, costs, methods for implementing suggestions, and so forth. As the writer, it is your obligation to flesh out the detail. After all, until you tell your readers what you want to tell them, they do not know.

There are many ways to gather data. In this chapter, and throughout the textbook, we provide options for gathering information. These planning techniques include

- Answering the reporter's questions
- Mind mapping
- Brainstorming or listing
- Outlining
- Storyboarding
- Creating organization charts
- Flowcharting
- Researching (online or at the library)

Each is discussed in greater detail on the following pages in Table 2.1 (except for research techniques which we discuss in Chapter 14). Table 2.2 lists some good Web sites for online research.

TABLE 2.1 PREWRITING TECHNIQUES		

Answering the Reporter's Questions	Sample Reporter's Questions	
By answering *who, what, when, where, why,* and *how*, you create the content of your correspondence.	*Who*	Joe Kingsberry, Sales Rep
	What	Need to know • what our discount is if we buy in quantities • what the guarantees are • if service is provided on-site • if the installers are certified and bonded • if Acme provides 24-hour shipping
	When	Need the information by July 9 to meet our proposal deadline
	Where	Acme Radiators 11245 Armour Blvd. Oklahoma City, Oklahoma 45233 Jkings@acmerad.com
	Why	As requested by my boss, John, to help us provide more information to prospective customers
	How	Either communicate with a letter or an e-mail. I can write an e-mail inquiry to save time, but I must tell Joe to respond in a letter with his signature to verify the information he provides.

Mind Mapping	
Envision a wheel. At the center is your topic. Radiating from this center, like spokes of the wheel, are different ideas about the topic. Mind mapping allows you to look at your topic from multiple perspectives and then cluster the similar ideas.	

Brainstorming or Listing	Improving Employee Morale
Performing either individually or with a group, you can randomly suggest ideas (brain-storming) and then make a list of these suggestions.	• Before meetings, ask employees for agenda items (that way, they can feel empowered) • Consider flextime • Review employee benefits packages • Hold yearly awards ceremony for best attendance, highest performance, most cold calls, lowest customer complaints, etc. • Offer employee sharing for unused personal days/sick leave days • Roll over personal days to next calendar year

This method, which works for almost all kinds of communication, is especially valuable for team projects.	• Include employees in decision-making process • Add more personal days (as a tradeoff for anticipated lower employee raises)

Outlining
 	
This traditional method of gathering and organizing information allows you to break a topic into major and minor components. This is a wonderful all-purpose planning tool.
	Topic Outline
	
	1.0 The Writing Process
	   1.1 Prewriting
	     • Planning Techniques
	   1.2 Writing
	     • All-Purpose Organizational Template
	     • Organizational Techniques
	   1.3 Rewriting
	2.0 Criteria for Effective Technical Writing
	   2.1 Clarity
	   2.2 Conciseness
	   2.3 Document Design
	   2.4 Audience Recognition
	   2.5 Accuracy

Storyboarding Storyboarding is a visual planning technique that lets you graphically sketch each page or screen of your text. This allows you to see what your document might look like.	**Brochure Storyboard** 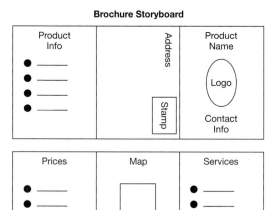
Creating Organization Charts This graphic allows you to see the overall organization of a document as well as the subdivisions to be discussed.	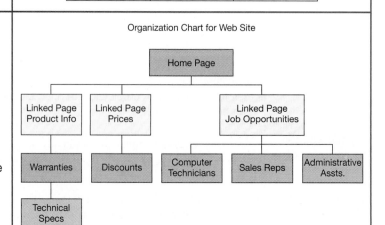

Flowcharting

Flowcharting is another visual technique for gathering data. Because flowcharting organizes content chronologically, it is especially useful for instructions. For example:

Stop/Start = ◯

Step = ▢

Decision = ◇

TABLE 2.2 INTERNET SEARCH ENGINES	
Purpose	Sites
Popular online search engines	Yahoo.com, Excite.com, Google.com, AltaVista.com, Lycos.com, alltheweb.com, Ask Jeeves, and HotBot.com.
Meta-search engines (multithreaded engines that search several major engines at once)	MetaCrawler.com, Dogpile.com, and Vivisimo.com.
Specialty search engines	Findlaw.com focuses on legal resources. Achoo.com lets you access health and medical sites.
Broad academic searches	Librarians' Index to the Internet (*http://lii.org*) and Infomine.ucr.edu.
Business search engines	ZDNet.com, EarthLink.net, Business Week Online (*http://www.businessweek.com*), and AbusinessResource.com. For information about business news in Great Britain, look at All Search Engines.com (*http://www.allsearchengines.co.uk/ business_list.htm*).
Government search sites	First Gov (*http://firstgov.gov/*) and Google's Uncle Sam (*http://www.google.com/unclesam*).
International search sites	Search Engine Colossus, Abyz News Links (international newspapers and magazines), and World Press Review (international perspectives on the United States).
Multipurpose search engines	All Search Engines.com gives you a one-stop search site for exactly what it says: *all search engines* (*http://www.allsearchengines.com/*).

Determine How the Content Will Be Provided

After you have determined your audience, your goals, and your content, the last stage in prewriting is to decide how best to convey your message. Will you write a letter, a memo, a report, an e-mail, a Web site, a proposal, an instructional procedure, a flier, or a brochure?

Single Sourcing

Maybe you will create content that will be used in many different ways simultaneously. A new phrase currently used in technical writing is *single sourcing*, the act of "producing documents designed to be recombined and reused across projects and various media" (Carter 2003, 317). In a constantly changing marketplace, you will need to communicate your content to many different audiences using many different types of communication channels. For instance, you might need to market your product or service using the Internet, a flier, a brochure, newsletters, and a sales letter. You might need to write hard-copy user manuals and develop online help screens. To ensure that content is reusable, the best approach would be to write a "single source of text" that will "generate multiple documents for different media" (Albers 2003, 337).

In Table 2.3, you can review the many channels or methods you may use for communicating your content.

TABLE 2.3 OPTIONS FOR PROVIDING CONTENT	
Communication Channels	**Good News/Bad News**
E-mail messages	**Benefits:** E-mail messages are quick and almost synchronous. You can have a real-time, electronic chat with one or more readers. Though e-mail messages should be very short (20 or so lines of text), you can attach documents, Web links, graphics, and sound and movie files for review. **Challenges:** E-mail tends to be less formal than other types of communication. E-mail might not be private (a company's network administrators can access your electronic communication).
Cell phone e-mail messages	**Benefits:** With the appropriate equipment, you can receive and send e-mail messages via your cell phone. The benefits are anyplace, anytime communication. **Challenges:** Cell phone monitors are as small as 1 × 1 inch, challenging readability. Cell phone e-mail costs you minutes or the price of the service. You can type an e-mail message only if you have a combination cell phone/personal digital assistant (PDA). You type the message using a very small stylus on a very small keypad. Attachments are virtually impossible on cell phone e-mails.
PDA e-mail messages	**Benefits:** Wireless PDAs with e-mail capabilities provide you all the benefits of e-mail and cell phone e-mail. A PDA monitor is larger than a cell phone, but smaller than your computer screen. **Challenges:** Readability and typing with a stylus are challenges. Attachments are very difficult on PDA e-mails.

Communication Channels	Good News/Bad News
Letters	**Benefits:** Typed on official corporate letterhead stationery, letters are formal correspondence to readers outside your company. **Challenges:** Letters are time consuming because they must be mailed physically. Although you can enclose documents, this might demand costly or bulky envelopes.
Memos	**Benefits:** Memos—internal correspondence to one or several coworkers—allow for greater privacy than e-mail (e-mail can be kept in corporate computer banks and observed by administrators within a company). Even though most memos are limited in length (one or two pages), you can attach or enclose documents. **Challenges:** Memos are both more time consuming than e-mail and less formal than letters.
Reports	**Benefits:** Reports, internal and external, are usually very formal. They can range in length, from one page to hundreds of pages (proposals and annual corporate reports to stakeholders, for example). Because of their length, reports are appropriate for extremely detailed information. **Challenges:** They can be time consuming to write.
Brochures	**Benefits:** Brochures are appropriate for informal informational and promotional communication to large audiences. **Challenges:** Most brochures are limited to six or so panels, the equivalent of a back and front hard copy. Thus, in-depth coverage of a topic will not occur.
Newsletters	**Benefits:** Newsletters are appropriate for formal and informal communication to large audiences, both within and outside a company. Newsletters, unlike brochures, can be long, thus allowing for very in-depth coverage of topics. **Challenges:** Because color and layout are key to newsletters, they can be costly and time consuming to create.
Web sites	**Benefits:** An Internet Web site can provide informal and public communication to the entire world—anytime, anyplace (with the appropriate technological connections). A company can have a firewall-protected intranet or extranet to allow more private communication for a large, selected audience. Web sites essentially have unlimited size, so you can provide lots of information, and the content can be updated instantaneously by Web designers. A Web site can include links to other sites, animation, graphics, and color. **Challenges:** Technology is a requirement.
Microsoft PowerPoint	**Benefits:** PowerPoint slides enhance written and oral communication, not only making correspondence look more professional but also aiding clarity. A pie chart, bar chart, line graph, or map within a PowerPoint presentation can make complex information more clear. **Challenges:** PowerPoint slides usually convey only key points or a synopsis, rather than very lengthy details.

WRITING

Writing lets you *package* your data. Once you have gathered your data, determined your objectives, and recognized your audience, the next step is writing the document. You need to package it (the draft) in such a way that your readers can follow your train of thought readily and can easily access your data. Writing the draft lets you *organize* your thoughts in some logical, easy-to-follow sequence. Writers usually know where they are going, but readers do not have this same insight. When readers pick up your document, they can read only one line at a time. They know what you are saying at the moment, but they don't know what your goals are. They can only hope that in your writing, you will lead them along logically and not get them lost in back alleys of unnecessary data or dead-end arguments.

Organization

To avoid leading your readers astray, you need to organize your thoughts. As with prewriting, you have many organizational options. In Chapter 3, we discuss organizing according to

- space (spatial organization)
- chronology
- importance
- comparison/contrast
- problem/solution

These organizational methods are not exclusive. Many of them can be used simultaneously within a memo, letter, or report to help your reader follow your train of thought.

Formatting

You also must *format* your text to allow for ease of access. In addition to organizing your ideas, you need to consider how the text looks on the page. If you give your readers a massive wall of words, they will file your document for future reading and look for the nearest exit. An unbroken page of text is not reader-friendly. To invite your readers into the document, to make them want to read the memo, letter, or report, you need to highlight key points and break up monotonous-looking text. You need to ensure that your information is accessible. (See Chapter 8 for more on formatting.)

REWRITING

Rewriting lets you *perfect* your writing. After you have prewritten (to overcome the blank page syndrome) and written your draft, your final step is to rewrite. There are no good writers, only good rewriters. People who write effective documents know that doing so requires a second or third write. Good writers fine-tune, hone, sculpt, and polish their drafts to make sure their final versions are perfect. To rewrite, you need to *revise*, *revise*, and *revise* again. Revision requires that you look over your draft and do the following:

- *Add* any missing detail for clarity.
- *Delete* dead words and phrases for conciseness.

- *Simplify* unnecessarily complex words and phrases to allow for easier understanding.
- *Move* around information (cut and paste) to ensure that your most important ideas are emphasized.
- *Reformat* (using highlighting techniques) to ensure reader-friendly ease of access.
- *Enhance* the tone and style of the text.
- *Correct* any errors to ensure accurate grammar and content.

We discuss each of these points in greater detail throughout the text.

Revision is possibly the most important stage in the writing process. If you prewrite effectively (gathering your data, determining your objectives, and recognizing your audience) and write an effective draft, you are off to a great start. However, if you then fail to rewrite your text, you run the risk of having wasted the time you spent prewriting and writing. Rewriting is the stage in which you make sure that everything is just right. Failure to do so not only can cause confusion for your readers but also can destroy your credibility.

Technology Tips

Using Word Processing Programs to Rewrite

Word processing programs help you rewrite your document in many ways:

- *Spell check*—whewn you misspell a word, often spell check will underline the error in red (as shown in this example). Spell check, unfortunately, will not catch all errors. If you use a word like *to* instead of *too*, spell check will not **no** the difference (of course, that should be "know" but spell check did not mark the error).

- *Grammar check*—word processors also can help you catch grammar errors. Grammar check underlines errors in green. When you right-click on the underlined text, the word processing package will provide an optional correction. Agreement error look like this one, for example.

error looks
errors look

- *Add/Delete*—word processing makes adding new content and deleting unneeded text very easy. All you need to do is place your cursor where you want to add/delete. Then, to add, you type. To delete, you hit the Backspace key or the Delete key.

- *Move*—the Copy, Cut, and Paste features of word processing allow you to move text with ease.

- *Enhance/Reformat*—In addition to changing the tone of your text, you also can enhance the visual appeal of your document at a keystroke. These include bullets, *italics*, **boldface**, FONT CHANGES, numbered lists, etc. Graphics (charts and figures) are easy to add also.

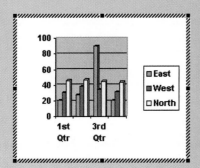

The process approach to writing—including prewriting, writing, and rewriting (usability testing)—can help you write successfully in any work environment or writing situation. In fact, the greatest benefit of process is that it is generic. Process is not geared to any one profession or type of correspondence. No author of a technical writing book can anticipate exactly where you will work, what type of document you will be required to write, or what your supervisors will expect in your writing. However, we *can* give you a methodology for tackling any writing activity. Writing as a process will help you write any kind of technical document, for any boss, in any work situation.

THE PROCESS IN PRACTICE

Following is a letter produced using the process approach to writing. The document was produced in the workplace by a senior transportation analyst for an international cosmetics firm. He had to write a problem/solution follow-up letter to a sales representative.

Prewriting

A senior transportation analyst received a phone call from a disgruntled sales representative. The sales rep had not received a shipment of goods on time, and the shipment was incomplete when it did arrive. While talking to the sales rep, the analyst jotted down notes, as shown in Figure 2.4 (using the listing method of prewriting).

In addition to listing, the transportation analyst used another prewriting technique—reporter's questions. The note tells us *who* the sales rep is (Beth); *what* her Social Security number, phone number, and sales area are; *what* her problem is (late and missing goods); *where* the shipment originated (Denver); *how* much was ordered ($700); and *when* the shipment was due (two weeks ago). By jotting down this list, the analyst is gathering data.

After concluding his discussion with Beth, the analyst contacted his manager to decide what to do next. This time, the analyst wrote down a list of objectives, as determined by his manager, as shown in Figure 2.5.

The list again answers the reporter's questions: *what* to do (write a letter), *who* gets a copy (manager), *what* to focus on in the letter (we understand your

Figure 2.4   Listing

> Send letter to sales rep.
>
> Send copy to manager.
>
> In letter
>
> > discuss problem encountered.
> >
> > show alternative method of shipment
> > for better control.
>
> Call manager for further help if needed.

Figure 2.5   Listing Objectives

problem; here is an alternative), and *why* to pursue the alternative (better control of shipment).

With data gathered and objectives determined, the analyst was ready to write.

Writing

First, the analyst wrote a rough draft (revising it as he wrote), as shown in Figure 2.6. In this draft, the analyst made subtle changes by adding new detail and deleting unnecessary words. However, he was unsatisfied with this draft, so he tried again (see Figure 2.7).

As is evident from the first two drafts (Figures 2.6 and 2.7), the analyst took the word *rough* seriously. When you draft, do not worry about errors or how the correspondence looks. It is meant to be rough, to free you from worry about making errors. You can correct errors when you revise.

Once the analyst drafted the letter, he typed a clean copy for his manager's approval (Figure 2.8). At this point, the manager added a dateline and added content to the second paragraph. He deleted wordiness in the second and third paragraphs. By deleting the entire fourth paragraph, the manager enhanced the tone of the document (see Figure 2.9).

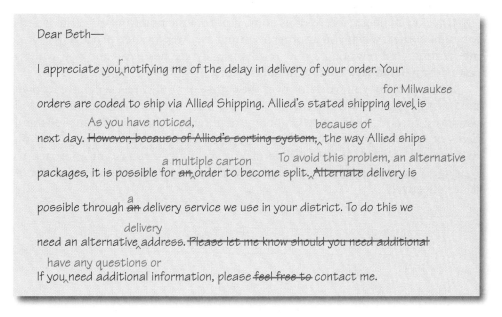

Figure 2.6   Rough Draft

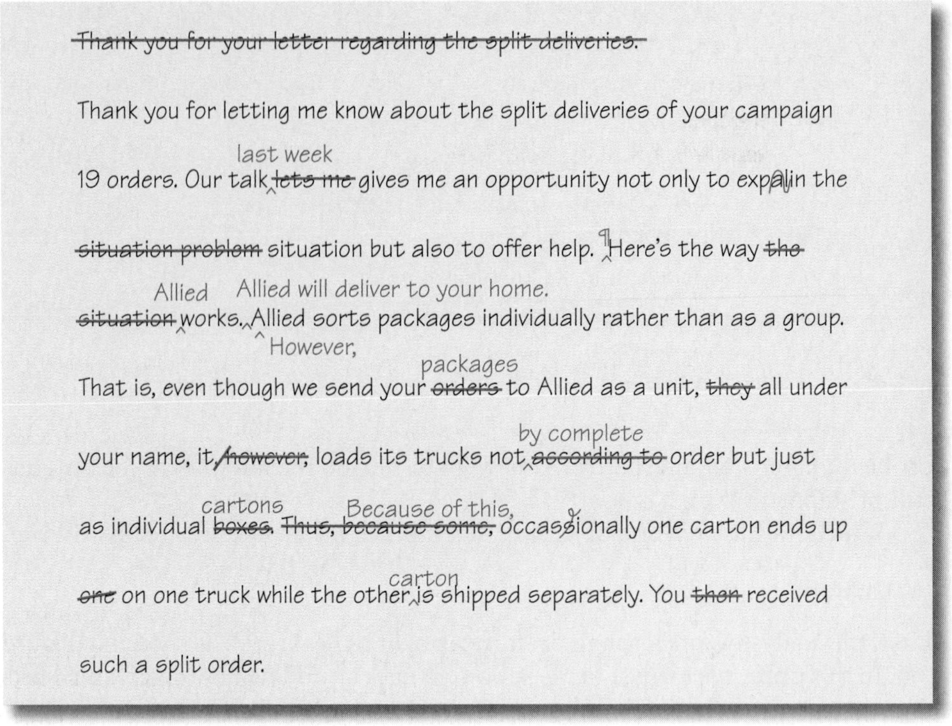

Thank you for your letter regarding the split deliveries.

Thank you for letting me know about the split deliveries of your campaign
19 orders. Our talk ~~lets me~~ *last week* gives me an opportunity not only to explain the
~~situation problem~~ situation but also to offer help. Here's the way ~~the~~
~~situation~~ works. *Allied* Allied will deliver to your home. Allied sorts packages individually rather than as a group.
However, That is, even though we send your ~~orders~~ *packages* to Allied as a unit, ~~they~~ all under
your name, it ~~however,~~ loads its trucks not ~~according to~~ *by complete* order but just
as individual ~~boxes~~ *cartons*. ~~Thus, because some,~~ *Because of this,* occasionally one carton ends up
~~one~~ on one truck while the other *carton* is shipped separately. You ~~then~~ received
such a split order.

Figure 2.7 Second Rough Draft

Rewriting

No writing is ever perfect. Every memo, letter, or report can be improved. Note how the manager improved the analyst's typed draft. When the senior transportation analyst received the revised letter from his manager, he typed and mailed the final version (Figure 2.10).

Once the manager received his copy, he wrote the note you see in the letter's top right corner. When you approach writing as a step-by-step process (prewriting, writing, and rewriting), your results usually are positive—and you will receive positive feedback from your supervisors.

Each company you work for over the course of your career will have its own unique approach to writing memos, letters, and reports. Your employers will want you to do it their way. Company requirements vary. Different jobs and fields of employment require different types of correspondence. However, you will succeed in tackling any writing task if you have a consistent approach to writing. A process approach to writing will allow you to write any correspondence effectively.

Mrs. Beth Fox
6078 Browntree
Milwaukee, WI 53131

Dear Mrs. Fox:

Thanks for letting me know about the split delivery of your Campaign 19 order. Our talk last week gives me an opportunity not only to explain the situation but also to offer help.

Here's the way Allied Shipping works. Allied will deliver to your home; however, Allied sorts packages individually rather than as a group. That is, even though we send your packages as a unit (all under your name), Allied loads its trucks *not by complete order, but just as individual cartons.* Because of this, occasionally, one carton ends up on one truck with another carton shipped separately. You received such a split order. This is an inherent flaw in Allied's system.

Because we understand this problem, we have an alternative delivery service for you. Here is our option. Free of charge, you can have your order delivered by our delivery agent, who does not split orders. Our agent, however, will deliver only within a designated area. All we need from you is an alternative address of a friend or relative in the designated delivery area.

I realize that neither of these options is perfect. Still, I wanted to share them with you. Your district manager now can help you decide which option is best for you.

Sincerely,

David L. Porter
Senior Transportation Analyst

— Date has been omitted

— Negative tone places blame on a vendor

— The writer avoids taking responsibility for the problem

Figure 2.8   Third Draft

Mrs. Beth Fox
6078 Browntree
Milwaukee, WI 53131

Dateline?

Dear Mrs. Fox:

Thanks for letting me know about the split delivery of your Campaign 19 order. Our talk last week gives me an opportunity not only to explain the situation but also to offer help.

~~Here's the way Allied Shipping works.~~ Allied will deliver to your home; however, Allied sorts packages individually rather than as a group. ~~That is,~~ even though we send your packages as a unit (all under your name), Allied loads its trucks *not by complete order, but ~~just as~~ individual cartons*. Because of this, occasionally, one carton ends up on one truck with another carton shipped separately. You received such a split order. This is an inherent flaw in Allied's system. ^

However, Allied constantly works with us to eliminate these service failures.

whose system has better control of orders

Because we understand this problem, we have an alternative delivery service for you. ~~Here is our option.~~ Free of charge, you can have your order delivered by our delivery agent, ~~who does not split orders.~~ Our agent, however, will deliver only within a designated area. ~~All we need from you is an alternative address of a friend or relative in the designated delivery area.~~ *If you would like this service, we will need an alternative delivery address in Milwaukee.*

~~I realize that neither of these options is perfect. Still, I wanted to share them with you. Your district manager now can help you decide which option is best for you.~~

Should you be unable to establish a different delivery address, we will still work with Allied to ensure that you receive home delivery of your complete orders.

Sincerely,

David L. Porter
Senior Transportation Analyst

Figure 2.9   Revised Draft

Carefree Cosmetics
83rd and Preen
Kansas City, MO 64141

September 21, 2005

Mrs. Beth Fox
6078 Browntree
Milwaukee, WI 53131

Great Job, David

Dear Mrs. Fox:

Thanks for letting me know about the split delivery of your Campaign 19 order. Our talk last week gives me an opportunity not only to explain the situation but also to offer help.

Allied will deliver to your home; however, Allied sorts packages individually rather than as a group. Even though we send your packages as a unit (all under your name), Allied loads its trucks not by complete order but by individual cartons. Because of this, occasionally one carton ends up on one truck with another carton shipped separately. You received such a split order. This is an inherent flaw in Allied's system; however, Allied constantly works with us to eliminate these service failures.

Because we understand this problem, we have an alternative delivery service for you. Free of charge, you can have your order delivered by our delivery agent, whose system has better control of orders. Our agent, however, delivers only within a designated area. If you would like this service, we will need an alternate delivery address in Milwaukee.

Should you be unable to establish a different delivery address, we will still work with Allied to ensure that you receive home delivery of your complete orders.

Sincerely,

David L. Porter

David L. Porter
Senior Transportation Analyst

pc: R. H. Handley

Figure 2.10   Finished Letter

CHAPTER HIGHLIGHTS

1. Writing effectively is a challenge for many people. Following the process approach to writing will help you meet this challenge.

2. Prewriting helps you determine your goals, consider your audience, gather your data, examine your purposes, and determine how content will be provided.

3. Prewriting techniques will help you get started. Try answering reporter's questions, mind mapping, brainstorming or listing, outlining, storyboarding, creating organization charts, flowcharting, or researching.

4. To begin writing a rough draft, organize your thoughts.

5. You can communicate content through e-mail messages, cell phone e-mail messages, PDA e-mail messages, letters, memos, reports, brochures, newsletters, Web sites, and PowerPoint presentations.

6. Perfect your text by rewriting: adding, deleting, simplifying, moving, reformatting, enhancing, and correcting your documents.

EXERCISES

The Writing Process

1. To practice prewriting, take one of the following topics. Then, using the suggested prewriting technique, gather data.

   a. *Reporter's questions.* To gather data for your resume, list answers to the reporter's questions for two recent jobs you have held and for your past and present educational experiences.

   b. *Mind mapping.* Create a mind map for your options for obtaining college financial aid.

   c. *Brainstorming or listing.* List five reasons why you have selected your degree program or why you have chosen the school you are attending.

   d. *Outlining.* Outline your reasons for liking or disliking a current or previous job.

   e. *Storyboarding.* If you have a personal Web site, use storyboarding to graphically depict the various screens. If you do not have such a site, use storyboarding to graphically depict what your site's screens would include.

   f. *Creating organizational charts.* What is the hierarchy of leadership or management at your job or college organization (fraternity, sorority, club, or team)? To graphically depict who is in charge of what and who reports to whom, create an organizational chart.

   g. *Flowcharting.* Create a flowchart of the steps you followed to register for classes, buy a car, or seek employment.

   h. *Researching.* Go online or find a hard copy of the *Occupational Outlook Handbook*. Then, research a career field that interests you. Reading the *Occupational Outlook Handbook*, find out the nature of the work, working conditions, employment opportunities, educational requirements, and pay scale.

2. Using the techniques illustrated in this chapter, edit, correct, and rewrite the following flawed memo.

DATE: April 3, 2005
TO: William Huddleston
FROM: Julie Schopper
SUBJECT: TRAINING CLASSES

Bill, our recent training budget has increased beyond our projections. We need to solve this problem. My project team has come up with several suggestions, you need to review these and then get back to us with your input. Here is what we have come up with.

We could reduce the number of training classes, fire several trainers, but increase the number of participants allowed per class. Thus we would keep the same amount of income from participants but save a significant amount of money due to the reduction of trainer salaries and benefits. The downside might be less effective training, once the trainer to participant ratio is increased. As another option, we could outsource our training. This way we could fire all our trainers which would mean that we would save money on benefits and salaries, as well as offer the same number of training sessions, which would keep our trainer to participant ratio low.

What do you think. We need your feedback before we can do anything so even if your busy, get on this right away. Please write me as soon as you can.

3. Choose the best prewriting techniques for each of the following communication needs and then justify your choice.

   a. You are planning an oral presentation about benefits and insurance options for newly hired personnel at your company.

   b. You must prepare a short proposal for marketing a new product, highlighting your product's purpose or the problems it will solve, the costs incurred, any new personnel required, facilities needed, and the benefits derived.

   c. You are writing a recommendation report for the purchase of computer printers. You must first decide on criteria, with the help of your project team.

   d. You are writing an instruction for testing electronic equipment on your company's recently purchased voltmeter.

   e. You are writing an incident report about an accident in the company's boiler room, which must include the date, location, time, personnel involved, causes, and financial ramifications.

   f. You are writing a six-panel brochure about a favorite vacation spot. Your goal is to sell this site's amenities as well as inform the public of the site's unique values.

   g. You are writing an e-mail message, which will be copied to a committee of 10 coworkers. This message lists the meeting's minutes and suggests future areas for discussion at the next meeting.

   h. You are creating a Web site. It will have screens dedicated to costs, product specifications, customer contacts, and directions to each of your company's office locations.

QUIZ QUESTIONS

1. What are the three main parts of the writing process?

2. What are four ways you can provide technical writing content?

3. What can you achieve by prewriting?

4. What is the difference between external and internal motivation?

5. Why should you consider your audience before you begin writing?

6. What are four different prewriting techniques?

7. Why do you consider format when you write a business document?

8. What are four rewriting techniques?

9. What happens when you fail to revise accurately?

10. How can the writing process help ensure that you become a successful writer?

11. What are four goals of technical writing?

12. What are three search engines?

13. How do reporters' questions differ from mind mapping?

14. What are benefits of sending e-mail messages?

15. How do memos differ from e-mail messages?

Objectives in Technical Writing

Chapter Preview

Clarity
Technical writing must be clearly worded and developed to avoid confusing its audience.

Conciseness
Concise technical writing saves time for both writers and readers.

Accuracy
Avoid grammatical errors by proofreading your work so that you will communicate effectively and appear professional.

Organization
Organize your thoughts to help your readers better understand your documents.

Ethics
Technical writing entails specific ethical and legal considerations.

Writing at Work

CompToday is a computer hardware company located in Boston, Massachusetts. Like all publicly traded companies, CompToday must meet the Sarbanes-Oxley Act, passed by Congress in response to accounting scandals following the Enron, Tyco, and WorldCom illegalities. As of June 14, 2004, this act makes all corporate executives responsible for their companies' accounting practices. Penalties for failing to abide by this act are severe, including heavy fines as well as the potential for prison sentences.

CompToday, like many other companies, realizes that maintaining ethics and legalities in business is not just a matter of being a good citizen in the community. Maintaining ethics in the work environment demands a great deal of written communication. A key component of the Sarbanes-Oxley Act, combating ethical problems in the workplace, is the need for extensive documentation, including the following:

- Written policies provided for all corporate employees, clarifying in low-tech terms the intent of the act as well as its standards
- Instructional procedures (hard-copy and online) guiding employees in the steps they must follow to abide by Sarbanes-Oxley
- Online help screens providing answers to FAQs (Frequently Asked Questions) relevant to Sarbanes-Oxley
- Monthly reports documenting key financial activities, such as accounts payables, cash disbursement, purchases, and earnings
- Annual reports submitted to and signed by outside auditors

To achieve this daunting goal, CompToday's chief executive officer William Huddleston has reassigned a number of technical writers from his Corporate Communication Department.

They had been working on other projects, such as the company's Web site, instructional manuals, and online help screens. Now, they are tasked with new responsibilities—creating the documentation required by CompToday to meet the Sarbanes-Oxley requirements.

This hurts the company's bottom-line profit margin. After all, new writers must be hired to complete unfinished writing projects while the reassigned technical writers work toward compliance with the act. Maintaining an ethical workplace can be costly. However, practicing good ethics and abiding by the law benefits everyone—employees, stockholders, clients, and vendors.

Source: Adapted from Harkness 2004, 16–18.

CLARITY

The ultimate goal of good technical writing is clarity. If you write a memo, letter, or report that is unclear to your readers, then what have you accomplished? You have wasted time. If your readers must write you a follow-up inquiry to determine your needs, this wastes *their* time. Once you receive the inquiry, you must rewrite your correspondence, trying to clarify your initial intentions. You have now written twice to accomplish the same goal. This wastes *your* time.

To avoid these time-consuming endeavors, write for clarity. But how do you do this?

Provide Specific Detail

One way to achieve clarity is by supplying specific, quantified information. If you write using vague, abstract adjectives or adverbs, such as *some* or *recently*, your readers will interpret these words in different ways. The adverb *recently* will mean thirty minutes ago to one reader, yesterday to another, and last week to a third reader. This adverb, therefore, is not clear. The same applies to an adjective like *some*. You write, "I need some information about the budget." Your readers can only guess what you mean by *some*. Do you want the desired budget increase for 2005, the budget expenditures for 2000, the allotted budget increase for 2006, the guidelines for implementing a budget increase, the budgeted allotment for travel, or the explanation for the budget decrease for training?

Look at the following example of vague writing caused by imprecise, unclear adjectives. (Vague words are underlined.)

> **Before**
>
> Our <u>latest</u> attempt at molding preform protectors has led to <u>some</u> positive results. We spent <u>several</u> hours in Dept. 15 trying different machine settings and techniques. <u>Several</u> good parts were molded using two different sheet thicknesses. Here's a summary of the findings.
>
> First, we tried the <u>thick</u> sheet material. At 240°F, this thickness worked well.
>
> Next, we tried the <u>thinner</u> sheet material. The <u>thinner</u> material is less forgiving, but after a <u>few</u> adjustments we were making good parts. Still, the <u>thin</u> material caused the most handling problems.

The engineer who wrote this report realized that it was unclear. To solve the problem, she rewrote the report, quantifying the vague adjectives.

During the week of 10/4/05, we spent approximately 12 hours in Dept. 15 trying different machine settings, techniques, and thicknesses to mold pre-form mold protectors. Here is a report on our findings.

<u>0.030" Thick Sheet</u>

At 240°F, this thickness worked well.

<u>0.015" Thick Sheet</u>

This material is less forgiving, but after decreasing the heat to 200°F, we could produce good parts. Still, material at 0.015" causes handling problems.

Your goal as a technical writer is to communicate clearly. To do so, state your exact meaning through specific, quantified word usage.

Answer the Reporter's Questions

A second way to write clearly is to answer the reporter's questions—who, what, when, where, why, and how. The best way we can emphasize the importance of answering these reporter's questions is by sharing with you the following memo, written by a highly placed executive, to a newly hired employee.

DATE: November 11, 2005
TO: Mary Jane Post
FROM: Don Goldenbaum
SUBJECT: TECHNICIAN MEETING

Please be prepared to plan a presentation on month-end reports. Please be sure that your explanations are very detailed. Thanks.

That's the entire memo. The questions are, "What doesn't the newly hired employee know?" "What additional information would that employee need to do the job?" "What needs clarifying?"

Simply, the employee needs answers to reporter's questions. For example, *What* is the subject of the presentation? If the answer is month-end reports, we still lack clarity. *Which* of the 12 month-end reports does Don want Mary Jane to discuss? *Who* is the audience? We read the word *technician* in the subject line, but is Don focusing on automotive technicians, electronic engineering technicians, or dental hygiene technicians? *Why* is this presentation being made? That is, what is the rationale or motivation for the meeting? *When* will the presentation be made, and *how* much detail is "very detailed"? *Where* will Mary Jane and the technicians meet for the presentation? Mary Jane has a right to ask for clarity on one more point: *What* exactly is she supposed to do? Is she being asked to make a presentation, plan a presentation, or merely prepare to plan a presentation?

This memo's lack of clarity, a result of its inability to answer reporter's questions, causes Mary Jane stress, anxiety, and tension. As a new hire, she wants to do her job well, but Don's unclear correspondence hinders her. You must write clearly to avoid causing your reader stress and to help your reader do the job effectively.

In contrast, the memo below achieves clarity by answering reporter's questions.

Phyllis knows *where* the meetings will be held (Conference Room C); *who* selected her for the committee (her manager); *what* she will accomplish (writing a proposal about changes to the CIA system); *when* the meetings will occur (Friday, Monday, and Wednesday); *why* she's involved (to assess and revise a current system); and *how* the work will be accomplished (the sequence of work during the three days of meetings).

Use Easily Understandable Words

Another key to clarity is using words that your readers can understand easily. Avoid *obscure words* and be careful when you use *acronyms, abbreviations,* and *jargon.*

DATE: September 5, 2005
TO: Phyllis Goldberg
FROM: Craig Mabrito
SUBJECT: CASH IN ADVANCE (CIA) PROCEDURE TASK FORCE

Who

Why

You have been chosen by your manager to be a member of the CIA Task Force. This committee will assess current CIA procedures and revise the system.

The following is your schedule for Task Force activities:

Friday, September 12: 8:30–10:30 A.M.
    Assess the current system and brainstorm new ideas.

When and how

Monday, September 15: 8:30–10:30 A.M.
    Review the suggested changes, add new ideas, and test the adjusted system.

Wednesday, September 17: 8:30–10:30 A.M.

What

    Review the adjusted system and write a proposal confirming your changes.

Where

All meetings will be held in Conference Room C. Thank you for your involvement. If you have any questions, please call me at ext. 1849.

Avoiding Obscure Words

A good rule of thumb is to *write to express, not to impress; write to communicate, not to confuse.* If your reader must use a dictionary, you are not writing clearly.

Try to make sense of the following examples of unclear writing.

> The following rules are to be used when determining whether or not to duplicate messages:
>
> • Do not duplicate nonduplicatable messages.
>
> • A message is considered nonduplicatable if it has already been duplicated.
>
> Your job duties will be to ensure that distributed application modifications will execute without abnormal termination through the creation of production JCL system testing.

These examples were written by businesspeople who were trying to communicate *something*. The examples are filled with outdated terms that are difficult to understand.

Following is a list of difficult, out-of-date terms and the modern alternatives.

Obscure Words	Alternative Words
aforementioned	already discussed
initial	first
in lieu of	instead of
accede	agree
as per your request	as you requested
issuance	send
this is to advise you	I'd like you to know
subsequent	later
inasmuch as	because
ascertain	find out
pursuant to	after
forward	mail
cognizant	know
endeavor	try
remittance	pay
disclose	show
attached herewith	attached
pertain to	about
supersede	replace
obtain	get

Impressive writing is correspondence we can understand easily. A modern thrust in technical writing is to write the way you speak—unless you speak poorly. Try to be casual, almost conversational.

Using Acronyms, Abbreviations, and Jargon

In addition to obscure words, a similar obstacle to readers is created by acronyms, abbreviations, and jargon.

We have all become familiar with common acronyms such as *scuba* (self-contained underwater breathing apparatus), *radar* (radio detecting and ranging), *NASA* (National Aeronautics and Space Administration), *FICA* (Federal

Insurance Contributions Act), and *MADD* (Mothers Against Drunk Driving)—single words created from the first letters of multiple words. We are comfortable with abbreviations like *FBI* (Federal Bureau of Investigation), *JFK* (John F. Kennedy), *NFL* (National Football League), *IBM* (International Business Machines), and *LA* (Los Angeles). Some jargon (in-house language) has become so common that we reject it as a cliché. Baseball jargon is a good example. It is hard to tolerate sportscasters who speak baseball jargon, describing line drives as "frozen ropes" and fast balls as "heaters."

However, more often than not, acronyms, abbreviations, and jargon cause problems, not because they are too common but because no one understands them. Your technical writing loses clarity if you depend on them. You might think your readers understand them, but do they?

Try to guess the meaning of the following abbreviations and jargon.

- *CIA*. Was your first guess Central Intelligence Agency? Wrong! By this abbreviation we mean *cash in advance*.

- *Spaghetti*. Did your first guess make you envision a romantic dinner at a candlelit Italian restaurant? Wrong! We are using fire prevention jargon. For firefighters, *spaghetti* is a term for various fire hoses and lines.

- *FIFO*. If you guessed that this is a natural, all-bran cereal, you're wrong. This is a confusing acronym, except for business office personnel, who immediately recognize it as meaning "first in, first out."

You have to decide when to use acronyms, abbreviations, and jargon and how to use them effectively. One simple rule is to define your terms. You can do so either parenthetically or in a glossary. Rather than just writing *CIA*, write *CIA* (*Cash in Advance*). Such parenthetical definitions, which are only used once per correspondence, don't take a lot of time and won't offend your readers. Instead, the result will be clarity.

If you use many potentially confusing acronyms or abbreviations, or if you need to use a great deal of technical jargon, then parenthetical definitions might be too cumbersome. In this case, supply a separate glossary. (This point is discussed in greater detail in Chapter 17.) A *glossary* is an alphabetized list of terms, followed by their definitions, as in the following example:

CPA	Certified Public Accountant
FICA	Federal Insurance Contributions Act (Social Security taxes)
Franchise	Official establishment of a corporation's existence
Gross pay	Pay before deductions
Line of credit	Amount of money that can be borrowed
Net pay	Pay after all deductions
Profit and loss statement	Report showing all incomes and expenses for a specified time

Use Verbs in the Active Voice Versus the Passive Voice

example

It has been decided that Joan Smith will head our Metrology Department.

The preceding sentence is written in the *passive voice* (the primary focus of the sentence, *Joan Smith*, is acted on rather than initiating the action). Passive voice causes two problems:

1. Passive constructions are often unclear. After reading the preceding sentence, our question is, *who* decided that Joan Smith will head the department? To solve this problem and to achieve clarity, replace the vague indefinite pronoun *it* with a precise noun: "Kin Norman decided that Joan Smith will head our Metrology Department."

2. Passive constructions are often wordy. Passive sentences always require helping verbs (such as *has been*). When we revise the sentence to read "Kin Norman decided that Joan Smith will head our Metrology Department," the helping verb *has been* disappears.

The revision ("Kin Norman decided that Joan Smith will head our Metrology Department") is written in the *active voice*. When you use the active voice, your subject *(Kin Norman)* initiates the action.

Another common problem with passive voice construction concerns prepositions. Look at the following example:

> ### example
> Overtime is favored by hourly workers. (passive voice)

Again, this sentence, written in the passive voice, uses a helping verb *(is)*, has the doer *(hourly workers)* acted on rather than initiating the action, and also includes a preposition *(by)*. The sentence is wordy. Revised, the sentence reads as follows:

> ### example
> Hourly workers favor overtime. (active voice)

We have omitted the helping verb *is;* we have deleted the preposition *by;* and the subject, *Hourly workers*, initiates the action. The sentence is less wordy and more precise.

Occasionally, it is appropriate to use the passive voice. When an individual is less important than an inanimate object, passive voice is appropriate. For example:

> ### example
> The new software was oversold by the salesperson, who guaranteed ease of use. (The individual, *salesperson*, is not as important as the inanimate object, *software*.)

You might also use passive voice when the individual is unknown.

> ### example
> The tax preparation software can be learned easily even when the user is new to our CPA firm. (The individual, *user*, is unnamed.)

tech link

Go to *http://www. plainlanguage.gov/cites/ memo.htm* to read the "Presidential Memorandum on Plain Language." In this memo, President Bill Clinton mandates plain language in government documents, focusing specifically on clarity and conciseness. Shortly after this mandate, the SEC published its booklet *A Plain English Handbook: How to Create Clear SEC Disclosure Documents.* (Washington, DC: U.S. Securities and Exchange Commission, 1998).

After clarity, your second major goal in technical writing is conciseness, providing detail in fewer words. Conciseness is important for at least three reasons.

Conciseness Saves Time

Remember how time consuming technical writing is in the work environment? American workers spend approximately 12 hours per week writing and additional time reading and revising others' writing. Conciseness in writing can help save some of this time. If you write concisely, you can save yourself time and take up less of your readers' time.

Conciseness Aids Clarity

Concise writing can aid comprehension. If you dump an enormous number of words on your readers, they might give up before finishing your correspondence or skip and skim so much that they miss a key concept. Wordy writing will lead your readers to think, "Oh no! I'll never be able to finish that. Maybe I can skim through it. I'll probably get enough information that way." Conciseness, on the other hand, makes your writing more appealing to your readers. They'll think, "Oh, that's not too bad; I can read it easily." If they can read your correspondence easily, they will read it with greater interest and involvement. This, of course, will aid their comprehension.

Technology Demands Conciseness

Technology is impacting the size of your technical writing. The size of the screen makes the difference, and screen sizes are shrinking. Thus, when you write, you need to consider the way in which technology limits your space. Today, more and more, effective technical writing must be concise enough to *fit in a box*.

Notice how the size of the "box" containing the following communication affects the way you package your content (see Figure 3.1 and Table 3.1).

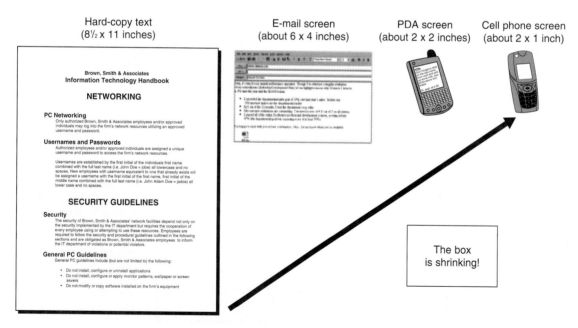

Figure 3.1   The Shrinking Size of Technical Writing

TABLE 3.1	SIZE CHARACTERISTICS OF COMMUNICATION CHOICES		
Communication Choice	Overall Size of the Screen	Lines per Page or Screen	Characters per Line
Hard-copy paper	8½ × 11 inches	55	70–80
E-mail screen	4 × 6 inches	20–22	60–70
PDA screen	2 × 2 inches	10–15	30–35
Cell phone screen	2 × 1 inch	4–6	10–15

Following are examples of technical writing that fit in various-sized boxes.

Resumes

A one-page resume is standard. One hard-copy page measures 8 1/2 × 11 inches. That is a box. In fact, this box (one page of text) allows for only about 55 lines of text, and each line of text allows for only about 70–80 characters. (A "character" is every letter, punctuation mark, or space.)

E-mail Messages

In contrast, much of today's written communication in the workplace is accomplished through e-mail messages. In Chapter 5 we quote statistics stating that 96 percent of business employees report using e-mail every day (Schulzrinne). Sixty-six percent of office workers say they are "e-mail only" users ("E-mail Statistics"). Another 45 percent say that e-mail has replaced their phone usage (Schulzrinne).

Because e-mail screens tend to be smaller (around only 4 × 6 inches) than the typical computer screen, e-mails should be brief and concise. Yes, you can scroll an e-mail message endlessly, but no one wants to do that. In fact, the reason that readers like a one-page resume is that one page allows for what is known as the "W-Y-S-I-W-Y-G" factor ("What You See Is What You Get"). Readers like to see what they will be getting in the correspondence. In contrast, if you make your readers scroll endlessly in e-mail, they do not see what they get. This causes problems. Therefore, a good e-mail message should fit in the box, letting the reader see the entire content at one glance. It should be limited to about 20 lines of text.

WebCT, an online educational tool for many colleges and universities, allows faculty to input course material online and students to access information by way of their password-protected personal identification number (PIN). WebCT also provides an online opportunity for faculty and classmates to communicate with each other by e-mail. In Figure 3.2, notice how WebCT provides a predefined box for its e-mail users.

PDAs and Cell Phones

The technological impact is even more dramatic when you consider the screen size for a handheld PDA (about 2 × 2 inches), the screen size for a cell phone (about 2 × 1 inch), and the screen size for a pager (often one line of text, equaling no more than 20 characters). In "Serving the Electronic Reader," Linda E. Moore says, "More and more e-readers are accessing documents using cell phones, PDAs, and other wireless devices. Although these applications are not yet widespread, they are making brevity even more critical" (2003, 17). Technical writers in the next decade will "have to figure out how to create content that works on a four-line cell phone screen" (Perlin 2001, 4–8).

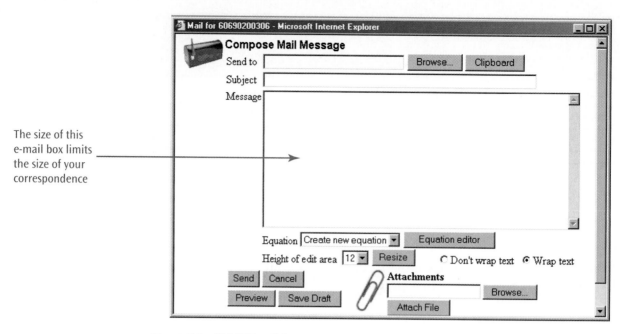

The size of this e-mail box limits the size of your correspondence

Figure 3.2   WebCT E-mail Screen

Online Order Forms

Similarly, when you need to fill out online forms, you will be given predetermined "fields" in which to write. Your text will be limited by the size of these boxes. You will find online forms for your banking, bill payments, stock exchanges, insurance claims, shopping, and many other uses. Figure 3.3 shows a sample online claim form for automobile insurance.

Online Help Screens

Another example of technical writing that must fit in a box is online help screens. If you work in information technology or computer sciences, you might need to create online help screens. If you do so, then your text will be limited by the size of these screens. Look at the Microsoft Word example in Figure 3.4. Notice how the text for "Display readability statistics" is provided in a vertical box aligned along the page's right margin. Then, if you click on the hypertext linked "ABC" check box, a smaller pop-up box emerges. What you are left with is boxes within boxes.

Microsoft PowerPoint

Finally, you also must fit your technical writing within a box (or boxes) when you use Microsoft PowerPoint software. Look at the New Slide dialog box that PowerPoint provides when choosing an AutoLayout for a presentation (Figure 3.5).

The bottom line is technology requires that you write concisely. If you are filling out an online form, you will be limited by the size of the form's fields. If you are writing correspondence that will be viewed as a PowerPoint presentation, an online help screen, or e-mail sent to someone's PDA or cell phone, you need to limit the size of your text. Therefore, as you write, consider the impact of technology and write concisely.

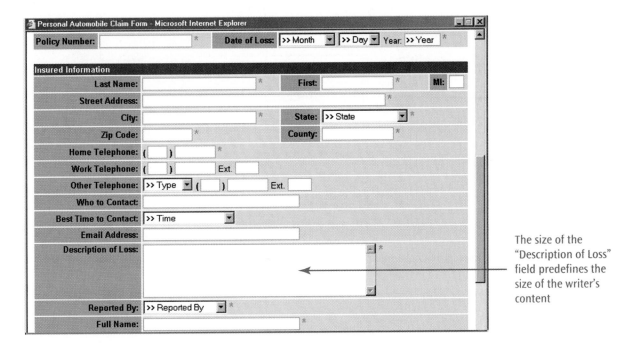

The size of the "Description of Loss" field predefines the size of the writer's content

Figure 3.3   Online Claim Form

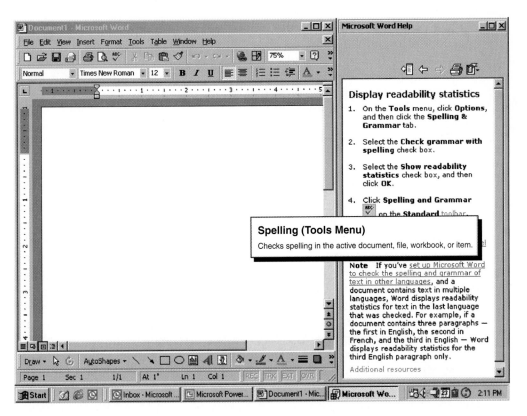

Figure 3.4   Online Help Screen with Pop-up Box

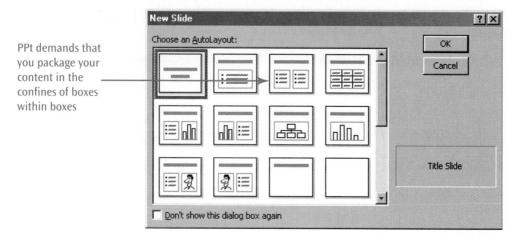

PPt demands that you package your content in the confines of boxes within boxes

Figure 3.5   Choosing an AutoLayout in PowerPoint

Limit Paragraph Length

Let's look at some poor writing—writing that is wordy, time consuming to read, and not easily comprehensible.

> **Before**
>
> Please prepare to supply a readout of your findings and recommendations to the officer of the Southwest Group at the completion of your study period. As we discussed, the undertaking of this project implies no currently known incidences of impropriety in the Southwest Group, nor is it designed specifically to find any. Rather, it is to assure ourselves of sufficient caution, control, and impartiality when dealing with an area laden with such potential vulnerability. I am confident that we will be better served as a company as a result of this effort.

Is that paragraph easy to understand? No, it is not. Why? What gets in your way? Do you have difficulty following it because you are an outsider and are not aware of the situation that generated it? That is only part of the problem. The reason you have difficulty understanding this paragraph is because it is poorly written. It causes difficulty for two reasons: (a) the paragraph is too long, and (b) the words and sentences in the paragraph are too long.

An excessively long paragraph is ineffective. In a long paragraph, you force your reader to wade through many words and digest large amounts of information. This hinders comprehension. In contrast, short, manageable paragraphs invite reading and help your reader understand your content.

As a rule of thumb, a paragraph in a technical document should consist of (a) no more than 4 to 6 typed lines, or (b) no more than 50 words. Sometimes you can accomplish these goals by cutting your paragraphs in half; find a logical place to stop a paragraph and then start a new one. Even the previous poorly written example can be improved in this way.

After

Please prepare to supply a readout of your findings and recommendations to the officer of the Southwest Group at the completion of your study period. As we discussed, the undertaking of this project implies no currently known incidences of impropriety in the Southwest Group, nor is it designed specifically to find any.

Rather, it is to assure ourselves of sufficient caution, control, and impartiality when dealing with an area laden with such potential vulnerability. I am confident that we will be better served as a company as a result of this effort.

The writing is still difficult to understand, but at least it is a bit more manageable. You can read the first paragraph, stop, and consider its implications. Then, once you have grasped its intent, you can read the next paragraph and try to tackle its content. The paragraph break gives you some room to breathe.

Limit Word and Sentence Length

In addition to the length of the example paragraph, the writing is flawed because the paragraph is filled with excessively long words and sentences. This writer has created an impenetrable wall of haze—the writing is foggy. In fact, we can determine how foggy this prose is by assessing it according to Robert Gunning's fog index.

Using the Fog Index

The following is his mathematical way of determining how foggy your writing is.

1. Count the number of words in successive sentences. Once you reach approximately 100 words, divide these words by the number of sentences. This will give you an average number of words per sentence.

2. Now count the number of long words within the sentences you have just reviewed. Long words are those with three or more syllables. You cannot count (a) proper names, like Leonardo DaVinci, Christopher Columbus, or Alexander DeToqueville; (b) long words that are created by combining shorter words, such as *chairperson* or *firefighter*; or (c) three-syllable verbs created by *–ed* or *–es* endings, such as *united* or *arranges*. Discounting these exceptions, count the remaining multisyllabic words. (A good example of a multisyllabic word is the word *mul-ti-syl-lab-ic*.)

3. Finally, to determine the fog index, add the number of words per sentence and the number of long words. Then multiply your total by 0.4.

Given this system, let's see how the original difficult paragraph (page 58) scores. The paragraph is composed of 92 words in 4 sentences. Thus, the average number of words per sentence is 23. The paragraph contains 16 multisyllabic words (*recommendations, officer, completion, period, undertaking, currently, incidences, impropriety, specifically, sufficient, impartiality, area, potential, vulnerability, confident,* and *company*).

$$
\begin{array}{rl}
23 & \text{(words per sentence)} \\
+16 & \text{(multisyllabic words)} \\
\hline
39 & \text{(total)}
\end{array}
$$

$$\begin{array}{rl} 39 & \text{(total)} \\ \times\ 0.4 & \text{(fog factor)} \\ \hline 15.6 & \text{(fog index)} \end{array}$$

What does a fog index of 15.6 mean? Look at Table 3.2. It shows that the paragraph is written at a level midway between college junior and senior, definitely above the danger line.

TABLE 3.2 FOG INDEX AND READING LEVEL			
	Fog Index	By Grade	By Magazine
	17	College graduate	No popular magazine
	16	College senior	scores this high.
	15	College junior	
	14	College sophomore	
Danger Line	13	College freshman	
	12	High school senior	*Atlantic Monthly*
	11	High school junior	*Time* and *Newsweek*
	10	High school sophomore	*Reader's Digest*
	9	High school freshman	*Good Housekeeping*
	8	Eighth grade	*Ladies' Home Journal*
	7	Seventh grade	Modern romances
	6	Sixth grade	Comics

Why is this level of writing considered dangerous? A fog index of 15.6 enters the danger zone for two reasons:

- Approximately 33 percent of Americans graduate from college. Thus, if you are writing at a college level, you could be alienating approximately 67 percent of your audience (Mollison 2001, 1).
- Studies show that college graduates read at approximately a tenth-grade level. Thus, even if you're writing to college graduates, you cannot assume college-level reading skills.

Given these facts, many businesses ask their employees to write at a sixth- to eighth-grade level. To accomplish this, you would have to strive for an average of approximately 15 words per sentence and no more than 5 multisyllabic words per 100 words.

$$\begin{array}{rl} 15 & \text{(words per sentence)} \\ +\ 5 & \text{(multisyllabic words)} \\ \hline 20 & \text{(total)} \end{array}$$

$$\begin{array}{rl} 20 & \text{(total)} \\ \times\ 0.4 & \text{(fog factor)} \\ \hline 8.0 & \text{(fog index level)} \end{array}$$

You cannot always avoid multisyllabic words. Scientists would find it impossible to write if they could never use words like *electromagnetism*, *nitroglycerine*, *telemetry*, or *trinitrotulene*. The purpose of a fog index is to make you aware that long words and sentences create reading problems. Therefore, although you cannot always avoid long words, you should be careful when using them. Similarly, you cannot always avoid lengthy sentences. However, try not to rely on sentences over 15 words long. Vary your sentence lengths, relying mostly on sentences less than 15 words long.

Here are ways to lower a potentially high fog index.

Use the Meat Cleaver Method of Revision

One way to limit the number of words per sentence is to cut the sentence in half or thirds. The following sentence, which contains 44 words, is too long.

> To maintain proper stock balances of respirators and canister elements and to ensure the identification of physical limitations that may negate an individual's previous fit-test, a GBC-16 Respirator Request and Issue Record will need to be submitted for each respirator requested for use.
>
> **Before**

If we use the meat cleaver approach, we can make this sentence more concise and easier to understand.

> Please submit a GBC-16 Respirator Request and Issue Record for each requested respirator. We then can maintain proper respirator and canister element stock balances. We also can identify physical limitations that may negate an individual's previous fit-test.
>
> **After**

The "before" sentence, now cut in thirds, is more digestible. Because we have less to swallow whole, we can understand the content more easily.

Avoid Shun Words

In the preceding examples, the original sentence contained 44 words: the revised version is composed of three sentences totaling 38 words. Where did the missing 6 words go?

One way to write more concisely is to shun words ending in –*tion* or –*sion*—words ending in a *shun* sound. For example, the original sentence reads "to ensure the identifica*tion* of physical limita*tions*." To revise this, we simply wrote "identify physical limitations." That's 3 words versus 7 in the original version. Deleting 4 words in this case reduces wordiness by over 50 percent. Shun words are almost always unnecessarily wordy.

Let's try another example. Instead of writing "I want you to take into considera*tion* the following," you could write "consider the following." That is 9 words versus 3, a 66.6 percent savings in word count.

Look at the following shun words and their concise versions:

Shun Words	Concise Versions
came to the conclu*sion*	concluded (or decided)
with the excep*tion* of	except for
make revi*sions*	revise
investiga*tion* of the	investigate
consider implementa*tion*	implement
utiliza*tion* of	use

Avoid Camouflaged Words

Camouflaged words are similar to shun words. In both instances, a key word is buried in the middle of surrounding words (usually helper verbs or unneeded prepositions). For example, in the phrase *with the exception of*, the key word *except* is camouflaged behind the unneeded *with*, *the*, *–tion*, and *of*. Once we prune away these unneeded words, the key word *except* is left, making the sentence less wordy.

Camouflaged words are common. Here are some examples and their concise versions.

Camouflaged Words	Concise Versions
make an *amend*ment to	amend
make an *adjust*ment of	adjust
have a *meet*ing	meet
*thank*ing *you* in advance	thank you
for the purpose of *discuss*ing	discuss
arrive at an *agree*ment	agree
at a *later* moment	later

Avoid the Expletive Pattern

Another way to write more concisely is to avoid the following expletives:

- *there* is, are, was, were, will be
- *it* is, was

Both these expletives (*there* and *it*) lead to wordy sentences. For example, consider the following sentence.

> **example**
>
> There are three people who will work for Acme.

This sentence can be revised to read, "Three people will work for Acme." The original sentence contains 9 words; the revision has 6. We've omitted 3 words by deleting the expletive *there*. Your response to this revision could be "So what, who cares? What's the point of deleting 3 words?" Deleting 3 words doesn't seem like much. However, the omission equals a 33 percent savings in word count, helping us achieve conciseness. In one sentence, that might be a minimal achievement, but if you can delete 3 words from every sentence, the benefits will add up.

The expletive *it* creates similar wordiness, as in the following sentence:

> **example**
>
> It has been decided that ten engineers will be hired.

If we delete the expletive *it*, the sentence reads, "Ten engineers will be hired." The original sentence contained 10 words; the revision has 5. We have achieved a 50 percent savings in word count.

Omit Redundancies

Redundancies are words that say the same thing. Conciseness is achieved by saying something once rather than twice. For example, in each of the following instances, the boldface words are redundant.

during **the year of** 2005

> (Obviously 2005 is a year; the words *the year of* are redundant.)

in **the month of** December

> (As in the preceding example, *the month of* is redundant; what else is December?)

needless to say

> (If it's needless to say, why say it?)

The computer will cost **the sum of** $1,000.

> (One thousand dollars *is* a sum.)

the results **so far achieved** prove

> (A result, by definition, is something that has been achieved.)

our **regular** monthly status reports require

> (Monthly status reports must occur every month; regularity is a prerequisite.)

We collaborated **together** on the project.

> (One can't collaborate alone!)

the **other** alternative is to

> (Every alternative presumes that some option exists.)

This is a **new** innovation.

> (As opposed to an old innovation!)

the consensus **of opinion** is to

> (The word *consensus* implies opinion.)

Avoid Wordy Phrases

Sentences may be wordy not because you have been redundant or because you have used shun words, camouflaged words, or expletives. Sometimes sentences are wordy simply because you've used wordy phrases.

Here are examples of wordy phrases and their concise revisions.

Wordy Phrases	Concise Revisions
in order to purchase	to buy
at a rapid rate	fast (or state the exact speed)
it is evident that	evidently
with regard to	about
in the first place	first
a great number of times	often (or state the number of times)
despite the fact that	although
is of the opinion that	thinks
due to the fact that	because
am in receipt of	received
enclosed please find	enclosed is

as soon as possible	by 11:30 A.M.
in accordance with	according to
in the near future	soon
at this present writing	now
in the likely event that	if
rendered completely inoperative	broken

ACCURACY

Clarity and conciseness are primary objectives of effective technical writing. However, if your writing is clear and concise but incorrect—grammatically or textually—then you have wasted your time and destroyed your credibility. To be effective, your technical writing must be *accurate*.

Accuracy in technical writing requires that you *proofread* your text. The examples of inaccurate technical writing below are caused by poor proofreading (we have underlined the errors to highlight them).

First City Federal Savings and Loan
1223 Main
Oak Park, Montana

October 12, 2005

Mr. and Mrs. David Harper
2447 N. Purdom
Oak Park, Montana

Dear Mr. and Mrs. Purdom:

Note that the savings and loan incorrectly typed the customer's street rather than the last name.

National Bank
1800 Commerce Street
Houston, TX

September 9, 2005

Adler's Dog and Oat Shop
8893 Southside
Bellaire, TX

Dear Sr.:
In response to your request, your account with us has been close out. We are submitted a check in the amount of $468.72 (your existing balance). If you have any questions, please fill free to conact us.

In addition to all the other errors, it should be "Dog and <u>C</u>at Shop," of course. The errors make the writer look incompetent.

To ensure accurate writing, use the following proofreading tips:

1. *Let someone else read it*—We miss errors in our own writing for two reasons. First, we make the error because we don't know any better. Second, we read what we think we wrote, not what we actually wrote. Another reader might help you catch errors.

2. *Use the gestation approach*—Let your correspondence sit for a while. Then, when you read it, you'll be more objective.

3. *Read backwards*—You can't do this for content. You should read backwards only to slow yourself down and to focus on one word at a time to catch typographical errors.

4. *Read one line at a time*—Use a ruler or scroll down your PC screen to isolate one line of text. Again, this slows you down for proofing.

5. *Read long words syllable by syllable*—How is the word *responsiblity* misspelled? You can catch this error if you read it one syllable at a time (re-spon-si-bl-i-ty).

6. *Use technology*—Computer spell checks are useful for catching most errors. They might miss proper names, homonyms (*their*, *they're*, or *there*) or incorrectly used words, such as *device* to mean *devise*.

7. *Check figures, scientific and technical equations, and abbreviations*—If you mean $400,000, don't write $40,000. Double-check any number or calculations. If you mean to say *HCl* (hydrochloric acid), don't write *HC* (a hydrocarbon).

8. *Read it out loud*—Sometimes we can hear errors that we cannot see. For example, we know that *a outline* is incorrect. It just sounds wrong. *An outline* sounds better and is correct.

9. *Try scattershot proofing*—Let your eyes roam around the page at random. Sometimes errors look wrong at a glance. If you wander around the page randomly reading, you often can isolate an error just by stumbling on it.

10. *Use a dictionary*—If you are uncertain, look it up.

If you commit errors in your technical writing, your readers will think one of two things about you and your company: (a) they will conclude that you are stupid, or (b) they will think that you are lazy. In either situation, you lose. Errors create a negative impression at best; at worst, a typographical error relaying false figures, calculations, amounts, equations, or scientific or medical data can be disastrous.

ORGANIZATION

If you are clear, concise, and accurate, but no one can follow your train of thought because your text rambles, you still haven't communicated effectively. Successful technical writing also must be well organized.

Here is an analogy to explain the importance of organization. Most artists cannot just dip a brush in paint and then splatter that paint on canvas. People want to make sense of what they see, and splattered images cause confusion. The same applies to technical writing. As the writer, you cannot haphazardly throw words on the page and expect readers to understand you clearly. In contrast, you should order that information on the page logically, allowing your readers to follow your train of thought.

No one method of organization always works. Following are five patterns of organization that you can use to help clarify content.

Spatial

If you are writing to describe the parts of a machine or a plot of ground, you might want to organize your text spatially. You would describe what you see as it appears in space—left to right, top to bottom, inside to outside, or clockwise. These spatial sequences help your readers visualize what you see and, therefore, better understand the physical qualities of the subject matter. They can envision the layout of the land you describe or the placement of each component within the machine.

For example, let's say you are a contractor describing how you will refinish a basement. Your text reads as follows:

example

At the basement's north wall, I will build a window seat 7' long by 2' wide by 2' high. To the right of this seat, on the east wall, I will build a desk 4' high by 5' long by 3' wide. On the south wall, to the left of the door, I will build an entertainment unit the height of the wall including four, 4' high by 4' wide by 2' deep shelving compartments. The west wall will contain no built-ins. You can use this space to display pictures and to place furniture.

Note how this text is written clockwise, uses points of the compass to orient the reader, and includes the transitional phrases "to the right" and "to the left" to help the reader visualize what you will build. That's spatial organization.

Chronological

Whereas you would use spatial organization to describe a place, you would use chronology to document time or the steps in an instruction. For example, an emergency medical technician (EMT) reporting services provided during an emergency call would document those activities chronologically.

example

At 1:15 P.M., we arrived at the site and assessed the patient's condition, taking vitals (pulse, respiration, etc.). At 1:17 P.M., after stabilizing the patient, we contacted the hospital and relayed the vitals. By 1:20 P.M., the patient was on an IV drip and en route to the hospital. Our vehicle arrived at the hospital at 1:35 P.M. and hospital staff took over the patient's care.

Chronology also would be used to document steps in an instruction. No times would be provided as in the EMT report. In contrast, the numbered steps would denote the chronological sequence a reader must follow.

Importance

Your page of text is like real estate. Certain areas of the page are more important than others—location, location, location. If you bury key data on the bottom of a page, your reader might not see the information. In contrast, content placed approximately one-third from the top of the page and two-thirds from the bottom (eye level) garners more attention. The same applies to a bulleted list of points. Readers will focus their attention on the first several points more than on the last few.

Knowing this, you can decide which ideas you want to emphasize and then place that information on the page accordingly. Organize your ideas by importance. Place the more important ideas above the less important ones.

The following agenda is incorrectly organized:

Agenda

- Miscellaneous ideas
- Questions from the audience
- Refreshments
- Location, date, and time
- Subject matter
- Guest speakers

Which of these points is most important? Certainly not the first two. A better list would be organized by importance, as follows:

Agenda

- Subject matter
- Guest speakers
- Location, date, and time
- Refreshments
- Questions from the audience
- Miscellaneous ideas

Comparison/Contrast

Many times in business you will need to document options and ways in which you surpass a competitor. These require that you organize your text by comparison/contrast. You compare similarities and contrast differences. For example, if you are writing a sales brochure, you might want to present your potential client alternatives regarding services, personnel, timetables, and fee structures. Table 3.3 shows how comparison/contrast provides the client options for cost and features.

TABLE 3.3 HOUSING COSTS		
Item	Features	Cost
The Broadmoor	4 bedrooms, 3 1/2 baths 2-car garage fully equipped kitchen	$200,000
The Aspen	4 bedrooms, 3 1/2 baths finished basement 3-car garage fully equipped kitchen	$240,000
The Regency	4 bedrooms, 3 1/2 baths patio deck finished basement with 1/2 bath 3-car garage fully equipped kitchen	$280,000

Each housing option is comparable in that the developer provides a 4-bedroom, 3-bath home with fully equipped kitchen. However, the homes contrast regarding garage size, deck availability, and basement. These options then affect the cost. The table, organized according to comparison/contrast, helps the reader understand these distinctions.

Problem/Solution

Every proposal and sales letter is problem/solution oriented. When you write a proposal, for instance, you are proposing a solution to an existing problem. If your proposal focuses on new facilities, your reader's current building must be flawed. If your proposal focuses on new procedures, your reader's current approach to doing business must need improvement. Similarly, if your sales letter promotes a new product, your customers will purchase it only if their current product is inferior.

To clarify the value of your product or service, therefore, you should emphasize the readers' need (their problem) and show how your product is the solution. Note how the following summary from a proposal is organized according to problem/solution.

> **example**
>
> Your city's 20-year-old wastewater treatment plant does not meet EPA requirements for toxic waste removal or ozone depletion regulations. This endangers your community and lessens property values in its neighborhoods.
>
> Anderson and Sons Engineering Company has a national reputation for upgrading wastewater treatment plants. Our staff of qualified engineers will work in partnership with your city's planning commission to modernize your facilities and protect your community's values.

The summary's first paragraph identifies the problem. The second paragraph promotes the solution. The problem/solution organization clarifies the writer's intent.

ETHICS

Here is the scenario. You are a technical writer responsible for producing a maintenance manual. Your boss tells you to include the following sentence:

> **example**
>
> NOTE: Our product has been tested for defects and safety by trained technicians.

When read literally, this sentence is true. The product has been tested, and the technicians are trained. However, you know that the product has been tested for only 24 hours by technicians trained on site without knowledge of international regulations.

So where's the problem? As a good employee, you are required to write what your boss told you. Right? Even though the statement is not completely true, legally you can include it in your manual. Correct?

The answer to both questions is no! Actually, you have an ethical responsibility to write the truth. Your customers expect it, and it is in the best interests of your company. Equally important is that including the sentence in your manual is illegal. Although the sentence is essentially true, it implies something that is false. Readers will assume that the product has been *thoroughly* tested by technicians who have been *correctly* trained. Thus, the sentence deceives the readers. Such comments are "actionable under law" if they lead to false impressions (Wilson 1987, WE-68). If you fail to properly disclose information, including dangers, warnings, cautions, or notes like the sentence in question, then your company is legally liable. (We discuss how to correctly write hazard alert messages in Chapter 12.)

Knowing this, however, does not make writing easy. Ethical dilemmas exist in corporations. The question is, what should you do when confronted with such problems?

One way to solve this dilemma is by checking your actions against these three concerns: legal, practical, and ethical. For example, if you plan to write operating instructions for a mechanism, will your text be

1. *legal*, focusing on liability, negligence, and consumer protection laws?

2. *practical*, because dishonest technical writing backfires and can cause the company to lose sales or to suffer legal expenses?

3. *ethical*, written to promote customer welfare and avoid deceiving the end user? (Bremer et al. 1987, 76–77)

These are not necessarily three separate issues. Each interacts with the other. Our laws are based on ethics and practical applications.

Legalities

If you're uncertain, that's what lawyers are for. When asked to write text that profits the company but deceives the customer, for instance, you might question where your loyalties lie. After all, the boss pays the bills, but your customers might also be your next-door neighbors. Such conflicts exist and challenge all employees. What do you do? You should trust your instincts and trust the laws. Laws are written to protect the customer, the company, and you—the employee. If you believe you are being asked to do something illegal that will harm your community, seek legal counsel.

Practicalities

Even though it might appear to be in the best interests of the company to hide potentially damaging information from customers, such is not the case. First, as a technical writer your goal is candor. That means you must be truthful, stating the facts. It also means you must not lie, keeping silent about facts that are potentially dangerous (Girill 1987, 178–79). Second, practically speaking, the best business approach is *good* business. The ultimate goal of a company is not just making a profit, but making money the right way—"good ethics is good business" (Guy 1990, 9). What good is it to earn money from a customer who will never buy from you again or who will sue for reparation? That is not practical.

Ethicalities

Here's the ultimate dilemma. Defining ethical standards has been challenging for professional organizations. As Shirley A. Anderson-Hancock, manager for the Society for Technical Communication (STC) Ethical Guidelines Committee, states, "Everyone who participates in discussions of ethical issues may offer a different perspective, and many viewpoints may be legitimate" (1995, 6). Due to

tech link

Go to *http://www.stc.org/* (the Society for Technical Communication) for additional information about the society's mission and goals, including ethical standards.

the difficulty of clearly defining what is and what is not ethical, the STC struggled for three years before publishing its guidelines in 1995 and updated them in 1998 (see Figure 3.6).

These ethical principles, as the STC Ethical Guidelines Committee admits, had to be "sufficiently broad to apply to different working environments for the next several years" (Anderson-Hancock 1995, 6).

One way to clarify these necessarily broad standards is by looking at the STC Code for Communicators, which reads as follows ("Code for Communicators" 2004).

As a technical communicator, I am the bridge between those who create ideas and those who use them. Because I recognize that the quality of my services directly affects how well ideas are understood, *I am committed to excellence in performance and the highest standards of ethical behavior.*

I value the worth of the ideas I am transmitting and the cost of developing and communicating those ideas. I also value the time and effort spent by those who read or see or hear my communication.

I therefore recognize my responsibility to communicate technical information truthfully, clearly, and economically.

My commitment to professional excellence and ethical behavior means that I will

- Use language and visuals with precision.
- Prefer simple, direct expression of ideas.
- Satisfy the audience's need for information, not my own need for self-expression.
- Hold myself responsible for how well my audience understands my message.
- Respect the work of colleagues, knowing that a communication problem may have more than one solution.
- Strive continually to improve my professional competence.
- Promote a climate that encourages the exercise of professional judgment and that attracts talented individuals to careers in technical communication.

Guide for Ethical Standards

Use language and visuals with precision.   In a recent survey comparing technical writers and teachers of technical writing, we discovered an amazing finding: professional technical writers rate grammar and mechanics higher than teachers do (Gerson and Gerson 1995). On a 5-point scale (5 equaling "very important"), writers rated grammar and mechanics 4.67, whereas teachers rated grammar and mechanics only 3.54. That equals a difference of 1.13, which represents a 22.6 percent divergence of opinion.

Given these numbers, would we be precise in writing "Teachers do not take grammar and mechanics as seriously as writers do"? The numbers accurately depict a difference of opinion, and the 22.6 percent divergence is substantial. However, these figures do not assert that teachers ignore grammar. To say so is imprecise and would constitute an ethical failure to present data accurately. Even though writers are expected to highlight their client's values and downplay their

Ethical Principles for Technical Communicators

As technical communicators, we observe the following ethical principles in our professional activities.

Legality

We observe the laws and regulations governing our profession. We meet the terms of contracts we undertake. We ensure that all terms are consistent with laws and regulations locally and globally, as applicable, and with STC ethical principles.

Honesty

We seek to promote the public good in our activities. To the best of our ability, we provide truthful and accurate communications. We also dedicate ourselves to conciseness, clarity, coherence, and creativity, striving to meet the needs of those who use our products and services. We alert our clients and employers when we believe that material is ambiguous. Before using another person's work, we obtain permission. We attribute authorship of material and ideas only to those who have made an original and substantive contribution. We do not perform work outside our job scope during hours compensated by clients or employers, except with their permission; nor do we use their facilities, equipment, or supplies without their approval. When we advertise our services, we do so truthfully.

Confidentiality

We respect the confidentiality of our clients, employers, and professional organizations. We disclose business-sensitive information only with their consent or when legally required to do so. We obtain releases from clients and employers before including any business-sensitive materials in our portfolios or commercial demonstrations or before using such materials for another client or employer.

Quality

We endeavor to produce excellence in our communication products. We negotiate realistic agreements with clients and employers on schedules, budgets, and deliverables during project planning. Then we strive to fulfill our obligations in a timely, responsible manner.

Fairness

We respect cultural variety and other aspects of diversity in our clients, employers, development teams, and audiences. We serve the business interests of our clients and employers as long as they are consistent with the public good. Whenever possible, we avoid conflicts of interest in fulfilling our professional responsibilities and activities. If we discern a conflict of interest, we disclose it to those concerned and obtain their approval before proceeding.

Professionalism

We evaluate communication products and services constructively and tactfully, and seek definitive assessments of our own professional performance. We advance technical communication through our integrity and excellence in performing each task we undertake. Additionally, we assist other persons in our profession through mentoring, networking, and instruction. We also pursue professional self-improvement, especially through courses and conferences.

Approved by the STC Board of Directors
September 1998.

Source: Reprinted with permission from the Society for Technical Communication, Arlington, VA, U.S.A.

Figure 3.6   STC Ethical Principles

client's shortcomings, technical writers ethically cannot skew numbers to accomplish these goals (Bowman and Walzer 1987). Information must be presented accurately, and writers must use language precisely.

Precision also is required when you use visuals to convey information. Look at Figures 3.7 and 3.8. Both figures show that XYZ's sales have risen. As with language, the writer is ethically responsible for presenting visual information precisely.

Another major consideration regarding the precise use of language and visuals involves the role of intellectual property laws as they relate to the Internet. If something is on the Internet, including graphics or text, we can just take it—right? We can download any graphic or just appropriate any text if it's online, right?

Of course, these assumptions are false. "Every element of a Web page—text, graphics, and HTML code—is protected by U.S. and international copyright laws, whether or not a formal copyright application has been submitted" (LeVie 2000, 20–21). In fact, if you and your company "borrow" from an existing Internet site, thus infringing upon that site's copyright, you can be assessed actual or statutory damages. From March 1997 to March 1998, at least eight companies suffered fines. A telecommunications company paid $100,000.00, a food la-

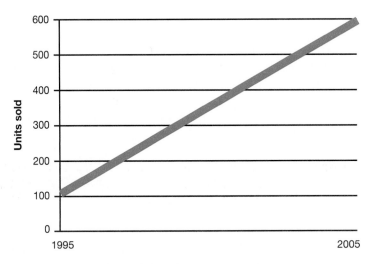

Figure 3.7   Imprecise Depiction of Company Growth

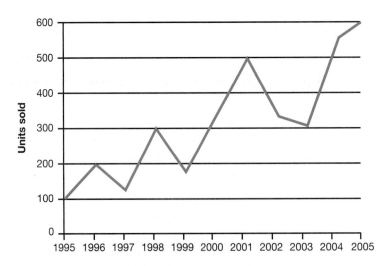

Figure 3.8   Precise Depiction of Company Growth

bel brokerage firm paid $400,000.00, an architectural firm paid $150,000.00, and an engineering company was assessed $140,000.00 (LeVie 2000, 21).

In addition to financial damages, your company, if violating intellectual property laws, could lose customers, damage its reputation, and lose future capital investments.

To solve these problems, and to protect your own property rights, you should

- Assume that any information on the Internet is covered under copyright protection laws unless proven otherwise.
- Obtain permission for use from the original creator of graphics or text.
- Cite the source of your information.
- Create your own graphics and text.
- Copyright any information you create.
- Place a copyright notice at the bottom of your Web site (Johnson 1999, 17).

Prefer simple, direct expression of ideas. While writing an instructional manual, you might be asked to use legal language to help your company avoid legal problems (Bowman and Walzer 1987). Here is a typically obtuse warranty that legally protects your company:

---

example

> Acme's liability for damages from any cause whatsoever, including fundamental breach, arising out of this Statement of Limited Warranty, or for any other claim related to this product, shall be limited to the greater of $10,000 or the amount paid for this product at the time of the original purchase, and shall not apply to claims for personal injury or damages to personal property caused by Acme's negligence, and in no event shall Acme be liable for any damages caused by your failure to perform your responsibilities under this Statement of Limited Warranty, or for loss of profits, lost savings, or other consequential damages, or for any third-party claims.

Can your audience easily grasp this warranty's lengthy sentence structure and difficult-to-understand words? No, the readers will become frustrated and will fail to recognize what is covered under such a warranty. Although you are following your boss's directions, you are not fulfilling one of your ethical requirements to use simple and direct language.

Writing can be legally binding *and* easy to understand. The first requirement does not negate the second. As a successful technical writer who values "the time and effort spent by those who read or see or hear [your] communication," you want to aid communication by using simple words and direct expression of ideas whenever possible ("Code for Communicators" 2004).

Satisfy the audience's need for information, not my own need for self-expression. The importance of simple language also encompasses the third point in STC's Code for Communicators: satisfying "the audience's need for information, not my own need for self-expression." The previously mentioned warranty might be considered poetic in its sentence structure and sophisticated in its word usage. It does not communicate what the audience needs, however. Such elaborate and convoluted writing will satisfy only one's need for self-expression: That is not the goal of effective technical writing.

Electronic mail (e-mail) presents another opportunity for unethical behavior. Because e-mail is so easy to use, many company employees use it too frequently. They write e-mail for business purposes, but they also use e-mail when writing to family and friends. Employees know not to abuse their company's metered

tech link

Go to *http://www.prenhall. com/gerson* for Web links, samples, and interactive activities.

mail by using corporate envelopes and stamps to send in their gas bills or write thank-you notes to Aunt Rose. These same employees, however, will abuse the company's e-mail system (on company time) by writing e-mail messages to relatives coast to coast. They are not satisfying their business colleagues' need for information; instead, they are using company-owned e-mail systems for their own self-expression (Hartman and Nantz 1995, 60). That is unethical.

Hold myself responsible for how well my audience understands my message. As a technical writer, you must place the audience first. When you write a precise proposal or instruction, using simple words and syntax, you can take credit for helping the reader understand your text. Conversely, you must also accept responsibility if the reader fails to understand.

Ethical standards for successful communication require the writer to always remember the readers—the real people who read manuals to put together their children's toys, who read corporate annual reports to understand how their stocks are doing, and who read proposals to determine whether to purchase a service. An ethical writer remembers that these real people will be frustrated by complex instructions, confused by inaccessible stock reports, and misled by inaccurate proposals. It is unethical to forget that your writing can frustrate people and even endanger them. It is unethical to use your writing to mislead or to confuse. In contrast, as an ethical writer, you need to treat these readers like neighbors, people you care about, and then write accordingly. Take the time to check your facts, present your information precisely, and communicate clearly so that your readers—your friends, coworkers, and clients—are safe and satisfied. You are responsible for your message (Barker et al. 1995).

Respect the work of colleagues. Each of the ethical considerations already discussed relates to your clients, the readers of your technical communication. This fifth point differs in two ways: First, it relates to coworkers or professional colleagues; second, it highlights the importance of online ethics, the ethical dilemmas presented by electronic communications on the Internet. We can discuss these two points simultaneously.

Much of today's technical writing takes place online, electronically (we discuss electronic communication in Chapter 13). This new venue for technical writing creates three unique reasons for an increased focus on ethics. As the technical writer, you must ethically consider confidentiality, courtesy, and copyright when working with the Internet or any private electronic network.

tech link

Go to Chapter 13, "Online Help and Web Sites," to learn more about the role of online help screens and Web sites in technical writing.

tech link

Go to *http://onlineethics.org/ codes/* for engineering ethical codes of conduct, including cases, scenarios, and essays.

- **Confidentiality.** The 1974 Privacy Act allows "individuals to control information about themselves and to prevent its use without consent" (Turner 1995, 59). The Electronic Communication Privacy Act of 1986, which applies all federal wiretap laws to electronic communication, states that e-mail messages can be disclosed only "with the consent of the senders or recipients" (Turner 1995, 60). However, both of these laws can be abused easily on the Internet. First, neither law specifically defines "consent." Data such as your credit records can be accessed without your knowledge by anyone with the right hardware and software. Your confidentiality can be breached easily.

  Second, the 1986 Electronic Communication Privacy Act fails to define "the sender." In the workplace, a company owns the e-mail system, just as it owns other more tangible items such as desks, computers, and file cabinets. Companies have been held liable for electronic messages sent by employees. Thus, many corporations consider the contents of one's e-mail and one's e-mailbox company property, not the property of the employee. With ownership comes the right to inspect an employee's messages. Although you might write an e-mail message assuming that your thoughts are confidential, this message can be monitored without your knowledge (Hartman and Nantz 61).

These instances might be legal, but are they ethical? Even though a company can eavesdrop on your e-mail or access your life's history through databases, that doesn't mean it should. As an employee or corporate manager, you should respect another's right to confidentiality. Ethically, you should avoid the temptation to read someone else's e-mail or to access data about an individual without her consent.

tech link

For ethical codes of conduct in many job fields, including business, communications, finance, and health care, go to *http://www.iit.edu/ departments/csep/ PublicWWW/codes/codes. html.*

- **Courtesy.** As already noted, e-mail is not as private as you might believe. Whatever you write in your e-mail correspondence can be read by others. Given this reality, you should be very careful about what you say in e-mail. Specifically, you don't want to offend coworkers by "flaming" (writing discourteous messages). Before you criticize a coworker's ideas or ability, and before you castigate your employer, remember that common courtesy, respect for others, is ethical (Adams et al. 1995, 328).

- **Copyright.** A final ethical consideration relates to copyright laws. Every English teacher you have ever had has told you to avoid *plagiarism* (stealing another writer's words and ideas). Plagiarism, unfortunately, is an even greater problem on the Internet. The Internet is perhaps the world's largest library without walls; it is almost a universal commons where anyone can set up a soap box and speak (or print) his opinion.

  This incredible ability to disseminate massive amounts of information presents problems. If you were an unethical technical writer, you could access data from the Internet and easily print it as your own. After all, it is hard to trace the identity of a writer on the Net, and Internet information can be downloaded by anyone (Adams et al. 1995, 328).

  An ethical writer will not fall prey to such temptation. If you have not written it, give the other author credit. Words are like any other possession. Taking words and ideas without attributing your source through a footnote or parenthetical citation is wrong. You should respect copyright laws.

Strive continually to improve my professional competence. Promote a climate that encourages the exercise of professional judgment. We've combined these last two tenets from the STC's Code for Communicators because together they sum up all of the preceding ethical considerations. Professional competence, for example, includes one's ability to avoid plagiarizing. Competence should also include professional courtesy, respect for another's right to confidentiality, a sense of responsibility for one's work, and the ability to write clearly and precisely. Ultimately, your competence is dependent upon professional judgment. Do you know the difference between right and wrong? Recognizing the distinction is what ethics is all about.

As a writer, you will always be confronted by a multitude of options, such as loyalty to your company, responsible citizenship, need for a salary, accountability to your client and your coworkers, and personal integrity. You must weigh the issues— ethically, legally, and practically—and then write according to your conscience.

Strategies for Making Ethical Decisions

When confronted with ethical challenges, try following these writing strategies (Guy 1990, 165).

a. *Define the problem.* Is the dilemma legal, practical, ethical, or a combination of all three?

b. *Determine your audience.* Who will be affected by the problem? Clients, coworkers, management? What is their involvement, what are their individual needs, and what is your responsibility—either to the company or to the community?

c. *Maximize values; minimize problems.* Ethical dilemmas always involve options. Your challenge is to select the option that promotes the greatest worth for all stakeholders involved. You won't be able to avoid all problems. The best you can hope for is to minimize those problems for both your company and your readers while you maximize the benefits for the same stakeholders.

d. *Consider the big picture.* Don't just focus on short-term benefits when making your ethical decisions. Don't just consider how much money the company will make now, how easy the text will be to write now. Focus on long-term consequences as well. Will what you write please your readers so that they will be clients for years to come? Will what you write have a long-term positive impact on the economy or the environment?

e. *Write your text.* Implement the decision by writing your memo, letter, proposal, manual, or report. When you write your text, remember to
   - Use precise language and visuals.
   - Use simple words and sentences.
   - Satisfy the audience's need for information, not your own need for self-expression.
   - Take responsibility for your content, remembering that real people will follow your instructions or make decisions based on your text.
   - Respect your colleagues' confidentiality, be courteous, and abide by copyright laws.
   - Promote professionalism and good judgment.

CHAPTER HIGHLIGHTS

1. If your technical writing is unclear, your reader may misunderstand you and then do a job wrong, damage equipment, or contact you for further explanations.

2. Use details to ensure reader understanding. Whenever possible, specify and quantify your information.

3. Answering *who, what, when, where, why*, and *how* (the reporter's questions) helps you determine which details to include.

4. For some audiences, you should avoid acronyms, abbreviations, and jargon.

5. Words that are not commonly used (legalisms, outdated terms, etc.) should be avoided.

6. Write to express, not impress—to communicate, not to confuse.

7. Avoid passive voice constructions, which tend to lengthen sentences and confuse readers.

8. Writing concisely helps save time for you and your readers.

9. Shorter paragraphs are easier to read, so they hold your reader's attention.

10. When possible, use short, simple words (always considering your reader's level of technical knowledge).

11. Apply readability formulas to determine your text's degree of difficulty.

12. Proofreading is essential to effective technical writing.

13. Well-organized documents are easy to follow.

14. Different organizational patterns—spatial, chronological, importance, comparison/contrast, and problem/solution—can help you explain material.

15. Consider whether or not your technical writing is legal, practical, and ethical.

CASE STUDIES
CompToday

CompToday computer hardware company must abide by the Sarbanes-Oxley Act, passed by Congress in response to accounting scandals. This act specifically mandates the following related to documentation standards:

- **Section 103: Auditing, Quality Control, and Independence Standards and Rules.** Companies must "prepare, and maintain for a period of not less than 7 years, audit work papers, and other information related to any audit report, in sufficient detail to support the conclusions reached in such report."

- **Section 401(a): Disclosures in Periodic Reports; Disclosures Required.** "Each annual and quarterly financial report . . . must be presented so as not to contain an untrue statement or omit to state a material fact necessary in order to make the pro forma financial information not misleading."

Beverly Warden, technical documentation specialist at CompToday, is responsible for managing the Sarbanes-Oxley reports. She is being confronted by the following ethical issues:

1. To help Beverly prepare the first annual report, her chief financial officer (CFO) has given her six months of audits (January through June). These prove that the company is meeting its accounting responsibilities. However, Beverly's report covers the entire year, including July through December. Section 103 states that the report must provide "sufficient detail to support the conclusions reached in [the] report."

Ethical Questions
Are the company's first six month of audits sufficient? If Beverly writes a report stating that her company is in compliance, is she abiding by her Society for Technical Communication Ethical Principles, which state that a technical writer's work is "consistent with laws and regulations"?

What are her ethical, practical, and legal responsibilities?

- Share your findings in an oral presentation.
- Write a letter, memo, report, or e-mail stating your opinion regarding this issue.

2. Beverly's CFO also has told her that during the year, the company fired an outside accounting firm and hired a new one to audit the company books. The first firm expressed concerns about several bookkeeping practices. The newly hired firm, providing a second opinion after reviewing the books, concluded that all bookkeeping practices were acceptable. The CFO sees no reason to mention the first firm.

Ethical Questions

Section 401(a) states that reports must contain no untrue statements or omit to state a material fact. The STC Ethical Principles also say that her writing must be truthful and accurate, to the best of her ability. Beverly can report factually that the new accounting firm finds no bookkeeping errors. Should she also report the first accounting firm's assessment? Is that a material fact? If she omits any mention of the first accounting firm, as her boss suggests, is she meeting both her STC technical writer's responsibilities and the needs of Sarbane-Oxley?

What are her ethical, practical, and legal responsibilities?

- Share your findings in an oral presentation.
- Write a letter, memo, report, or e-mail stating your opinion regarding this issue.

Blue Valley Wastewater Treatment Plant

The Blue Valley Wastewater Treatment Plant processes water running to and from Frog Creek, a water reservoir that passes through the North Upton community. This water is usually characterized by low alkalinity (generally, <30 mg/l), low hardness (generally, <40 mg/l), and minimal water discoloration. Inorganic fertilizer nutrients (phosphorus and nitrogen) are also generally low, with limited algae growth.

Despite the normal low readings, algae-related tastes and odors occur occasionally. Although threshold odors range from 3 to 6, they have risen to 10 in summer months. Alkalinity rises to <50 mg/l, hardness to <60 mg/l, and discoloration intensifies. Taste and odor problems can be controlled by powdered activated carbon (PAC); nutrient-related algae growth can be controlled by filtrated ammonia. Both options are costly.

The odors and tastes are disturbing North Upton residents. The odors are especially bothersome to outdoor enthusiasts who use the trails bordering Frog Creek for biking and hiking. Residents also worry about the impact of increased alkalinity on fish and turtles, many of which are dying, further creating an odor nuisance. Frog Creek is treasured for its wildlife and recreational opportunities.

The Blue Valley Wastewater Treatment Plant is under no legal obligation to solve these problems. The alkalinity, hardness, color, and nutrient readings are all within regulated legal ranges. However, community residents are insistent that their voices be heard and that restorative steps be taken to improve the environment around their homes.

This is an ethical and practical dilemma. In response to this dilemma, divide into small groups to write one of the following documents.

- You are Blue Valley Wastewater Treatment Plant's director of public relations. Write a letter to the City Commission stating your plant's point of view. (We discuss letters in Chapter 6.)
- You are North Upton community's resident representative. Write a letter to the City Commission stating your community's point of view.

- You are an employee for the Blue Valley Wastewater Treatment Plant. Write a memo to the plant director suggesting ways in which the plant could solve this environmental and community relations problem. (We discuss memos in Chapter 5.)
- You are the Blue Valley Wastewater Treatment Plant's director. Write a memo to the engineering supervisor (your subordinate) stating how to solve this environmental and community relations problem.

These documents could be organized by the problem/solution or comparison/contrast methods. Whichever method you choose, consider the strategies for making ethical decisions, discussed in this chapter.

Exercises

Writing Precisely

The following sentences are vague and imprecise. They will be interpreted differently by different readers. Revise these sentences—replace the vague, impressionistic words with more specific information.

1. We need this information as soon as possible.
2. The machinery will replace a flawed piece of equipment in our department.
3. Failure to purchase this will have a negative impact.
4. Weather problems in the area resulted in damage to the computer systems.
5. The most recent occurrences were caused by insufficient personnel.
6. Fire in the office caused substantial losses.
7. If we can't solve this problem soon, we will lose a large percentage of our business.
8. The automobile has a smaller turning radius than last year's model.
9. Several employees commended her for her expertise.
10. Make your explanations very detailed.

Avoiding Obscure Words

Obscure words make the following sentences difficult to understand. Improve the sentences by revising the difficult words and making them more easily understood.

1. As Very Large Scale Integration (VLSI) continues to develop, a proliferation of specialized circuit simulators will be utilized.
2. As you requested at the commencement of the year, I am forwarding my regular quarterly missive.
3. Though Randolph was cognizant of his responsibility to advise you of any employment aberrations, he failed to abide by this mandate.
4. Herewith is an explanation of our rationale for proferring services, pursuant to your request.
5. Please be advised that the sale constitutes a successful closure.

6. Inasmuch as we have endeavored to determine the causes of the dilemma without success, we are terminating this fact-finding operation immediately.

7. To facilitate the initiation of this activity, we have assigned the ensuing job responsibilities.

8. Can you assist us in ascertaining the causes pertaining to yesterday's mechanism malfunction?

9. I wonder if you would be so kind as to avail yourself of this opportunity to respond accordingly to our questionnaire.

10. In lieu of further discussion, we want to state in the affirmative that what transpired was due to the fact that the vehicle had insufficient braking capabilities to avoid the collision.

Using the Active Voice Versus the Passive Voice

Use of the passive voice often leads to vague, wordy sentences. Revise the following sentences by writing them in the active voice.

1. Implementation of this procedure is to be carried out by the Metrology Department.

2. Benefits derived by attending the conference were twofold.

3. The information was demonstrated and explained in great detail by the training supervisor.

4. Discussions were held with representatives from Allied, who supplied analytical equipment for automatic upgrades.

5. Also attended was the symposium on polymerization.

6. Process control systems for foam encapsulation were reviewed with vendors.

7. Effort should be expended to reduce overtime. Overtime in excess of eight hours should be closely monitored.

8. The reassignment of this activity was the result of changes requested by manufacturing.

9. Installation of the fiber optic networking is estimated to occur early next month.

10. Misapplication of a dry film lubricant has been the primary cause of defectiveness.

Limiting Paragraph Length

You can achieve clarity and conciseness if you limit the length of your paragraphs. An excessively long paragraph (beyond six typed lines) requires too much work for your reader. Revise the following paragraph to make it more reader-friendly.

As you know, we use electronics to process freight and documentation. We are in the process of having terminals placed in the export departments of some of our major customers around the country so they may keep track of all their shipments within our system. I would like to propose a similar tracking mechanism for your company. We could handle all of your export traffic from your locations around the country and monitor these exports with a terminal located in your home office. This could have many advantages for you. You could generate an export invoice in your export department, which could be transmitted via the computer to our office. You could trace your shipments more readily. This would allow you to determine rating fees more accurately. Finally, your accounting department would benefit. All in all, your export operations would achieve greater efficiency.

Reducing Word Length

Multisyllabic words can create long sentences. To limit sentence length, limit word length. Find shorter words to replace the following words:

1. advise	6. endeavor	11. prohibit
2. anticipate	7. inconvenience	12. residence
3. ascertain	8. indicate	13. subsequent
4. cooperate	9. initially	14. sufficient
5. determine	10. presently	15. terminate

Reducing Sentence Length

Each of the following sentences is too long. Revise them using the techniques suggested in this chapter: Use the meat cleaver approach; avoid shun words, camouflaged words, and expletives; omit redundancies; and delete wordy phrases.

1. In regard to the progress reports, they should be absolutely complete by the fifteenth of each month.

2. I wonder if you would be so kind as to answer a few questions about your proposal.

3. I am in receipt of your memo requesting an increase in pay and am of the opinion that it is not merited at this time due to the fact that you have worked here for only one month.

4. On two different occasions, I have made an investigation of your residence, and I believe that your sump pump might result in damage to your neighbor's adjacent property. I have come to the conclusion that you must take action to rectify this potential dilemma, or your neighbor might seek to sue you in a court of law.

5. In this meeting, our intention is to acquire a familiarization with this equipment so that we might standardize the replacement of obsolete machinery throughout our entire work environment.

6. It is evident that the company's request for electrical equipment to be placed in the laboratory has become rather important inasmuch as this need is prioritized in our CEO's most recent letter.

7. It is anticipated that these changes will lead to a reduction in the failure rate.

8. There is the possibility that we will implement these suggestions early next month.

9. New personnel will be assessed when brought on board and then tested on a yearly basis in order to ensure their continued successful job prowess.

10. If there are any questions that you might have, please feel free to contact me by phone.

Using Organization

1. *Spatial:*
   a. Using spatial organization, write a paragraph describing your classroom, your office, your work environment, your dorm room, your apartment, or any room in your house.
   b. Using spatial organization, write an advertisement describing the interior of a car, the exterior of a mechanism or tool, a piece of clothing, or a motorcycle.

2. *Chronological:* Organizing your text chronologically, write a report documenting your drive to school or work, your activities accomplished in class or at work, your discoveries at a conference or vacation, or your activities at a sporting event.

3. *Importance:* A fashion merchandising retailer asked her buyers to purchase a new line of clothing. In her memo, she provided them the following list to help them accomplish their task. Reorganize the list by importance, and justify your decisions.

   DATE: January 15, 2005
   TO: Buyers
   FROM: Sharon Baker
   SUBJECT: CLOTHING PURCHASES

   It is time again for our spring purchases. This year, let's consider a new line of clothing. When you go to the clothing market, focus on the following:

   • Colors
   • Materials
   • Our customers' buying habits
   • Price versus markup potential
   • Quantity discounts
   • Wholesaler delivery schedules

   Good luck. Your purchases at the market are what make our annual sales successful.

4. *Comparison/contrast:* Visit two auto dealerships, two clothing stores, two restaurants, two music shops, two prospective employers, two colleges, and so on. Based on your discoveries, write a report using comparison/contrast to make a value judgment. Which of the two cars would you buy, which of the two restaurants would you frequent, and at which of the two music shops would you purchase CDs?

5. *Problem/solution case study:*   You work for Acme Electronics as an electrical engineer. Carol Haley, your boss, informs you that your department's electronic scales are measuring tolerances inaccurately. You are asked to study the problem and determine solutions. In your study, you find that one scale (ID #1893) is measuring within 90 percent of tolerance; another scale (ID #1887) is measuring within 75 percent of tolerance; a third scale (ID #1890) is measuring within 60 percent of tolerance; a final scale (ID #1885) is measuring within 80 percent of tolerance. Standards suggest that 80 percent is acceptable. To solve this problem, the company could purchase new scales ($2,000 per scale); reduce the vibration on the scales by mounting them to the floor ($1,500 per scale); or reduce the vibration around the scales by enclosing the scales in plexiglass boxes ($1,000 per scale).

   Write a memo to your boss detailing your findings (the problems) and suggesting the solutions.

Considering Ethics

1. The Society for Technical Communication constantly is trying to redefine its Ethical Principles. As a class, how would you define the word *ethics?* Brainstorm new definitions as they apply to technical communication and come to a class consensus. Then in small groups, based on your class's definition of *ethics,*
   - List five or more examples of ethical responsibilities you believe technical writers should have when writing memos, letters, reports, proposals, or instructions (other than those already discussed in this chapter).
   - List five or more examples of failures to abide by ethical responsibilities you have seen either in writing or in other types of media (recordings, movies, television programs, newspapers, magazines, news reports, etc.).
   - Technical writing is factual, as are newspaper, magazine, radio, and television news reports; recordings, movies, and television programs are art forms. Do the same ethical considerations apply for all types of media? If there are differences, explain your answer.

2. Bring to class examples of ethically flawed communication. These could include poorly written warranties (which are too difficult to understand) or dangers, warnings, and cautions (which do not clearly identify the potential for harm). You might find misleading annual reports or unethical advertisements. Poor examples could even include graphics that are visually misleading. Then in small groups, rewrite these types of communication or redraw the graphics to make them ethical.

3. Every day, the Internet is presenting ethical dilemmas to lawmakers and to the populace. Congress is currently debating laws to curb unethical practices online. These unethical practices include hacking, pornography, solicitation, infringements on confidentiality, hate mail, inappropriate advertisements, spam, and unauthorized viewing of e-mail.

   To update your knowledge of ethics problems online, research any of the topics listed. Then, present your findings as follows:

   - Write an individual or group report on your findings (we discuss reports in Unit 6).
   - Give an individual or group oral presentation on your findings (we discuss oral presentations in Chapter 18).
   - Write a summary of the article(s) you've researched (we discuss summaries in Chapter 15).

QUIZ QUESTIONS

1. Why is clarity the ultimate goal of technical writing?

2. How can you achieve clarity?

3. How will answering the reporter's questions help you achieve clarity?

4. Why should you avoid using acronyms, abbreviations, and jargon in your technical writing?

5. In what two places in a technical document can you define a term?

6. Why should you strive for conciseness in your technical writing?

7. Why is a long paragraph ineffective in your technical documents?

8. What is a fog index?

9. Why should you use shorter sentences?

10. What are three causes of wordy sentences?

11. What are four proofreading tips?

12. When will you organize a document spatially?

13. Why would you organize chronologically?

14. What is the effect of organization by importance?

15. When is comparison/contrast used effectively in a business document?

16. How can you use problem/solution in a document?

17. What do you consider when you focus on the legal aspect of a document?

18. What do you consider when you focus on the practical aspect of a document?

19. What do you consider when you focus on the ethical aspect of a document?

20. What are three things you consider when you follow the guide for ethical standards?

Audience Recognition and Involvement

Writing at Work

Home and Business Mortgage

Home and Business Mortgage (HBM), a Phoenix, Arizona, company, prides itself on being a good neighbor, an asset to the community. Its goal is to ensure that the company meets all governmental personnel regulations, including those required by the Equal Employment Opportunity Commission (EEOC), the Family Medical Leave Act (FMLA), and the Americans with Disabilities Act (ADA). HBM is committed to achieving diversity in its workplace.

In fact, the company is so focused on its appreciation of and commitment to diversity that HBM's technical writers have created the following:

- Bilingual brochures, newsletters, and annual reports
- Brochures geared toward different stakeholders—senior citizens (age 55 and up), African-American, Hispanic, and Native American homebuyers, women, and young singles (18–25)
- A Web site with photographs depicting Phoenix's diverse population

HBM's Web site is unique in another important way. It has sought to abide by W3C WAI mandates. The World Wide Web Consortium (W3C) hosts the Web Accessibility Initiative (WAI). This initiative states that "the power of the Web is its universality. Access by everyone regardless of disability is an essential aspect." HBM's webmasters, in agreement with the WAI, realize that graphic-laden Web sites, as well as those with audio components, are not accessible to people with disabilities, such as hearing and visual impairment.

To meet this WAI platform, HBM's webmasters have achieved the following at the company's Web site:

- Avoided online audio.
- Limited online video and used the **Alt** attribute to describe each visual's function.

85

- Increased the font of Web text to 20-point versus the standard 12- to 14-point font.
- Avoided all designer fonts, like cursive, which are hard to read online.
- Avoided red and green for color emphasis because these colors are hard to read for audience's with color blindness.
- Avoided frames that not only fail to load on all computers but also create readability challenges online.

HBM is honored to serve all of its customers, regardless of race, age, or religion. One way to achieve the company's commitment to diversity is through its technical communication.

When you write a memo, letter, or report, someone reads it. That individual or group of readers is your audience. To compose effective technical writing, you should achieve (a) audience recognition and (b) audience involvement.

AUDIENCE RECOGNITION

In the business world, you will never write or speak in a vacuum. When you write your memo, e-mail, letter, report, or brochure, *someone* will read it. When you give an oral briefing, convene a meeting, communicate with customers in a salesroom, or make a speech at a conference, *someone* will be listening. The question is, "Who?"

- Who is your audience?
- What does this reader or listener know?
- What does this reader or listener not know?
- What must you write or say to ensure that your audience understands your point?
- How do you communicate to more than one person (multiple audiences)?
- What is his or her position in relation to your job title?
- What diversity issues (gender, sexual orientation, cultural, multicultural) must you consider?

If you do not know the answers to these questions, your communication might miss the mark. Your letter may contain jargon or acronyms the reader will not understand. The tone of the memo may be inappropriate for management (too dictatorial) or for your subordinates (too relaxed). Your verbal communication might not factor in your audience's unique culture and language. To communicate successfully, you must recognize your audience's level of understanding. You also must factor in your audience's unique traits, which could have an impact on your communication success. These could include many variables as shown in Table 4.1.

Knowledge of the Subject Matter

What does your audience know about the subject matter? Does he or she work closely with you on the project? That would make the audience a *high-tech peer*. Does the audience have general knowledge of the subject matter, but his or her area of expertise is elsewhere? That would make the audience a *low-tech peer*. Is the audience totally uninvolved in the subject matter? That would make him or her a *lay* audience. Finally, could your audience be a combination of these types? Then, you would be confronted with a *multiple audience*.

TABLE 4.1	AUDIENCE VARIABLES

Knowledge of the Subject Matter

- High tech
- Low tech
- Lay
- Multiple

Issues of Diversity

- Age
- Gender
- Race and/or religion
- Sexual orientation
- Language and/or culture of origin—*multicultural* or *cross-cultural*

High-Tech Audience

High-tech readers work in your field of expertise. They might work directly with you in your department, or they might work in a similar capacity for another company. Wherever they work, they are your colleagues because they share your educational background, work experience, or level of understanding.

If you are a computer programmer, for example, another computer programmer who is working on the same system is your high-tech peer. If you are an environmental engineer working with hazardous wastes, other environmental engineers focusing on the same concerns are your high-tech peers.

Once you recognize that your reader is high tech, what does this tell you? High-tech readers have the following characteristics:

- They are experts in the field you are writing about. If you write an e-mail message to one of your department colleagues about a project you two are working on, your associate is a high-tech peer. If you write a letter to a vendor requesting specifications for a system she markets, that reader is a high-tech expert. If you write a journal article geared toward your colleagues, they are high tech.

- Because their work experience or education are comparable to yours, high-tech readers share your level of understanding. Therefore, they will understand high-tech jargon, acronyms, and abbreviations. You do not have to explain to an electronics technician, for example, what *MHz* means. Defining *megahertz* for this high-tech reader would be unnecessary and even offensive. We recently read a procedure written by a manager to his engineers detailing how to write departmental technical reports. In the memo, the boss specified that the engineers should use "metric SI units in their reports." Because the boss recognized that his audience was high tech, he did not have to define *SI* as "System International."

- High-tech readers require minimal detail regarding standard procedures or scientific, mathematical, or technical theories. Two physicists, when discussing the inflation theory of the universe's creation, would not labor over the importance of quarks and leptons (subatomic particles). Although this jargon is unintelligible to most of us, to high-tech peers no explanations are required.

- High-tech peers read to discover new technical knowledge or for updates regarding the status of a project.
- High-tech readers need little background information regarding a project's history or objectives unless the specific subject matter of the correspondence is new to them. If, for example, you are writing a status report to your first-line supervisor, who has been involved in a project since its inception, then you will not need to flesh out the history of the project. On the other hand, if you have a new supervisor or if you are updating a colleague new to your department, even though these readers are high tech, you will need to provide background data.

If you work in an environment in which you write only to high-tech peers, you are a rare and lucky individual. Writing to high-tech readers is rather easy because you can use acronyms, abbreviations, jargon, complex graphics, and so on. You usually do not have to define terms or provide background information. Writing within a technical environment, however, also requires that you write to low-tech readers.

Low-Tech Audience

Low-tech readers include your coworkers in other departments. Low-tech readers also might include your bosses, your subordinates, or your colleagues who work for other companies. For instance, if you are a biomedical equipment technician, the accountant or personnel director or graphic artists in your company are low-tech peers. These individuals have worked around your company's equipment and, therefore, are familiar with your technology. However, they do not understand the intricacies of this technology.

Your bosses are often low tech because they no longer work closely with the equipment. Although they might have been technicians at one time, as they moved more and more into management they moved further and further away from technology. Your subordinates might be low tech because their levels of education or work experience are less than yours.

Finally, although your colleagues at other companies have your level of education or work experience, they could be low tech if they are not familiar with your company's procedures or in-house jargon, acronyms, and abbreviations.

When you write memos, letters, and reports to low-tech readers, remember that they share the following characteristics:

- Low-tech readers are familiar with the technology you are writing about, but their job responsibilities are peripheral to the subject matter. They either work in another department, manage you, work under your supervision, or work outside your company.
- Because low-tech readers are familiar with your subject matter, they understand *some* abbreviations, jargon, and technical concepts. To ensure that readers understand your content, therefore, define your terms. An abbreviation like *VLSI* can't stand alone. Define it parenthetically: VLSI (very large scale integration). Similarly, technical jargon like *silicon foundry* needs a follow-up explanation. You should write instead, "A silicon foundry, as the name implies, is a factory or manufacturer that casts into silicon an integrated chip based on a customer's specifications."

  In addition, technical concepts must be defined for low-tech readers. For example, whereas high-tech readers understand the function of pressure transducers, a low-tech reader needs further information, such as "Pressure transducers: Solid-state components sense proximal pressure in the patient tubing circuit. The transducers convert this pressure value into a proportional voltage for the control system."

- Since the low-tech reader is not in your normal writing "loop"—that is, not someone to whom you write often regarding your field of expertise—you need to provide more background information. When you submit a status report to upper-level management, for example, you can not just begin with work accomplished. You need to explain why you are working on the project (objectives, history), who is involved (other personnel), when the project began and its scheduled end date, and how you are accomplishing your goals. Low-tech readers understand the basic concepts of your work, but they have not been involved in it daily. Fill them in on past history.

Lay Audience

Customers and clients who neither work for your company nor have any knowledge about your field of expertise are your lay audience. If you work in telecommunications for a telephone company, for example, and you write a letter to a client regarding a problem with the company's phone line, your audience is a lay reader. If your field of expertise is biomedical equipment and you write a procedures manual for the patient end user, you are writing to a lay audience.

If you are an automotive technician writing a service report for a customer, that customer is a lay reader. Although you understand your technology, your reader, who uses the phone or the medical equipment or the car, is not an expert in the field. These readers are using your equipment or require your services, but the technology you are writing about is not within their daily realm of experience.

This makes writing for a lay audience difficult. It is easy to write to a high-tech reader who thoroughly understands the technology you are discussing. However, writing to a low-tech coworker or a lay audience totally outside your field of expertise is demanding.

To write successfully to a lay audience, remember that these readers share the following characteristics:

- Lay readers are unfamiliar with your subject matter. They do not understand your technology. Therefore, you should write simply. That's not to say that you should insult your lay reader with a remedial discussion or with a patronizing tone. You must, however, explain your topic clearly. You achieve clarity through precise word usage, depth of detail, and simple graphics.

- Because your lay readers do not understand your technology or work environment, they won't understand any of your in-house jargon, abbreviations, or acronyms. Avoid high-tech terms or define them thoroughly.

- Lay readers will need background information. If you leap into a discussion about a procedure without explaining to your lay readers why they should perform each step, they will not understand the causes or rationale. High-tech and possibly even low-tech readers might not need such explanations, but the lay reader needs you to clarify.

    For example, look at the maintenance procedure, which is part of a user's manual provided for the purchaser of an audio recorder (Figure 4.1). The introductory paragraph clarifies for the lay reader why she should clean the head section, instead of incorrectly assuming that

MAINTENANCE

Cleaning the Head Section

The heads, capstan, and pinch rollers get dirty easily. If this head section becomes dirty, the high-frequency sound will not be reproduced and the stereo balance will be impaired. This hurts your system's sound quality.

To avoid these problems, clean your system's head section regularly by following these simple steps:

1. Push the STOP/EJECT button to open the cassette door.
2. Dip a cleansing swab into the cleaning fluid.
3. Wipe the heads, capstan, and pinch rollers with the swab.
4. Allow 30 seconds to dry.

Note:
- Do not hold screwdrivers, metal objects, or magnets close to the heads.
- When demagnetizing the heads, be sure the unit's POWER switch is in the OFF position.

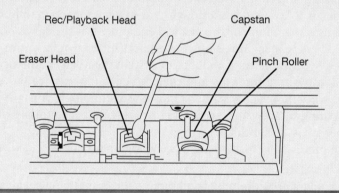

Figure 4.1 Cleaning the Head Section

the reader will understand without being told. This procedure also uses a simple graphic to clearly depict what action is required.

On the other hand, look at the procedure for cable preparation from an installation manual geared toward high-tech readers (Figure 4.2). It provides no background about why to perform the action, no clarity about how to perform the steps, and no graphics to help the reader visualize the procedure.

A lay reader would have no idea what is going on here. Why are we performing this act? What's the purpose? What will be achieved? Perhaps these questions are irrelevant for the technician, who knows the background or rationale, but a lay audience will be confused without an introductory overview. In addition, a lay audience will not know what a gasket, a cable dielectric, or a center conductor is. These terms are high-tech jargon, which a lay reader will not understand. Finally, what dimensions are we supposed to achieve? The high-tech reader, who is familiar with this operation, knows what is meant by *proper*

1. Place nut and gasket over cable and cut jacket to dimension shown.
2. Comb out braid and fold out. Cut cable dielectric to dimension shown. Tin center conductor.
3. Pull braid wires forward and taper toward center conductor. Place clamp over braid and push back against cable jacket.
4. Fold back braid wires as shown, trim to proper length (D), and form over clamp as shown. Solder contact to center conductor.
5. Insert cable and parts into connector body. Make sure sharp edge of clamp seats properly in gasket. Tighten nut.

Figure 4.2 Cable Preparation

length, but a lay audience won't. And, when is a clamp seated "properly in gasket"? Again, a lay audience needs these points clarified.

Composing effective technical writing requires that you recognize the differences among high-tech, low-tech, and lay audiences. If you incorrectly assume that all readers are experts in your field, you will create problems for yourself as well as for your readers. If you write using high-tech terms to low-tech or lay audiences, your readers will be confused and anxious. You will waste time on the phone clarifying the points that you did not make clear in the technical document.

Multiple Audiences

Correspondence is not always sent to just one type of audience. Sometimes your correspondence has multiple audiences. Sometimes you write to *all* of the aforementioned audiences *simultaneously.*

For example, when writing a report, most people assume that the first-line supervisor will be the only reader. This might not be the case, however. The first-line supervisor could send a copy of your report to the manager, who could then submit the same report to the executive officer. Similarly, your first-line supervisor might send the report to your colleagues or to your subordinates. Your report might be sent out to other lateral departments. Figure 4.3 shows the possibilities.

Writing correspondence for multiple readers with different levels of understanding and different reasons for reading creates a challenge for you. When you add the necessity of using a tone that will be appropriate for all of these varied readers, the writing challenge becomes even greater.

How do you meet such a challenge? The first key to success is recognizing that multiple audiences exist and that they share the following characteristics.

- Your intended audience will not necessarily be your only readers. Others might receive copies of your correspondence.

- Some of the multiple readers will be unfamiliar with the subject matter. You will have to provide background data (objectives, overviews) to clarify the history of the report for these readers. In a short letter, memo, report, or e-mail message, this background information can't be too elaborate. Often, all you do is provide a reference line suggesting where the readers can find out more about the subject matter if they

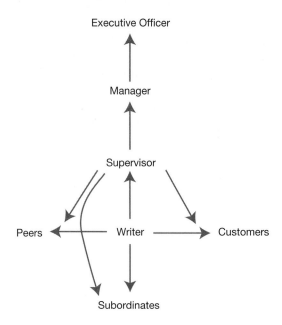

Figure 4.3 Examples of Possible Audiences for Correspondence

wish—"Reference: Operations Procedure 321 dated 9/21/03." In longer reports, background data will appear in the summary or abstract, as well as in the report's introduction.

Summaries, abstracts, introductions, and references are especially valuable when you consider one more fact regarding technical writing: reports, memos, and letters are kept on file. When you write the correspondence initially, you can assume that your reader or readers have knowledge of the subject matter. But months (or years) later when the report is retrieved from the files, will your readers still be familiar with the topic? Will you still have the same readers? Many people, some of whom didn't even work for your company at the time of the original writing, will read your correspondence. These future multiple readers need background information.

tech link

Go to *http://www.prenhall. com/gerson* for Web links, samples, and interactive activities.

- Multiple readers have diverse understandings of your technology. Some will be high tech; others will be low tech. This requires that you define jargon, abbreviations, and acronyms. As mentioned earlier in this chapter, you can define your terms either parenthetically within your text or in a glossary, depending on the length of the correspondence. Short e-mail messages, memos, or letters allow for only parenthetical definitions. Long reports, Web sites, and manuals allow for glossaries and extended definitions.

- Correspondence geared toward multiple readers must have a matter-of-fact, businesslike tone. You should not be too authoritative because upper-level management might read the memo, letter, or report. You should not be too ingratiating because lower level subordinates might also read the correspondence.

The memo in Figure 4.4 is written to a multiple audience. "To: Distribution" is a common way to direct correspondence to numerous readers. The multiple audience, then, affects the memo's content.

High-tech terms such as *operating procedure*, *engineering notice*, and *product quality requirements* are followed by parenthetical abbreviations—*(OP)*, *(EN)*, and *(PQR)*. If these abbreviations alone

DATE: July 7, 2005
TO: Distribution — Designation for multiple readers
FROM: Rochelle Kroft
SUBJECT: REVISION OF OPERATING PROCEDURE (OP) 354 DATED 5/31/05

The reissue of this procedure was the result of extensive changes requested by Engineering, Manufacturing, and Quality Assurance. These procedural changes will be implemented immediately according to Engineering Notice (EN) 185. — Parenthetical abbreviation

Some important changes are as follows:

1. *Substitutions.* An asterisk (*) can no longer be used to identify substitutable items in requirement lists. Engineering will authorize substitutions in work orders (WOs). Substitutions also will be specified by item numbers rather than by genetic description names.
2. *Product Quality Requirements (PQRs).* These will be included in WOs either by stating the requirements or by referring to previous PQRs.
3. *Oral Instructions.* When oral instructions for process adjustments are given, the engineer must be present in the department.

All managers will review OP 354 and EN 185. Next, the managers will review the changes with their supervisors to make sure that each supervisor is aware of his or her responsibilities. These reviews will occur immediately. — Abbreviation after prior parenthetical definition

— Objective tone (passive voice), versus a directive (active voice) that could offend management

Distribution:

Rob Harken   Manuel Ramos
Julie Burrton   Jeannie Kort
Hal Lang   Jan Hunt
Sharon Myers   Earl Eddings

Figure 4.4   Effective Memo for Multiple Audiences

had been presented, certain high-tech readers would have understood them. Other readers in the distribution list, however, might not have known what *OP, EN,* or *PQR* meant. To cover all readers in the multiple audience, the writer correctly used both the written-out terms and the abbreviations.

Finally, the tone of the memo is appropriate for readers with different levels of responsibility. Although it contains no direct commands, which might have been offensive to management, the memo is still assertive. The matter-of-fact tone simultaneously suggests to management that these actions will be carried out and informs subordinates to do so.

DEFINING TERMS FOR DIFFERENT AUDIENCE LEVELS

Communicating with audiences at different levels, including high-tech, low-tech, lay, and multiple readers, is challenging because each industry has its own specialized vocabulary. Usually your high-tech readers work most closely with you and share your expertise; colleagues understand your high-tech jargon. Therefore, you can use acronyms and abbreviations for these high-tech readers without providing definitions. However, it is your responsibility as a technical writer to define terms that might be unfamiliar to audiences at other levels. You can accomplish this goal by using any or all of the following:

- Words or phrases that define your terms, presented either parenthetically or in a glossary
- Sentences that define your terms, either in a glossary or following the unfamiliar terms
- Extended definitions of one or more paragraphs

Low-tech readers at your company have different areas of expertise. They are unfamiliar with your high-tech acronyms and abbreviations, but they are familiar with your work environment. Therefore, parenthetical definitions or brief definitions in a glossary should suffice. You could define *ATM* parenthetically as *(asynchronous transfer mode)*. Then, your low-tech reader will not misconstrue *ATM* as *automatic teller machine*. You could also use a glossary.

> **example**
>
> | HTTPS | Hypertext Transfer Protocol, Secure |
> | TDD | telecommunication device for the deaf |
> | TTY | teletypewriter |

Lay readers are further removed from your immediate work world. Customers, members of a city council, end users of your equipment, and vendors have no knowledge of your high-tech jargon and won't be helped if you merely define your terms. These readers need more information. You could provide either a follow-up sentence or an extended definition. If you provide a sentence definition, include the following:

Term + Type + Distinguishing characteristics

The **term** consists of the words, abbreviations, or acronyms you are using. The **type** explains what class your term fits into. For example, a "car" is a "thing," but the word *thing* applies to too many types of "things," including bananas, hammers, and computers. You need to classify your term more precisely. What type of thing is a car? "Vehicle" is a more precise classification.

But a motorcycle, a dirigible balloon, and a submarine are also types of vehicles. That's when the **distinguishing characteristics** become important. How does a car differ from other vehicles? Perhaps the distinguishing characteristics of a car would include the facts that cars are land-driven, four-wheeled vehicles. Thus, the definition would read as follows:

> **example**
>
> A car is a *vehicle* that contains *four wheels* and is *driven on land*.
> *Term + Type + Distinguishing characteristics*

Using a sentence to define *HTTP*, you would write

example

> *Hypertext Transfer Protocol* is a computer access code that provides secure communications on the Internet, an intranet, or an extranet.

When you need to provide an extended definition of a paragraph or more, in addition to providing the term, type, and distinguishing characteristics, also consider including examples, procedures, and descriptions. Look at the following definition of a voltmeter:

example

> The voltmeter is an instrument used to measure voltage. The voltmeter usually consists of a magnet, a moving coil, a resistor, and control springs. Types of voltmeters include the microvoltmeter, millivoltmeter, and kilovoltmeter, which measure voltages with a span of 1 billion to 1. By connecting between the points of a circuit, voltmeters measure potential difference.

There are other ways to define your terms.

- Place the definition in the front matter or an appendix of a document (such as a manual).
- Use endnotes in a report with other researched material.
- Use pop-up screens for online help incorporated into your Internet, intranet, or extranet site. The reader can access the definitions by clicking on "hot buttons" or hypertext links. (See Figure 4.5.)

Use a gutter margin

1. On the **File** menu, click **Page Setup,** and then click the **Margins** tab.

2. In the **Gutter** box, enter a value for the gutter margin.

3. Under **Gutter position,** click **Left** or **Top.**

Tip   To set gutter margins for part of a document, select the text, and then change the gutter margins as usual. In the **Apply to** box, click **Selected text.** Word automatically inserts section breaks before and after ⟵————————— Hypertext link to pop-up

> **section break**
> A mark you insert to show the end of a section. A section break stores the section formatting elements, such as the margins, page orientation, headers and footers, and sequence of page numbers. A section break appears as a double dotted line that contains the words "Section Break."

Pop-up box, defining the term ("section break"), the type ("a mark"), and the unique aspects of the term ("stores the section formatting elements . . ."), etc.

Figure 4.5   Pop-up Definition

TABLE 4.2	HOW TO COMMUNICATE TO DIFFERENT AUDIENCE LEVELS	
Audience Level	How To Communicate	Sample
High-tech	Use jargon, acronym, or abbreviation alone.	The wastewater is being treated for DBPs.
Low-tech	Use jargon, acronym, or abbreviation with a parenthetical definition.	The wastewater is being treated for DBPs (disinfection by-products).
Lay	Use jargon, acronym, or abbreviation with a parenthetical definition *and* a brief explanation or extended definition.	The wastewater is being treated for DBPs (disinfection by-products), such as acid, methane, chlorine, and ammonia.

Table 4.2 shows techniques for communicating with audiences with different levels of technical knowledge. These various techniques for defining terms are neither foolproof nor mandatory. You are always the final judge of how much information to provide for your intended readers. However, remember that if your readers fail to understand your content, then you have failed to communicate. No one will complain if you define your terms, but readers who are uncertain about your meaning either will call you for assistance, wasting your time, or will perform tasks incorrectly, wasting their time. It is better to provide too much information than not enough.

BIASED LANGUAGE—ISSUES OF DIVERSITY

Your audience will not be composed of people just like you. In contrast, your readers or listeners will more than likely be diverse. Diversity includes "gender, race/ethnicity, religion, age, sexual orientation, class, physical and mental characteristics, language, family issues, [and] departmental diversity" (Grimes and Richard 2003, 8).

Think of it this way: You work for a city government and your audience is the citizenry. Who comprises your city's populace? Or, you work in a hospital. Who will visit your facility? Or, you work in a retail establishment. Who will shop there? Or, you work in a company, any company. Who are your coworkers? They are people of many different interests, levels of knowledge, and backgrounds—and they are all valuable to your business.

Why should you be concerned about a diverse audience?

1. *Diversity is protected by the law*—Prejudicial behavior and discrimination on the job will not be tolerated. "Non-compliance with Equal Opportunity or Affirmative Action legislation can result in fines and/or loss of contracts" (McInnes 2003).

2. *Respecting diversity is the right thing to do*—People should be treated equally, regardless of their age, gender, sexual orientation, culture, or religion.

3. *Diversity is good for business*—"An environment where all employees feel included and valued yields greater commitment and motivation" ("What Is The 'Business Case' For Diversity?" 2003). Clients obviously would prefer shopping in an environment devoid of prejudice. In addition, "buying power . . . is represented by people from all walks of

life. . . . To ensure that . . . products and services are designed to appeal to this diverse customer base, 'smart' companies are hiring people from those walks of life—for their specialized insights and knowledge" (McInnes 2003).

4. *A diverse workforce keeps companies competitive*—Talent does not come in one color, nationality, or belief system. Instead, talent is "represented by people from a vast array of backgrounds and life experiences. Competitive companies cannot allow discriminatory preferences and practices to impede them from attracting the best available talent" (McInnes 2003).

Diversity management is such an important concern that "American businesses, educational institutions, and governmental agencies have spent untold millions on multicultural awareness and diversity training. The intended outcome: organizations that value and celebrate differences as well as similarities, thereby creating a more harmonious and productive work/study environment" (Jordan 1999, 1).

This is further verified in "the results of the 1998 *Society for Human Resource Management Survey of Diversity Initiatives*, [which stated that] 84 percent of human resource professionals at Fortune 500 companies say their top-level executives think diversity management is important. At organizations outside of the Fortune 500, 67 percent of human resource professionals said diversity management is important to their organizations' high-level executives" ("What Is The 'Business Case' For Diversity?" 2003).

> **tech link**
> Go to *http://www.shrm.org/ diversity/* (the Society for Human Resource Management) for additional information about diversity in the workplace, including definitions, links, and articles about diversity in the news.

MULTICULTURALISM

Multicultural Communication

Another feature of diversity in business communication is today's global economy. Your company will market its products or services worldwide. International business requires multicultural communication, the sharing of written and oral information between businesspeople from many different countries (Nethery 2003). Who is doing business internationally? Almost everyone!

The Global Economy

Examples of companies doing work globally abound. Read Table 4.3 to see how production and communication at various major corporations are affected by the global economy.

The Challenges of Multicultural Communication

The Internet and e-mail affect global communication and global commerce constantly. With these technologies, companies can market their products internationally and communicate with multicultural clients and coworkers at a keystroke. An international market is great for companies because a global economy increases sales opportunities. However, international commerce also creates written and oral communication challenges. Companies that work internationally must communicate with all of their employees and clients; therefore, communication must be multilingual.

Look at this one startling example to prove our point. Medtronic, a leading medical technology company, does business in 120 countries. Many of those countries mandate that product documentation be written in the local language.

TABLE 4.3	GLOBAL COMMUNICATION
Company	Impact
Coca Cola	Coca Cola produces 300 different brands in 200 countries. Seventy percent of their income is generated from outside the United States ("Around the World" 2003).
Microsoft	Microsoft realizes that "increasing numbers of companies need to have their computing systems support multiple languages. [Thus,] the Microsoft Windows XP Professional operating system [addresses] a multilingual environment by providing different versions and options . . . for all . . . users regardless of their location or language." To accommodate this multicultural need, Windows XP Professional is written in 25 different languages ("Multilingual Features in Windows XP Professional" 2001).
General Motors	General Motors, which employs about 355,000 people around the world, realizes that "integrity transcends borders, language, and culture." Thus, it seeks to "establish common goals, monitor progress and share best practice across [its] global operations" (General Motors 2003).
Black & Veatch	Black & Veatch, a global engineering company, has locations in Argentina, Australia, Botswana, Brazil, China, the Czech Republic, Egypt, Germany, Hong Kong, India, Indonesia, Korea, Kuwait, Malaysia, Mexico, Philippines, Poland, Saudi Arabia, Singapore, South Africa, Swaziland, Taiwan, Thailand, Turkey, United Arab Emirates, United Kingdom, United States, Vietnam, Zambia, and Zimbabwe.
Hallmark Cards	Margaret Keating, vice president-operations for Hallmark Cards Inc., oversees Hallmark's North American manufacturing and distribution, graphics, global procurement and global operations. She says, "communication is critical with a group [of 6,000 employees] geographically dispersed." "She spends a lot of time making communication clear, concise and compelling. That it has to be translated into different languages makes efficient communication even more important." Business communication at Hallmark "'has to make sense to the audience, and the audiences are very different'" (Cardarella 2003).

tech link

To learn more about these companies' global communication concerns, visit their Web sites:
- *http://www2.coca-cola. com/ourcompany/ aroundworld.html*
- *http://www.microsoft. com/Office/previous/xp/ multilingual/*
- *http://www2.bv.com/ locations/index.htm*
- *http://www.hallmark.com* ("About Hallmark"— "Hallmark International")

To meet these countries' demands, Medtronic translates its manuals into 11 languages: French, Italian, German, Spanish, Swedish, Dutch, Danish, Greek, Portuguese, Japanese, and Chinese (Walmer 1999, 230).

Multilingual reports, for example, create unique communication challenges, as John K. Courtis and Salleh Hassan note. Will each language version be identical "in terms of reading ease" and "in terms of content"? "Does the official first language version signal any advantages to investors over second or third language versions? Are different language versions prepared with the same attention to tone, style, and emphasis?" "Does the external audit firm read each language version and test for accuracy, comparability and completeness?" (2002, 395).

Multicultural Team Projects

What about international, multilingual project work teams? If, for example, your U.S. company is planning to build a power plant in China, you will work with Chinese engineers, financial planners, and regulatory officials. To do so effectively, you will need to understand that country's

- Verbal and nonverbal communication norms
- Management styles
- Decision-making procedures
- Sense of time and place
- Local values, beliefs, and attitudes

Our natural instinct is to evaluate people and situations according to our sense of values, our cultural perspectives. That is called ethnocentrism—a belief that one's own culture represents the norm. Such is not the case. The world's citizenry does not share the same perspectives, beliefs, values, political systems, social orders, languages, or habits. Successful business communication takes into consideration language differences, nonverbal communication differences, and cultural differences.

Due to the multicultural makeup of your audience, you must ensure that your writing, speaking, and nonverbal communication skills accommodate language barriers and cultural customs. The classic example of one company's failure to recognize the importance of translation concerns a car that was named Nova. In English, *nova* is defined as a star that spectacularly flares up. In contrast, *no va* in Spanish is translated as "no go," a poor advertisement for an automobile.

Communicating Globally . . . in Your Neighborhood

Cross-Cultural Communication

Multiculturalism will affect you not just when you communicate globally. You will be confronted with multicultural communication challenges even in your own city and state. "Marlene Fine . . . argues that 'the challenge posed by the increasing cultural diversity of the U.S. workforce is perhaps the most pressing challenge of our times'" (Grimes and Richard 2003, 9). Another term for this challenge is *cross-cultural communication,* writing and speaking between businesspeople of two or more different cultures within the same country (Nethery 2003).

How big a challenge is this? After all, everyone in the United States understands English—right? Wrong! "About 19 million people in the United States are not proficient in English" (Sanchez 2003, A1). In addition, look at the following statistics regarding America's melting pot, as shown in Figure 4.6.

These numbers are estimated to change significantly. Though the U.S. population of African Americans will stay at 14 percent through 2050, the U.S. Hispanic

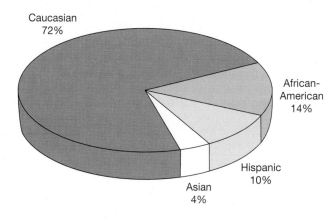

Figure 4.6 U.S. Occupational Employment in Private Industry by Race/Ethnic Group in 2000

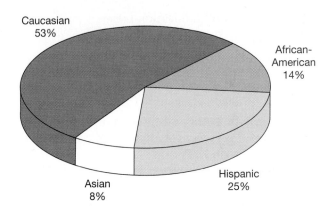

Figure 4.7 U.S. Occupational Employment in Private Industry by Race/Ethnic Group in 2050

population will grow from 10 percent in 2000 to 14 percent in 2010, 16 percent in 2020, and 25 percent in 2050. The Asian population in the U.S. will grow more slowly: 5 percent in 2010, 6 percent in 2020, and 8 percent in 2050 (Skettino 2004, 13–15).

In 2000, the U.S. workplace is predominately Caucasian, representing 70 percent of the workers. In 2050, the U.S. workplace will look very different. Figure 4.7 shows what this new workplace is projected to look like.

What is generally referred to as "minority groups"—African-American, Hispanic, and Asian—will actually grow to represent almost 50 percent of the total U.S. population. One of the challenges presented by the increasingly multicultural nature of our society and workplace is language. Language barriers are an especially challenging situation for hospitals, police and fire personnel, and governmental agencies where failure to communicate effectively can have dangerous repercussions.

One hospital reported that it has "13 staff members supplying Spanish, Arabic and Somali translations. . . . In 2001, nearly 21,000 Spanish interpretations were performed at [the hospital]. The numbers for 2002 . . . exceeded 29,000 interpretations." However, this hospital's successful use of translators to help with doctor-patient communication is rare. "Only about 14 percent of U.S. hospitals provide training for volunteer translators." Most hospitals depend on the patient's relatives. "In one case studied, an 11-year-old sibling translated. The child made 58 mistakes" (Sanchez 2003, A4). Imagine how such errors can negatively impact healthcare and medical records.

The communication challenges are not just evident for employees in healthcare and other community infrastructures. Industries as diverse as banking, hospitality (restaurants and hotels), construction, agriculture, meat production and packing, and insurance also face difficulties when communicating with clients and employees for whom English is a second language.

To communicate effectively to a multicultural audience, follow these guidelines.

Define Acronyms and Abbreviations

Acronyms and abbreviations cause most readers a problem. Although you and your immediate colleagues might understand such high-tech usage, many readers won't. This is especially true when your audience is not native to the United States.

To avoid communication problems, define your acronyms and abbreviations parenthetically the first time you use them. This applies even to acronyms and abbreviations that most people take for granted. Do you know what *FYI* means?

You probably do, but would your readers in Japan, France, Mexico, and China understand this commonly used acronym? Why take the risk? Just define it parenthetically: *FYI (for your information)*.

A current e-mail acronym is *LOL*. When this is placed at the end of a document or sentence, the writer is telling the readers not to take the message seriously. *LOL* is deined as *laughing out loud*. How many readers abroad will recognize this acronym?

Here's another example. Corporate employees often abbreviate the job title *system manager* as *sysmgr*. However, in German, the title *system manager* is called the *system leiter*; in French, it's *le responsable*. The abbreviation *sysmgr*. would make no sense in either of these countries (Swenson 1987, WE-193).

Avoid Jargon and Idioms

The same dilemma applies to *jargon* and *idioms*, words and phrases that are common expressions in English but that could be meaningless outside our borders. Every day in the United States, we use *on the other hand* as a transitional phrase and *in the black* or *in the red* to denote financial status. What will these idioms mean in a global market?

Similarly, the computer industry says, "The system crashed so we rebooted." A literal translation of this jargon into Chinese, German, or French will only confuse the readers (Swenson 1987, WE-194).

Distinguish Between Nouns and Verbs

Many words in English act as both nouns and verbs. This is especially true with computer terms, such as *file*, *scroll*, *paste*, *code*, and *help*. If your text will be translated, make sure that your reader can tell whether you're using the word as a noun or a verb (Rains 1994).

Watch for Cultural Biases/Expectations

Your text will include words and graphics. As a technical writer, you need to realize that many colors and images that connote one thing in the United States will have different meanings elsewhere.

For example, the idioms *in the red* and *in the black* will not necessarily communicate your intent when they are translated. Even worse, the colors black and red have different meanings in different cultures. Red in the United States connotes danger; therefore, *in the red* suggests a financial problem. In China, however, the word *red* has a positive connotation, which would skew your intended meaning. The word *black* often implies death and danger, yet *in the black* suggests financial stability. Such contradictions could confuse readers in various countries.

Moreover, images like those of an apple, a grim reaper, or a police officer might represent such concepts as wholesomeness, death, and safety in the United States. These same images, however, are not universal. In various countries abroad, the image of a police officer could be perceived as the symbol of totalitarian oppression. The grim reaper would be meaningless to a Hindu or Buddhist. Biblically, apples represent sin, not American apple pie. Hands are an even greater problem. For example, the "thumbs up" gesture is considered obscene in many countries. Even an image as simple as an electrical plug presents multicultural problems, because these plugs vary from country to country.

Animals represent another multicultural challenge. In the United States, we say you're a "turkey" if you make a mistake, but success will make you "soar like an eagle." The same meanings don't translate in other cultures. Take the friendly piggy bank, for example. It represents a perfect image for savings accounts in the United States, but pork is a negative symbol in the Mideast. If you are "cowed" by your competition in the United States, you lose. Cows, in contrast, represent a positive and sacred image in India (Horton 1993, 686–93).

Be Careful When Using Slash Marks

Our fourth consideration, "Watch for cultural biases/expectations," uses a slash mark(/). What do we mean by this typographic character? Does the slash mark mean "and," "or" or both "and/or"? The word *and* means "both," but the word *or* means "one or the other," not necessarily both. If your text will be translated to another language, will the translator know what you meant by using a slash mark? To avoid this problem, determine what you want to say and then say it (Rains 1994).

Avoid Humor and Puns

Humor is not universal. In the United States, people talk about regional humor. If a joke is good in the South but not in the North, how could that same joke be effective overseas? Microsoft's software package Excel is promoted by a logo that looks like an *X* superimposed over an *L*. This visual pun works in the United States because we pronounce the letters *X* and *L* just as we would the names of the software package. If your readers are not familiar with English, however, they might miss this clever sound-alike image (Horton 1993, 686).

Realize That Translations Might Take More or Less Space

Paper Size   If your writing will be conveyed not on paper but on disk or on the Internet, you must consider software's line-length and screen-length restrictions. For example, a page of hard-copy text in the United States will consist of approximately 55 lines that average 80 characters per line. How could this present a problem? The standard sheet of paper in the United States measures $8\frac{1}{2} \times 11$ inches. In contrast, the norm in Europe for standard-sized paper is A4—210×297 millimeters, or 8.27×11.69 inches.

Why is the size of paper important? This has nothing to do with language barriers, right? Here is the problem: If you format your text and graphics for an $8\frac{1}{2} \times 11$-inch piece of paper in Atlanta, travel to London, download the files on a computer there, and hit Print, you might find that what you get is not what you hoped for. The line breaks will not be the same. You will not be able to three-hole punch and bind the text. The margins will be off, and so will the spacing on your flowchart or table (Scott 2000, 20–21).

Web Sites   An even greater problem occurs when you are writing for the Internet. On a Web site, you will provide a navigation bar and several frames. Why is this a problem? A page of English text translates to the same length in any language, right? The answer is no. The word count of a document written in English will expand more than 30 percent when translated into some European languages. For example, the four-letter word *user* has 12 characters in German (Hussey and Homnack 1990, RT-46). In Table 4.4, notice how English words become longer when translated into other languages (Horton 1993, 691).

The Swiss government is trying to curb what it defines as "the encroachment of English." To do so, the government's French Linguistics Service is asking its citizens to avoid using the word *spam,* instead opting for *courier de masse non sollicite,* meaning "unsolicited bulk mail" ("Swiss fight encroachment of English" 2002, A16).

TABLE 4.4	TRANSLATIONS INCREASE WORD LENGTH
English Word	Translations into Other Languages
Print	*Impression* (French)
File	*Archivo* (Spanish)
View	*Visualizzare* (Italian)
Help	*Assistance* (French)
E-mail	*Courriel* (Swiss)

Whereas the above examples add length to a document, "Chinese character writing captures the subject matter more compactly than English. Consequently, the length of the Chinese translation is usually about 60% of that needed by the English version" (Courtis and Hassan 2002, 397).

You must consider length to accommodate translations. This is especially important when you realize the "surge of international business" and the impact of the Internet for communication purposes (St. Amant "Web Sites" 2003, 16). International access to the Internet has grown at an astounding rate. Look at China, for example. In 1999, approximately 2 million users in China accessed the Internet. This number has grown to over 25 million users today (St. Amant "Web Sites" 2003, 15). Facts like this alone suggest that a company must assume that all online communication projects will be translated.

Avoid Figurative Language

Many of us use sports images to figuratively illustrate our points. We "tackle" a chore; in business, a "good defense is the best offense"; we "huddle" to make decisions; if a sale isn't made, you might have "booted" the job; if a sale is made, you "hit a home run"; if you know you are not going to succeed at a task, you just "bail." Each of these sports images might mean something to native speakers, but will they communicate worldwide? We doubt it. Instead, say what you mean, using precise words (Weiss 1998, 14).

Cultural differences and language barriers present the technical writer many challenges. You might even want to ignore these problems as being too time consuming. Creating text that can be translated successfully will cost your company time and money—up front. The payoff comes later, however, when you ensure your company's reputation and avoid potential lawsuits. By recognizing the multicultural basis of your audience, you can protect your company from unintentional cultural offenses. Similarly, by considering the needs of multicultural readers, you can create text that is easy to follow and understand. This will minimize misreadings that could lead to mechanism failures or physical harm—each of which could cost your company in legal fees.

Be Careful with Numbers, Measurements, Dates, and Times

Numbers and Measurements   For example, if your text uses measurements, you are probably using standard American inches, feet, and yards. However, most of the world measures in metrics. Thus, if you write 18 high × 20 wide × 30 deep, what are the measurements? There is a huge difference between $18 \times 20 \times 30$ inches and $18 \times 20 \times 30$ millimeters.

TABLE 4.5	DIFFERENT WAYS OF UNDERSTANDING AND WRITING THE U.S. DATE 05/03/05
Country	Date
United States	May 3, 2005
United Kingdom	March 5, 2005
France	5 mars 2005
Germany	5. Marz 2005
Sweden	05–05–03
Italy	5.3.05

TABLE 4.6	DIFFERENT WAYS OF WRITING THE U.S. TIME 5:15 P.M.
Country	Time
United States	5:15 P.M.
France	17:15
Germany	17.15
Quebec, Canada	17 h 15

Dates   In the United States, we tend to abbreviate dates as MM/DD/YY: 05/03/05. In the United Kingdom, however, this could be perceived as March 5, 2005, instead of May 3, 2005. See Table 4.5 for additional examples.

Time   Time is another challenge. Table 4.6 shows how different countries write times.

In addition to different ways of writing time, you must also remember that even within the United States, 1:00 P.M. does not mean the same thing to everyone. Is that central time, Pacific time, mountain time, or eastern time? Add to this the problems with world time zones, and the challenge increases. You must define time zones clearly.

Another challenge with time occurs when we incorrectly assume that everyone everywhere abides by the same work hours. In the United States, the average workweek is 40 hours, and the typical workday is from 8:00 A.M. to 5:00 P.M. However, this is not the norm globally. French laws have reduced the workweek to 35 hours. Many Middle Eastern countries close work "for all or part of Friday," the beginning of Sabbath. Offices in parts of southern Europe shut down for a traditional two-hour lunch (noon to 2 P.M.) (St. Amant "Virtual" 2003, 28). Therefore, if you write an e-mail telling a coworker in Spain or Jordan that you will call at 2:00 P.M. his or her time, that could be an inappropriate time for your audience.

Finally, even simple words like *today, yesterday,* or *tomorrow* can cause problems. Japan is 14 hours ahead of U.S. eastern standard time. Thus, if you need a report "tomorrow," do you mean tomorrow—the next day *your time*— or tomorrow—two days from your reader's time?

To solve these problems, determine your audience and make changes accordingly. That might mean

- Writing out the date completely (January 12, 2005)

- Telling the reader what standard of measurement you will use ("This document provides all measurements in metrics.")
- Telling the reader what scheme of time presentation you will use ("This document relates time using a 24-hour clock rather than a 12-hour clock.")
- Using multiple formats ("Let's meet at 2:30 P.M./14:30.")
- Avoiding vague words like *today, tomorrow,* or *yesterday*
- Recognizing that people have different work schedules globally

Use Stylized Graphics to Represent People

A photograph or realistic drawing of people will probably offend someone and create a cultural conflict. You want to avoid depicting race, skin color, hairstyles, and even gender. To solve this problem, avoid shades of skin color, choosing instead pure white or black to represent generic skin. Use simple, abstract, even stick figures to represent people. Stylize hands so they are neither male nor female—and show a right hand rather than a left hand, if possible (a left hand is perceived as "unclean" in some countries) (Flint 1991, 241).

Recognizing the importance of the global marketplace is smart business and a wise move on the part of the technical writer.

Following is an example of poor communication for a multilingual audience.

Before

DATE: February 10, 2005
TO: Jose Guerrero, Mexico City, Mexico Office; Yong Kim, Hong Kong Office; Hans Rittmaster, Berlin Office
FROM: Leonard Liss, New York Office
SUBJECT: AGENDA FOR TELECONFERENCE

Time to wrap up that deal, guys. If we don't finish the project soon, we're all behind the eight ball. So, here's what I'm planning for the 03/07/05, 12:00 discussion:

- Restructured design—rather than build the part at 8 × 10 × 23, let's consider a smaller design.

- Shipping method—let's use a new carrier/vendor. We've not had good luck with Flyrite Overnight. I'm open to your suggestions. Let's think outside the box and just wing it.

Brainstorm with your men before our teleconferences so we can tackle this topic pronto. My boss, the old "ball and chain," needs our suggestions ASAP, so I need it even sooner. I know you'll come through with flying colors.

This e-mail fails for many reasons:

1. *Date*—Does "03/07/05" mean March 7, 2005 or July 3, 2005?
2. *Time*—"12:00" not only fails to specify the time zone, but creates the question whether midnight or noon is intended.
3. *Figurative language*—"Wrap up the deal," "behind the eight ball," "think outside the box," "just wing it," "tackle the topic," "ball and chain," and "flying colors" all are idiomatic phrases. Though these phrases might communicate within the United States, they do not translate internationally.

4. *Informal tone*—Informality is common in most stateside businesses, but this is not true in all countries. Germany and Japan, for example, are far more formal in their business communication expectations. Thus, "guys" and the contractions will cause problems.

5. *Sexist language*—"Guys" and "men" should be avoided when writing any correspondence. The implication is that only men will be involved in the discussions, which of course is completely erroneous.

6. *Measurements*—"8 × 10 × 23" is confusing. Is the writer discussing inches, feet, or meters?

7. *Slash marks*—"New carrier/vendor" could mean either "new carrier *or* vendor," or "new carrier *and* vendor." Which is it?

8. *Undefined abbreviations or acronyms*—"ASAP" must be defined. Though all readers within the United States might understand this to mean "as soon as possible," you cannot assume the same level of understanding internationally.

9. *Cultural sensitivity*—"Brainstorming" is the norm in the United States. However, in Germany, such flexibility is not as common. In the United States, we might see humor in referring to a boss as "the old ball and chain" (though this is highly doubtful). In Japan, however, where saving face is a cultural norm, potentially offending a superior would be an egregious flaw. Finally, using the word "pronto" is cowboy slang at best, and offensive at worst to some readers. To communicate internationally, you must consider each country's cultural norms—and be careful to avoid offense.

In contrast, the following example corrects these communication problems:

After

DATE: February 10, 2005
TO: Jose Guerrero, Mexico City, Mexico Office; Yong Kim, Hong Kong Office; Hans Rittmaster, Berlin Office
FROM: Leonard Liss, New York Office
SUBJECT: AGENDA FOR TELECONFERENCE

We need to complete our team project. Doing so will allow our respective companies to meet our client's deadline. A teleconference is our best way to communicate, given everyone's diverse locations. I have made all the technical arrangements. The teleconference is scheduled for March 7, 2005, at 12:00 noon, Pacific standard time. This is the only time that is at least somewhat convenient for all of us. During the teleconference, we will discuss the following:

- Restructured design—rather than build the part at 8" × 10" × 23" as planned (all dimensions in inches), we should consider a smaller design.

- Shipping method—a new carrier or vendor might save us money and time. Flyrite Overnight has increased its shipping fees by 25 percent. What do you think? I am open to your creative ideas.

Brainstorm with your coworkers before our teleconference so we can use our time effectively. My boss, Sue Cottrell, needs our suggestions by March 8, 2005, 4:00 P.M. central standard time. With your help, I know our company will make the correct decisions.

Sexist Language

Many of your readers will be women. This does not constitute a separate audience category. Women readers will be high tech or low tech, management or subordinate. Thus, you don't need to evaluate a woman's level of understanding or position in the chain of command any differently than you do for readers in general.

Recognize, however, that women constitute over half the workforce. As such, when you write, you should avoid *sexist language*, which is offensive to all readers.

Let's focus specifically on ways in which sexism is expressed and techniques for avoiding this problem. Sexism creates problems through omission, unequal treatment, and stereotyping, as well as through word choice.

Omission

When your writing ignores women or refers to them as secondary, you are expressing sexist sentiments. The following are examples of biased comments and their nonsexist alternatives:

Biased	Unbiased
Radium was discovered by a woman, Marie Curie.	Radium was discovered by Marie Curie.
When setting up his experiment, the researcher must always check for errors.	When setting up experiments, the researcher must always check for errors.
As we acquired scientific knowledge, men began to examine long-held ideas more critically.	As we acquired scientific knowledge, people began to examine long-held ideas more critically.

Unequal Treatment

Modifiers that describe women in physical terms not applied to men are patronizing.

Biased	Unbiased
The poor women could no longer go on; the exhausted men . . .	The exhausted men and women could no longer go on.
Mrs. Acton, a statuesque blonde, is Joe Granger's assistant.	Jan Acton is Joe Granger's assistant.

Stereotyping

If your writing implies that only men do one kind of job and only women do another kind of job, you are stereotyping. For example, if all management jobs are held by men and all subordinate positions are held by women, this is sexist stereotyping.

Biased	Unbiased
Current tax regulations allow a head of household to deduct for the support of his children.	Current tax regulations allow a head of household to deduct for child support.
The manager is responsible for the productivity of his department; the foreman is responsible for the work of his linemen.	Management is responsible for departmental productivity. Supervisors are responsible for their personnel.

Biased	Unbiased
The secretary brought her boss his coffee.	The secretary brought the boss's coffee.
The teacher must be sure her lesson plans are filed.	The teacher must file all lesson plans.

Sexist language disappears when you use pronouns and nouns that treat all people equally.

Pronouns

Pronouns such as *he, him,* or *his* are masculine. Sometimes you read disclaimers by manufacturers stating that although these masculine pronouns are used, they are not intended to be sexist. They're only used for convenience. This is an unacceptable statement. When *he, him,* and *his* are used, a masculine image is created, whether or not such companies want to admit it.

To avoid this sexist image, avoid masculine pronouns. Instead, use the plural, generic *they* or *their.* You also can use *he or she* and *his or her. (S)he* is not a good compromise; it's just too odd looking. Sometimes you can solve the problem by omitting all pronouns.

Biased	Unbiased
Sometimes the doctor calls on his patients in their homes.	Sometimes the doctor calls on patients in their homes.
The typical child does his homework after school.	Most children do their homework after school.
A good lawyer will make sure that his clients are aware of their rights.	A good lawyer will make sure that clients are aware of their rights.

Nouns

Use nouns that are nonsexist. To achieve this, avoid nouns that exclude women and denote that only men are involved.

Biased	Unbiased
mankind	people
manpower	workers, personnel
the common man	the average citizen
wise men	leaders
businessmen	businesspeople
policemen	police officers
firemen	firefighters
foreman	supervisor
chairman	chairperson, chair
stewardess/steward	flight attendant
waitress/waiter	server

Consider the following examples of sexist and nonsexist writing in a letter advertising landscaping services. In the first letter the sexist writing is highlighted, whereas in the second example the sexist terms are replaced with nonsexist boldfaced words.

SEXIST WRITING

DiBono and Sons Landscaping
1349 Elm
Oakland Hills, IA 20981
April 2, 2005

Owner
TurboCharge, Inc.
2691 Sommers Rd.
Iron Horse, IA 20992

Dear Sir:

Interested in beautifying your company property, with no care or worries? We have been in business for over 25 years helping men just like you. Here's what we can offer:

1. Shrub Care—your workmen will no longer need to water shrubs. We'll take care of that with a sprinkler system geared toward your business's unique needs.

2. Seasonal System Checks—don't worry about your foreman having to ask his workers to turn on the sprinkler system in the spring or turn it off in the winter. Our repairmen take care of that for you as part of our contract.

3. New Annual Plantings—we plant rose bushes and daffodils that are so pretty, your secretary will want to leave her desk and pick a bunch for her office. Of course, you might want to do that for her yourself ☺.

4. Trees—we don't just plant annuals. We can add shade to your entire property. Just think how beautiful your parking lot will look with elegant elms, oaks, and maples. Then, when you have that company picnic, your employees and their wives will be surrounded by nature.

Call today for prices. Let our skilled craftsmen work for you!

Sincerely,

Richard DiBono

Sexist salutation, implying only a man will read the letter

Sexist word usage

Sexist "gender-tagged word"

Stereotyping

Stereotyping

Sexist "gender-tagged word"

NONSEXIST REVISION

DiBono and Sons Landscaping
1349 Elm
Oakland Hills, IA 20981
April 2, 2005

Owner
TurboCharge, Inc.
2691 Sommers Rd.
Iron Horse, IA 20992

Subject: Sales Information

Interested in beautifying your company property, with no care or worries? We have been in business for over 25 years helping **business owners** just like you. Here's what we can offer:

continued

continued

- Shrub Care—your **employees** will no longer need to water shrubs. We'll take care of that with a sprinkler system geared toward your business's unique needs.

- Seasonal System Checks—don't worry about your **supervisor** having to **ask his or her** workers to turn on the sprinkler system in the spring on turn it off in the winter. Our **staff** takes care of that for you as part of our contract.

- New Annual Plantings—we plant rose bushes and daffodils that are so pretty, **all of your employees** will want vase-filled flowers on **their** desks.

- Trees—we don't just plant annuals. We can add shade to your entire property. Just think how beautiful your parking lot will look with elegant elms, oaks, and maples. Then, when you have that company picnic, your employees and **their families and friends** will be surrounded by nature.

Call today for prices. Let our skilled **experts** work for you!

Sincerely,

Richard DiBono

AUDIENCE INVOLVEMENT

In addition to audience recognition, effective technical writing demands audience involvement. You not only need to know whom you are writing to in your technical correspondence (audience recognition), but also you need to involve your readers—lure them into your writing and keep them interested. Achieving audience involvement requires that you strive for (a) personalized tone and (b) reader benefit.

Personalized Tone

Companies do not write to companies; people write to people. A major thrust in modern technical writing is person-to-person communication. Remember that when you write your memo, letter, report, or procedure, another person will read it. As such, you want to achieve a personalized tone to involve your reader. This personalization can be accomplished in any of the following ways:

- pronoun usage
- names
- contractions

Pronouns

The best way to personalize correspondence is through pronoun usage. If you omit pronouns in your technical writing, as some old schools of thought advocated, the text will read as if it has been computer generated, devoid of human contact. On the other hand, when you use pronouns in your technical writing, you humanize the text. You reveal that the memo, letter, report, or procedure is written by people, for people.

The generally accepted hierarchy of pronoun usage is as follows:

Pronoun		Focus
You Your	⟶	The reader
We Us Our	⟶	The team
I Me My	⟶	The ego

The first group of pronouns, *you/your*, is the most preferred. When you use *you* or *your*, you are speaking directly to your reader(s) on a one-to-one basis. The readers, whoever they are, read the word *you* or *your* and see themselves in the pronoun, envisioning that they are being spoken to, focused on, and singled out. In other words, *you* or *your* appeals to the reader's sense of self-worth. We all like to be thought of as special and worthy of the writer's attention. By focusing on the pronouns *you* or *your* in technical writing, you make your readers feel special and involved in the writing transaction.

The second group of pronouns (*we, us,* and *our*) uses team words to connote camaraderie and group involvement. These pronouns are especially valuable when writing to multiple audiences or when writing to subordinates. In either instance, *we, us,* or *our* implies to the readers that "we're all in this together." Such a team concept helps motivate by making the readers feel an integral part of the whole.

The third group (*I, me,* and *my*) denotes the writer's involvement. If overused, however, these pronouns can connote egocentricity ("All I care about is *me, me, me*"). Because of this potential danger, emphasize *you* and *your* and downplay *I, me,* and *my*. The rule of thumb is to strive for a two-to-one ratio. For every *I, me,* or *my,* double your use of *you* or *your*. We are not saying that you should avoid using first-person pronouns. You cannot write without involving yourself through *I* or *my*. We are just saying that an overuse of these first-person pronouns creates an egocentric image, whereas an abundance of *you* orientation achieves audience involvement.

Let's look at an automobile manufacturer's user manual, which omits pronouns and, thus, reads as if it's computer generated.

CLAIMS PROCEDURE

To obtain service under the Emissions Performance Warranty, take the vehicle to the company dealer as soon as possible after it fails an I/M test along with documentation showing that the vehicle failed an EPA-approved emissions test.

Before

Compare this flat, dry, dehumanized example with the more personalized revision.

CLAIMS PROCEDURE

How do you get service under the Emissions Performance Warranty? To get service under this warranty, take your car to the dealer as soon as possible after it has failed an EPA-approved test. Be sure to bring along the document that shows your car failed the test.

Without any pronouns, the first version has no personality. It's dull and stiff, and it reads as if no human beings are involved. In contrast, when pronouns are added in the second version, the writing involves the reader, who is being spoken to on a person-to-person basis.

Many writers believe that the first version is more professional. However, professionalism does not require that you write without personality or deny the existence of your reader. The pronoun-based writing in the second version is more friendly. Friendliness and humanism are positive attributes in technical writing.

The second version also ensures reader involvement. As a professional, you want your readers (customers, clients, colleagues) to be involved, for without these readers no transaction occurs. You end up writing in a vacuum denying your audience, which is not the goal of technical writing.

This desire for reader involvement through pronoun usage is evident in other examples of technical writing. Here are a few lines from a legal contract, which again shows that professional writers are striving to involve readers through pronoun usage.

example

Throughout this policy, *you* and *your* refer to the "named insured" shown in the Declaration. *We, us,* and *our* refer to the company providing the insurance.

Using pronouns takes a conscious effort. Technical writers want to erase the old-fashioned style of writing that denies the reader's existence or depersonalizes the reader through stiff euphemisms such as the "named insured." Instead, writers want to involve the readers and humanize the text. Pronouns soften the technical edge of technical writing.

The following is an example of dry, depersonalized correspondence and a more friendly revision.

example

Depersonalized	Friendly
Dear Sir:	Dear Mr. Riojas:
With regard to lost policy 123, enclosed is a lost policy form. Complete this form and return it ASAP. Upon receipt, the company will issue a replacement policy.	Your lost policy 123 can be replaced easily. I've enclosed a form to help us replace it for you. All you need to do is fill it out for us. As soon as we get it, we'll send you your replacement policy.

Names

Another way to personalize and achieve audience involvement is through use of names—incorporating the reader's name in your technical writing. By doing so, you create a friendly reading environment in which you speak directly to your reader. Some examples follow.

DATE: June 22, 2005
TO: Steve McMann
FROM: Maureen Pierce
SUBJECT: CABLE PURCHASE ORDERS

Steve, attached are requisitions for cable replacement for California. These are held pending our analysis and your approval:

Requisition No.	Amount	Location
1045	$3,126	Waconia
1825	$4,900	Chaska
2561	$3,829	Chanhassen

When you sign the attached requisition forms, Steve, we'll get right on the purchases.

Pfeiffer Consulting

February 13, 2005

Jennifer Shanmugan
Home Health Care
1238 Roe
Phoenix, AZ 34143

Subject: Proposed Business Writing Seminar

Thank you, Ms. Shanmugan, for inviting us to help your colleagues with their business writing. We're looking forward to working with you.

Here's a reminder of our agreed arrangements:
Date: February 22, 2005
Time: 8:00 A.M.–5:00 P.M.
Site: Cedar Inn

If we can provide any other information, please let us know.

Note that the reader's first name is used in the first example and the reader's last name in the second example. When do you use first names versus last names? The decision is based on your closeness to or familiarity with your reader. If you know your reader well, have worked with him or her for a while, and know your reader will not be offended, then use the first name. However, if you do not know the reader well, have not worked with this reader before, or worry that the reader might be offended, then use the surname instead. In either instance, calling a reader by name will involve that reader and personalize your writing.

Contractions

One way to achieve conversationalism in technical writing is to use contractions occasionally (*we're, let's, here's, it's, can't,* etc.). Contractions now are accepted in composition classes as well as in technical writing. We all use them every day, and they make technical writing more approachable and more friendly.

The following examples prove our point. The first is from a user manual; the second is from a memo written by a CEO and distributed to an entire company.

CONNECTING THE DG-40 TO YOUR COMPUTER

Now that your printer is working, it's time to hook it up to your computer. It's best to turn off both the computer and the printer when you hook them up.

Remember that each computer communicates differently with a printer. If your computer communicates through a parallel interface, all you'll need is a cable. If your computer requires another kind of interface, then you'll need an interface board.

If you don't know what a parallel interface is, your computer manual or dealer will tell you.

In addition to the contractions, note that the preceding text achieves audience involvement through pronouns.

Dear Friends,

I can't tell you how proud I am of the extraordinary effort each of you has put forth toward reaching our goal. As you've probably heard by now, we attained 99.89 percent of our March 31, 2005, ship schedule.

We did not quite reach our gold-ring goal of 100 percent ship performance, but we are so close that we're going to celebrate as if we did. We are committed to continuous improvement. I'd like to invite you to a special celebration this Tuesday.

Let's take a moment to congratulate ourselves before we get back to work. We've got more units to ship and more gold rings to reach. When we work together as a team, we can make significant achievements.

This letter epitomizes motivation through personalization. The text involves the readers by making them feel a part of the team (through pronouns) and softens the technical edge of the writing (through conversational contractions).

Unfortunately, most technical writing is impersonal and alienates readers by its stiffness. The foregoing examples break that mold. Most modern technical writing strives to engage readers, not deny them. (At least one exception to this would be lab reports, in which impersonal objectivity is preferred.) Why will contractions help involve readers? Contractions, which are conversational, help make technical writing less threatening. Often, technical writing reads too legalistically. This style deters most of us. In contrast, technical writing that uses contractions feels more casual and accessible. Contractions appeal to us because they duplicate our normal speech patterns.

Reader Benefit

A final way to achieve audience involvement is to motivate your readers by giving them what they want or need. We're not saying that you should make false promises. Instead, you should show your audience how they will benefit from your technical writing. You can do this one of two ways: (a) explain the benefit and (b) use positive words and verbs.

Explain the Benefit

Until you tell your readers how they'll benefit, they don't know. Therefore, in your letter, memo, report, or manual, state the benefit clearly. You can do this anywhere, but you're probably wise to place the statement of benefit either early (first paragraph or abstract/summary) or late (last paragraph or conclusion). Placing the benefit early in the writing will interest readers and help ensure that they read on, remaining alert throughout the rest of the document. Placing the benefit at the end could provide a motivational close, leaving readers with a positive impression when they finish.

For example, when you write a procedure, you want your readers to know that by performing the steps appropriately, they will avoid mechanism breakdowns, maintain tolerance requirements, achieve a proper fit, or ensure successful equipment operation. By following the procedures, they will reap a benefit.

The following examples of procedures show how reader benefit is conveyed.

---

example

Instructions for Poured Foundations

A poured foundation will provide a level surface for mounting both the pump and motor. Carefully aligned equipment will provide you a longer and more easily maintained operation.

---

example

Instructions for Drive Belt Installation

When installing your Amox Drive Belt, you need to achieve a belt tension of 3/8" deflection for a 9-lb. to 13-lb. load applied at the center span. Check tension frequently during the first 30 hours of operation. Correct belt tension is important for maximum belt life in a large drive system.

MEMORANDUM

DATE: December 2, 2005
TO: Fred Mittleman
FROM: Renata Shuklaper
SUBJECT: IMPROVED BILLING CONTROL PROCEDURES

Three times in the last quarter, one of the billing cycles excluded franchise taxes. This was a result of the CRT input being entered incorrectly.

To strengthen internal controls and reduce the chance of future errors, the following steps will be implemented, effective January 1, 2006.

1. Input   The control screen entry form will be completed by the accounting clerk.
2. Entry   The CRT entry clerk will enter the control screen input and use local print to produce a hard copy of the actual input.
3. Verification   The junior accountant will compare the hard copy to the control screen and initial if correct.
4. Notification   The junior accountant will tell data processing to start the billing process.

The new procedure will provide more timely and accurate billing to our customers. Customer service will be happier with accounting, and we will have more pride in our contributions.

Figure 4.8   Memo Achieving Audience Involvement

Memos can also develop reader benefit to involve audiences. Figure 4.8 achieves audience involvement through reader benefit in many ways.

In this instance, the reader benefit is emphasized in the last paragraph. The pronouns *we* and *our* in the last sentence involve the readers in the activity. A team concept is implied. The benefit, however, is evident in the positive words sprinkled throughout the memo—*improved, strengthen, reduce, timely, accurate, happier, pride,* and *contributions*. These words motivate readers by making them perceive a benefit from their actions.

Use Positive Words and Verbs

In all of the preceding examples geared toward reader benefit, the motivation or value is revealed through positive words or verbs. A sure way to involve your audience is to sprinkle positive words throughout your correspondence or to inject verbs (power writing) into your text. Positive words give your writing a warm glow; verbs give it punch. Positive words and verbs can sell the reader on the benefits of your subject matter.

Positive Words

advantage	effective	happy	profitable	successful
asset	efficient	please	satisfied	thank you
benefit	enjoyable	pleased	succeed	value
confident	favorable	pleasure	success	

To clarify how important such words are, let's look at some examples of negative writing, followed by positive revisions.

Negative

- We cannot process your request. You failed to follow the printed instructions.

- The error is your fault. You keep your books incorrectly and cannot complain about our deliveries. If you would cooperate with us, we could solve your problem.

- We have received your letter complaining about our services.

- Your bill is now three weeks overdue. Failure to pay immediately will result in lower credit ratings.

- The invoice you sent was useless by the time it arrived. You wasted my time and money.

Positive

- So that we may process your request rapidly, please fill in line 6 on the printed form.

- To ensure prompt deliveries, let's get together to review our bookkeeping practices. Would next Tuesday be convenient?

- Thank you for writing to us about our services.

- If you're as busy as we are, you've probably misplaced our recent bill (mailed three weeks ago). Please send it in soon to maintain your high credit ratings.

- We're proud of our ability to maintain schedules. But we need your help. When you return invoices by the fifteenth, we save time and you save money.

Making something positive out of something negative is a challenge, but the rewards for doing so are great. If you attack your readers with negatives, you lose. When you involve your readers through positive words, you motivate them to work with you and for you.

Verbs also motivate your readers. The following verbs will motivate your readers to action.

accomplish	develop	insure	raise
achieve	direct	lead	recommend
advise	educate	maintain	reduce
analyze	ensure	manage	reestablish
assess	establish	monitor	serve
assist	guide	negotiate	supervise
build	help	operate	support
conduct	implement	organize	target
construct	improve	plan	train
control	increase	prepare	use
coordinate	initiate	produce	
create	install	promote	

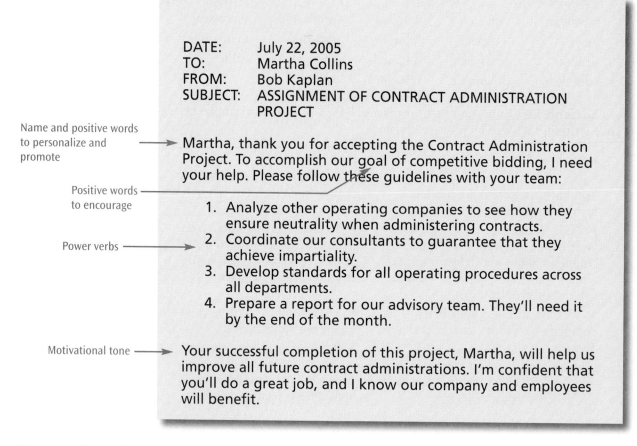

Name and positive words to personalize and promote →

Positive words to encourage →

Power verbs →

DATE: July 22, 2005
TO: Martha Collins
FROM: Bob Kaplan
SUBJECT: ASSIGNMENT OF CONTRACT ADMINISTRATION PROJECT

Martha, thank you for accepting the Contract Administration Project. To accomplish our goal of competitive bidding, I need your help. Please follow these guidelines with your team:

1. Analyze other operating companies to see how they ensure neutrality when administering contracts.
2. Coordinate our consultants to guarantee that they achieve impartiality.
3. Develop standards for all operating procedures across all departments.
4. Prepare a report for our advisory team. They'll need it by the end of the month.

Motivational tone →

Your successful completion of this project, Martha, will help us improve all future contract administrations. I'm confident that you'll do a great job, and I know our company and employees will benefit.

Figure 4.9   Memo Using Power Verbs

The memo in Figure 4.9 uses power verbs to motivate the reader. This memo brings together all of the techniques we've been discussing. It achieves audience involvement by speaking directly to the reader ("Martha"), by using pronouns, by including contractions to soften the technology, by using verbs that give the memo punch, and by employing positive words to motivate the reader to action.

Audience Checklist

To communicate effectively with your readers, ask yourself the following questions. Then circle the appropriate responses or fill in the needed information.

1. What is my audience's level of understanding regarding the subject matter?
   - High-tech
   - Low-tech
   - Lay
   - Multiple

2. Given my audience's level of understanding, have I written accordingly?
   - Have I defined my acronyms, abbreviations, and jargon?
   - Have I supplied enough background data?
   - Have I used the appropriate types of graphics?

3. Who else might read my correspondence?
   - How many people?
   - What are their levels of understanding?
     - High-tech
     - Low-tech
     - Lay
   - Will my audience be multicultural?

4. What is my role in relation to my audience?
   - Do I work for the reader?
   - Does the reader work for me?
   - Is the reader a peer?
   - Is the reader a client?
5. What response do I want from my audience? Do I want my audience to act, respond, confirm, consider, decide, or file the information for future reference?
6. Will my audience act according to my wishes? What is my audience's attitude toward the subject (and me)?
   - Negative
   - Positive
   - Noncommittal
   - Uninformed
7. Is my audience in a position of authority to act according to my wishes (can he or she make the final decision)? If not, who will make the decision?
8. What are my audience's personality traits?
   - Slow to act
   - Eager
   - Receptive
   - Questioning
   - Organized
   - Disorganized
   - Reticent
9. Have I motivated my audience to act?
   - Have I involved my reader in the correspondence by using pronouns and his or her name?
   - Have I shown my reader the benefit of the proposal by using positive words and verbs?
   - Have I considered any questions or objections my audience might have?

10. Have I considered my audience's preferences regarding style?
    - Will my reader accept contractions?
    - Is use of a first name appropriate?
    - Should I use my reader's last name?
11. Have I avoided sexist language?
    - Have I used *their* or *his or her* to avoid sexist singular pronouns like *his*?
    - Have I used generic words such as *police officer* versus sexist words like *policeman*?
    - Have I avoided excluding women, writing sentences such as "All present voted to accept the proposal" versus the sexist sentence "All the men voted to accept the proposal"?
    - Have I avoided patronizing, writing sentences such as "Mr. Smith and Mrs. Brown wrote the proposal" versus the sexist sentence "Mr. Smith and Judy wrote the proposal"?
12. Have I considered diversity?
    - Have I avoided any language that could offend various age groups, people of different sexual orientations, people with disabilities, or people of different cultures and religions?
    - Have I considered that people from different countries and people for whom English is a second language will be involved in the communication? This might mean that I should
      - Clarify time and measurements
      - Define abbreviations and acronyms
      - Avoid figurative language and idiomatic phrases, unique to one culture
      - Avoid humor and puns
      - Consider each country's cultural norms

CHAPTER HIGHLIGHTS

1. High-tech readers understand acronyms, abbreviations, and jargon. However, not all of your readers will be high tech.

2. Low-tech readers need glossaries or parenthetical definitions of technical terms.

3. Lay readers usually will not understand technical terms. Consider providing these readers with extended definitions.

4. In addition to considering high-tech, low-tech, and lay audience levels, you also should consider issues of diversity, such as gender, race, religion, age, and sexual orientation.

5. Multiple audiences have different levels of technical knowledge. This fact affects the amount of technical content and terms you should include in your document.

6. The definition of a technical term usually includes its type and distinguishing characteristics but can also include examples, descriptions, and procedures.

7. For multicultural audiences, define terms, avoid jargon and idioms, and consider the context of the words you use. Avoid cultural biases, avoid complicated punctuation, be careful with humor, and allow space for translation.

8. More than 50 percent of your audience will be female, so you should avoid sexist language.

9. To get your audience involved in the text, personalize it by using pronouns, the reader's name, and contractions.

10. Show your reader how he or she will benefit from your message or proposal.

CASE STUDY

Home and Business Mortgage

Home and Business Mortgage

Home and Business Mortgage (HBM) has suffered several lawsuits recently. A former employee sued the company, contending that it practiced "ageism" by promoting a younger employee over him. In an unrelated case, another employee contended that she was denied a raise due to her ethnicity. To combat these concerns, HBM has instituted new Human Resources practices. The goal is to ensure that the company meets all governmental personnel regulations, including those required by the Equal Employment Opportunity Commission (EEOC), the Family Medical Leave Act (FMLA), and the Americans with Disabilities Act (ADA). HBM is committed to achieving diversity in its workplace.

HBM now needs to hire a new office manager for one of its branch operations. It has three outstanding candidates. Following are their credentials:

- **Carlos Gutierrez**—Carlos is a 27-year-old recent recipient of an MBA (masters in business administration) from an acclaimed business school. At this university, he learned many modern business applications, including capital budgeting, human resource management, organizational behavior, diversity management, accounting and marketing management, and team management strategies. He has been out of graduate school for only two years, but his work during that time has been outstanding. As an employee at one of HBM's branches, he has already impressed his bosses by increasing the branch's market share by 28 percent through innovative marketing strategies he learned in college. In addition, his colleagues enjoy working with him and praise his team-building skills. Carlos has never managed a staff, but he is filled with promise.

- **Cheryl Huff**—Cheryl is a 37-year-old employee of a rival mortgage company, Farm-Ranch Equity. She has a BGS (bachelor of general studies) from a local university, which she acquired while working full time in the mortgage/real estate business. At Farm-Ranch, where she has worked for over 15 years, she has moved up the ladder, acquiring many new levels of responsibility. After working as an executive assistant for 5 years, she then became a mortgage payoff clerk for 2 years, a loan processor for 2 years, a mortgage sales manager for 4 years, and, most recently, a residential mortgage closing coordinator for the last 2 years. In the last two positions, she managed a staff of five employees.

According to her references, she has been dependable on the job. Her references also suggest that she is a forceful taskmaster. Though she has accomplished much on the job, her subordinates have chaffed at her demanding expectations. Her references suggest, perhaps, that she could profit from some improved people skills.

- **Rose Massin**—Rose is 47 years old and has been out of the workforce for 12 years. During that absence, she raised a family of three children. Now, the youngest child is in school and Rose wants to reenter the job market. She has a bachelor's degree in business, and was the former office manager of this HBM branch. Thus, she has management experience, as well as mortgage experience.

  While she was branch manager, she did an outstanding job. This included increasing business, working well with employees and clients, and maintaining excellent relationships with lending banks and realtors. She was a highly respected employee and was involved in civic activities and community volunteerism. In fact, she is immediate past-president of the local Rotary club. Though she has been out of the workforce for years, she has kept active in the city and has maintained excellent business contacts. Still, she's a bit rusty on modern business practices.

Assignment

Whom will you hire from these three candidates? Based on the information provided, make your hiring decision. This is a judgment call. Be sure to substantiate your decision with as much proof as possible. Then share this finding as follows:

- **Oral Presentations**—give a 3–5 minute briefing to share with your colleagues the results of your decision.
- **Written**—write an e-mail, memo, or report about your findings.

EXERCISE

Avoiding Sexist Language

1. Revise the following sentences to avoid sexist language.
   a. All the software development specialists and their wives attended the conference.
   b. The foremen met to discuss techniques for handling union grievances.
   c. Every technician must keep accurate records for his monthly activity reports.
   d. The president of the corporation, a woman, met with her sales staff.
   e. Throughout the history of mankind, each scientist has tried to make his mark with a discovery of significant intellectual worth.
   f. The chairman of the meeting handled the debate well.
   g. All manmade components were checked for microscopic cracks.
   h. The supervisors have always brought the girls flowers on Mother's Day.

i. Mr. Smith (the CEO) introduced his staff: Mr. Jones (the vice president), Mr. Brown (the chief engineer), and Judy (the executive secretary).

j. The policemen and firemen rushed to the scene of the accident.

Achieving Audience Involvement

The following sentences are either dull, dry, and impersonal or egocentric.

1. Revise them to achieve audience involvement through personalization, adding pronouns, names, or contractions. Strive for a two-to-one ratio, using *you* twice for every *I*.

   a. The company will require further information before processing this request.

   b. It has been decided that a new procedure must be implemented to avoid further mechanical failures.

   c. The department supervisor wants to extend a heartfelt thanks for the fine efforts expended.

   d. I think you have done a great job. I want you to know that you have surpassed this month's quota by 12 percent. I believe I can speak for the entire department by saying thank you.

   e. If the computer overloads, simultaneously press Reset and Control. Wait for the screen command. If it reads "Data Recovered," continue operations. If it reads "I/O Error," call the computer resource center.

   f. Ampex Corporation announces the opening of a new office in the Fairway Village area.

   g. Telech's new communication system can help meet business telecommunication needs. This new system offers multiparty extensions, call forwarding, call waiting, intercom, and call accounting.

   h. There are three ways to solve the problem: line checks, system checks, and random checks.

   i. I want a member from each department to hear my speech so I can receive feedback on what I can do to improve my content and my delivery.

   j. Here is how to number pages. To number from page one, press Page Layout and then the number 1. To number from any other page, press Page Layout and then the required number, whatever it might be.

2. Revise the following sentences to achieve reader benefit, using positive or power verbs.

   a. We cannot lay your cable until you sign the attached waiver.

   b. John, don't purchase the wrong program. If we continue to keep inefficient records, our customers will continue to complain.

   c. You have not paid your bill yet. Failure to do so might result in termination of services.

   d. If you incorrectly quote and paraphrase, you will receive an *F* on the assignment.

   e. Send me the requested information by January 12.

   f. Poor input of information delays shipments, which forces us to pay unneeded expenses. These costs eventually come out of your year-end raises.

   g. Thank you for your recent call complaining about our product.

   h. Your team has lost 6 of their last 12 games.

   i. Your memo suggesting an improvement for the system has been rejected. The reconfigurations you suggest are too large for the area specifications. We need you to resubmit if you can solve your problem with calculations.

3. Rewrite the following flawed correspondence. Its tone is too negative and commanding, regardless of the audience. If the audience is composed of subordinates, the tone is too dictatorial. Rather than encouraging team building, the correspondence borders on officiousness. If the audience is composed of coworkers or upper-level management, then the tone creates an even greater offense. Soften the tone to achieve better audience involvement and motivation.

DATE: October 15, 2005
TO: Distribution
FROM: Darryl Kennedy
SUBJECT: FOURTH QUARTER GOALS

Due to a severe lack of discipline, the company failed to meet third quarter goals. To avoid repeating this disaster for the fourth quarter, this is what I think you all must do—**ASAP.**

1. Demand that the sales department increase cold calls by 15 percent.
2. Require weekly progress reports by all sales staff.
3. Penalize employees when reports are not provided on time.
4. Tell managers to keep on top of their staff, prodding them to meet these goals.

Remember, when one link is weak in the chain, the entire company suffers. **DON'T BE THE WEAK LINK!**

Defining Terms for Different Audience Levels

a. Find examples of definitions provided in computer word processing programs, e-mail packages, Internet dictionaries, and online help screens. Determine whether these examples provide the *term*, its *type*, and its *distinguishing characteristics*. Are the definitions effective? If so, explain why. If not, rewrite the definitions for clarity.

b. Find examples of definitions in manuals such as your car's user manual, a manual that accompanied your computer, or manuals packaged with your radio or alarm clock, coffeemaker, VCR, or lawn mower. Note where these definitions are. Are they placed parenthetically within the text or in a glossary? Next, determine whether the definitions provide the *term*, its *type*, and its *distinguishing characteristics*. Are the definitions effective? If so, explain why. If not, rewrite the definitions for clarity.

c. Find examples of definitions in your textbooks. Where are these definitions placed—parenthetically within the text or in a glossary? Next, determine whether the definitions provide the *term*, its *type*, and its *distinguishing characteristics*. Are the definitions effective? If so, explain why. If not, rewrite the definitions for clarity.

d. Select 5 to 10 terms, abbreviations, or acronyms from your area of expertise or interest (telecommunications, accounting, electronic engineering, automotive technology, biomedical records, data processing, etc.). First, define each term using words and phrases. Next, define each term in a sentence, conveying the *term*, its *type*, and its *distinguishing characteristics*. Finally, define each term using a longer paragraph. In these longer definitions, include examples, descriptions, or processes.

Recognizing Issues of Diversity

1. Rewrite the following sentences for multicultural, cross-cultural audiences.
   a. Let's meet at 8:30 P.M.
   b. The best size for this new component is $16 \times 23 \times 41$.
   c. To keep us out of the red, we need to round up employees who can put their pedal to the medal and get us out of this hole.
   d. We need to produce fliers/brochures to increase business.
   e. The meeting is planned for 07/09/05.

2. Rewrite the following flawed correspondence. Be sure to take into consideration your multicultural or cross-cultural audience's needs.

TO: Andre Castro, Barcelona; Sunyun Wang, Singapore; Nachman Sumani, Tel Aviv
FROM: Ron Schaefer, New York
SUBJECT: Brainstorming

I need to pick your brains, fellows. We've got a big one coming up, a killer deal with a major European player. Before I can make the pitch, however, let's brainstorm solutions. The client needs a proposal by 12/11/05, so I need your input before that date. Give me your ideas about the following:

1. What should we charge for our product, if the client buys in bulk?
2. What's our turnaround time for production? I know that your people tend to work slowly, so can you hurry the team up if I promise delivery in six weeks?

E-mail me your feedback by tomorrow, 1:00 P.M. my time, at the latest. Trust me, guys. If we boot this one, everyone's bonuses will be in a sling.

---

3. Visit local banks, hospitals, police or fire stations, or city officials. Ask employees at these sites what diversity challenges they have encountered and how they have tried to solve these problems. These challenges could be language barriers presented by multiculturalism or cross-culturalism. Perhaps the issue is diversity hiring to avoid ageism or sexism. Maybe you will discover problems in the workplace created by diversity hiring, such as incompatible salaries or meeting Equal Opportunity regulations. Report on your findings either orally or in writing.

Team Projects

1. Rewrite the following flawed correspondence. Be sure to achieve effective audience understanding and involvement. To do so, avoid sexist language and define high-tech terms. Search online to find any abbreviations or acronyms that need defining. Remember: Though the immediate readers might understand all of the high-tech terminology, other readers might not. These additional audiences could include lawyers, employees from other departments, or low-tech clients.

West Central Auditors
"Your Technical Engineering and Energy Resource Experts"
1890 River
Pocato, Idaho 89022

March 12, 2005

Marks-McGraw, Inc.
2145 Oceanview
Clackamas, Oregon

Gentlemen:

We will visit your plant next week for the TEA. To ensure that our visit goes smoothly, we plan this procedure:

- Your plant foreman will have his engineer assemble his CAD drawings for the power plant at Brush Prairie.
- Our men will review these drawings against TE specs, such as specific cost estimates for energy usage and energy data.
- Our ER group will then help your men plan and prepare your facilities for technology changes mandated by the DOE's office of the EERE.
- We also want to review your engineer's plans for any building envelopes. Make sure he brings all relevant correspondence he has had with his clients.

This audit should take approximately seven man-hours. It's a man-sized job. If we have to go off site for lunch or breaks, the job will take even longer. Therefore, please ask your secretary to provide drinks and food for our six representatives. Tell her that we have no dietary concerns, so anything she provides will be greatly appreciated.

Sincerely,

Jim Wynn, Team Manager

2. Individually, find examples of writing geared to high-tech, low-tech, and lay audiences. To do so, read professional journals, find procedures and instructions, look at marketing brochures, read trade magazines, or ask your colleagues and coworkers for memos, letters, or reports.

    Once you have found these examples, bring them to class. In small groups, discuss your findings to determine whether they are written for high-tech, low-tech, or lay readers. Use the Audience Evaluation Form below to record your decisions. Share these with all class members.

Audience Evaluation Form					
EXAMPLE NUMBER	1	2	3	4	5
TYPE **(CIRCLE ONE)**	HIGH LOW LAY	HIGH LOW LAY	HIGH LOW LAY	HIGH LOW LAY	HIGH LOW LAY

Criteria
Language
 • Abbreviations
 • Acronyms
 • Jargon (defined?)
Content
 • General
 • Specific
 • Background
Tone
 • Formal (high)
 • less formal (low)
 • least formal (lay)
Format
 • Highlighting
 • Graphics
  —type
  —complexity
  —color
  —labeling					

3. Once you've made your decisions regarding team project 1, select one of the high-tech examples and rewrite it for a low-tech or lay audience. To do so, work with your group to define your high-tech terms, improve the tone through pronouns and positive words, enhance the page layout through highlighting techniques, and add appropriate graphics.

4. In small groups composed of individuals from like majors, list 10 high-tech terms (jargon, acronyms, or abbreviations) unique to your degree programs. Then, envisioning a lay audience, parenthetically define and briefly explain these terms.

   To test the success of your communication abilities, orally share these high-tech terms with other students who have different majors. First, state the high-tech term to see if they understand it. If they don't, provide the parenthetical definition. How much does this help? Do they understand now? If not, add the third step—the brief explanation.

   How much information do the readers need before they understand your high-tech terms?

5. In small groups, read the following two examples and the flowchart to determine whether they are written to a high-tech or low-tech audience. To justify your answer, consider the use of acronyms, abbreviations, or jargon; the presentation of background data; the length of sentences and paragraphs; and the way in which the text has been formatted (how it looks on the page).

---

example

This documentation summarizes the items to be tested and the methodology utilized in the testing routine. The test plan tested four new 8536 Switcher features: OIM channel commands, OIM disconnect commands, OIM dump commands, and OIM log-off messages.

The patch files associated with the base must disable the features; consequently, only testing to ensure that this is operational will be addressed.

Software development personnel, after testing the features, will deliver copies of this report and product acceptance releases to program management, who are then responsible for writing functional specifications and submitting them to administration for manufacturing and marketing.

---

example

Purpose. This document identifies the time to be tested and the methods used for testing. The tests center on the 8536 Switcher (a network consisting of two nodes) and CN (connect node) bases. The 8536 Switcher and CN bases perform the same function as our currently marketed G9 and G12 systems.

OIM (Operator Interface Module) Features Tested:
- OIM channel commands
- OIM disconnect commands
- OIM dump commands
- OIM log-off messages

Testing Focus. To work appropriately, these features must be disabled by our patch files if an error occurs. Therefore, our testing focus will be the disabling capabilities.

Responsibilities. After testing the features, our software development group will send copies of this report and product acceptance releases to program management. They then will write the functional specifications and send them to administration. Administration is responsible for manufacturing and marketing.

The flowchart on the following page graphically depicts this sequence.

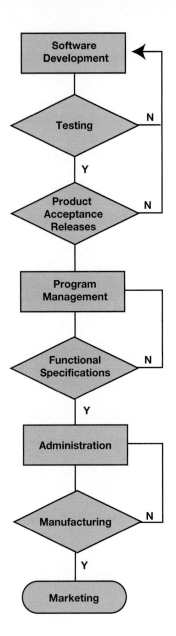

QUIZ QUESTIONS

1. What are the four main types of audiences?

2. What are the differences among the types of audiences?

3. What are three places for definitions in technical documents?

4. What are the main parts of a definition?

5. Why should you consider a multicultural audience when you write a technical document?

6. What guidelines should you follow to write to a multicultural audience?

7. What is sexist language?

8. What are three ways to avoid sexist writing?

9. How can you achieve audience involvement?

10. What are two things you can do to express reader benefit?

CORRESPONDENCE

5 Memos and E-mail

Chapter Preview

The Differences Between Memos and E-mail Messages
Memos and e-mail messages have differences and similarities in relation to their destination, format, audience, topics, tone, speed or delivery time, attachments, length, and security.

Memos
Memos are an important part of on-the-job communication. Using an all-purpose template will help you format and write memos.

Sample Memos
The samples provide you with guidelines for memo components, organization, and writing style.

Effective Memo Checklist
The memo checklist will help you evaluate your memos, ensuring that you have met all of your goals.

E-mail
E-mail is an extremely important part of your on-the-job communication. More e-mail messages are written at work than any other type of correspondence.

Why Is E-mail Important?
E-mail is important because it can save time, it is convenient, it is affordable, and it allows for expanded discussion of topics by numerous people.

Writing at Work

Barney Allis Stores, home based in Seattle, Washington, has stores in all 50 states and in the following international cities: Toronto, Vancouver, London, Paris, Berlin, Barcelona, Rome, Beijing, Hong Kong, Sidney, Mexico City, and Rio de Janeiro.

BAS is an all-purpose department store, offering clothing, high-end electronics, and furniture. The BAS electronics department includes

- computer hardware, software, and accessories
- cameras and video equipment
- DVD and CD players, televisions, and audio equipment
- Cell phones, PDAs, and hybrids

Not only does BAS do business all over the world, but also it has a worldwide network of vendors. This allows BAS to maintain an inventory of state-of-the-art electronics equipment at cost-effective prices.

Communicating with its international vendors, purchasing agents, and employees would be impossible without e-mail. Letters ("snail mail") would take days; telephoning long distance not only would be costly but employees and vendors would only end up playing "telephone tag":

- the manager calls a vendor
- the vendor is out of the office
- the manager leaves a voice-mail message
- the vendor, perhaps hours or days later, listens to the message and returns the call
- the manager is now out of the office
- the vendor leaves a voice-mail message

E-mail messages sent internationally are more cost effective and less time consuming. In addition, BAS's international vendors and employees work in different time zones. A phone call from the Seattle home office to an employee in Rome, Berlin, or Barcelona is fruitless—no one will be there.

E-mail, in contrast, makes good business sense. By communicating through e-mail,

- Managers, vendors, and employees can develop their content more thoroughly and thoughtfully.
- The company has a record of the transactions.
- Multiple audiences can be involved through cc's (complimentary copies).
- The interested parties can attach files (forms, memos, GIF and JPEG visuals, and hypertext links).

THE DIFFERENCES BETWEEN MEMOS AND E-MAIL MESSAGES

This chapter focuses on memos and e-mail messages. To give you an overview of the differences and similarities between them, look at Table 5.1.

MEMOS

Purposes

Memos are an important means by which employees communicate with each other. First, you will write memos often on the job. A Pitney Bowes study suggests that you might write as many as 18 memos a day ("Pitney Bowes Study" 2000). Next, you will write memos to a wide range of readers (Figure 5.1). This includes your supervisors, coworkers, subordinates, and multiple combinations of these audiences.

Memos are flexible and can be written for different purposes, including the following:

- **Documentation**—report on expenses, incidents, accidents, problems encountered, projected costs, study findings, hiring, firings, and reallocations of staff or equipment.
- **Cover/transmittal**—tell the reader you have attached a document.
- **Confirmation**—tell a reader about a meeting agenda, date, time, and location; decisions to purchase or sell; topics for discussion at upcoming teleconferences; conclusions arrived at; and fees, costs, or expenditures.
- **Procedures**—explain how to set up accounts, research on the company intranet, operate new machinery, use new software, apply online for job opportunities through the company intranet, create a new company Web site, or solve a problem.
- **Recommendations**—provide reasons to purchase new equipment, fire or hire personnel, contract with new providers, merge with other companies, revise current practices, and renew contracts.

E-mail Challenges
Though an important part of your business communication, e-mail presents problems. Computers have limitations, e-mail does not guarantee privacy, misunderstandings can occur, and e-mail messages written too casually can be seen as unprofessional.

Techniques for Writing Effective E-mail Messages
Consider the helpful tips when you write e-mail messages.

Sample E-mail Messages
The samples provide you with guidelines for writing effective e-mail messages that give instructions, provide information, and recommend changes persuasively.

Effective E-mail Message Checklist
The e-mail message checklist will help you evaluate your e-mail, ensuring that you have met all of your goals.

The Communication Process
Follow the writing process to create your memos and e-mail messages: prewrite, write, and rewrite.

Process Log
The process log takes you through the step-by-step sequence of creating and writing a memo.

TABLE 5.1 MEMOS VERSUS E-MAIL MESSAGES

Characteristics	Memos	E-mail Messages
Destination	Internal: correspondence written to coworkers within a company.	Internal *and* external: correspondence written to friends and acquaintances, coworkers within a company, and clients and vendors.
Format	Identification lines: Date, To, From, Subject. Options include cc (complimentary copy), Ref (reference), and Distribution (other recipients of the memo).	Identification lines: To and Subject. The Date and From are computer generated. Options include cc (complimentary copy), Ref (reference), and Distribution (other recipients of the e-mail message).
Audience	Generally specialists and semi-specialists within the company.	Multiple readers due to the internal and external nature of e-mail.
Topics	Company related, allowing for abbreviations and acronyms due to well-informed audience.	A wide range of diverse topics determined by the audience.
Tone	Determined by audience: can be informal when written to coworkers; might be more formal when written to management.	A wide range of tones due to diverse audiences. Usually informal when written to friends, informal to coworkers, more formal to management. The largest challenge involves multicultural readers.
Speed or Delivery Time	Determined by a company's in-house mail procedures. Memos could be delivered within hours or days.	Often instantaneous, usually within minutes. Delays can be caused by system malfunctions or excessively large attachments.
Attachments	Hard-copy attachments can be stapled to a memo or enclosed within an envelope. Size is not an issue. Complimentary copies can be sent to other readers by way of the company's mail system.	Computer word processing files, HTML files and Web links, PDF files, RTF files, or downloadable graphics can be attached to e-mail. Complimentary copies can be sent to other readers. Size of these files is an issue, because large documents can crash a reader's system. A good rule of thumb is to limit files to 750 Kilobytes (K).
Length	A typical memo is no longer than one printed page (8½ × 11 inches).	An effective e-mail message is limited to one viewable screen (requiring no scrolling).
Security	If a company's mail delivery system is reliable, the memo will be secure. Memos can be delivered within a sealed envelope. No one else will see the memo unless the reader or writer shares it.	E-mail systems are not secure. E-mail can be tampered with, read by others, and sent to many people. E-mail stays within a company's computer backup system and is the property of the company. Therefore, e-mail is not private.

- **Feasibility**—study the possibility of changes in the workplace (practices, procedures, locations, staffing, equipment, missions, or visions).
- **Status**—provide a daily, weekly, monthly, quarterly, biannual, or yearly progress report about sales, staffing, travel, practices, procedures, and finances.

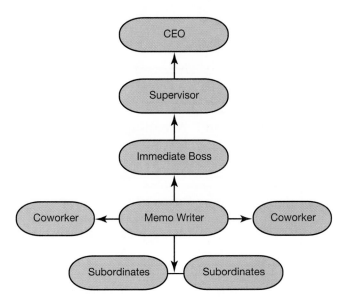

Figure 5.1   Multiple Memo Readers

- **Directive (delegation of responsibilities)**—inform subordinates of their designated tasks.
- **Inquiry**—ask questions about upcoming processes, procedures, or assignments.

Criteria

Your audiences and purposes will change from memo to memo. Also, memos written at different companies will not look the same. Every company will have its own, unique corporate style or template for writing memos. Even within a company, you will find differences. The accounting supervisor on the fifth floor wants his employees to write memos this way; the sales manager on the third floor wants her employees to write memos that way. However, memos should contain the following key components:

- Memo ID lines
- Introduction
- Discussion
- Conclusion
- Audience recognition

Figure 5.2 shows an all-purpose organizational template that works well for memos.

Subject Line

One hundred percent of your readers read the subject line (after all, we have to start reading any text at its beginning. The beginning of a memo is the subject line). The subject line, that can be typed in all caps, is where you begin talking to your reader(s). Therefore, you want it to work for you. One-word subject lines don't communicate effectively, as in the flawed subject line on page 134.

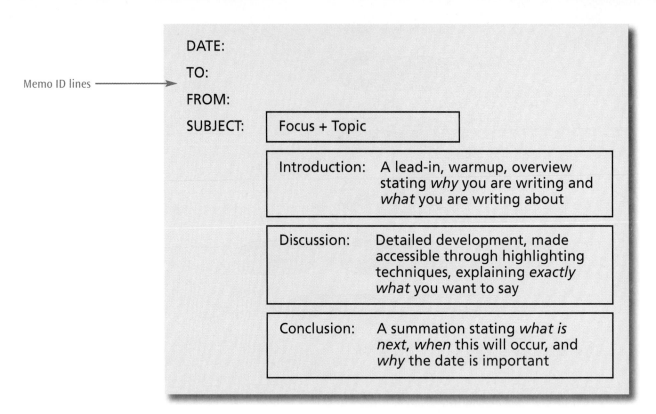

DATE:

TO:

FROM:

SUBJECT: | Focus + Topic |

| Introduction: | A lead-in, warmup, overview stating *why* you are writing and *what* you are writing about |

| Discussion: | Detailed development, made accessible through highlighting techniques, explaining *exactly what* you want to say |

| Conclusion: | A summation stating *what is next*, *when* this will occur, and *why* the date is important |

Memo ID lines →

Figure 5.2   All-Purpose Memo Template

Subject: COMPTROLLERS

Before

We've got a *topic* (a *what*), but we're missing a *focus* (a *what* about the *what*). An improved subject line would read as follows:

Subject: SALARY INCREASES FOR COMPTROLLERS

After

This subject line gives us the *topic*—comptrollers (the *what*)—plus a focus—salary increases (the *what* about the *what*)—all linked by a preposition.

Such a sequence works in all subject lines. For instance, in the following examples, note how the focus alters the meaning of the subject line's topic.

example

Subject: Focus	plus	Topic
Termination	of	Comptrollers
Hiring Procedures	for	Comptrollers
Vacation Schedules	for	Comptrollers
Training Seminars	for	Comptrollers

Although the topic stays the same, the focus changes and clarifies for the reader the actual subject matter of the memo.

Introduction

Once you've communicated your intent in the subject line, you want to get to the point in the introductory sentence(s). Readers are busy and don't want your memo to slow down their work. To avoid any delays for your audience, you want your first sentence or sentences to communicate immediately. A goal is to write one or two clear introductory sentences that tell your readers *what you want* and *why you are writing*. Remember, until you tell them, they don't know.

example

In the third of our series of quality control meetings this quarter, I'd like to get together again to determine if improvements have been made.

This example invites the reader to a meeting, thereby communicating what the writer's intentions are. It also tells the reader that the meeting is one of a series of meetings, thus communicating why the meeting is being called.

example

As a follow-up to our phone conversation yesterday (8/12/05), I have met with our VP regarding your suggestions. He'd like to meet with you to discuss the following ideas in more detail.

This introduction reminds the reader why this memo is being written—as a follow-up—and tells the reader what will happen next.

Discussion

The discussion section allows you to develop your content specifically. You want to respond to the reporter's questions mentioned in Chapter 3 (*who*, *what*, *when*, *why*, *where*, *how*), but you also want to make your information accessible. Because very few readers read every line of your memo (tending instead to skip and skim), traditional blocks of data (paragraphing) are not effective visually. The longer the paragraph, the more likely your audience is to avoid reading. Instead, try to make your text more reader-friendly by applying some of the highlighting techniques discussed in Chapters 8 and 9: (a) itemization, (b) white space, (c) boldface type, (d) headings, (e) columns, and (f) graphics.

Note the difference between the following examples: the first is reader-unfriendly, and the second is reader-friendly.

READER-UNFRIENDLY TEXT

This year began with an increase, as we sold 4.5 million units in January compared to 3.7 for January 2005. In February we continued to improve with 4.6, compared with 3.6 for same time in 2005. March was not quite so good, as we sold 4.3 against the March 2005 figure of 3.9. April was about the same with 4.2, compared to 3.8 for April 2005.

READER-FRIENDLY TEXT

Comparative Quarterly Sales (in Millions)

	2005	2006	Increase/Decrease
Jan.	3.7	4.5	0.8+
Feb.	3.6	4.6	1.0+
Mar.	3.9	4.3	0.4+
Apr.	3.8	4.2	0.4+

The first block of data is unappealing to read and hard to follow because of its complexity. You can't see the relationship between figures and years clearly, nor can you see the variance from year to year easily. On the other hand, the table in the second example is accessible at a glance. It's clear and concise—perfect technical writing.

Conclusion

Conclude your memo with a *complimentary close* or a *directive close*. A complimentary close motivates your readers and leaves them happy, as in the following example:

> **example**
>
> If our quarterly sales continue to improve at this rate, we will double our sales expectations by 2005. Congratulations!

A directive close tells your readers exactly what you want them to do next or what your plans are (and provides dated action).

> **example**
>
> Next Wednesday (12/22/05), Mr. Jones will provide each of you a timetable of events and a summary of accomplishments.

Why is a conclusion important? Without it, the reader senses a lack of closure and does not know what to do next or why the actions requested are important. Without a conclusion, the reader's response is going to be "OK, but now what?" or "OK, but so what?" To write an effective memo, sum up and provide closure.

Audience

Another criterion for effective memo writing is audience recognition. In memos, audience is both easier and more complex than in letters. Because letters go outside your company, your audience is usually low tech or lay, demanding that you define your terms more specifically. In memos, on the other hand, your in-house audience is easier to define (usually low tech or high tech). Thus, you often can use more acronyms and internal abbreviations.

However, whereas your audience for letters is usually singular—one reader—your audience for memos might be multiple. You might be writing simultaneously to your immediate supervisor (high tech), to his or her boss (low tech), to your colleagues (high tech), and even to a CEO (low tech).

This diverse readership presents a problem. How do you communicate to a large and varied audience? How do you avoid offending your high-tech readers with seemingly unnecessary data, which your low-tech readers need?

The problem is not easy to solve. Sometimes you can use parenthetical definitions. For instance, do not just write *CIA,* which to most people means Central Intelligence Agency. Provide a parenthetical definition—*CIA (cash in advance)*—if your usage differs from what most people will automatically assume. You are always better off saying too much.

Style

The appropriate style for memos is the same technical writing style discussed in earlier chapters, depending on conciseness, clarity, and accessibility. Use simple words, readable sentences, specific detail, and highlighting techniques.

In addition, strive for an informal, friendly tone. Memos are part of your interpersonal communication abilities. Just as you should be concerned with how you speak to others on the telephone or at the water cooler, you should assess your tone within a memo. Do you sound natural, normal, informal, and friendly? Does your memo sound like it is written by the kind of person you would want to work with or for? If not, you might be creating a negative image.

Decide which of the following examples has the more positive "feel." Which boss would you prefer working with?

---

example

We will have a meeting next Tuesday, Jan. 11, 2005. Exert every effort to attend this meeting. Plan to make intelligent comments regarding the new quarter projections.

---

example

Let's meet next Tuesday (Jan. 11, 2005). Even if you're late, I'd appreciate your attending. By doing so you can have an opportunity to make an impact on the new quarter projections. I'm looking forward to hearing your comments.

Isn't the latter more personal, informal, appealing, and motivational? This tone is achieved through audience involvement (*you* usage), contractions, and positive words, aspects of effective writing discussed in Chapter 4.

When you write a memo, you are writing to someone with whom you work every day. The Golden Rule is write as you would like to be written to.

Grammar

Abide by all grammatical conventions when writing memos. Poor grammar or typographical errors destroy your credibility.

Wizards and Templates

Word processing **wizards** and **templates** also can help you format and organize your memos. Look at the memo templates provided by Microsoft Word's wizards. If you click on "File" and "New" within Microsoft Word, you will find memo templates for "elegant memos," "contemporary memos," and "professional memos."

Each of these templates differs in small ways. The contemporary memo provides identification lines in this order: "To," "CC" (for carbon copy or complimentary copy), "From," "Date," and then "Re" (the abbreviation for *Regarding*).

In contrast, the elegant memo template lists these identification elements as follows: "to," "from," "subject," "date," and "cc." Here the order has changed, "subject" has replaced "re," and the words are typed in lowercase rather than initial capitals. The professional memo template lists "To," "From," "CC," "Date," and "Re." The order, typing style, and word usage have changed. There is no one way to order these identification lines.

Memo Templates

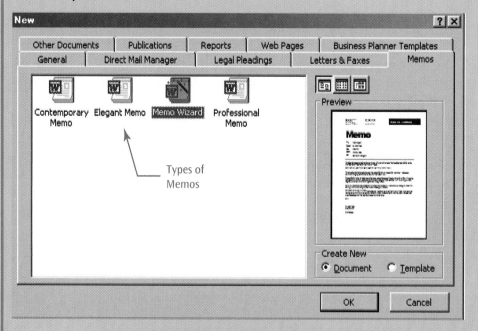

Wizards allow you to customize your memo in a variety of ways. Through a wizard, you can

- Select the style: Professional, Contemporary, or Elegant

- Include a memo title: "Interoffice Memo" or one of your choice

- Choose heading fields: "Date," "From," "Subject," or "Priority"

- Choose who should receive the memo: "To," "CC," or a separate distribution list

- Select closing fields: "Writer's Initials," "Typist's Initials," "Enclosure," or "Attachments"

- Select headers: "Date," "Topic," "Page #," or "Confidential"

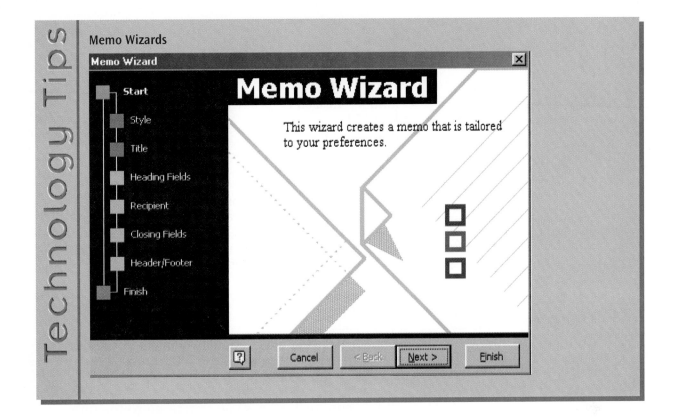

SAMPLE MEMOS

On the following pages are additional examples of both student and professionally written memos (complete with our evaluations of the memos' successes and needed improvements). The first, a student-written memo (Figure 5.3), is well organized and developed, but it has several stylistic flaws.

Successes

Let's discuss the successes first. We like the way the memo is formatted (indented body for accessibility). It has an introduction, discussion, and conclusion, following our template. The writer also has developed her ideas thoroughly. The introductory paragraph answers the reporter's questions *who* and *what;* the body discusses *why;* and the conclusion focuses on *when.* Finally, the tone of the memo is appropriate for management (especially in the last paragraph).

Failures

However, the memo is flawed stylistically. Look how long that first sentence is! This type of writing requires a high degree of patience on the reader's part. Instead, the first paragraph could read as follows: "Our product development staff has designed a new phone answering machine. This breakthrough product screens calls and, by voice activation, routes the calls to an appropriate message. This message gives a busy signal when the person doesn't want to be bothered." In addition, the writer has made two punctuation errors. No period is needed in the subject line. The colon after *because* in the body is incorrect. Either no punctuation is needed, or the sentence should be rewritten so a colon could be used. For example, if the sentence reads " . . . it would benefit our company for the following reasons," then you could insert a colon after *reasons.* (Chapter 19 covers this grammatical rule.)

MEMO

DATE: January 19, 2005
TO: Tom Lisk
FROM: Juliet Kincaid
SUBJECT: PRODUCING THE ANNOY-NO-MORE CALL SCREENING MACHINE.

Subject line should not end with periods

Our product development staff has come up with a breakthrough new product, a phone answering machine that answers the phone, screens calls, and, by voice activation, routes the call to an appropriate message and gives a busy signal when the person doesn't want to be bothered.

Good technical writing should be concise—this sentence is too long

We have researched this product fully and believe it would benefit our company because:

Grammatically, this colon is used incorrectly

1. The product has a relatively simple design that occupies no more space than a regular phone and can be designed for wall or desk mounting.

2. The cost for manufacturing is low compared to the retail value. The unit costs approximately $100.00 to produce and retails for $999.99.

3. The unit can be interchanged easily with all phone networks including MBB and Shout.

"around" is vague word usage

We would like to begin mass production around April 1, 2005. Our Westport facility could handle the production with the addition of 10 technicians and engineers to aid in production. We hope you give this product consideration and endorse it fully. Thanks.

Figure 5.3 Student-Written Memo

The example in Figure 5.4 is professionally written. It too has successes and failures.

Successes

This memo succeeds in the following ways. As with the earlier student-written example, this memo uses the introduction-body-conclusion template. It ends with an excellent series of directives, specifying what is to occur next. The body is formatted for reader-friendly ease of access. The memo answers the reporter's questions *what, who, why,* and *when.*

Failures

The memo is flawed, however, in several ways. First, the subject line has a topic (operating procedure) but no focus. What about this procedure? Will the memo

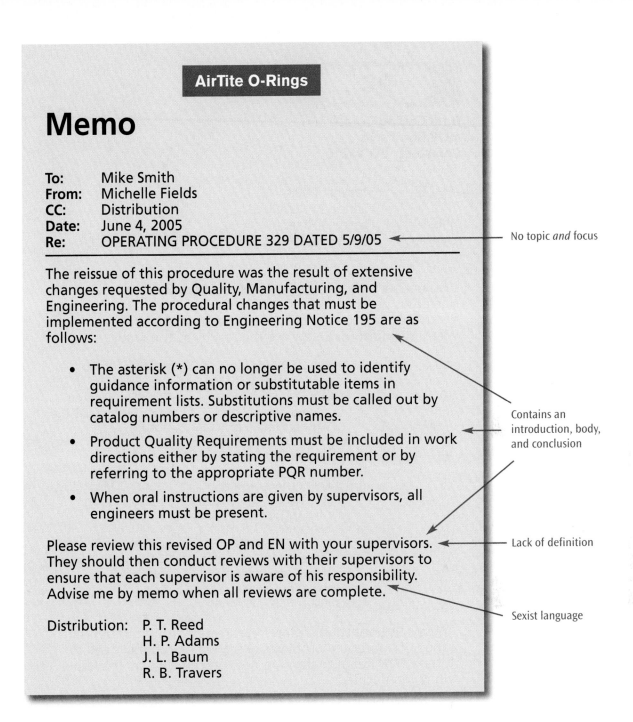

Figure 5.4 Professionally Written Memo

discuss the termination of the procedure, the writing of a new procedure, or changes in the procedure? A subject line such as "Revision of Operating Procedure 329 dated 1-19-05" would provide the necessary focus.

Next, note that this memo is written to a multiple audience (Distribution). Will each reader recognize that *PQR*, *OP*, and *EN* are abbreviations for *product quality requirement, operating procedure,* and *engineering notice?* You would have to guess at an answer, and guessing gets you in trouble in technical writing. When in doubt, spell it out. To avoid any problems with audience recognition, add your abbreviation parenthetically. For example, write, "operating procedure (OP) 329." Doing so takes very little time but eliminates any audience doubt.

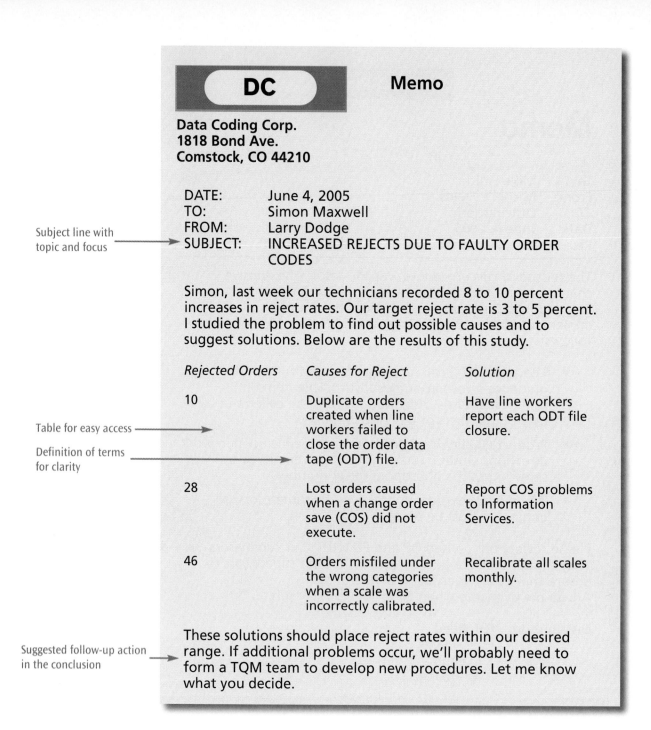

DC **Memo**

Data Coding Corp.
1818 Bond Ave.
Comstock, CO 44210

DATE: June 4, 2005
TO: Simon Maxwell
FROM: Larry Dodge
SUBJECT: INCREASED REJECTS DUE TO FAULTY ORDER
          CODES

Simon, last week our technicians recorded 8 to 10 percent
increases in reject rates. Our target reject rate is 3 to 5 percent.
I studied the problem to find out possible causes and to
suggest solutions. Below are the results of this study.

Rejected Orders	Causes for Reject	Solution
10	Duplicate orders created when line workers failed to close the order data tape (ODT) file.	Have line workers report each ODT file closure.
28	Lost orders caused when a change order save (COS) did not execute.	Report COS problems to Information Services.
46	Orders misfiled under the wrong categories when a scale was incorrectly calibrated.	Recalibrate all scales monthly.

These solutions should place reject rates within our desired
range. If additional problems occur, we'll probably need to
form a TQM team to develop new procedures. Let me know
what you decide.

Figure 5.5   Professionally Written Problem-Solution Memo

Finally, the memo uses sexist language in the sentence "They should then
conduct reviews with their supervisors to ensure that each supervisor is aware of
his responsibility." It must read "*his or her* responsibility."

The two professionally written memos in Figures 5.5 and 5.6 succeed in all
ways: organization, visual appeal, development, grammar, and tone. Note how
the memos have short introductory paragraphs, accessible bodies, and concise
conclusions. The writers have used specific details so the readers can clearly understand the content.

tech link

Go to *http://www.prenhall.*
com/gerson for Web links,
samples, and interactive
activities.

Diamond Delivery
101 Starburst Dr.
Antelope, GA 54210

MEMORANDUM

DATE: December 12, 2005
TO: Distribution
FROM: Luann Brunson
SUBJECT: REPLACEMENT OF MAINTENANCE RADIOS ◄——————— Effective subject line
                                                      with topic and focus

On December 5, the manufacturing department supervisor informed the purchasing department that our company's maintenance radios were malfunctioning. Purchasing was asked to evaluate three radio options (the RPAD, XPO 1690, and MX16 radios). Based on my findings, I have issued a purchase order for 12 RPAD radios.

The following points summarize my findings.

1. *Performance*
   During a one-week test period, I found that the RPAD outperformed our current XPO's reception. The RPAD could send and receive within a range of 5 miles with minimal interference. The XPO's range was limited to 2 miles, and transmissions from distant parts of our building broke up due to electrical interference.

2. *Specifications*
   Both the RPAD and the MX16 were easier to carry, because of their reduced weight and size, than our current XPO 1690s.

	RPAD	XPO 1690	MX16
Weight	1 lb	2 lb	1 lb
Size	5" x 2"	8" x 4"	6" x 1"	◄—— Details for clarity

3. *Cost*
   The RPAD is our most cost-effective option because of quantity cost breaks and maintenance guarantees.

	RPAD	XPO 1690	MX16
Cost per unit	$70	$215	$100
Cost per doz.	$750	$2,580	$1,100
Guarantees	1 year	6 months	1 year

Purchase of the RPAD will give us improved performance and comfort. In addition, we can buy 12 RPAD radios for approximately the cost of 4 EXPOs. If I can provide you with additional information, please call. I'd be happy to meet with you at your convenience.

Distribution: M. Ellis   M. Rhinehart   T. Schroeder
                 P. Michelson   R. Travers   R. Xidis

Figure 5.6   Professionally Written Compare-and-Contrast Memo

1. Have you used the correct memo format, including date, to, from, and subject lines?

2. Is your subject line correct?
   - Is it typed in all caps?
   - Does it contain a topic and a focus?

3. Does your introduction tell why you are writing and what you are writing about?

4. Does the body explain exactly what you want to say?

5. Does the conclusion tell what's next, providing either a complimentary or a directive close?

6. Is your page layout reader-friendly?
   - Do you use highlighting techniques for accessibility, such as bullets, boldface, underlining, and white space?

   Refer to Chapter 8 for help.

7. Is your writing style concise?
   - Do you limit the number of words per sentence, the number of syllables per word, and the number of lines per paragraph?

   Refer to Chapter 3 for help.

8. Is your writing clear?
   - Do you answer reporter's questions?
   - Do you avoid vague words, such as *some, several, many,* or *few,* specifying instead?

   Refer to Chapter 3 for help.

9. Have you written appropriately to your audience?
   - Have you defined high-tech terms where necessary for low-tech or lay readers?
   - Have you achieved the correct tone and personalized your memo, based on your audience and purpose?

   Refer to Chapter 4 for help.

10. Are errors eliminated? Remember, a memo with grammatical or mechanical errors will destroy your credibility. You must prooofread to catch erros like the ones in this sentence— prooofread and erros.

E-Mail

tech link

For an excellent article on the shift to electronic mail and its revolutionary effect on business, see *http://www.emc.com/information_generation/in_depth_archive/12112000_posthaste.jsp.*

Memos have been and will continue to be an important aspect of your business communication. However, e-mail is rapidly becoming an even more important business communication channel.

The use of e-mail has exploded in recent years:

- 50 e-mail messages are sent and received daily by business employees ("Pitney Bowes Study" 2000).
- 66 percent of business employees say they are e-mail only users.
- 96 percent of business employees report using e-mail every day.
  - 80 percent of business employees say that e-mail has replaced posted mail.
  - 72.5 percent of business employees say that e-mail has replaced faxing.
  - 45 percent of business employees say that e-mail has replaced telephone calls (Schulzrinne).
  - 2.1 billion personal and business messages are sent daily.
  - E-mail messages outnumber first-class letters 30:1 ("E-mail Statistics") (Schwabach D13) (Wickman D1).

Note:
Number of years to reach 50 million users:
- Radio: 38 years
- Personal Computers: 16 years
- E-mail/Internet: 4 years

Source: From MacIntyre 2002. (MacIntyre)

John MacIntyre of EMC Corporation states, "In the business world, e-mail has revolutionized communication, often replacing the phone, the fax, the postal service and even face-to-face meetings."

Mark Levitt, an IDC analyst, reports that "the number of e-mails sent will balloon from 2.6 trillion in 2000 to 9.2 trillion by 2005. Soon there will be more e-mail addresses than phone numbers in the world" (MacIntyre 2002).

Why Is E-mail Important?

Why is e-mail proving to be such a valuable communication tool? Many companies are "geared to operate with e-mail," creating what the Harvard Business School calls "e-mail cultures" for the following reasons ("The Transition to General Management" 1998).

Time

"Everything is driven by time. You have to use what is most efficient" (Miller et al. 10). The primary driving force behind e-mail's prominence is time. E-mail is quick. You can send an e-mail message around the world in seconds compared to "snail mail." In addition, communicating by e-mail saves time within the office. Instead of walking down the hall or up several flights, business employees communicate with each other in the same office building via e-mail.

Convenience

"You can't spill coffee all over it" (Miller et al. 15). With wireless communication, you can send e-mail from notebooks to handhelds. Current communication systems combine a voice phone, a personal digital assistant, and e-mail into a package that you can slip into a pocket or purse. This lets you e-mail coworkers, clients, and vendors from any location, achieving instantaneous communication.

Internal/External

Memos are internal communication within a company. E-mail allows you to communicate internally to coworkers and externally to customers and vendors.

Cost

E-mail is cost effective because it is paper-free. E-mail currently requires no postage. However, this might change. In January 2004, at the World Economic Forum in Switzerland, Bill Gates discussed postage fees for e-mail (Jesdanun 2004, B6) .

With an ability to attach files, you can send many kinds of documentation without paying shipping fees. This is especially valuable when considering international business.

Documentation

Your company's computer system administrator maintains electronic files to back up your e-mail. E-mail provides an additional value when it comes to documentation. Because so many writers merely respond to earlier e-mail messages, what you end up with is a "virtual paper trail" (Miller et al. 15). When e-mail is printed out, often the printout will contain dozens of e-mail messages, representing an entire string of dialogue. This provides a company an extensive record for future reference.

E-mail Challenges

Despite e-mail's many benefits, this form of correspondence is not foolproof. It presents the writer with numerous problems.

The Rise of Instant Messaging

E-mail could be too slow for today's fast-paced workplace. In fact, e-mail might be the next snail mail.

Instant messaging (IM) could replace e-mail in the workplace within the next five years. IM pop-ups are already providing many businesses these benefits:

• Increased speed of communication

• Improved efficiency for geographically dispersed workgroups

• Collaboration by multiple users in different locations

However, IM used for business purposes is so new that corporate standards have not been formalized. Software companies have not yet redesigned IM home versions for the workplace. This leads to problems with security, archiving, monitoring, and employee misuse.

To solve potential problems, consider these five suggestions:

1. *Choose the correct communication channel.* Use IM for speed and convenience. If you need length and detail, other options—e-mail, memos, reports, letters—are better choices. In addition, sensitive topics should never be handled through IM. These deserve the personal attention provided by telephone calls or face-to-face meetings.

2. *Document important information.* For future reference, you must archive key text. IM does not allow for this. Therefore, you will need to copy and paste IM text into a word processing tool for long-term documentation.

3. *Summarize decisions.* IM is great for collaboration. However, all team members might not be online when decisions are made. Once conclusions have been reached that affect the entire team, the designated team leader should e-mail everyone involved. In this e-mail, the team leader can summarize the key points, editorial decisions, timetables, and responsibilities.

4. *Tune in, or turn off.* The moment you log on, IM software tells everyone who is active online. Immediately, your IM buddies can start sending messages. IM pop-ups can be distracting. Sometimes, in order to get your work done, you might need to turn off your IM system. Your IM product might give you status options, such as "on the phone," "away from my desk," or "busy." Turning on IM could infringe upon your privacy and time. Turning off might be the answer.

5. *Limit personal use.* Your company owns the instant messaging in the workplace. IM should be used for business purposes only.

Source: From Hoffman *2004*, 16–17.

Computer Limitations

Hardware and software are not universally compatible. The computer commands you use for your e-mail will not necessarily mean the same thing on someone else's computer. If you have used a certain command to create textual enhancements (underlining, italics, boldface, or color—as you can with memos), these commands might be interpreted differently on your reader's computer.

Even a typed word might not transmit as you would assume. Take the simple word *it's*, for example. People have received e-mail in which this word is displayed as *it96s*. Why? At a minimum, all computers support 128 American Standard Code for Information Interchange (ASCII) characters. ASCII is "the standard way that printable . . . characters are represented in the USA" (Shipman 1).

Other computer systems, however, have additional ASCII codes, called extended ASCII. If the writer's computer has extended ASCII and the reader's computer does not, there is a problem—the technologies do not interface. In the example regarding *it's*, the technology challenge is created by the apostrophe. ASCII apostrophes look like this ('), whereas extended ASCII supports apostrophes that look like this ('). If the e-mail systems are not compatible, the reader can see *it96s*, the computer's representation of the writer's apostrophe.

Lack of Privacy

Traditional mail is confidential. Not only is a letter sealed in an envelope, but also privacy laws protect your mail. In contrast, e-mail is not confidential. "Employers have the right to keep track of how their employees spend their work hours. A recent survey of 722 companies by the Society of Human Resource Management found that almost three quarters monitored employee Internet use and periodically checked employee e-mail" (Rodenbough 2003, 29).

How do companies check employee e-mail? A company's systems "operator is allowed to intercept messages during transmission. The operator is allowed to disclose user messages to anyone [he or she] chooses" (Rose 172). Deleting e-mail does not ensure confidentiality. Deleted e-mail can be stored in backup systems or recycle bins (Kim 52).

More important, employees do not own their e-mail. The 1986 Electronic Communications Privacy Act asserts that the company owns an employee's e-mail. Sent e-mail may be saved internally, and archived e-mail messages are officially company documents. In fact, "in some countries (including the U.S.), e-mail may be legally monitored and even subpoenaed in a lawsuit."

Finally, an e-mail message's content, after receipt by another company, may be reused without the writer's knowledge or consent. Once a reader receives your e-mail, your "messages may be printed, distributed, forwarded, or saved by the recipient for later dissemination" (Munter et al. 2003, 29).

To combat the problem of lack of confidentiality, many companies, municipalities, or organizations have started to append a disclaimer to their corporate e-mail.

tech link

For the full text of the Electronic Communications Privacy Act, Title 18, Part I, Chapter 121, Sec. 2701, "Unlawful access to stored communications," go to *http://www4.law.cornell.edu/uscode/18/2701.html*

Applebee's International, Inc. Confidentiality Notice

PRIVILEGED/CONFIDENTIAL INFORMATION may be contained in this message or any attachments. This information is strictly confidential and may be subject to attorney-client privilege. This message is intended only for the use of the named addressee. If you are not the intended recipient of this message, unauthorized forwarding, printing, copying, distributing, or using such information is strictly prohibited and may be unlawful. If you have received this in error, you should kindly notify the sender by reply e-mail and immediately destroy this message. Unauthorized interception of this e-mail is a violation of federal criminal law. Applebee's International, Inc. reserves the right to monitor and review the content of all messages sent to or from this e-mail address. Messages sent to or from this e-mail address may be stored on the Applebee's International, Inc. e-mail system.

tech link

See *http://www.cmac.state.ct.us/policies/emailcon.htm* for the State of Connecticut's "Electronic Mail Acceptable Use Policy," which focuses on privacy, security, e-mail management, and user responsibilities.

Olathe, Kansas, School District

CONFIDENTIALITY NOTICE: This message is from the Olathe District Schools. The message and any attachments may be confidential or privileged and are intended only for the individual or entity identified above as the addressee. If you are not the addressee, or if this message has been addressed to you in error, you are not authorized to read, copy, or distribute this message or any attachments. We ask that you please delete this message and any attachments and notify the sender by return e-mail or by phone.

Misunderstandings or Erroneous Messages

When you talk to others, either face-to-face or on the telephone, you occasionally employ verbal sarcasm and humor. Through voice inflections and gestures, you help your phone audience or face-to-face listener recognize your sarcastic or humorous intent.

E-mail, however, leaves no room for gesture or inflection. Will your reader recognize your sarcasm? Let's say that you received a memo asking whether your company should select XYZ Corporation as a hardware vendor. You think the idea is terrible, so you sarcastically write an e-mail responding, "Great idea." What will the reader think? Will he or she recognize the sarcasm? If the reader does not and then signs a contract with the vendor, you have made a terrible mistake by letting e-mail's inherent casualness affect your correspondence ("E-mail Etiquette").

Casual, Unprofessional Tone

Because you can "talk" back and forth with your e-mail reader, e-mail tends to become much friendlier than memos or letters—even chatty. This casual tone can lead to problems.

For example, e-mail users occasionally rely on the following abbreviations:

- B/C   Because
- C-Ya   See ya'
- D/L   Download
- ESP   Especially
- IC   I see
- B4   Before
- UR   You're
- 2   To/too
- IOW   In other words
- TIA   Thanks in advance
- BTW   By the way

(*Source:* From Cobbs 2002, F5)

Other casual e-mail writing conventions include numerous periods or dots to suggest ongoing thoughts, instead of traditional periods to end sentences. Many writers have stopped capitalizing. Finally, casual e-mail writers use "emoticons"—faces created by colons, parentheses, semicolons, and hyphens.

People use them to express the attitudes commonly heard in speech but which cannot be heard online.

:)	happy face to show agreement
:(	sad face to show unhappiness
;)	winking face to show sarcasm
:-0	startled shock or amazement
%-)	confusion
:-X	lips sealed for confidentiality

If your e-mail message is sent to a boss, client, or vendor and will be printed for future reference, a conversational tone could lead to trouble. This failure to recognize audience is highlighted in e-mail because e-mail addresses rarely list the reader's title. You never know if the e-mail recipient is a salesperson, purchasing agent, or vice president, unless you are writing to an acquaintance.

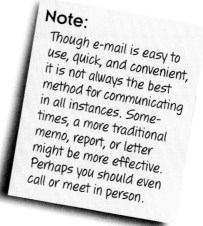

Note:
Though e-mail is easy to use, quick, and convenient, it is not always the best method for communicating in all instances. Sometimes, a more traditional memo, report, or letter might be more effective. Perhaps you should even call or meet in person.

Sample of Casual, Unprofessional E-mail

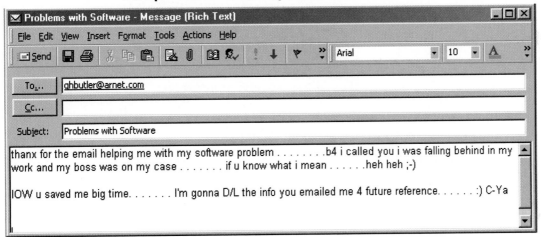

✉ Problems with Software - Message (Rich Text)

File Edit View Insert Format Tools Actions Help

Send | Arial | 10 | A

To...: ghbutler@arnet.com
Cc...:
Subject: Problems with Software

thanx for the email helping me with my software problem . . . . .b4 i called you i was falling behind in my work and my boss was on my case . . . . if u know what i mean . . . . . .heh heh ;-)

IOW u saved me big time. . . . . . . I'm gonna D/L the info you emailed me 4 future reference. . . . . . :) C-Ya

TECHNIQUES FOR WRITING EFFECTIVE E-MAIL MESSAGES

To combat the challenges e-mail presents, follow these tips for writing effective e-mail.

Recognize Your Audience

E-mail messages can be sent to many people at the same time—throughout the world. Thus, your reader's culture and language might differ from your own. Remember that abbreviations and acronyms are not universal. Dates, times, measurements, and monetary figures differ from country to country (see our discussion of multiculturalism in Chapter 3). In addition, your reader's e-mail system might not have the same features or capabilities that you have. Therefore, to communicate effectively to a diverse audience, recognize your reader's unique language and technology needs.

Identify Yourself

This is especially important given the problem with e-mail viruses, like the "love bug" that crashed computer systems worldwide in May 2000. Corporations constantly warn their employees about opening e-mail messages from unfamiliar writers.

To solve this problem and present a professional first impression, identify yourself by name, affiliation, or title. You can accomplish this either in the "from" line of your e-mail or by creating a signature file or ".sig file." This .sig file acts like an online business card. Once this identification is complete, readers will be able to open your e-mail without fear of corrupting their computer systems ("E-mail Etiquette").

For example, in the following graphic, note how Microsoft Outlook allows you to select from "Insert" a "Signature." Once you scroll to this command to open it, you can customize the signature to your choosing.

By left-clicking on "Signature," you can access options for personalizing a ".sig file."

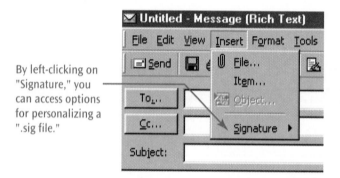

Use the Correct E-mail Address

A common e-mail address looks like this: name@arpnet.uh.hou.tx.us. You can imagine how easy it is to omit any part of this address or to punctuate it incorrectly. To connect with your intended recipient, confirm the e-mail address before you hit the "Send" button.

Provide an Effective Subject Line

Memos and e-mail messages have similarities and differences (see Table 5.1). As with memos, an effective subject line will consist of a topic and a focus. Here is a successful subject line:

Subject: Meeting Dates for Tech Prep Conference

The topic of this e-mail message is a "Tech Prep Conference." If you had received only these words in the subject line, your question would have been, "What about the conference?" "Meeting Dates" provides the focus.

Effective subject lines are more important in e-mail messages than in memos. When a reader picks up a hard-copy memo, the content is obvious at a glance. In contrast, when readers open their e-mail packages, they do not see the entire message. All they might see is their in-box display. Then, based on this in-box, the readers will "decide if and when to read [the] message" (Munter et al. 2003, 30). Given readers' unwillingness to open unsolicited or unknown e-mail, the subject line has special importance.

Keep Your E-mail Message Brief

"Apply the 'top of the screen' test. Assume that your readers will look at the first screen of your message only" (Munter et al. 2003, 31). Limit your message to one screen (if possible). Readers do not want to scroll through several screens to read your message.

Organize Your E-mail Message

Successful writing usually contains an introductory paragraph, a discussion paragraph, and a conclusion. Your e-mail message also should contain these components. Use the introduction to tell the reader why you are writing and what you are writing about. In the discussion, clarify your points thoroughly. Use the conclusion to tell the reader what is next, possibly explaining when a follow-up is required and why that date is important.

Use Highlighting Techniques Sparingly

Many e-mail packages will let you use highlighting techniques, such as **bold-face**, *italics*, <u>underlining</u>, computer-generated bullets and numbers, centering, font color highlighting, and font color changes, as follows:

Other e-mail packages, however, do not support these highlighting techniques. Therefore, even if your system allows you to highlight, avoid doing so. The commands required for these operations might not be compatible with your end user's computer. In that case, the enhanced text you send could "be displayed on the receiver's computer as an unreadable mess!" (Christopher 2).

Try the following options to make your e-mail message accessible:

- Use the Tab key or space bar to indent the e-mail discussion unit (body).
- Set off the itemized points in this discussion either with an asterisk (*), hyphens (-), or typed numbers. These are keystrokes on your keypad, as opposed to computer-generated highlighting techniques.
- Use headings as a lead-in to new paragraphs.
- Type headings in all capital letters.
- Double-space between paragraphs.

The added white space will help readers access your text. The use of upper-case letters will also make a key word or phrase emphatic, but *do not use all caps throughout the text.* An entire e-mail message composed of capitalized letters is difficult to read and creates "flaming" (discussed later in this chapter).

Proofread Your E-mail Message

E-mail can be written and transmitted rapidly. That's good news . . . and bad news. It is valuable to have electronic communication abilities that let you correspond with others around the world, in seconds. Rapid writing also can lead to mistakes, unfortunately. Whereas most word processing packages have spell checks and grammar checks, many e-mail systems do not. Sometimes, even if a computer's e-mail system has spell checks, writers do not know how to set these tools. This means that you can only avoid errors in your e-mail message the old-fashioned way—read it and reread it. Errors will undermine your credibility.

Even if you have spell check or grammar check in your e-mail system, following are tips for more effective proofreading:

1. Type your text first in a word processing package, like Microsoft Word.

2. Print it out. Sometimes it is easier to read hard-copy text than on-screen text. Also, your word processing package, with its spell check and grammar check, will help you proofread your writing.

3. Once you have completed these two steps (writing in Word or WordPerfect and printing out the hard-copy text), copy and paste the text from your word processing file into your e-mail message.

Make Hard Copies for Future Reference

Like memos, letters, and reports, e-mail messages can communicate important business transactions or document significant decisions. Though many companies would like to become paper-free, you still should make hard copies of correspondence for future reference. Your company might need your message next week, next month, or next year. Save your e-mail printouts.

Be Careful When Sending Attachments

Another e-mail netiquette consideration involves attachments. When you send attachments, follow this procedure:

- Tell your reader within the body of the e-mail message that you have attached a file.

- Specify the file name of your attachment and the software application that you have used (HTML, PowerPoint, PDF, RTF—rich text format—Word, or Works). By telling the reader what software you have used for your attachment, she can then determine if the file can be opened.

- Limit the size of the attached file(s). If you try to send an excessively large file or number of attachments, the file(s) might not be delivered or it could take forever to download on your reader's system.

  - Limit your attachments to files no larger than 750 Kilobytes (K) ("E-mail Etiquette Guidelines").

  - Use compression (ZIP) files.

Do Not Automatically Reply to All Recipients

Often you will receive e-mail messages that have been sent to additional readers. When you respond to the message, you have two choices: You can respond only to the writer, or you can choose to reply to all the recipients. When you choose the latter option, everyone who received the original e-mail also will receive your response.

If you reply to all recipients without considering the consequences, not only might you clutter up other readers' e-mail boxes, but also you might be distributing confidential information. Therefore, "check the original distribution list and consider which individuals really want, need, or even should receive your message" (Munter et al. 2003, 36).

An Angry E-mail Sends Stock Plummeting

After an executive's e-mail message to managers was posted on the Internet, the executive watched the share price of his company's stock plummet. In his e-mail, he accused his staff of being lazy, threatened to fire them, and gave them a two-week ultimatum to shape up.

Within three days of his e-mail being posted on the Internet, stock in the company dropped 22 percent because of concerns about employee morale.

The e-mail read: "We are getting less than 40 hours of work from a large number of our employees. The parking lot is sparsely used at 8 A.M.; likewise at 5 P.M. As managers, you either do not know what your EMPLOYEES are doing or you do not CARE. In either case, you have a problem and you will fix it or I will replace you."

The e-mail continued, "NEVER in my career have I allowed a team which worked for me to think they had a 40-hour job. I have allowed YOU to create a culture which is permitting this. NO LONGER." He also said that "hell would freeze over" before he would increase employee benefits, and the car park should be nearly full by 7:30 A.M. and half full on weekends. He gave the managers two weeks, "Tick, tock."

Source: From Broughton 2001, online.

Practice Netiquette

When you write your e-mail messages, observe the rules of "netiquette":

- *Be courteous.* Do not let the instantaneous quality of e-mail negate your need to be calm, cool, deliberate, and professional.

- *Avoid abusive, angry e-mail messages.* Because of its quick turnaround abilities, e-mail can lead to negative correspondence called flaming. Flaming is sending angry e-mail, often typed in all caps.

tech link

Go to *http://www.library.yale.edu/training/netiquette/* for Yale University's guidelines for sending and responding to e-mail.

When angry e-mail is sent negative results usually occur, as in the Spotlight feature.

The executive in the memo assumed that his remarks were confidential. However, within a week, the boss's "private" e-mail appeared on a Yahoo financial message board—for the world to review. If your company e-mail leads to a loss of corporate income, you know you have made a huge mistake.

"Business professors and market analysts . . . [criticized not only the] . . . angry tone, but also [the] mode of communication." The executive "ran afoul of two cardinal rules for modern managers. . . . Never try to hold large-scale discussions over e-mail. And never, ever, use the company's e-mail system to convey sensitive information or controversial ideas to more than a handful of trusted lieutenants" (Wong 2001).

E-mail messages can be written for unlimited reasons. You might want to write e-mail messages that persuade, instruct, inform, or build trust. Following are examples of successful e-mail messages that give instructions, provide information, and recommend changes persuasively.

Instructions—Correcting Online Account Problem

Introduction explaining why the e-mail has been written

Itemized, step-by-step text for easy access

Conclusion with follow-up information

Positive tone for audience involvement and to build trust

"Sig." lines, a "virtual" business card, giving the name and contact information of the e-mail writer

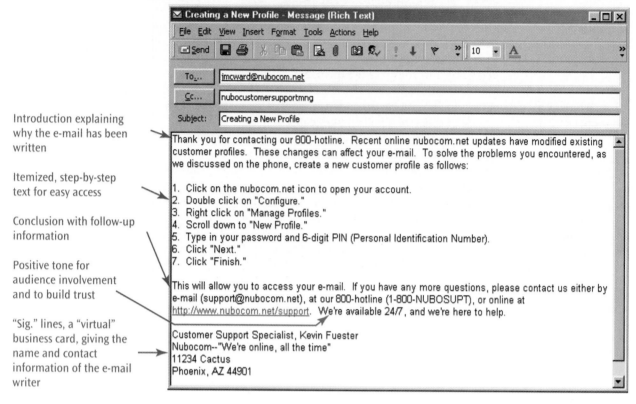

Creating a New Profile - Message (Rich Text)

File Edit View Insert Format Tools Actions Help

Send | 10 | A

To... jmcward@nubocom.net

Cc... nubocustomersupportmng

Subject: Creating a New Profile

Thank you for contacting our 800-hotline. Recent online nubocom.net updates have modified existing customer profiles. These changes can affect your e-mail. To solve the problems you encountered, as we discussed on the phone, create a new customer profile as follows:

1. Click on the nubocom.net icon to open your account.
2. Double click on "Configure."
3. Right click on "Manage Profiles."
4. Scroll down to "New Profile."
5. Type in your password and 6-digit PIN (Personal Identification Number).
6. Click "Next."
7. Click "Finish."

This will allow you to access your e-mail. If you have any more questions, please contact us either by e-mail (support@nubocom.net), at our 800-hotline (1-800-NUBOSUPT), or online at http://www.nubocom.net/support. We're available 24/7, and we're here to help.

Customer Support Specialist, Kevin Fuester
Nubocom--"We're online, all the time"
11234 Cactus
Phoenix, AZ 44901

Information—Confirming an Upcoming Meeting

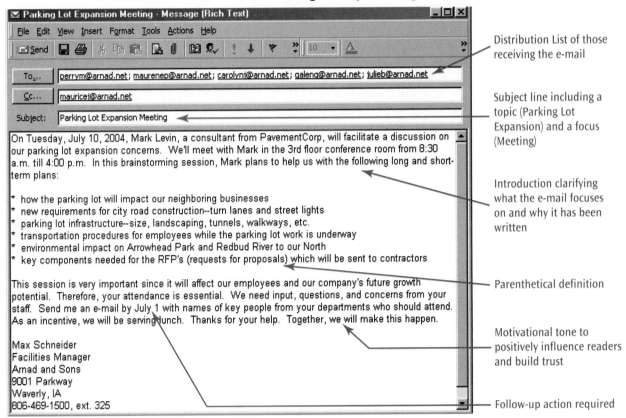

Distribution List of those receiving the e-mail

To... perrym@arnad.net; maurenep@arnad.net; carolyni@arnad.net; galeng@arnad.net; julieb@arnad.net

Cc... mauricej@arnad.net

Subject: Parking Lot Expansion Meeting

Subject line including a topic (Parking Lot Expansion) and a focus (Meeting)

On Tuesday, July 10, 2004, Mark Levin, a consultant from PavementCorp, will facilitate a discussion on our parking lot expansion concerns. We'll meet with Mark in the 3rd floor conference room from 8:30 a.m. till 4:00 p.m. In this brainstorming session, Mark plans to help us with the following long and short-term plans:

Introduction clarifying what the e-mail focuses on and why it has been written

* how the parking lot will impact our neighboring businesses
* new requirements for city road construction--turn lanes and street lights
* parking lot infrastructure--size, landscaping, tunnels, walkways, etc.
* transportation procedures for employees while the parking lot work is underway
* environmental impact on Arrowhead Park and Redbud River to our North
* key components needed for the RFP's (requests for proposals) which will be sent to contractors

Parenthetical definition

This session is very important since it will affect our employees and our company's future growth potential. Therefore, your attendance is essential. We need input, questions, and concerns from your staff. Send me an e-mail by July 1 with names of key people from your departments who should attend. As an incentive, we will be serving lunch. Thanks for your help. Together, we will make this happen.

Motivational tone to positively influence readers and build trust

Max Schneider
Facilities Manager
Arnad and Sons
9001 Parkway
Waverly, IA
806-469-1500, ext. 325

Follow-up action required

Persuasive—Directive E-mail Requiring Follow-up Action

Subject line with topic (Digital Switch Quote) and a focus (Request)

Audience involvement with first name

Body with accessible, specific text. Highlighting techniques depend on keyboard strokes, versus computer-generated bullets or numbers, which might not be readable on all computers

Conclusion with due date and reason why

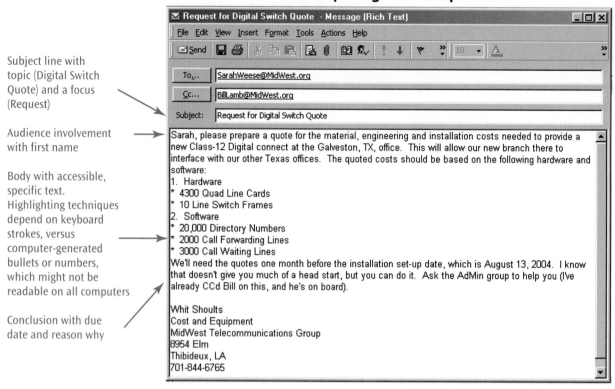

Effective E-mail Message Checklist

___ 1. Does the e-mail use the correct address?

___ 2. Have you identified yourself? Provide a "Sig." (signature) line.

___ 3. Did you provide an effective subject line? Include a *topic* and a *focus*.

___ 4. Have you effectively organized your e-mail message? Include the following:
   - an introductory paragraph telling *why* you are writing and *what* you are writing about.
   - a body unit with itemized points telling *what exactly* the message is discussing.
   - a conclusion, *summing up* your e-mail message.

___ 5. Have you used highlighting techniques sparingly?
   - Avoid **boldface**, *italics,* color, or underlining.
   - Use asterisks (*) for bullets, numbers, all caps, and double-spacing for access.

___ 6. Did you practice netiquette?
   - Be polite, courteous, and professional.
   - Do not flame.

___ 7. Is the e-mail concise?

___ 8. Did you identify and limit the size of attachments?
   - Tell your reader(s) if you have attached files and what types of files are attached (PPt, PDF, RTF, Word, etc.).
   - Limit the files to 750 K.

___ 9. Does the memo recognize audience,
   - Defining acronyms or abbreviations where necessary?
   - Considering a diverse audience (factoring in issues, such as multiculturalism or gender)?

___ 10. Did you avoid grammatical errors?

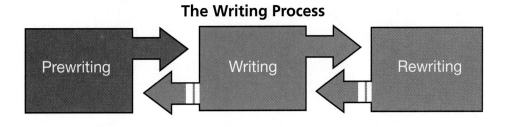

The Writing Process

The Writing Process		
Prewriting	Writing	Rewriting
• Examine your purposes • Determine your goals • Consider your audience • Gather your data • Determine how the content will be provided	• Organize the draft according to some logical sequence that your readers can follow easily • Format the content to allow for ease of access	• Revise 　° Add missing details 　° Delete wordiness 　° Simplify word usage 　° Enhance the tone of your communication 　° Reformat your text for ease of access 　° Practice the speech or review the text • Proofread 　° Correct errors

PROCESS

Now that you know what to include in a memo or an e-mail message, follow the three-step process to create the correspondence.

- Prewriting
- Writing
- Rewriting

The writing process is dynamic, with the three steps frequently interwoven.

Prewriting

No single method of prewriting is more effective than another, so let's try clustering or mind mapping.

The type of prewriting you prefer depends on your unique learning style. Clustering or mind mapping is visual and free-form. Figure 5.7 shows a typical mind-map pattern.

The idea of clustering is to start with your main topic, whatever that might be, and then determine what that topic is composed of. For example, if your main topic is a company picnic, the next question is to ask what that picnic should entail. If you are planning such a picnic, you will want to consider food, entertainment, and a schedule for activities. These items represent the major components and are diagrammed in Figure 5.8.

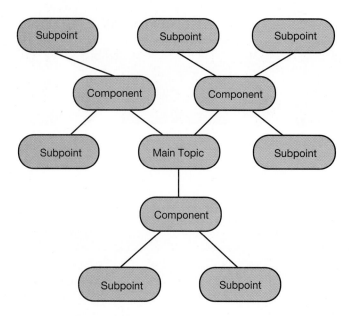

Figure 5.7   Clustering or Mind Mapping

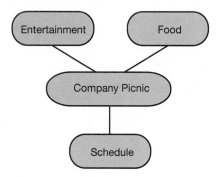

Figure 5.8   Major Components of a Company Picnic Mind Map

Writing always requires that you develop your points so you are clear and detailed. Thus, the next step in clustering is to provide more detail through the addition of subpoints. Subpoints clustered around the food component might include the entree (BBQ beef), condiments (pickles, relish, potato salad), drinks (soda and lemonade), and desserts (apple pie). Subpoints clustered around the entertainment component might include a live band and games for both adults and children.

Figure 5.9 shows a mind map including subpoints.

Clustering/mind mapping is valuable because it allows you to sketch your ideas freely without falling prey to the rigid structure of an outline (which often deters some writers' creativity). Clustering/mind mapping also allows you to see graphically the relationship between subpoints and components of an idea. This helps you determine if you have omitted any key concerns or included any irrelevant ideas. If so, you can develop your ideas further by adding new subpoints or delete irrelevancies simply by scratching off one of your subpoint balloons. Thus, clustering or mind mapping is an excellent prewriting technique because it easily allows you to gather information and organize your thoughts.

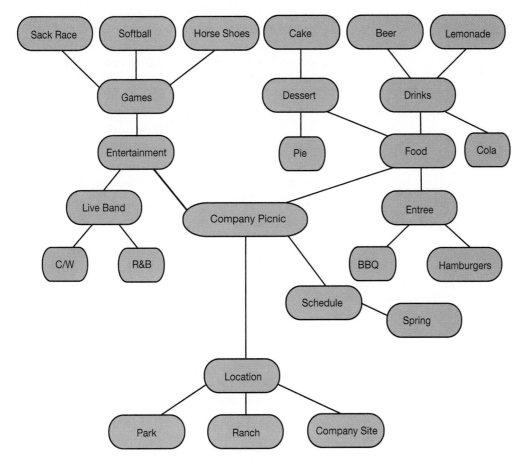

Figure 5.9   Company Picnic Mind Map, Including Subpoints

Writing

Once you have gathered your data and determined your objectives in prewriting, your next step is to draft your memo or your e-mail message. In doing so, consider the following drafting techniques.

Review Your Prewriting

Before you begin writing, look back over your clustering to see if you have missed anything important. This review also provides one final reminder of what you want to say (the content of the memo) and why you are going to say it (your objectives).

Determine Your Focus

What is the topic sentence, thesis statement, or objective of the memo or e-mail message? Prior to drafting, write down your objective for this correspondence in one or two sentences. For example, regarding the company picnic, you might want to write, "My plan for the picnic includes food, entertainment, and a schedule for activities." Such a sentence provides you with a clear focus for your draft. It helps keep you from straying off course in your memo or e-mail.

Clarify Your Audience

Again, before writing the draft, be sure that you know whom you are writing to. Is your audience high tech, low tech, or lay? Is the audience composed of management, subordinates, or lateral colleagues? Is the audience multiple, consisting of many of the above levels? Is the audience multicultural or cross-cultural? Make this determination and then write accordingly.

Review Your Criteria

Remember what a memo or e-mail entails. In addition to listing date, to, from, and subject lines, try to use the template shown earlier (introduction, body, and conclusion).

Organize Your Ideas

If your supporting details are presented randomly, your audience will be confused. As a writer, you are obligated to develop your content in a manner that will allow your readers to follow your train of thought easily. Therefore, when you draft your memo or e-mail, choose a method of organization that will help your readers understand your objectives. In Chapter 3, we discussed organizing by importance. You may want to use this method of organization.

Another method of organization is *chronology*. A chronological organization follows a time sequence, from first to last, from 8:00 A.M. to 5:00 P.M., from Monday through Friday, from first quarter to fourth quarter, from step 1 through step 27. This method of organization is useful if your memo or e-mail focuses on any of the following:

Past Histories	Future Events
Incidents or accidents	Instructions
Meeting minutes	Deadlines
Performance appraisals	Agendas
Customer complaints	Itineraries

The memo shown in Figure 5.10 is organized chronologically.

Write the Draft

Once you have decided on content (by reviewing your prewriting), determined your focus, clarified your audience, reviewed your criteria for effective memos and e-mail messages, and determined how best to organize your thoughts, all that is left is to draft the text.

Rewriting

To revise your memo or e-mail message and make it as good as it can be, follow these revision techniques:

- *Add New Detail for Clarity.* Reread your draft. If you have omitted any information stemming from the reporter's questions *(who, what, when, why, where, how)*, insert answers to these questions.
- *Delete Dead Words and Phrases for Conciseness.* Refer to Chapter 3 for help.
- *Simplify Words and Phrases.* For example, *in lieu of* might confuse someone. Why not simplify this to *instead of*? Do not "write down" to your audience. A good rule of thumb is to write as you speak, unless

MEMORANDUM

DATE: May 1, 2005
TO: Planning Committee
FROM: Steve Janasz
SUBJECT: PLANNING AGENDA FOR JULY 4 COMPANY PICNIC

Congratulations! We had a series of outstanding meetings with excellent input from all team members. I appreciate your involvement.

To confirm our decisions reached yesterday at the final planning committee meeting, here's the agenda for everyone's July 4 Company Picnic responsibilities:

Entertainment
> Mary, you'll need to focus on two topics: music and games. By *May 10,* select and book a country-western band. The team members' first choice is the band Rattlesnake, but they also would be happy with Texas River. By *May 15,* make arrangements with the local community college to lend us softball equipment (balls, bats, and gloves).

Food
> Tasha, you're in charge of food. We've already chosen the menu (hamburgers, hot dogs, potato salad, baked beans, cole slaw, chocolate cake, and vanilla ice cream). By *May 20,* select a caterer. Remember, budget is important!

Location
> Galen, because the committee members voted on softball for our company entertainment, you'll need to reserve a ball field. By *May 25,* contact the city Parks and Recreation Department and reserve field #2 at Pease Park. This field is the only one with covered eating facilities nearby.

Invitations
> Jordan, you'll need to write and mail invitations by *June 10.* Use our desktop publishing software to produce the mailers. Be sure to include where we'll hold the picnic, when to arrive, when the festivities plan to end, costs per family, and the agenda for events.

Thank you for all your support on this project. Together we'll make this team effort a success. July 4 will be a company festival guaranteed to raise corporate morale.

Figure 5.10   Directive Memo Organized Chronologically

you speak poorly. Be more casual. Chapters 3 and 4 will help you achieve these goals.

- *Move Information from Top to Bottom or Bottom to Top for Emphasis.* Place your most important information first, or move information around to maintain an effective chronological order.

- *Reformat for Access.* Use highlighting techniques (white space, bullets, numbers, headings, etc.) for reader-friendly appeal. (See Chapter 8 for information on document design.)

- *Enhance the Tone and Style of Your Memo or E-mail.* If you want to sound tough, then revise your memo or e-mail to sound that way. If you want to tone down your correspondence, now is the time to do so. If you want to personalize the document, then add pronouns or names to give it a more person-to-person tone. Chapter 4 discusses this important aspect of writing.

- *Correct for Accuracy.* Proofread!

- *Avoid Biased Language.* All readers need to feel included equally.

Process Log

The following process log illustrates a student's prewriting, writing, and rewriting of a memo. In this log, you can see how the writing process works. To prewrite, the student gathers information for the memo. In writing, she drafts her text. Finally, with the help of peer evaluations, she revises her memo in the rewriting stage.

Prewriting

Figure 5.11 shows one student's prewriting, written in response to a request that all students write a memo telling how to write a memo.

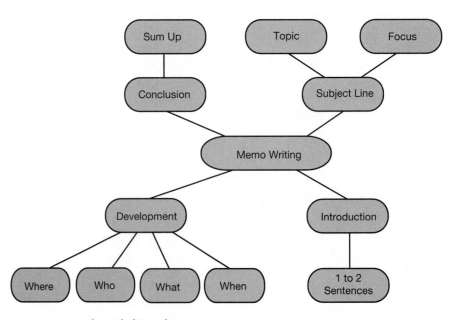

Figure 5.11   Student Mind Mapping

Writing

Following is the student's rough draft (Figure 5.12).

Subject: Memo Writting

Following are ways to write better memos.

Include a Subject Line, which provides a topic and a focus. Then have an Introduction to tell your readers what he needs to know. This Introduction could be about one or two sentences. Next, develope your ideas in a Discussion. Finally, sum it up in a Conclusion.

These suggestions should help you write better memos.

Figure 5.12   Rought Draft

Rewriting

Student peer evaluators suggested the revisions in Figure 5.13. The finished copy is shown in Figure 5.14.

Add Date, To, From

Subject: Memo Writing
SP

(No Focus - what about it?)
and needs to be typed
in all caps
^

Following are ways to write better memos.

Add why!
What's their motivation?

Reformat

Include a Subject Line, which provides a topic and a focus.

agreement?

Reformat

Then have an Introduction to tell your readers what he

needs to know. This Introduction could be about one or

Reformat

SP

two sentences. Next, develop your ideas in a Discussion.

Finally, sum it up in a Conclusion.

Reformat the text—make
it more accessible through
highlighting

Your entire
memo needs
more development

These suggestions should help you write better memos.

Poor tone—motivate & tell them
when (dated action)

Figure 5.13   Suggested Revisions

DATE: October 19, 2005
TO: George Calvert
FROM: Candice Millard
SUBJECT: SUGGESTIONS FOR IMPROVED MEMO WRITING ◀—— Subject line includes topic and focus

George, I've just returned from a training seminar titled "Better Business Writing." Our department manager asked me to share tips I've learned about writing effective memos. ◀—— Introduction explaining why the memo is written
Following are the improved memo-writing techniques I learned at this training seminar.

1. *Subject Line*   One hundred percent of readers read the subject line. Thus, to make this first line of communication effective, include a topic ("memo writing") and a focus ("suggestions for").   ◀—— Effective use of numbered points and headings for access

2. *Introduction*   Limit your introduction to one or two sentences. State why you're writing and what you're writing about.

3. *Discussion*   In the body, develop your points specifically (answering *who, what, where,* and *why*). Then itemize your points for easy access.

4. *Conclusion*   End by telling the reader what to do next and *when* to accomplish this task.

By using these techniques, you'll be able to communicate more successfully. If you have any questions, please drop by my office. I'd be happy to share more information with you. ◀—— Positive words and pronouns successfully used to achieve a personalized tone

Figure 5.14   Revised Memo

CHAPTER HIGHLIGHTS

1. Memos and e-mail messages are an important part of your interpersonal communication on the job.
2. Memos and e-mail messages differ in many ways, such as destination, format, audience, tone, delivery time, and security.
3. Subject lines are read by 100 percent of your audience in memos and e-mail.
4. Use a topic and focus in your subject line.
5. In the introduction, state what you want and why you are writing.
6. In the discussion section, state the details.

7. Conclude by telling the reader what you plan to do next or what you expect him or her to do next. You might also want to date this action.

8. Consider the level of your audience when you write a memo or e-mail.

9. Wizards allow you to customize your memos in a variety of ways.

10. E-mail messages represent 24 percent of the approximately 206 messages you will send and receive daily.

CASE STUDIES

Apex Corporation

Apex Corporation is experiencing repeated failures with its Maxwell air compressors. The uncoated radiators tend to corrode, which causes pinhole leaks that reduce air pressure and volume. Because the radiators were built before current energy-saving devices were mandated, the Maxwells use excess electricity as well. This forces compressors to make up for lost air. Finally, the Maxwells, which are now 10 years old, have a high failure rate (24 percent downtime). This leads to increased maintenance costs and a lack of dependability.

Cheryl Huff, the department supervisor for maintenance engineering, needs to write a memo to the plant supervisor, Mark Levin, suggesting a solution. Her options are to retrofit the current Maxwells with modern energy-saving devices, to coat the Maxwells to prevent additional corrosion, or to purchase new replacement radiators. The cost for replacements would be approximately $15,000. Although that is high, the cost for continued maintenance and downtime is also high. Cheryl received this $15,000 bid contingent upon ordering before a November 15 deadline. After that date, the costs will be $16,750 because of a year-end price increase. Retrofitting and coating the radiators are stop-gap solutions that can be undertaken by in-house maintenance crews. Whichever choice Cheryl promotes must happen soon. Winter is fast approaching, and the air demand will increase.

Make the decision for Cheryl, based on the information provided (you can invent any additional information you choose). Then write the memo or e-mail. Abide by the criteria provided in this chapter.

Barney Allis Stores

Barney Allis Stores, home based in Seattle, Washington, has department stores in all 50 states and in the following international cities: Toronto, Vancouver, London, Paris, Berlin, Barcelona, Rome, Beijing, Hong Kong, Sidney, Mexico City, and Rio de Janeiro.

'Tis the Season to Be Jolly! It's time for BAS's marketing department to inform all BAS stores of what's being "pushed" for the Christmas season sales. This year, BAS has created a new product line, called "BASK"—Barney Allis Stores-Kids. The focus will be on winter and spring children's wear, for ages Pre–K through fifth grade. Specifically, marketing needs to highlight three item lines:

- Corduroy pants.
- Reversible jackets.
- Long-sleeve T-shirts. These will be customized to each country or locale with slogans (in the country's language) or cartoon characters decorating the shirtfronts. The cartoon characters will also be customized, stemming from each country or region's Saturday morning cartoon television shows.

Assignment

As BAS director of marketing, your job is to write an e-mail message to all state-side and international marketing managers. The e-mail will not only provide the above information but also act as a cover correspondence, transmitting an attached product catalog. Plus, the e-mail must direct the marketing managers to create print, radio, and television advertisements for their local markets. To help them, direct them to the BAS intranet site *http://www.bas.com/BASK/salestalk.html*. They should already know their passwords.

In writing this e-mail, abide by the criteria provided in this chapter.

EXERCISES

Writing Memos or E-mail Messages

1. Write a memo or e-mail on one of the following topics. To do so, first prewrite (using mind mapping or clustering), then write a draft, and finally rewrite (revising your draft).
   a. Your work environment is experiencing a problem (with scheduling, layoffs, turnover, production, quality, morale, etc.). Your boss has asked you to write a memo or e-mail noting the problem and suggesting solutions.
   b. An employee under your supervision sees a problem in his or her work environment and has written a memo suggesting a solution. You do not believe the suggested solution will work. Write a memo or e-mail providing your response.
   c. A major project is being introduced at work. Write a directive memo or e-mail informing your work team of their individual responsibilities and schedules.
   d. Your department needs a new piece of equipment to perform work. Write a memo or e-mail requesting this equipment. Justify the need for the equipment and give the date when the equipment is needed.
   e. You work in the purchasing department and must buy a new piece of equipment. You must first compare bids. You've done so and now must write a comparison/contrast memo or e-mail explaining why you plan to purchase one piece of equipment versus another.
   f. It's time for your quality circle team to meet again (or any committee you chair). Write a memo or e-mail calling the meeting. Provide an agenda.

2. Practice mind mapping or clustering. Choose any of the topics presented in activity 1 and create a mind map/cluster.

Team Projects

1. Choose one of the topics listed in activity 1. In groups of three to five students, write the correspondence requested. Once the group documents are written, evaluate them. You can do this as a class by projecting transparencies, or you could exchange group work and evaluate them in peer review groups. Decide which documents are successful and explain why. Decide which memos or e-mail messages are flawed and explain why.

2. Gather memos or e-mail messages from your work site (or the work site of your friends or relatives). In small groups of three to five, discuss whether these documents are successful or unsuccessful. To make your judgments, use the criteria discussed in this chapter. Explain to the class why the good documents succeed and why the flawed correspondence fails. Then revise the flawed memos and e-mail messages to improve them.

3. In groups of three to five students, revise the following flawed memo. Consider the criteria for good memos and rewriting techniques presented in this chapter. You can invent any information you consider necessary to improve the memo.

DATE: April 6, 2005
TO: Jan Pascal
FROM: Wilkes Berry
SUBJECT: Engineering

To solve the recent problems with sewer design, the engineering shift will perform some studies on corrosion, water seepage, pipe decay, alignment, depth, soil composite, and so forth. We'll need to do this soon. If you have any questions, let me know.

4. Most city, county, and state governments have written guidelines or policies for their employees' e-mail. For example, see "Electronic and Voice Mail: Connecticut's Management And Retention Guide For State And Municipal Government Agencies" *(http://www.cslib.org/email.htm)*. In addition, see the e-mail policy provided by Hennepin County, Minnesota (*http://www.co. hennepin.mn.us/wemail.html)*.

   Research city, county, and state e-mail policies. Determine what they say, what they have in common, and how they differ from municipality to municipality.

   Find out if your city, county, or state has a policy regarding e-mail usage.

5. Most colleges and universities have written guidelines or policies for their employees' or students' e-mail use. For example, see what Florida Gulf Coast University says about e-mail use *(http://admin.fgcu.edu/compservices/ policies3.htm)*. Also look at *http://www.dbs.umd.edu/records/emailpolicies. html,* which lists the e-mail policies at many universities.

   Research college and university e-mail policies to determine what they say, what they have in common, and how they differ from school to school. Find out if your college or university has a policy regarding e-mail usage.

6. Write either an e-mail or memo to accomplish any of the following purposes (you can pick any topic you would like to complete the assignment). Whatever topic you write about, and regardless of the purpose, your correspondence should include an introduction (telling why you are writing and what you are writing about); a body (itemizing the key points); and a conclusion asking for or suggesting a follow-up.
   - *Directive*—inform subordinates of their designated tasks.
   - *Cover/transmittal*—tell a reader that you have attached a document.
   - *Documentation*—report on expenses, incidents, accidents, problems encountered, projected costs, study findings, hirings, firings, reallocations of staff or equipment.
   - *Confirmation*—tell a reader about a meeting agenda, date, time, and location; decisions to purchase or sell; topics for discussion at upcoming teleconferences; conclusions arrived at; and fees, costs, or expenditures.
   - *Procedures*—explain how to set up accounts, research on the company intranet, operate new machinery, use new software, apply online for job op-

portunities through the company intranet, create a new company Web site, or solve a problem.

- *Recommendations*—provide reasons to purchase new equipment, fire or hire personnel, contract with new providers, merge with other companies, revise current practices, and renew contracts.
- *Feasibility*—study the possibility of changes in the workplace (practices, procedures, locations, staffing, equipment, missions or visions).
- *Status*—provide a daily, weekly, monthly, quarterly, biannual, or yearly progress report about sales, staffing, travel, practices, procedures, and finances.
- *Inquiry*—ask questions about upcoming processes, procedures, or assignments.

7. For some situations you need to write a memo, whereas for others you need to write an e-mail message. Sometimes, you are better off writing neither of these two types of communication, choosing instead to have a face-to-face meeting. Perhaps an in-person meeting also will be followed by correspondence documenting what was discussed. Analyze the following scenarios and decide whether a memo, an e-mail, or a person-to-person meeting might be best. Then write an e-mail to your instructor to explain your answers.

- An employee (John) attended a work-related conference in another city. He received company financial support for this job-related travel. Another work colleague (Allan), who also attended the conference, claimed that John had not attended all of the conference sessions, using his time to visit the city's tourist attractions instead. John's boss, Ellen, had to address this accusation.

  How should she handle the situation? Should she write a memo to John, an e-mail to John, or visit with John personally? Write an e-mail or memo to your instructor explaining your answer. Or, make a brief oral presentation. Be sure to give reasons for and against each option.

- You are the manager of a Human Resources Department. You are planning a quarterly meeting with your staff (training facilitators, benefits employees, personnel directors, and company counselors). The staff works in 3 different cities and 12 different offices. You need to accomplish three goals: get their input regarding agenda items, invite them to the meeting, and provide the final agenda.

  How should you communicate to them? Should you write a memo, an e-mail, or should you telephone them? Write an e-mail or memo to your instructor explaining your answer. Or, make a brief oral presentation. Be sure to give reasons for and against each option.

- As the Manufacturing manager, you have just met with an employee (Sarah) regarding her year-end evaluation. You discussed her efficiency rating, cost-effective goals, relationship with coworkers, and continuing education options. The meeting went fairly well. She and you agreed on ways to achieve more effective efficiency and cost effectiveness. Sarah is looking forward to attending more training classes to meet her continuing education requirement. However, she and you disagreed about her relationship with coworkers. You reported to her about a few complaints you had received from colleagues, but Sarah felt the comments were inaccurate. This is an important meeting because you will use it to determine her annual raise.

  It is time to document the meeting. How should you share your comments with Sarah—a memo or an e-mail? Write an e-mail or memo to

your instructor explaining your answer. Or, make a brief oral presentation. Be sure to give reasons for and against each option.

- You are an independent consultant, providing a unique service to companies (assessment of their international business opportunities). You just met with a team of eight managers within a company to discuss ways you could help them. Due to your expertise, they hired you on the spot. When you got back to your office, you thought of several more ideas to benefit their company. Now you need to share these new visions.

  How should you communicate to the eight managers? Should you write a memo, an e-mail, or telephone them? Write an e-mail or memo to your instructor explaining your answer. Or, make a brief oral presentation. Be sure to give reasons for and against each option.

WEB WORKSHOP

1. Access an Internet search engine (Google, Yahoo, Excite, Lycos, etc.) and type in a phrase like "government e-mail policies." You will find local, state, and national governmental sites specifying their e-mail policies. In addition, you will find private companies marketing their e-mail management and monitoring systems ("e-policy" hardware/software and consulting), as well as sites denying the validity of such e-mail management.

   Review a sampling of these sites and either make an oral presentation on your findings or write a memo documenting the diverse aspects of governmental e-mail policies.

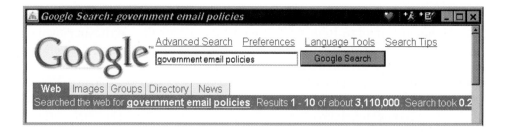

2. Access an Internet search engine (Google, Yahoo, Excite, Lycos, etc.) and type in a phrase like "corporate e-mail policies." You will find the e-mail policies of individual corporations. In addition, you also will find the reasons why corporations have created e-policies (including lawsuits resulting from inappropriate e-mail communications as well as corporate e-mail systems crashing due to inappropriate employee usage). Furthermore, you can find online articles regarding statutes used by companies to legitimize their policies.

   Review a sampling of these sites and either make an oral presentation on your findings or write a memo documenting the diverse aspects of corporate e-mail policies.

3. Many universities and colleges provide faculty, staff, and student e-mail opportunities. Thus, the schools usually require adherence to standards and provide specific guidelines.

   Access an Internet search engine (Google, Yahoo, Excite, Lycos, etc.) and type in a phrase like "university e-mail policies" or "college e-mail policies." Does your school have an e-mail policy for faculty, staff, and students? If so, what is it?

Review a sampling of these sites and either make an oral presentation on your findings or write a memo documenting the diverse aspects of e-mail policies provided by colleges and universities.

Quiz Questions

1. What are five distinctions between memos and e-mail messages?

2. Why is a subject line in a memo or e-mail important?

3. What are the components of a subject line?

4. What do you hope to accomplish in the introduction to a memo or e-mail?

5. Why should you be concerned with document design in a memo or e-mail?

6. What are three highlighting techniques you can use in a memo or e-mail?

7. What should you include in a conclusion to a memo or e-mail?

8. Who is the audience of a memo or e-mail?

9. What are the characteristic elements of writing style you should strive for in a memo or e-mail?

10. How can clustering or mind mapping be valuable to you as a form of prewriting a memo or e-mail?

6 Letters

Chapter Preview

Letter Components
A letter should include eight essential components: the writer's address, the date, an inside address for the recipient, a salutation, the body of the letter, a complimentary close, and the writer's signed and typed names.

Letter Wizards and Templates
You can use software templates to help design a letter.

Letter Formats
Standard letter formats include full block, full block with subject line, modified block, and simplified format.

Criteria for Different Types of Letters
Follow our criteria for writing different types of letters: letters of inquiry, cover letters, good-news letters, bad-news letters, complaint letters, adjustment letters, and sales letters.

Process
Following a letter of inquiry through the writing process will help you write your letters.

Writing at Work

CompuRam **CompuRam,** a wholesale provider of computer hardware, is home based in Reno, Nevada. CompuRam manufactures and sells monitors, printers, CPUs, cables, and other computer accessories. The company is expanding into a new market—biotechnology equipment—under the business operating name **CompuMed**. In this capacity, CompuMed plans to capitalize on emerging nanotechnology to manufacture and sell the following:

- Extremely lightweight and portable heart monitors and ventilators.
- Pacemakers and hearing aids, $\frac{1}{10}$ the size of current products on the market.
- Microscopic biorobotics that can be injected in the body to manage, monitor, and destroy blood clots, metastatic activities, arterial blockages, alveoli damage due to carcinogens or pollutants, and scar tissues creating muscular or skeletal immobility. CompuMed also is researching nanotechnology biorobotics that will be programmed to "eat" away targeted fat cells, replacing the need for invasive surgeries such as liposuction. A promising nanotechnology use of microscopic robotics could be to "knit" or fuse ruptured blood vessels, at the capillary level, again without the need for surgery.

CompuRam/CompuMed is a growing company with over 5,000 employees located in two dozen cities and three states. To manage this business, supervisors and employees must write on average over 20 letters a day. These letters are written to many different audiences and serve various purposes. CompuRam must write letters for employee files, to customers, to job applicants, to outside auditors, to governmental agencies involved in biotechnology regulation, to insurance companies, and more. Specifically, the company writes the following:

- *Sales letters* to computer and biotechnology retailers.
- *Responses to letters of inquiry* from retailers seeking product information (technical specifications, pricing, warranties, guarantees, credentials of service staff, etc.).
- *Good-news and bad-news responses* to letters from clients documenting problems encountered with CompuRam's product line.
- *Good-news and bad-news letters* to employees and potential employees. For employees, these letters document promotions, demotions, and lay-offs. For potential employees, these letters are written either to offer employment or to reject job applicants.
- *Cover letters* prefacing CompuRam's proposals.
- *Complaint letters* written to parts manufacturers if and when faulty equipment and materials are received in shipping.
- *Adjustment letters* to compensate retailers when problems occur.

Letters sent to agencies, auditors, and regulators must explain practices and procedures. CompuRam's letters inform, instruct, persuade, and build trust.

Letters are external correspondence that you send from your company to a colleague working at another company, to a vendor, to a prospective client, to an agency, or to a friend who lives around the corner or across the continent. Letters leave your work site (as opposed to memos, which stay within the company).

Because letters are sent to readers in other locations, you must write them effectively. Your letters not only reflect your communication abilities but also are a reflection of your company. If your letter is sloppy, is marred with grammatical errors, or creates a negative tone, you will look bad and so will your company. Even worse, if your letter communicates incorrect information regarding prices, guarantees, due-date promises, or equipment and labor commitments, your company will be held legally responsible.

When you write on your company's letterhead stationery and sign it, your letter constitutes a legally binding contract. Therefore, no matter what type of letter you are writing, take care to follow accepted letter formats, maintain the proper tone, and avoid errors.

This chapter helps you write the following kinds of letters:

- inquiry
- cover (transmittal)
- good news
- bad news
- complaint
- adjustment
- sales

So that you can write these letters effectively, we provide letter components and formats, examples, criteria for writing different types of letters, and a process log to follow.

LETTER COMPONENTS

Essential Components

Your letter should be typed or printed on $8\frac{1}{2} \times 11$-inch unlined paper. Leave 1 to $1\frac{1}{2}$-inch margins at the top and on both sides.

Your letter should contain the following components (see Figure 6.1).

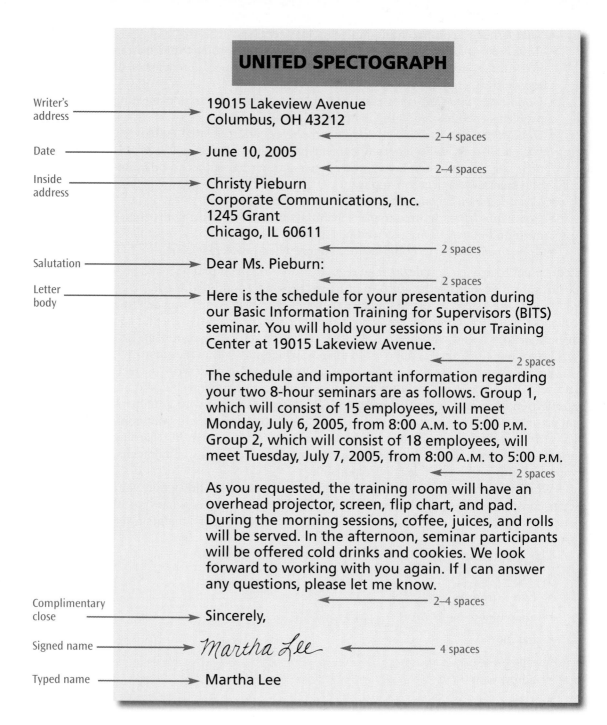

Figure 6.1   Essential Letter Components

Writer's Address

This section contains either your personal address or your company's address (you may be writing on printed, letterhead stationery).

If the heading consists of your address, you will include (a) your street address (do not abbreviate the words *street, avenue, road, drive,* etc.) and (b) the city, state, and zip code. Do not include your name. The state may be abbreviated with the appropriate two-letter abbreviation (see Chapter 19).

If the heading consists of your company's address, you will include (a) the company's name; (b) the street address; and (c) the city, state, and zip code.

Date

Document the month, day, and year when you write your letter. You can set up your date in one of two ways: May 31, 2005, or 31 May 2005.

Place the date one or two spaces below the writer's address (if the letter is long) or three to eight spaces below the writer's address (for shorter letters). For a short letter, leave more space so the spacing on the letter is balanced.

Inside Address

This is the address of the person or people to whom you are writing. The inside address contains these elements in the following order.

- Your reader's name (If you do not know the name of this person, begin the reader's address with a job title or the name of the department.)
- Your reader's title (optional if you include the name)
- The company name
- The company street address
- The company city, state, and zip code

Place this information two lines below the date.

Salutation

The traditional salutation, placed two spaces beneath the inside address, is your reader's last name, preceded by *Dear* and followed by a colon.

> **example**
>
> Dear Mr. Smith:

You can also address your reader by his or her first name if you are on a first-name basis with this person.

> **example**
>
> Dear John:

If you are writing to a woman and are unfamiliar with her marital status, address the letter as follows:

> **example**
>
> Dear Ms. Jones:

However, if you know the woman's marital status, you can address the letter accordingly.

> **example**
>
> Dear Miss Jones:
>   or
> Dear Mrs. Jones:

Occasionally, you will write to someone whom you do not know and who has a non-gender-specific first name, such as Pat, Kim, Kelly, Stacy, or Chris. In this case, avoid either *Mr., Miss, Mrs.,* or *Ms.,* and write as follows:

example

> Dear Chris Evans:

Often, when writing to another company, you will not know your reader's name. In this instance, you have limited options. You could address the reader by his or her title *(Dear Vice President of Operations:)* or address the letter to the reader's department *(Accounting Department:)*. If you address the letter to the department, *Dear* would be inappropriate and should be omitted.

Each of these salutations is acceptable. However, several formerly common salutations are no longer useful in business communications. For example, as discussed in Chapter 4, sexist salutations such as *Dear Sir* and *Gentlemen* cannot be used (unless you know beyond all doubt that each of your readers is a man; however, we suggest that you eliminate the problem by avoiding the sexist salutation).

Other awkward salutations include *Good morning* (because mail is delivered twice daily at many corporations), *To Whom It May Concern* (which is trite and imprecise), and *Greetings* (which either has a seasonal ring to it or a military connotation). *Dear Sir/Madam* is also ineffective and should be avoided.

To avoid these pitfalls in salutations, you might want to call the company to which you are writing and ask for the name of your intended reader (and the correct spelling of the name).

tech link

Go to *http://www.prenhall. com/gerson* for Web links, samples, and interactive activities.

Letter Body

Begin the body of the letter two spaces below the salutation. The body includes your introductory paragraph, discussion paragraph(s), and a concluding paragraph. The body should be single-spaced with double-spacing between paragraphs. Whether you indent the beginning of each paragraph or leave them flush with the left margin is determined by the letter format you employ. We discuss this issue later in this chapter.

Complimentary Close

Place the complimentary close, followed by a comma, two spaces below the concluding paragraph. Although several different complimentary closes are acceptable, such as *Yours truly* and *Sincerely yours,* we suggest that you limit your close to *Sincerely*.

Signed Name

Sign your name legibly beneath the complimentary close.

Typed Name

Type your name four spaces below the complimentary close. If you wish, you may type your title one space beneath your typed name.

Sincerely,

Henry Marshall

Henry Marshall
Comptroller

Optional Letter Components

In addition to the letter essentials, you can include the following optional components.

Subject Line

A subject line, which is mandatory in memos (see Chapter 5), is also applicable in letters. You can type the subject line in all capital letters two spaces below the inside address and two spaces above the salutation.

Dr. Ron Schaefer
Linguistics Department
Southern Illinois University
Edwardsville, IL 66205

Subject: LINGUISTICS CONFERENCE REGISTRATION PAYMENT

Dear Dr. Schaefer:

You also could use a subject line instead of a salutation.

Linguistics Department
Southern Illinois University
Edwardsville, IL 66205

Subject: LINGUISTICS CONFERENCE REGISTRATION PAYMENT

A subject line not only helps readers understand the letter's intent but also (if you are uncertain of your reader's name) helps you avoid such flawed salutations as *To Whom It May Concern, Dear Sirs,* and *Ladies and Gentlemen.*

In the simplified letter, both the salutation and the complimentary close are omitted, and a subject line is included.

New-Page Notations

If your letter is longer than one page, you will need to cite your name, the page number, and the date on all pages after page 1. Place this notation either flush with the left margin at the top of subsequent pages or across the top of subsequent pages. (You must have at least two lines of text on the next page to justify another page.)

example

Left, Margin, Subsequent Page Notation

Mabel Tinjaca
Page 2
May 31, 2005

example

Across Top of Subsequent Pages

Mabel Tinjaca   2   May 31, 2005

Writer's and Typist's Initials

If the letter has been typed by someone other than the writer, include both the writer's and the typist's initials two spaces below the typed signature. The writer's initials are capitalized, the typist's initials are typed in lowercase, and the two sets of initials are separated by a colon. If the typist and the writer are the same person, this notation is not necessary.

example

Sincerely,

W. T. Winnery
W. T. Winnery

WTW:mm

Enclosure Notation

If your letter prefaces enclosed information, such as an invoice, a report, or graphics, mention this enclosure in the letter and then type an enclosure notation two spaces below the typed signature (or two spaces below the writer and typist initials). The enclosure notation can be abbreviated *Enc.*; written out—*Enclosure*; show the number of enclosures, such as *Enclosures (2)*; or specify what has been enclosed—*Enclosure: January Invoice*.

Copy Notation

If you have made a complimentary copy or a photocopy of your letter, show this in a copy notation. A complimentary copy is designated by a lowercase *cc*: whereas a photocopy is designated by a lowercase *pc*. Type the copy notation two spaces below the typed signature or two spaces below either the writer's and typist's initials or the enclosure notation.

Sincerely,

Brian Altman

Brian Altman

Enclosure: August Status Report

pc

If you are sending copies of the letter to other readers, list these readers' names following the copy notation.

Sincerely,

Brian Altman

Brian Altman

Enclosure: August Status Report

pc: Marcia Rittmaster
Erica Nochlin

LETTER WIZARDS AND TEMPLATES

Most word processing packages provide you with templates for letters. In Microsoft Word, for example, you click on "File" and "New" to get templates for what they designate as "Contemporary Letters," "Elegant Letters," and "Professional Letters." Each of these templates gives you an already-designed letter format, complete with spacing, font selection, and layout. In addition, these templates provide fields in which you merely need to type the appropriate information (address, company name, date, salutation, complimentary close, your name and title, etc.). Then, the template tells you to "Type your letter here" (see Figure 6.2 on page 180).

Microsoft Word also provides a letter wizard, which differs from the templates. The wizard walks you through a four-step sequence, prompting you to fill in a number of blank fields.

- Step 1, "Letter Format," lets you pick the type of letter format you want (contemporary, elegant, or professional) and whether you want this letter to be printed in full block, modified block, or semiblock. Once you have made these choices, you click on "Next," which takes you to the next step.

- Step 2, titled "Recipient Info," provides fields that let you type in the recipient's name, delivery address, and a choice of salutations: informal ("Dear Joe,"), formal ("Dear Mr. Jones,"), or business ("Dear Mr. Jones:"). Again, you click on "Next" to move on.

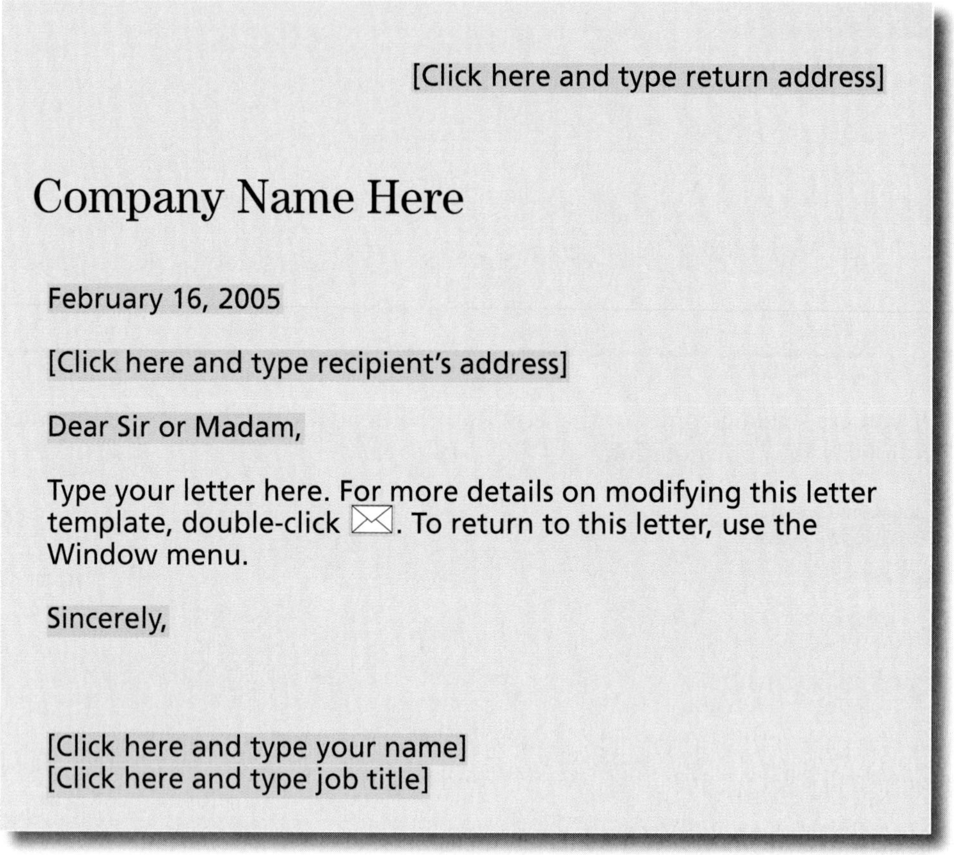

[Click here and type return address]

Company Name Here

February 16, 2005

[Click here and type recipient's address]

Dear Sir or Madam,

Type your letter here. For more details on modifying this letter template, double-click ✉. To return to this letter, use the Window menu.

Sincerely,

[Click here and type your name]
[Click here and type job title]

Figure 6.2   Microsoft Word's Template for "Contemporary Letters"
Screen shot reprinted by permission from Microsoft Corporation.

- Step 3, "Other Elements," directs you to add a reference line, mailing instructions (personal, registered, etc.), an attention line, a subject line, and courtesy copies. Finally, after clicking on "Next," you arrive at your final destination.
- Step 4, "Sender Info," gives you the options to add your sender's name, return address, and select from 13 complimentary closes, including "take care," "best wishes," and "love." Once you click on "Finish," your template is provided with whatever information you have included in the appropriate fields.

These templates and wizards are both good and bad. They remind you of which components can be included in a letter, they make it easy for you to include these components, and they let you choose ready-made formats. However, the templates also can create some problems. First, they are somewhat limiting in that they dictate where you will put information (such as the return address). The placement of this information might contradict your teacher's or boss's requirements. Second, the templates are prescriptive, limiting your choice of font sizes and types. Third, some of the information these templates offer as valid options, such as the salutation *Dear Sir or Madam,* is old-fashioned. In contrast, we would never suggest that you use this salutation. Finally, the templates are too inclusive, suggesting that you should insert all of the information they provide (such as mailing instructions, courtesy copies, and attention lines). Our advice would be to use these templates and wizards with caution.

LETTER FORMATS

Type your letter using any of the formats shown in Figures 6.3 through 6.6.

Full Block Format

In the *full block format,* (Figures 6.3 and 6.4), you type all information at the left margin—without indenting paragraphs, the date, the complimentary close, or the signatures. Figure 6.4 differs from Figure 6.3 only in the inclusion of a subject line.

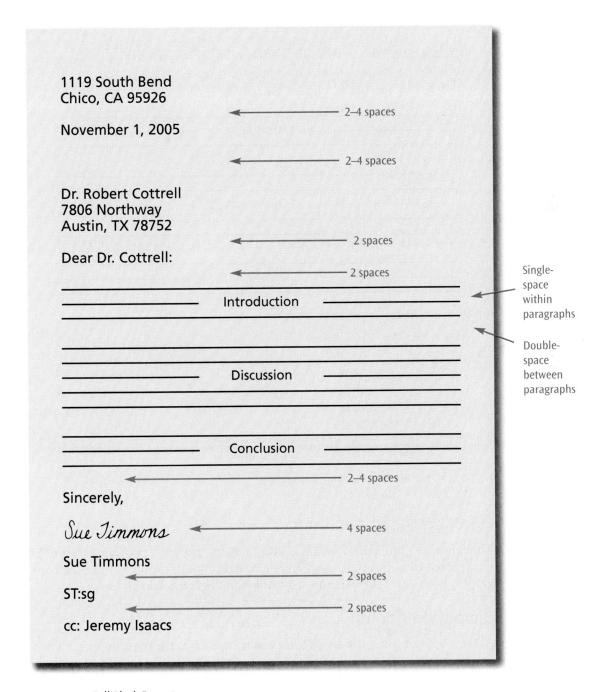

Figure 6.3 Full Block Format

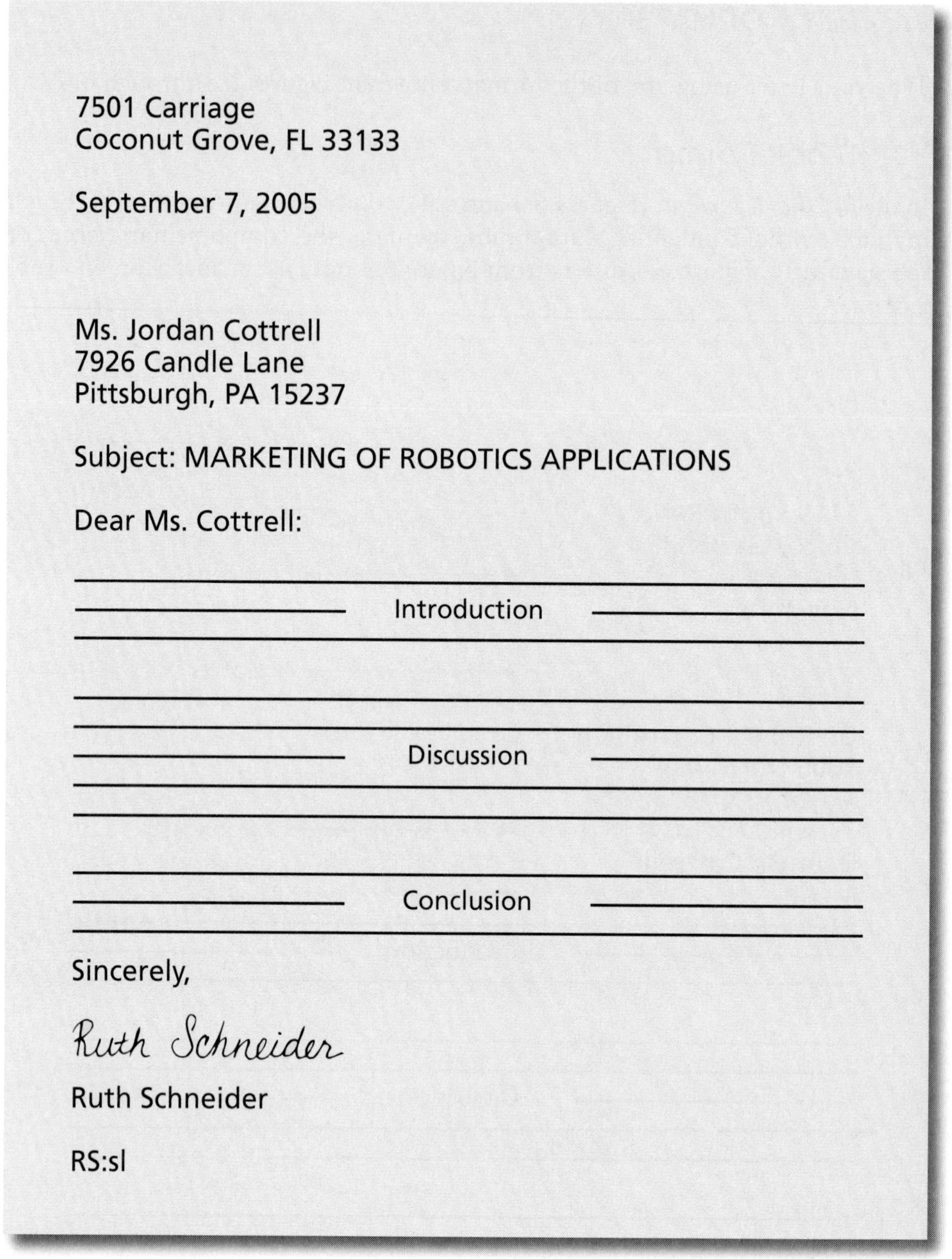

7501 Carriage
Coconut Grove, FL 33133

September 7, 2005

Ms. Jordan Cottrell
7926 Candle Lane
Pittsburgh, PA 15237

Subject: MARKETING OF ROBOTICS APPLICATIONS

Dear Ms. Cottrell:

——————————— Introduction ———————————

——————————— Discussion ———————————

——————————— Conclusion ———————————

Sincerely,

Ruth Schneider

Ruth Schneider

RS:sl

Figure 6.4   Full Block Format with Subject Line

Modified Block Format

Though once common in business and industry, the *modified block format* is now perceived as outdated (see Figure 6.5). Its primary feature is indented paragraphs, which give the letter an old-fashioned, less professional look.

Simplified Format

A letter written using the *simplified format* is perhaps the most unique, controversial, but useful format (see Figure 6.6). This type of format is similar to the full block letter in that all text is typed margin left. However, note the three significant changes: no salutation (*Dear _____:*), an added subject line, and no complimentary close (*Sincerely,*).

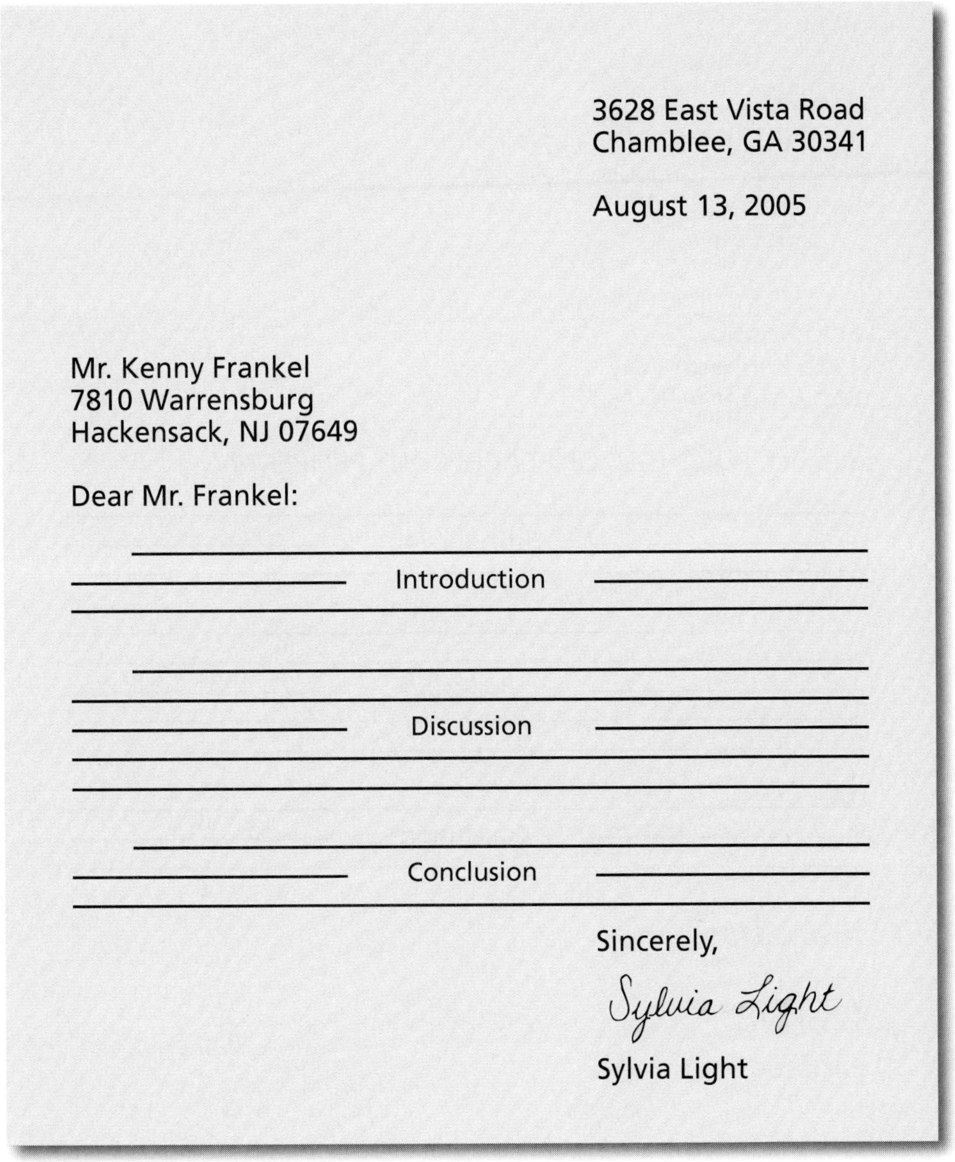

3628 East Vista Road
Chamblee, GA 30341

August 13, 2005

Mr. Kenny Frankel
7810 Warrensburg
Hackensack, NJ 07649

Dear Mr. Frankel:

—————————————— Introduction ——————————————
——————————————

——————————————
—————————————— Discussion ——————————————
——————————————

—————————————— Conclusion ——————————————

Sincerely,

Sylvia Light

Sylvia Light

Figure 6.5 Modified Block Format

Dropping the salutation is a valuable option for two reasons:

- **Cliché salutations**—If you do not know your reader's name, you might be tempted to start your letter with the trite and ineffective salutation, *To whom it may concern.*

- **Non-gender-specific names**—You know the reader's name, but it is nongender specific. For example, names like Jesse, Jordan, Morgan, Alex, Tracy, Chris, Stacy, Pat, Alex, and Dakota can be either men's or women's names. Then what do you do? You will be uncertain as to whether the letter's salutation should read *Dear Mr.* or *Dear Mrs.*

The Administrative Management Society (AMS) suggests that you replace the salutation with a subject line. This is an excellent idea. As we related in Chapter 5, subject lines aid clarity. Whereas a salutation only repeats what has been provided in the letter's inside address, the subject line gives your reader a clear overview of what the letter will discuss.

935 West Hermosa
Oakland, CA 94610

September 5, 2005

Edie Kreisler
1126 Ranleigh Way
San Antonio, TX 78213

Subject: PURCHASE OF BEACHFRONT PROPERTY

————————————————————————————————
—————————————————— Introduction ——————————————————
————————————————————————————————

————————————————————————————————
—————————————————— Discussion ——————————————————
————————————————————————————————
————————————————————————————————

————————————————————————————————
—————————————————— Conclusion ——————————————————
————————————————————————————————

Walt McDonald

Walt McDonald

Enclosures (3)

Figure 6.6   Simplified Format

Note:
The simplified format makes a letter look more like a memo or e-mail. Remember, memos and e-mail messages do not begin with *Dear* nor do they end with *Sincerely*.

Finally, the AMS suggests that when you omit the salutation, you also should omit the complimentary close *Sincerely*. This is a controversial issue. We have heard people say that deleting the salutation and the complimentary close makes the letter cold and unfriendly. However, the AMS claims that if your letter is warm and friendly, these omissions will not be missed. More importantly, if your letter is negative, beginning with *Dear* and ending with *Sincerely* will not improve the letter's tone.

CRITERIA FOR DIFFERENT TYPES OF LETTERS

Letters of Inquiry

If you want information about degree requirements, equipment costs, performance records, turnaround time, employee credentials, or any other matter of interest to you, you write a letter requesting that data. Letters of inquiry require that you be specific and precise. For example, if you write, "Please send me any information you have on the XYZ transformer," you're in trouble. You will either receive any information the reader chooses to dispose of or none at all. Look at the following letter of inquiry.

> Dear Mr. Jernigan:
>
> Please send us information about the following filter pools:
>
> 1. East Lime Pool
> 2. West Sulphate Pool
> 3. East Aggregate Pool
>
> Thank you.

The disgruntled worker replied as follows:

> Dear Mr. Scholl:
>
> I would be happy to provide you with any information you would like. However, you need to tell me *what* information you require about the pools.
>
> I look forward to your response.

To receive the information you need, write an effective letter of inquiry. Successful letters of inquiry contain all the letter essentials, maintain an effective technical writing style, achieve audience involvement through pronoun usage, and avoid grammatical and mechanical errors. In addition, you must accomplish the following tasks.

Introduction

Clarify your intent in the introduction. Until you tell your readers why you are writing, they do not know. It's your responsibility to clarify your intent and explain your rationale for writing. Also tell your reader immediately what you are writing about (the subject matter of your inquiry). You can state your intent and subject matter in one to three sentences, as follows:

> **example**
>
> My company is planning to purchase 30 new tractors by the end of the year. Your 2005 tractor with the onboard computer might be the answer to our problems. In addition to sending us a picture brochure of the CP95 tractor, could you also answer the following questions about your vehicle?

In this example, the first two sentences explain why you are writing. The last sentence explains what you want.

Discussion

Specify your needs in the discussion. To ensure that you get the response you want and need, ask precise questions or list specific topics of inquiry. You must quantify. For example, rather than vaguely asking about machinery specifications, you should ask more precisely about "specifications for the 12R403B Copier." Rather than asking, "Will the roofing material cover a large surface?" you need to quantify—"Will the roofing material cover 150×180 feet?"

Conclusion

Conclude precisely. First, explain when you need a response (because until you tell the reader, he does not know). Do not write, "Please respond as soon as possible." Provide dated action—tell the reader exactly when you need your answers. Dated action does not mean you will get the answers when you want them, but your chances are better. Second, to sell your readers on the importance of this date, explain why you need answers by the date given.

> **example**
>
> Please send the above information by August 15. This will allow my shop supervisor to study the manuals and arrange for a test drive prior to our board of directors meeting on September 1.

Figure 6.7 will help you understand the requirements for effective letters of inquiry.

Cover Letters

In business, you are often required to send information to a client, vendor, or colleague. You might send multipage copies of reports, invoices, drawings, maps, letters, memos, specifications, instructions, questionnaires, or proposals. The reader may have requested this information. Maybe you are sending the data of your own accord. Whatever the situation, if you merely submit the data without a cover letter, your readers will be overwhelmed with information that they must wade through.

A cover letter accomplishes two goals. First, it lets you tell readers up front what they are receiving. Second, it helps you focus your readers' attention on key points within the enclosures. Thus, the cover letter is a reader-friendly gesture geared toward assisting your audience.

1020 Sunshine Road
Jamaica Plain, MA 02130

July 23, 2005

Mr. Leonard Liss
Director of Marketing
Johnson Motor Company
4718 West 91st Street
Tacoma, WA 82492

Dear Mr. Liss:

Thank you for your sales letter regarding the Model CP95 tractor with fuel-saving devices. My company plans to purchase 30 new tractors by the end of the year. Your tractors may be what we need.

Before we decide, however, we would like more information. Please send us a brochure of your CP95, along with answers to these questions:

1. What is the cost per tractor?
2. What are your financing options for quantity purchases?
3. Do you provide both a paper-bound instructional manual and an intranet manual with online help screens?
4. What are your tractor's comfort features?
5. When can you deliver 30 tractors?

Please send this information by August 15. Then my shop supervisor and I can study the material and arrange a test drive. I plan to present a proposal for purchase to our board of directors on September 1. I look forward to your response.

Sincerely,

Kathy Brewington

Kathy Brewington

Begin with positive word usage to create a pleasant tone and build rapport

Use itemized lists to improve accessibility

Include dated action to encourage a timely response

Figure 6.7   Letter of Inquiry

As with the previously discussed letters, successful cover letters include the letter essentials, maintain an effective technical writing style, avoid grammatical and mechanical errors, and achieve audience recognition and involvement. In addition, a well-written cover letter contains an introduction, a discussion, and a conclusion.

Introduction

In the introductory paragraph, tell your reader *why* you are writing and *what* you are writing about. An appropriate introduction providing your reader a rationale for your writing might read like this: "To help you prepare for next week's audit, we have enclosed the following necessary forms."

What if the reader has asked you to send the documentation? Do you still need to explain why you are writing? The answer is yes. Although the reader requested the data, time has passed, other correspondence has been written, and your reader might have forgotten the initial request. Your job is to remind her. Write something like, "In response to your request, enclosed are the completed questionnaires from our engineers." These introductory sentences provide the reader information about *why* you are writing and *what* you are sending. Note that neither exceeds one sentence. Clarity of intent can be provided *concisely*.

Discussion

In the body of the letter, you want to accomplish two things. First, you either want to tell your reader *exactly what you have enclosed* or *exactly what of value is within the enclosures*. In both instances, you should provide an itemized list. For example, you can itemize enclosures as follows:

example

To help you prepare for next week's audit, we have enclosed the following necessary forms:

- Year-to-date sales figures
- Fiscal year sales figures
- Year-to-date expenditures
- Fiscal year expenditures

Similarly, you can itemize the important facts within the enclosures as follows:

example

In response to your request, enclosed are the completed questionnaires from our engineers.

Of special interest to you within the questionnaires are their answers to questions regarding the following:

- Easement dimensions . . . . page 1
- Sewer construction materials . . . . page 3
- Residential/commercial ratios . . . . page 5

This example shows where important information can be found within the enclosure. Page numbers are a friendly gesture toward your audience. Instead of just dumping the data on your reader and saying "Best of luck—hope you find what you want," you are helping the reader locate the important information. In so doing, you are achieving audience recognition and involvement.

However, including page numbers has a greater benefit than audience involvement. These page numbers also allow you to focus your reader's attention on what you want him to see. In other words, if you merely provide the enclosure without a cover letter, then you leave it up to readers to sift through the information and decide on their own what is important. In contrast, by providing an itemized list with page numbers, you direct the reader's attention. You can emphasize, for your benefit, the points in the enclosure that you consider to be important.

Conclusion

Your conclusion should tell your readers *what* you want to happen next, *when* you want this to happen, and *why* the data are important. Without such a conclusion, your reader will review the documentation and say, "OK, so what? What do you want me to do with the data?" or "OK, but what are your future plans?"

The *what* clarifies your intentions. The *when* specifies the date. The *why* either sells the importance of this date, possibly hinting at the urgency, or suggests your conscientiousness.

Here are examples of cover letters, complete with conclusions.

> To help you prepare for next week's audit, we have enclosed the following necessary forms:
>
> - Year-to-date sales figures
> - Fiscal year sales figures
> - Year-to-date expenditures
> - Fiscal year expenditures
>
> Please fill out these forms before our Wednesday arrival. Doing so will facilitate the audit, thereby allowing you and your colleagues to return to your regular work activities more rapidly.

> In response to your request, enclosed are the completed questionnaires from our engineers.
>
> Of special interest to you within the questionnaires are their answers to questions regarding the following:
>
> - Easement dimensions . . . . page 1
> - Sewer construction materials . . . . page 3
> - Residential/commercial ratios . . . . page 5
>
> We would be happy to meet with you at your convenience to discuss this issue in greater detail. After you have had a chance to review these enclosures, please call my office at 952-469-8500 to set up a meeting. We want to answer your questions regarding our construction planning.

The cover letter seen in Figure 6.8 contains letter essentials.

AMERICAN
HEALTHCARE

1401 Laurel Drive
Denton, TX 76201

November 11, 2005

Jan Pascal
Director of Outpatient Care
St. Michael's Hospital
Westlake Village, CA 91362

Dear Ms. Pascal:

Thank you for your recent request for information about our specialized outpatient care equipment. *American Healthcare's* stair lifts, bath lifts, and vertical wheelchair lifts can help your patients. To show how we can serve you, we've enclosed a brochure including the following information:

Maintenance, warranty, and guarantee information 1–3
Technical specifications for our products, including
    sizes, weight limitations, colors, and installation
    instructions 4–6
Visuals and price lists for our products 7–8
An *American Healthcare* order form 9
Our 24-hour hotline for immediate service 10

Early next month, I'll call to make an appointment at your convenience. Then we can discuss any questions you might have. Thank you for considering *American Healthcare*, a company that has provided exceptional outpatient care for over 30 years.

Sincerely,

Toby Sommers

Toby Sommers

Enclosure

In the introduction include why *you are writing and* what *you are writing about*

By listing page numbers, your reader will be able to access text easily

Figure 6.8 Cover Letter

Good-News Letters

You will often have the opportunity to write good-news letters. For example, you might write a letter promoting an employee or offering an individual a job at your corporation. You might write to commend a colleague for a job well done. Maybe you will write to tell a customer that her request for a refund was justified.

Introduction

As in other letters discussed in this chapter, the introduction should immediately explain *why* you're writing and tell *what* you're writing about. The point of these letters is good news. Furthermore, because your goal is to convey good news, you should begin with positive word usage, as in the following examples:

> **example**
>
> **Commending a Colleague (introduction)**
>
> Judy, you have proved to be indispensable again. Your work for the textbook committee has made all of our jobs easier.

> **example**
>
> **Promoting an Employee (introduction)**
>
> Congratulations! We are proud to offer you early promotion, Jan.

Discussion

Once you have explained why you are writing and what you are writing about, the next step is to provide the detail—to explain *exactly what* has justified the commendation or the promotion.

> **example**
>
> **Commending a Colleague (introduction and discussion)**
>
> Judy, you have proved to be indispensable again. Your work for the textbook committee has made all of our jobs easier.
>
> Thank you for performing the following services:
> - Meeting with all the sales reps to convey our departmental requirements
> - Reviewing those texts with computer-aided design components
> - Screening the textbook options and selecting the three most suited to our needs

> **example**
>
> **Promoting an Employee (introduction and discussion)**
>
> Congratulations! We are proud to offer you early promotion, Jan.
>
> You have earned a grade raise to E30 for the following reasons:
> 1. *Productivity.* Your line personnel produced 2,000 units per month throughout this quarter.
> 2. *Efficiency.* You maintained a 95 percent manufacturing efficiency rating.
> 3. *Supervisory skills.* You received only four grievances; your annual performance appraisals showed that your subordinates appreciated your motivational management techniques.

Conclusion

Once you have clarified exactly what has resulted in the good news, your last paragraph should state *what* you plan next (or expect from your reader next), *when* this action will occur, and *why* the date is important. The following good-news letters are complete with introductions, discussions, and conclusions.

Judy, you have proved to be indispensable again. Your work for the textbook committee has made all of our jobs easier.

Thank you for performing the following services:

- Meeting with all the sales reps to convey our departmental requirements
- Reviewing those texts with computer-aided design components
- Screening the textbook options and selecting the three most suited to our needs

Due to your assistance, we have decided on the Wilkes text and plan to place our orders this April. That will allow us to stock the bookstore for the fall semester. Your work has made a difference, and we appreciate your efforts—another job well done!

Congratulations! We are proud to offer you early promotion, Jan.

You have earned a grade raise to E30 for the following reasons:

1. *Productivity*	Your line personnel produced 2,000 units per month throughout this quarter.
2. *Efficiency*	You maintained a 95 percent manufacturing efficiency rating.
3. *Supervisory Skills*	You received only four grievances; your annual performance appraisals showed that your subordinates appreciated your motivational management techniques.

Because of your excellent work, you will receive your pay increase the first of next month. You deserve it. Good work, Jan.

Bad-News Letters

Unfortunately, you occasionally will be required to write bad-news letters. These letters might reject a job applicant, deny an employee a raise, tell a vendor that his or her company's proposal has not been accepted, or reject a customer's request for a refund. Maybe you will have to write to a corporation to report that its manufacturing is not meeting your company's specifications. Maybe you will need to write a letter to a union documenting a grievance. You might even have to write a bad-news letter to fire an employee.

In any of these instances, tact is required. You cannot berate a customer or client. You should not reject a job applicant offensively. Even your grievance must be worded carefully to avoid future problems.

Because the point of a bad-news letter is bad news, you will need to structure your correspondence carefully to avoid offending your reader.

Introduction

We have suggested throughout this chapter that your introduction should explain why you're writing and what you're writing about. However, such conciseness and clarity in a bad-news letter would be harsh and abrupt. Therefore, to avoid these lapses in diplomacy, begin your bad-news letter with a buffer. Start your letter with information that your reader can accept as valid but that will sway your reader to accept the bad news to come.

---

example

Rejecting a Job Applicant (introduction)

Thank you for your recent letter of application. As you can imagine, we received many letters from highly qualified applicants.

---

example

Terminating a Client/Vendor Relationship (introduction)

John, as you know, our business demands exact tolerances and precision work. Because of these requirements and the reputation your company has for quality production, we were happy to pursue a long-term contract with you.

Discussion

Once you have provided the buffer, swaying your audience to your point of view, you can no longer delay the inevitable. The discussion paragraph states the bad news. However, to ensure that the reader accepts the bad news, preface your assertions with quantifiable proof.

---

example

Rejecting a Job Applicant (introduction and discussion)

Thank you for your recent letter of application. As you can imagine, we received many letters from highly qualified applicants.

Although we appreciate your interest in Acme, the advertisement specifically required that all applicants have an MS in computer science and at least five years of experience in telecommunications. We also suggested that a knowledge of fiber optics would be preferred. Your degree meets our criteria successfully. However, your years of experience fall below our requirements, and your resume does not mention fiber optics expertise. Therefore, we must reject your application.

---

example

Terminating a Client/Vendor Relationship (introduction and discussion)

John, as you know, our business demands exact tolerances and precision work. Because of these requirements and the reputation your company has for quality production, we were happy to pursue a long-term contract with you.

However, your last two shipments contained flawed goods. In fact, we found these problems:

- 37 percent of your shipped components were off tolerance by 0.025 mm.
- Your O-rings suffered stress fractures when under 2,000 lb of pressure.

Because of these failures, we are returning the products.

Conclusion

If you end your bad-news letter with the bad news, then you leave your reader feeling defeated and without hope. You want to maintain a good customer-client, supervisor-subordinate, or employer-employee relationship. Therefore, you need to conclude your letter by giving your readers an opportunity for future success.

Provide your readers options which will allow them to get back in your good graces, seek employment in the future, or reapply for the refund you have denied. Then, to leave your readers feeling as happy as possible, given the circumstances, end upbeat and positively.

The following examples are complete with introductions, discussions, and conclusions.

> Thank you for your recent letter of application. As you can imagine, we received many letters from highly qualified applicants.
>
> Although we appreciate your interest in Acme, the advertisement specifically required that all applicants have an MS in computer science and at least five years of experience in telecommunications. We also suggested that a knowledge of fiber optics would be preferred. Your degree met our criteria. However, your years of experience fell below our requirements, and your resume did not mention fiber optics expertise. Therefore, we must reject your application.
>
> If you have fiber optics knowledge or have acquired additional job experiences that pertain to our work requirements, we would be happy to reconsider your application. In any case, we will keep your letter on file. When new positions open up, your letter will be reassessed. Good luck in your job search.

> John, as you know, our business demands exact tolerances and precision work. Because of these requirements and the reputation your company has for quality production, we were happy to pursue a long-term contract with you. However, your last two shipments contained flawed goods. In fact, we found these problems:
>
> - 37 percent of your shipped components were off tolerance by 0.025 mm.
> - Your O-rings suffered stress fractures when under 2,000 lb of pressure.
>
> Because of these failures, we are returning the products.
>
> If you can correct these problems and document to our satisfaction that the errors have been eliminated, we would be willing to reconsider our stance. We have enjoyed working with you, John, and look forward to the possibility of future contracts.

Although the preceding two letters convey bad news, they couch the negatives in positive terms. For example, although the first letter rejects a job applicant, the writer uses words and phrases such as *thank you, appreciate, happy,*

State of Mind Insurance
11031 Bellbrook Drive
Stamford, CT 23091
213-333-8989

September 7, 2005

John Chavez
4249 Uvalde
Stamford, CT 23091

Dear Mr. Chavez:

Thank you for letting us provide you and your family with insurance for the last 10 years. We have appreciated your business. ←——— Begin with a buffer statement in a bad-news letter

However, according to our records, you have filed three claims in the past three years:

1. Damage to your basement due to a failed sump pump
2. Flooring damage due to a broken water seal in your second-floor bathroom
3. Fender repair coverage for a no-fault automobile accident ←——— In the body, explain what happened and then state the bad news

SMI's policy stipulates that no more than two claims may be filed by a client within a three-year period. Therefore, we must cancel your policy as of October 15, 2005.

If you have any questions about this cancellation, please call our 24-hour assistance line (913-482-0000). Our transition team can help you find new coverage. Thank you for your patronage and your understanding. ←——— In the conclusion, give your reader follow-up options

Sincerely,

Darryl Kennedy

Darryl Kennedy

Figure 6.9   Bad-News Letter

and *good luck*. The second letter does the same: *quality, happy, enjoyed,* and *look forward to*. As the writer of the bad-news letter, you are in charge of the tone. Why not make it positive?

The example in Figure 6.9, which is from an insurance company cancelling an insurance policy, shows how a bad-news letter should be constructed.

Complaint Letters

You are purchasing director at an electronics firm. Although you ordinarily receive excellent products and support from a local manufacturing firm, two of your recent orders have been filled incorrectly and included defective merchandise. You do not want to have to hunt for a new supplier.

How can you improve your relations with the company and receive the service you want? Write a letter of complaint.

Introduction

In the introduction, *politely* state the problem. Although you might be angry over the service you have received, you want to suppress that anger. Blatantly negative comments don't lead to communication; they lead to combat. Because angry readers will not necessarily go out of their way to help you, your best approach is diplomacy.

To strengthen your assertions, in the introduction, include supporting documents, such as the following:

- Serial numbers
- Dates of purchase
- Invoice numbers
- Check numbers
- Names of salespeople involved in the purchase

Also state that copies of these documents are enclosed.

example

On August 15, you shipped to my company 36 XYZ digital oscilloscopes (copies of invoice #3492 are enclosed). Several of our customers have since complained that the o-scopes are malfunctioning.

Discussion

In the discussion paragraph(s), explain in detail the problems experienced. This could include dates, contact names, information about shipping, breakage information, or an itemized listing of defects. Be specific. Generalized information won't sway your readers to accept your point of view. In a complaint letter, you suffer the burden of proof. Help your audience understand the extent of the problem. After documenting your claims, state what you want done and why.

example

The following occurred after we delivered the o-scopes to three of our customers.

1. August 20—AAA Electronics stated that two of its five o-scopes were malfunctioning, giving incorrect readings.
2. August 27—Richards Electronics, Inc., said three of its o-scopes were incorrectly calibrated.
3. September 5—Five of ABC Computers' o-scopes would not interface with its computers.

These 10 o-scopes need to be repaired. You also should send service technicians to troubleshoot our clients' other o-scopes to avoid future problems.

Conclusion

End your letter positively. Remember, you want to ensure cooperation with the vendor, and you want to be courteous, reflecting your company's professionalism. Include your phone number and the time you can best be reached.

> You can reach me at 612-469-8200 from 8:00 A.M. to 5:00 P.M., Monday through Friday. Thank you for your help. After working with your company for the past eight years, I have been impressed with the way you stand behind your products. I know you will help us with these defective o-scopes.

Figure 6.10 is an example of a complaint letter.

Adjustment Letters

Responses to letters of complaint, also called adjustment letters, can take three different forms.

- 100 percent yes—you could agree 100 percent with the writer of the complaint letter.
- 100 percent no—you could disagree 100 percent with the writer of the complaint letter.
- Partial adjustment—you could agree with some of the writer's complaints but disagree with other aspects of the complaint.

100 Percent Yes Letter

This letter is like the good-news letter discussed earlier in the chapter. Because you have got good news for the reader, there is no reason to delay the message. In the introductory paragraph, state that you agree with your reader's complaint and will honor her recommendations for adjustment.

> Thank you for calling our attention to your problem. We will be happy to replace your defective oscilloscopes.

In the discussion paragraph(s), explain what happened, why the problem occurred, and how the problem will be avoided in the future.

> I have looked into your order and discovered the cause of the problem. The o-scopes were damaged during shipment. They were stacked too high, which led to breakage while handling. I have met with the carrier, who agrees that the packages were handled improperly. I assure you that none of these problems will occur again. Also, I am sending a service technician to check on your other o-scopes from this order.

In your conclusion paragraph, resell to maintain customer satisfaction. End the letter upbeat.

> Acme has worked with your company for many years, and I look forward to continuing this business relationship. I apologize for the inconvenience. Please call me at 434-392-8270 if you have any other questions.

1234 18th Street
Galveston, TX 77001

May 10, 2005

Mr. Holbert Lang
Customer Service Manager
Gulfstream Auto
1101 21st Street
Galveston, TX 77001

Dear Mr. Lang:

On February 12, 2005, I purchased two shock absorbers in your automotive department. Enclosed are copies of the receipt and the warranty for that purchase. One of those shocks has since proved defective.

I attempted to exchange the defective shock at your store on May 2, 2005. The mechanic on duty, Vernon Blanton, informed me that the warranty was invalid because your service staff did not install the part. I believe that your company should honor the warranty agreement and replace the part for the following reasons:

1. The warranty states that the shock is covered for 48 months and 48,000 miles.
2. The warranty does not state that installation by someone other than the dealership will result in warranty invalidation.
3. The defective shock absorber is causing potentially expensive damage to my tire and suspension system.

I can be reached between 1 P.M. and 6 P.M. on weekdays at 651-763-9280 or at 651-763-9821 anytime on weekends. I look forward to hearing from you. Thank you for helping me with this misunderstanding.

Sincerely,

Carlos De La Torre

Carlos De La Torre

Enclosures (2)

Include enough details such as date of purchase to establish context for your complaint

In the body explain what happened, state what you want, and justify what you want

Figure 6.10   Letter of Complaint

Unfortunately, sometimes you have to say no to a letter of complaint. If the customer is wrong, you have no alternative but to say so. However, you still want to maintain positive customer relationships. Although you will deny the customer's request for adjustment, you want to keep that customer. To do so, follow the pattern discussed earlier in this chapter for a bad-news letter.

In the introductory paragraph, begin with a buffer. This should include a positive statement and facts that all readers can accept, regardless of the situation.

> **example**
>
> Thank you for your letter regarding the XYZ oscilloscopes. We were sorry to hear about the malfunctions your customers are experiencing. Although such occurrences are rare, as you know, products occasionally are damaged in shipment.

In the discussion paragraph(s), precisely explain what happened, providing facts, figures, and incontestable detail. Then state the bad news.

> **example**
>
> After researching the matter thoroughly, we found that the problem was created by the shipping company. They stacked the boxes too high, which led to damage during handling. Because the damage was caused by the carriers and was not due to our negligence during manufacturing, we cannot replace the o-scopes.

In the conclusion, end on an upbeat note. To do so, use positive words and provide the reader an alternative.

> **example**
>
> To assist you in solving this problem, we have contacted Archie Cox, the shipping manager at ABC Freight (501-535-2020, ext. 345). He expects to hear from you soon. If we can help you in any other ways, please call.

Partial Adjustment

Life is complex. Sometimes you can agree with part of a writer's complaint but disagree with other parts. If you are not going to agree completely or disagree completely, then you will want to write a partial adjustment letter.

In the introductory paragraph, state the good news first. As always, you want to be diplomatic, winning the reader to your point of view.

> **example**
>
> Thank you for your letter regarding the XYZ oscilloscopes. I was sorry to hear about the malfunctions your customers are experiencing. Because we pride ourselves on our products' quality, I am happy to replace the 10 defective o-scopes.

In the discussion paragraph(s), precisely explain what happened, and then state the bad news.

> **example**
>
> The 10 o-scopes, which were damaged during shipment due to packaging problems, will be replaced within six to eight weeks. I am unable to send a service technician to troubleshoot the remaining 26 o-scopes. That service is not part of our contract with you. (Please see section 4.1 of your service contract: "On-site service will not be provided in case of malfunction.")

In the conclusion, resell. Your reader will not be completely happy with you. You have provided some relief but not all that was requested. The last paragraph is your opportunity to win back a bit of the reader's good faith in your company. After all, you want to keep the customer for future contracts.

> **example**
>
> If any of the other o-scopes malfunction, please call us immediately. We stand by our products, as you know from having been one of our satisfied customers for over a decade. I'll be happy to help you in any way I can.

Figure 6.11 is an example of a 100 percent yes adjustment letter.

Sales Letters

You have just manufactured a new product (an electronic testing device, a fuel injection mechanism, a fiber optic cable, or a high-tech, state-of-the-art computer). Perhaps you have just created a new service (computer maintenance, automotive diagnosis, home repair, or telecommunications networking). Congratulations!

However, if your product just sits in your basement gathering dust or your service exists only in your imagination, what have you accomplished? To benefit from your labors, you must market your invention. Connect with your end users. Let the public know that you exist. The question is, how? If you have unlimited time, money, and personnel, you can try telemarketing or door-to-door cold calls. However, a more time-efficient and cost-effective way to market your product or service is to write a sales letter. You write a sales letter once (which saves money) and then mass mail it (which saves time).

To write your sales letter, you will have to include the letter essentials discussed earlier in this chapter, maintain an effective technical writing style with a low fog index, and avoid grammatical and mechanical errors. You will also have to accomplish the following objectives.

Arouse Reader Interest

The introductory paragraph of your sales letter tells your readers *why* you are writing (you want to increase their happiness or reduce their anxieties). You also tell your readers *what* you are writing about (the product or service you are marketing). However, studies tell us that you have only about 30 seconds to grab your readers' attention, after which they will lose interest and toss out your letter. To lure readers into your correspondence, you must arouse their interest imaginatively in the first few sentences. Try any of the following lead-ins.

Gulfstream Auto

1101 21st Street
Galveston, TX 77001
712-451-1010

May 31, 2005

Mr. Carlos De La Torre
1234 18th Street
Galveston, TX 77001

Dear Mr. De La Torre:

Thank you for your recent letter. I am pleased to inform you that Gulfstream will happily replace your defective shock absorber according to the warranty agreement. ◄—— Begin with the good news

The Trailhandler Performance XT shock absorbers that you purchased were discontinued in October 2004. Mr. Blanton, the mechanic to whom you spoke, incorrectly assumed that Gulfstream was no longer honoring the warranty on that product. Because we no longer carry that product, we either ◄—— Explain what happened and tell your reader what to do next
will replace it with a comparable model or refund the purchase price—you make the decision. Just ask to see Mrs. Brandsgard at the automotive desk on your next visit to our store. She is expecting you and will handle the exchange.

We appreciate your business, Mr. De La Torre. I'm glad you brought this problem to my attention. If I can help you in the future, please contact me.

Sincerely,

Holbert Lang

Holbert Lang

Figure 6.11   100 Percent Yes Adjustment Letter

- **An anecdote**—a brief, dramatic story.

example

> It's late at night, the service stations are closed, and you have just had a blowout on Highway 35. Don't worry. Our new Tire-Right will solve your problems!

- **A question**—to make your readers read on for an answer.

> **example**
>
> "Where will I get the money for my kid's college education?" "How am I going to afford to retire?" "Will my insurance cover all the medical bills?" You have asked yourself these questions. Our estate planning video has the answers.

- **A quotation**—to give your letter the credibility of authority.

> **example**
>
> "Omit needless words," technical writers say. If you can't, let us help you. *Write Now,* our new office communications service, can help you write your in-house newsletters—clearly and concisely.

- **Data**—researched information, again ensuring your credibility.

> **example**
>
> You are not alone. In fact, if you are at least 25 years old, you are in the majority. Today, 51 percent of our students are older than 25. So why not enroll now? Our college offers the nontraditional student many benefits.

Develop Your Assertions

In the discussion paragraph(s), specify exactly what you offer to benefit your audience or how you will solve your readers' problems. You can do this in a traditional paragraph. In contrast, you might want to itemize your suggestions in a numerical or bulletized list. Whichever option you choose, the discussion should accomplish any of the following:

- *Provide data* to document your assertions.

> **example**
>
> Eighty-five percent of the homeowners contend that . . .
> or
> Seven out of ten buyers said they would . . .
> or
> Ten thousand retired metalworkers can't be wrong!

- *Give testimony* from satisfied customers.

> **example**
>
> The Job Corps of Blue Valley, Titan Co., Amex Inc., and the Southwestern Parish swear by our product.

- *Document your credentials*—years in business, certification of employees, number of items sold, or satisfied customers (this last point overlaps with our suggestions regarding data).

In business since 1992.

or

Each of our mechanics is ASE certified.

or

We've sold over 2 billion widgets.

or

Over 3,000 corporate managers and supervisors have benefited from our training services.

Make Your Readers Act

If your conclusion says "We hope to hear from you soon," you have made a mistake. Such a weak conclusion forces you to stare at the phone, twiddle your thumbs, and hope for a customer response. The concluding paragraph of a sales letter cannot let the reader off the hook. Do something to make the reader act. Here are some suggestions.

- Give directions (with a map) to your business location.
- Provide a tear-out to send back for further information.
- Supply a self-addressed, stamped envelope for customer response.
- Offer a discount if the customer responds within a given period of time.
- Give your name or a customer-contact name and a phone number (toll free if possible).

Use an Appealing Style

Your sales letter should not only convey the necessary information but also present an appealing style. Consider the following requirements:

- Use verbs. Verb usage, or power writing, will give your letter punch.
- Format for reader-friendly ease of access. Highlight through white space, underlining, boldface, bullets, numbers, and so on.
- Achieve audience recognition and involvement through *you* orientation.
- Show ease of use. Sprinkle the words *easy* and/or *simple* throughout your correspondence.
- Imply urgency. Use the words *now, today, soon,* or *don't delay* to force the reader to immediate action.

Figure 6.12 shows an example of a successful sales letter.

PROCESS

Now that you know what letters of inquiry, cover letters, good-news letters, bad-news letters, and sales letters entail, the next question is, "How do I write these letters?" If you merely get out a blank piece of paper, or turn on your computer screen, and then fill it with words, assuming that they will be good enough, you are making a mistake. No one can write a successful letter the first time through. In contrast, you need to approach writing more systematically. As we discussed in Chapter 2, to construct effective correspondence, approach writing as a process.

2629 Westport   Oswego, ND 55290   405-555-2121

October 9, 2005

Jane Golub
Media Publishing
10901 College Blvd.
Mountainview, ND 55292

Dear Ms. Golub:

> *In the introduction, emphasize a reader-related problem and show how the product or service can help*

"Do you have the software I use?" "Can you print my PC job?" If you have asked these questions, let **Image Graphics** help. **Image Graphics** has the printing technology, accessibility, and knowledge that you are looking for.

> *The body shows how the product or service can benefit the reader*

⇨ **Image Graphics** is the most technologically advanced graphic design firm in the Midwest, according to *Printer's Wheel Weekly*.
⇨ We are the only design firm in your area with an in-house, 24-hour service bureau.
⇨ Our service bureau can print any job to any media with no color or resolution limitations.
⇨ We support all brands of software.
⇨ Each member of our support staff has a master's degree in graphic technology.

> *Encourage the reader to act*

I am glad I could introduce you to **Image Graphics**, Ms. Golub. Look at the enclosed brochure and pricing sheet. Then call me personally (ext. 2055) this week for a ***50 percent*** discount on your first output order.

Sincerely,

Macy Hart

Macy Hart

Enclosure

Figure 6.12   Successful Sales Letter

You should do the following:

- Prewrite (to gather your data and to determine your objectives).
- Write a draft (to organize and format your text).
- Rewrite (to revise your correspondence, making it as perfect as it can be).

The writing process is dynamic with the three steps frequently overlapping.

The Writing Process

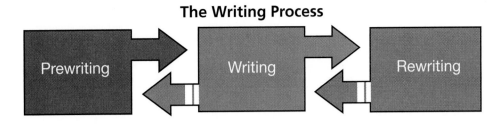

The Writing Process		
Prewriting	Writing	Rewriting

• Examine your purposes
• Determine your goals
• Consider your audience
• Gather your data
• Determine how the content will be provided
	• Organize the draft according to some logical sequence that your readers can follow easily
	• Format the content to allow for ease of access
		• Revise
		  ° Add missing details
		  ° Delete wordiness
		  ° Simplify word usage
		  ° Enhance the tone of your communication
		  ° Reformat your text for ease of access
		  ° Practice the speech or review the text
		• Proofread
		  ° Correct errors

Prewriting

To overcome the blank page syndrome, or writer's block, spend some time *before* writing your letter gathering as much information as you can about your subject matter. After all, how can you expect to write an effective letter if you are not sure what you want to say? In addition, determine your objectives. Ask yourself why you are writing the letter and what you want to achieve.

No single way of prewriting is more effective than another. One prewriting technique helpful to many writers is answering reporter's questions. By providing answers to *who, what, when, why, where,* and *how,* you can help yourself write more effective letters. Answers to these questions provide you content for your correspondence as well as focus in your writing objectives.

Following is a Reporter's Questions Checklist for a Letter of Inquiry.

Reporter's Questions Checklist for a Letter of Inquiry

___ **1.** *Who* is your audience?
- High-tech peer
- Low-tech peer
- Lay reader
- Management
- Subordinate
- Multiple audiences

___ **2.** *Why* are you writing?

___ **3.** *What* is the general topic of your request?

___ **4.** *What exactly* do you want to know (the specifics itemized in the discussion paragraph of your letter)?

___ **5.** *What* do you want the reader to do next?

___ **6.** *When* do you want the reader to act (dated action)?

___ **7.** *Why* is this date important?

Writing

Once you have gathered your data and determined your objectives, the next step in the process is to begin your rough draft. Doing so requires the following:

Study the Letter Criteria

By studying the specific criteria for the type of letter you'll write, you can remind yourself of what information should be included in each paragraph.

Review Your Prewriting

Now that you have reminded yourself of what each paragraph should include, reviewing the prewriting will help you determine whether, in fact, you have provided the correct details. Have you omitted any significant information? If so, now is the time to add missing content. Have you included unnecessary information? If so, now is the time to delete these irrelevancies.

Organize the Data for Your Discussion Paragraph

One organizational pattern especially effective for most letters is *importance*. When you organize by importance, you place the most important information first and less important ideas later. To do so, you can use the inverted journalist's pyramid shown in Figure 6.13.

This method works well for itemizing the discussion paragraph in your sales letters, letters of inquiry, and cover letters. You also can use importance to organize the body paragraph in your good-news letter.

However, because a bad-news letter leads up to the bad news rather than begins with it, importance would not be an effective organizational pattern. Instead, you will want to organize your ideas inductively (indirectly)—specifics to general result. For instance, in a bad-news letter, you will begin with the specific causes leading to bad news (recession, depression, changes in the market, unfavorable work habits, etc.). Then, after setting the stage with these specifics, you will relate the general conclusion reached—the bad news (layoffs, refunds denied, promotions rejected, etc.). The bad-news letter examples in this chapter reveal this inductive organizational pattern.

Draft Your Correspondence

Finally, after you have organized your information, the last step is to write your rough draft. Once you know what you want to say, you should just say it—and

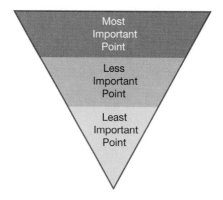

Figure 6.13 Inverted Journalist's Pyramid

let it suffice for the moment. Write a rapid rough draft, focusing on content and organization, not on grammar, mechanics, or style. Your primary goal in a rough draft is to get words on the paper or screen. The time for fine tuning comes later in rewriting. During rewriting, you will be able to check your spelling, consider your punctuation, and perfect your tone. Drafting, in contrast, should be a stage in the writing process still free of those fears caused by grammar anxiety. After all, if you are worrying about the correctness of every word while you write it, you'll never complete a sentence.

Rewriting

After you have drafted your correspondence, the final step in the writing process and, in many people's opinion, the most important step is to revise your rough copy. Rewriting is the step that most mediocre writers omit. Poor writers merely let the rough draft suffice as their finished version. They assume that their draft is sufficient. It isn't. Successful writers, in contrast, realize that an effective letter must be as perfect as it can be. To achieve this level of expertise, good writers revise their rough drafts—fine-tuning, honing, sculpting, molding, and polishing so what leaves their hands is something they can be proud of.

To revise your letters, consider the following revision criteria:

Add for Clarity and Correctness

Look back over your rough draft. In doing so, ask yourself what you might need to add for clarity. This might entail reviewing the reporter's questions. For example, has your letter clarified *who* will do the work (personnel), act as a contact person (liaison), manage the system (management), and so on? If not, then you need to provide further information by answering the reporter's question *who*. Have you correctly stated *when* you will follow up, when you need the information, or when the bad news will occur? If not, add these dates. The same applies for all the reporter's questions. Add any missing information regarding *what, why, where,* or *how*.

In addition to adding answers to the reporter's questions, review your draft and add any missing letter essentials: letterhead address, date, inside address, salutation, complimentary close, typed or signed signature, and optional inclusions such as enclosure or copy notations.

Delete for Conciseness

Deleting relates to style and content. Review your rough draft to delete any dead words and phrases which will raise your fog index (discussed in Chapter 3). Delete irrelevant content. This might include background sentences in the introductory paragraph that delay your main point. Writers often feel compelled to lead in to their thesis. This is a mistaken notion in technical writing. All the reader really wants to know is why you are writing and what you are writing about. Get to your point immediately; delete the unneeded background data.

Simplify to Aid Easy Understanding

Chapter 3 stated that you should avoid old-fashioned words and phrases, such as *pursuant, accede, supersede,* and *in lieu of*. Rather than using such pompous words, simplify. Just write naturally. Why not say *after, accept, replace,* and *instead of?* This simplified style not only affects your fog index positively but also helps your readers understand your content, and that's the goal of good writing.

Move Information for Emphasis

This involves cutting and pasting—moving information within your letter. For example, let's say you've read your rough draft and realized that the itemized discussion paragraph is not organized according to importance. In fact, you have buried a key point deep in the body where readers might miss it. To focus your reader's attention on this idea, you should move it from the bottom of the list to the top. Doing so makes the idea more emphatic.

Reformat for Reader-Friendly Ease of Access

Review your letter's format. How does the letter look on the page? The letter's appearance affects your readers before they read one word. If the letter is open and appealing, then you're off to a good start. If the letter is blocky and inaccessible, then you've already done yourself a disservice.

Before you type your final copy, answer the following questions:

- Have you used enough (or too much) white space?
- Should you underline a key word or phrase?
- Are bullets appropriate in your discussion paragraph, or should you use numbers instead?
- Would boldfacing help you draw attention to a key concern?

Enhance the Letter's Tone

Letters are not the inanimate objects they appear to be. They do not just dump data. Your letter is a reflection of your interpersonal communication skills and your company's attitudes. Reread your rough draft; is it revealing the personality or attitude that you want it to reveal? If the letter is stuffy, impersonal, high-handed, abrupt, offensive, or mean-spirited, then you must alter the tone.

To do so, add more pronouns, especially *you*-oriented ones; add verbs for punch; use contractions for a conversational tone; or add positive words and phrases to give your letter an upbeat tone. Using such personalization techniques (as discussed in Chapter 4) ensures customer and client cooperation.

Correct Errors

Finally, before your letter leaves your office, correct any errors you've committed. The most important step of rewriting may be proofreading. Check and double-check your grammar; check and double-check your mathematical computations and scientific notations. Your letter is a legal document. A mistake can cost your company money and you your job. Proofreading is boring, but reading the want ads to find a new job is worse. We presented several proofreading tips in Chapter 3. Save yourself future problems by proofing your letters.

PROCESS LOG

To give you a better idea of how this three-step process approach to letter writing works, we provide the following student-written process log. It includes the student's answers to the reporter's questions (prewriting), her rough draft (writing), her revisions (rewriting), and her finished copy.

Prewriting

First, the student filled in the Reporter's Questions Checklist for a Letter of Inquiry (see Figure 6.14).

1. *Who* is your audience?
   - High-tech peer __Vendor__
   - Low-tech peer _____
   - Lay reader _____
   - Management _____
   - Subordinate _____
   - Multiple audiences _____

2. *Why* are you writing? __In response to sales letter—to request information__

   What is the general topic of your request? __Information about the personal shopping service__

3. *What exactly* do you want to know (the specifics itemized in the discussion paragraph of your letter)?
   - __cost of services__
   - __number of employees needed__
   - __commission__
   - __what department to work with__
   - _____

4. *What* do you want the reader to do next? __Respond to questions__

5. *When* do you want the reader to act (dated action)? __March 1__

6. *Why* is this date important? __For meeting with management__

Figure 6.14   Reporter's Questions Checklist for a Letter of Inquiry

Writing

After the student gathered her data, she wrote a rough draft (Figure 6.15).

Rewriting

Next, with the assistance of a peer review group consisting of two other students, the writer revised the letter (Figure 6.16).

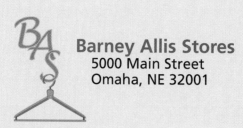

Barney Allis Stores
5000 Main Street
Omaha, NE 32001

February 15, 2005

Fashion Pace
100 Eby Drive
New York, NY 01034

Dear Ms. Pace:

Thank you for the information about your personal shopping service. I am interested in Fashion Pace and would like to receive the following information:

1. Sales figures.
2. Customers serviced.
3. Requirements and costs to start service.

I would appreciate your response by March 1, 2005. Because I will be attending a meeting of the board on March 10, 2005 your early response will allow me to present your service at this time.

Sincerely,

Shirley Chandley

Shirley Chandley, Manager

Figure 6.15   Rough Draft of Letter of Inquiry

Once the student made the suggested revisions, she submitted the finished copy (Figure 6.17). Notice how the changes both the student made and those suggested by her peers correspond to our suggestions regarding revision:

- **Add**—to clarify her requests, the student, with the help of her peers, not only added more questions but also made these questions more precise. For example, instead of merely writing "Sales figures," the student rewrote this to read "What are your sales figures for the last two years?" The noun phrase "sales figures" leaves too much room for guesswork. In contrast, the revision is more specific.
- **Move**—several items have been moved in the text. These include Shirley Chandley's title ("manager") and the second sentence of the rough draft's first paragraph. By moving this second sentence, the writer now uses it as a transitional lead into the body's numbered list.

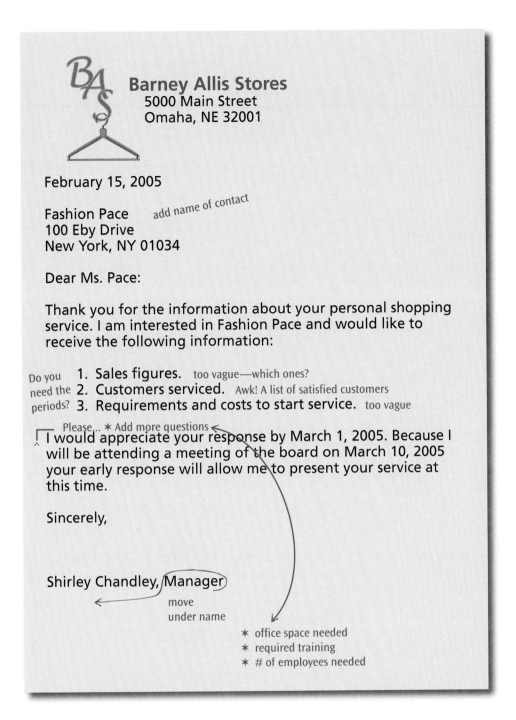

Barney Allis Stores
5000 Main Street
Omaha, NE 32001

February 15, 2005

Fashion Pace   *add name of contact*
100 Eby Drive
New York, NY 01034

Dear Ms. Pace:

Thank you for the information about your personal shopping service. I am interested in Fashion Pace and would like to receive the following information:

Do you need the periods?
1. Sales figures.   *too vague—which ones?*
2. Customers serviced.   *Awk! A list of satisfied customers*
3. Requirements and costs to start service.   *too vague*

Please... ✳ Add more questions

I would appreciate your response by March 1, 2005. Because I will be attending a meeting of the board on March 10, 2005 your early response will allow me to present your service at this time.

Sincerely,

Shirley Chandley, Manager

move under name

✳ office space needed
✳ required training
✳ # of employees needed

Figure 6.16   Revision of Letter of Inquiry

- **Enhance**—a very subtle change is made in the last paragraph. Initially, the writer says, "I would appreciate your response by March 1, 2005." Though the tone of this sentence is pleasant, it lacks conviction. By rewriting this to say, "Please send answers to these questions by March 1, 2005," the writer has achieved a more aggressive tone.
- **Correct**—the writer makes two corrections for the final draft. First, she adds the reader's name. This not only makes the letter more personal but also now the correct recipient of the letter will receive it. In

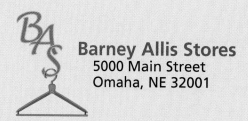

Barney Allis Stores
5000 Main Street
Omaha, NE 32001

February 15, 2005

Ms. Jackie Pace
Fashion Pace
100 Eby Drive
New York, NY 01034

Dear Ms. Pace:

Thank you for the information about your personal shopping service.

I am interested in Fashion Pace and would like to receive answers to the following questions:

1. What are your sales figures for the last two years?
2. Would you send me a satisfied customer list?
3. How much office space will your consultants need?
4. How many employees will you need to run your service?
5. What training is required for these employees?
6. What will you charge customers for your service?

Please send answers to these questions by March 1, 2005. Your early response will allow me to present your proposal at our March 10 sales meeting. I look forward to hearing from you.

Sincerely,

Shirley Chandley

Shirley Chandley
Manager

Figure 6.17   Letter of Inquiry Final Copy

addition, one small grammatical error is corrected. The unnecessary periods following the itemized body points are removed and replaced by more complete questions.

Chapter Highlights

1. Letters are external correspondence.

2. When writing a letter, include these essential components:
    *   Writer's address
    *   Date
    *   Inside address
    *   Salutation
    *   Body of the letter
    *   Complimentary close
    *   Your signed name
    *   Your typed name

3. You could include optional letter components such as
    *   A subject line
    *   New-page notations
    *   Writer's and typist's initials
    *   Enclosure notation
    *   Complimentary copy notation

4. In a letter of inquiry, you are asking for information.

5. In a cover letter, tell your reader what he is receiving in the enclosure, such as a questionnaire or a proposal.

6. Good-news letters provide your reader with positive information.

7. Bad-news letters require tact and politeness on your part as the writer.

8. Complaint letters should politely state what you want and explain why your request is justified.

9. You can adjust (or answer) a letter of complaint in three ways:
    *   100 percent yes—you agree.
    *   100 percent no—you disagree.
    *   Partial adjustment—you agree in part.

10. Sales letters have to arouse interest, so explain how your product or service will benefit the reader.

CASE STUDIES

After reading the following case studies, write the appropriate letters required for each assignment.

1. Vivian Davis, who lives at 2939 Cactus in Santa Clara, CA 95054, has invented a new product, the VAST (a voice-activated speaker telephone). This telephone can be "dialed" without the use of one's hands, simply by saying names or numbers into the system's speaker. Vivian believes that such a machine could benefit people with disabilities, people who are elderly or home bound, homemakers, and businesspeople. She can sell this phone for $25 to electronic stores, hobby shops, general-purpose retailers, and so on, who then can market it for $50. Write a sales letter for Vivian based on the preceding information.

2. Mark Shabbot works for Apex, Inc., at 1919 W. 23rd Street, Denver, CO 80204. Apex, a retailer of electronic equipment, wants to purchase 125 new oscilloscopes from a vendor, Omnico, located at 30467 Sheraton, Phoenix, AZ 85023. The oscilloscopes will be sold to a college in Denver (Northwest Hills Vocational-Technical College). However, before Apex purchases these oscilloscopes, Mark needs information regarding bulk rates, shipping schedules, maintenance agreements, equipment specifications, and machinery capabilities. Northwest Hills needs this equipment before the new term (August 15). Write a letter of inquiry for Mark based on the preceding information.

3. Sharon Baker works as a technical writer for Prismatic Consulting Engineering, 123 Park, Boston, MA 01755. In response to an RFP (request for proposal), she has written a proposal to the Oceanview City Council, 457 E. Cypress Street, Oceanview, MA 01759. The proposal suggests ways in which Oceanview can improve its flood control. Now Sharon needs to write a cover letter prefacing her proposal. In this cover letter, she wants to call her readers' attention to key concerns within the proposal: suggested costs, time frames, problems that could occur if the proposed suggestions are not implemented, ways in which the proposal will solve these problems, and Prismatic's credentials. Once the Oceanview City Council receives the proposal, Prismatic representatives will schedule follow-up discussions. Write Sharon's cover letter based on the preceding information.

4. Bob Ward, a lineworker at HomeCare Health Equipment, 8025 Industrial Parkway, Ashley, NC 27709, deserves a letter of commendation for his excellent job record. He has not missed a day of work in five years. In addition, his production line has achieved a 5 percent error reading (7 percent is considered acceptable). He has also trained new hires. Most importantly, he made six suggestions for improvements, three of which saved the company money. The company president, Peter Tsui, based at HomeCare's home office at 4791 Research Avenue, Wasa, MN 55900, wants to award Bob with a plaque at the annual awards dinner on September 7, 2005. Write Peter Tsui's good-news letter based on the preceding information.

5. Stacy Helgoe works as a service technician for EEE Electronic Servicing, 11201 Blanco, Santa Fe, NM 88004. Yesterday, she went to Schoss-

McGraw Associates, 1628 W. 18th Street, Taos, NM 88003, to service their computer systems. She billed them $75.00 for her time, but she did not bill them for parts because the machinery was supposedly under warranty. However, when she returned to EEE, Stacy's manager, Marilyn Hoover, informed Stacy that the machinery was not under warranty and that Schoss-McGraw would have to be billed an additional $45.87 for parts. Schoss-McGraw is an excellent client, so Marilyn wants Stacy to be especially tactful in requesting the additional money. Based on the preceding information, write Stacy's bad-news letter.

6. Diane Waisner (who lives at 1439 87th Street, Monroe, LA 67054) purchased a VCR on August 13 from Smiley's TV Town (8201 Magnolia, Monroe, LA 67056). The VCR came with a "ninety-day warranty against all defects" and a guarantee for "in-home free repairs and labor." On October 30, Diane's VCR showed a horizontal line across the screen when she replayed tapes. She called the store manager, Jill Miller, and explained the problem. Jill said the horizontal lines were caused by a dirty head and told Diane to bring in the VCR for cleaning. Jill also told Diane that she would be charged for this service because dirty VCR heads were basic wear and, therefore, not covered by the warranty. Diane was angered by this response from the store manager and decided to write a letter of complaint. Based on the information provided, write Diane's letter of complaint.

7. Gregory Peña (121 Mockingbird Lane, San Marcos, NV 87900) has written a letter of complaint to Donya Kahlili, the manager of CompuRam (4236 Silicon Dr., Reno, NV 87601). Gregory purchased a computer from a CompuRam outlet in San Marcos. The San Marcos *Tattler* advertised that the computer "came loaded with all the software you'll need to write effective letters and perform basic accounting functions." (Gregory has a copy of this advertisement.) When Gregory booted up his computer, he expected to access word production software, multiple fonts, a graphics package, a grammar check, and a spreadsheet. All he got was a word processing package and a spreadsheet. Gregory wants Ms. Kahlili to upgrade his software to include fonts, graphics, and a grammar check; he wants a computer technician from CompuRam to load the software on his computer; and he wants CompuRam to reimburse him $400 (the full price of the software) for his trouble.

Ms. Kahlili agrees that the advertisement is misleading and will provide Gregory software including the fonts, graphics, and grammar check (complete with instructions for loading the software). However, she refuses his other two requests.

Write Ms. Kahlili's letter of partial adjustment to Gregory based on the information provided.

WEB WORKSHOP

Access any Internet search engine (Yahoo, Google, Lycos, Excite, etc.) and research a type of letter (sales, inquiry, good news, bad news, cover). You will find samples, guidelines, checklists, templates, and articles providing detailed analyses of these letters' pros and cons.

Review a sampling of these sites. Then, analyze your findings. How do the letters and guidelines compare to those discussed in this textbook?

a. Write a letter to your instructor detailing your findings.

b. If you find sample letters that you dislike, rewrite them according to the criteria provided in this textbook.

EXERCISES

Writing Letters

1. Write a sales letter. To do so, imagine what product or service you could provide to a potential client. Select one from your area of expertise or academic major. Write your letter according to the criteria for sales letters and the writing process techniques discussed in this chapter.

2. Write a letter of inquiry. You might want to write to a college or university requesting information about a degree program or to a manufacturer for information about a product or service. Whatever the subject matter, be specific in your request. Follow the criteria for writing letters of inquiry and the writing process techniques.

3. Write a cover letter. Perhaps your cover letter will preface a lab report you are working on in school, a report you are writing at work, or documentation you will need to send to a client. Use the criteria for writing cover letters and the writing process techniques.

4. Write a good-news letter. Commend a coworker for a job well done. Congratulate a subordinate for her promotion. Better yet, write a thank-you letter to a teacher showing your appreciation for her professionalism and dedication. Follow the criteria for writing a good-news letter and the writing process techniques.

5. Write a bad-news letter. Inform a job candidate that he has not been offered the job, tell a customer that her request for a refund is denied, or explain to employees why they are to be laid off from their work. Use the criteria for writing bad-news letters and the writing process techniques.

6. Write a complaint letter. Perhaps you have purchased a product that has malfunctioned, received poor service from a salesperson, ordered one item but received another, failed to have a warranty honored, or received damaged or broken equipment in shipment. You want to write a letter of complaint to solve these problems. To do so, write your letter by following the suggestions provided in this chapter.

7. Write an adjustment letter. Envision that a client has written a complaint letter about a problem he has encountered with your product or service. Write a letter in response to the complaint. This could either be a 100 percent yes letter, a 100 percent no letter, or a partial adjustment letter. Follow the suggestions provided in this chapter when writing your letter.

Team Projects

1. Find business letters in your home or at work and bring them to class. In small groups of three to five students, categorize these letters as either sales

letters, letters of inquiry, cover letters, good-news letters, bad-news letters, complaint letters, or adjustment letters. Then, in your groups, decide whether the letters are successful or unsuccessful, based on the criteria provided in this chapter. If the letters are good, be prepared to explain why, either in writing or orally. If the letters fail, discuss how they could be improved.

2. In small groups of three to five students, select any of the flawed letters from activity 1 and rewrite them. Use the revision techniques discussed in this chapter.

Quiz Questions

1. Is a letter external or internal company correspondence?

2. What are the eight essential letter components?

3. What are four optional letter components?

4. Why would you write a letter of inquiry?

5. What are the goals of a cover letter?

6. What are two illustrations of good-news letters?

7. Why would you begin a bad-news letter with a buffer?

8. What are three ways to adjust, or respond to, a letter of complaint?

9. How can you arouse reader interest in a sales letter?

10. What are characteristics of style of writing in a sales letter?

ACTIVITIES

Chapter Preview

How to Find Job Openings
Resources include networking, classified advertisements, and the Internet.

Criteria for Effective Resumes
Well-written resumes can open the door for you to get an interview.

Optional Resume Components
Personal data and references are not essential.

Style
How you write and design your resume is key to your success.

Methods of Delivery
You can send resumes by mail, post them on the Internet, or send through e-mail.

Criteria for Effective Letters of Application
A resume is generic, but a letter of application personalizes your job search.

Techniques for Interviewing Effectively
Consider our helpful tips before you interview for a job.

Samples
Examine sample resumes, letters of application, and follow-up letters.

Checklists
Our checklists will assist you in your job search.

Writing at Work

The job search involves at least two people—the applicant and the individual making the hiring decision. Usually more than two people are involved, however, because companies typically hire based on a committee's decision. That is the case at **DiskServe**. This St. Louis-based company is hoping to hire a Customer Service Representative for its computer technology department. DiskServe is eager to hire a new employee because one of its best workers has just advanced to a new position in the company. DiskServe asked applicants to apply using e-mail (the quickest means of communication). Thus, the applicants submitted an e-mail letter of application and an attached ASCII-scannable resume.

DiskServe advertised this opening in the career placement centers at local colleges, in the city newspaper, and online at its Web site: *http://www.DiskServe.com*.

In addition to DiskServe's chief executive officer (CEO), Harold Irving, the hiring committee will consist of two managers from other DiskServe departments, the former employee whose job is being filled, and two coworkers in the computer technology department.

Ten candidates were considered for the position. All candidates first had teleconference interviews. While Harold talked with the candidates, the other hiring committee members listened on a speaker phone. After the telephone interviews, five candidates were invited to DiskServe's work site for personal interviews—Macy Heart, Aaron Brown, Rosemary Lopez, Quisha Southerland, and Robin Scott.

Harold Irving, who has worked hard to create a family-oriented environment at DiskServe, values three traits in his employees: technology know-how, teamwork, and a positive attitude toward customers and coworkers. When the candidates arrived at DiskServe, Harold gave them a tour of the facilities, introducing them to many employees. Then the interviews began.

Each job candidate was asked a series of questions that included the following:

- What is your greatest strength? Give an example of how this reveals itself on the job.
- What did you like most about your past job?
- How have you handled customer complaints in the past?
- Where do you see yourself in five years?

Then, each candidate was taken to the computer repair lab and confronted with an actual hardware or software problem. The candidates were asked to solve the problem, and their work was timed. Finally, the applicants were allowed to ask questions about DiskServe and their job responsibilities.

Harold Irving is a stickler for good manners and business protocol. He waited 48 hours after the final interview to make his hiring decision. The wait time allowed him to check references. More important, he wanted to see which of the candidates wrote follow-up thank-you notes, and he planned to assess the quality of their communication.

Harold Irving takes the hiring process seriously. He wants to hire the best people because he hopes those employees will stay with the company a long time. Hiring well is a good corporate investment.

OBJECTIVES

A job search is demanding. You will have to know how to look for a job, write a resume and letter of application, interview for the job, and write a follow-up letter.

In this chapter, we help with your job search by giving you

- Ways to find job openings
- Criteria for effective resumes
- Criteria for effective letters of application
- Techniques for interviewing effectively
- Criteria for effective follow-up letters
- Sample resumes, letters of application, and follow-up letters

HOW TO FIND JOB OPENINGS

When it's time for you to look for a job, how will you begin your search? You know you can't just wander up and down the street, knocking on doors randomly. That would be time consuming, exhausting, and counterproductive. Instead, you must approach the job search more systematically.

Visit Your College or University Job Placement Center

This is an excellent place to begin a job search, for several reasons. First, your school's job placement service might have job counselors who will counsel you regarding your skills and job options. Second, your job placement center can give you helpful hints on preparing resumes, letters of application, and follow-up letters. Third, the center will post job openings within your community and possibly in other cities. Fourth, the center will be able to tell you when companies will visit campus for job recruiting. Finally, the service can keep on file your letters of recommendation or portfolio. The job placement center will send these out to interested companies upon your request.

Talk to Your Instructors

Whether in your major field or not, instructors can be excellent job sources. They will have contacts in business, industry, and education. They may know of job openings or people who might be helpful in your job search. Furthermore, because your instructors obviously know a great deal about you (having spent a semester or more working closely with you), they will know which types of jobs or work environments might best suit you.

Network with Friends and Past Employers

The latest studies tell us that networking is the best way to find jobs. Over 60 percent of jobs are found by talking to people you know. A July 2003 *Smart Money* magazine article reports that 62 percent of job searchers find employment through "face-to-face networking" (Bloch 2003, 12). A study performed by Drake Beam Morin confirms this, stating that "64 percent of . . . almost 7,500 people surveyed said they found their jobs through socializing and meeting people" (Drakeley 2003, 5).

Why is networking so important? It is simple math. Your friends and past employers know people. Those people know people. By visiting acquaintances, you can network with numerous individuals. The more people you talk to about your job quest, the more job opportunities you will discover.

Get Involved in Your Community

There are many ways to network. In addition to talking to your family or past employers, you can network by getting involved in your community. Consider volunteering for a community committee, pursuing religious affiliations, joining community clubs, or participating in fund-raising events. Join Toastmasters. Take classes in accounting, HTML, RoboHelp, or Flash at your local community college. Each instance not only teaches you a new skill but also provides an opportunity to meet new people. Then, "when a job comes open, you'll be on their radar screen" (Drakeley 2003, 6).

Check Your Professional Affiliations and Publications

If you belong to a professional organization, this could be a source of employment in three ways. First, your organization's sponsors or board members might be aware of job openings. Second, your organization might publish a listing of job openings. Finally, your organization's publications might list job opportunities.

Read the Want Ads

Check the classified sections of your local newspapers or in newspapers in cities you might like to live and work in. These want ads list job openings, requirements, and salary ranges.

Read Newspapers or Business Journals to Find Growing Businesses and Business Sectors

Which companies are receiving grants, building new sites, winning awards, or creating new service or product lines? Which companies have just gained new clients or received expanded contracts? Newspapers and journals report this kind of news, and a growing company or business sector might be good news for you. If a company is expanding, "they'll most likely be hiring resources next" (Drakeley 2003, 5). This means more job opportunities for you to pursue.

Take a "Temp" Job

Temporary ("temp") jobs, accessed through staffing agencies, are great for several reasons. First, you are getting paid while you look for a job. Second, you are acquiring new skills. Third, temp jobs help you network. Finally, many staffing agencies provide "temp-to-perm job opportunities" (Drakeley 2003, 6).

Get an Internship

Some internships are paid, but most are unpaid. That is the bad news. However, the good news is that an internship will provide you outstanding job preparedness skills, help you meet new people for networking, and improve your resume. Meanwhile, an unpaid internship in your preferred work area might lead to full-time employment. An internship "gives you the opportunity to show your skills, work ethic, positive attitude, and passion for your work." By interning, you can prove that you should be "the next employee the company hires" (Drakeley 2003, 7).

Research the Internet

In the mid-to-late 1990s, the Internet was the preferred means by which job seekers found employment. That has changed drastically. *Newsweek* magazine calls the Internet "a time waster." Quoting the head of a Chicago outplacement firm, *Newsweek* writes that some of the Internet's popular job sites are "big black holes"—your resume goes in, but you never hear from anyone again (Stern 2003, 67).

Others are equally pessimistic about the Internet's value as a source for jobs. Only 10 percent of technical or computer-related jobs are found from electronic job searches, "about 13 percent of interviews for managerial-level jobs result from responding to an online posting," and a mere 4 percent of jobs overall are found through the Internet (Bloch 2003, 12).

Nonetheless, you should make the Internet part of your job search strategy. If it is not the best place to find a job, the Internet still provides numerous benefits. Internet job search engines provide excellent job search resources. Table 7.1 lists numerous job search sites.

These sites and many more provide the following career information resources:

- **Resumes**—explaining the difference between resumes and curriculum vitae (CV), how to address gaps in your career history, avoiding the top 10 resume mistakes, and 15 tips for writing winning resumes.
- **Interviews**—interviewing to get the job and handling illegal questions.
- **Cover letters and thank-you letters**—sample cover letter techniques and ten ways to build a better thank-you letter.
- **Job search tips**—leaping to a twenty-first-century technology career and employing the correct netiquette.

TABLE 7.1	SITES FOR INTERNET JOB SEARCHES

- *http://www.Monster.com*—lets you post resumes and search for jobs, and provides career advice
- *http://www.CareerJournal.com*—the *Wall Street Journal's* career search site; provides salary and hiring information, a resume database, and job hunting advice
- *http://www.FlipDog.com*—provides national job listings in diverse fields by job title, location, and date of listing
- *http://www.HotJobs.yahoo.com*—lets you search for jobs by keyword, city, and state
- *http://www.WantedJobs.com*—lists jobs in the United States and Canada
- *http://www.WetFeet.com*—gives company and industry profiles, resume help, city profiles, and international job sites
- *http://www.CareerLab.com*—provides a cover letter library
- *http://www.CareerCity.com*—offers a detailed discussion and samples of functional versus chronological resumes
- *http://www.JobOptions.com*—lists job opportunities in the fields of accounting, customer service, engineering, human resources, sales, and technology
- *http://www.CareerMag.com*—allows for online job searches by keyword
- *http://www.JobLocator.com*—offers access to many other Internet job search engines
- *http://www.CareerShop.com*—helps you post and edit resumes, offers career advice, and lists hiring employers
- *http://www.CareerBuilder.com*—lets you search for jobs by company, industry, and job type

CRITERIA FOR EFFECTIVE RESUMES

Once you have found a job that interests you, it is time to apply. Your job application will start when you send the prospective employer your resume. Resumes are usually the first impression you make on a prospective employer. If your resume is effective, you have opened the door to possible employment—you have given yourself the opportunity to sell your skills during an interview. If, in contrast, you write an ineffective resume, you have closed the door to opportunity.

Your resume should present an objective, easily accessible, detailed biographical sketch. However, do not try to include your entire history. Because the primary goal of your resume, together with your letter of application, is to get an interview, you can use your interview to explain in more detail any pertinent information that does not appear on your resume.

When writing a resume, you have two optional approaches. You can write either a reverse chronological resume or a functional resume.

Reverse Chronological Resume

Write a reverse chronological resume if you

- Are a traditional job applicant (a recent high school or college graduate, aged 18 to 25)
- Hope to enter the profession in which you have received college training or certification
- Have made steady progress in one profession (promotions or salary increases)
- Plan to stay in your present profession

Functional Resume

Write a functional resume if you

- Are a nontraditional job applicant (are returning to the workforce after a lengthy absence, are older, are not a recent high school or college graduate)
- Plan to enter a profession in which you have not received formal college training or certification
- Have changed jobs frequently
- Plan to enter a new profession

Key Resume Components

Whatever type of resume you write, you should include these elements that will be explained in detail in the following sections.

- Identification
- Career objectives
- Employment
- Education
- Professional skills
- Military experience (if applicable)
- Professional affiliations

Identification

Begin your resume with the following:

- Your name (full first name, middle initial, and last name). Your name can be in boldface and printed in a larger type size (14-point, 16-point, etc.).
- Your address. Beneath your name, type your street address, your city, state (use the correct two-letter abbreviation provided in Chapter 19), and zip code. If you are attending college or serving in the armed forces, you might also want to include a permanent address. By including alternative addresses, you will enable your prospective employer to contact you more easily. What's the point of sending a resume if you can't be reached?
- Your area code and phone numbers. As with addresses, you can provide alternative phone numbers, such as cell phone numbers. However, limit yourself to two phone numbers, and don't provide a work phone. Having prospective employers call you at your present job is not wise. First, your current employer will not appreciate your receiving this sort of personal call; second, your future employer might believe that you often receive personal calls at work—and will continue to do so if he or she hires you. You can also include your e-mail or Web site address and fax number.

Career Objectives

A career objectives line is not mandatory. Some employers like you to state your career objectives; others do not. You will have to decide whether you want to include this section in your resume.

The career objectives line is like a subject line in a memo or report. Just as the subject line clarifies your memo or report's intent, your career objective informs the reader of your resume's focus.

Be sure your career objective is precise. Too often, career objectives are so generic that their vagueness does more harm than good, as in the following example:

example

Flawed Career Objective

Career Objective: Seeking employment in a business environment offering an opportunity for professional growth

This poorly constructed career objective provides no focus. What kind of business? What kind of opportunities for professional growth? Employers don't want to hire people who have only nebulous notions about their skills and objectives. Instead, look at the following clearly focused career objectives.

example

Good Career Objective

Career Objective: An entry-level position as a computer programmer

example

Better Career Objective

Career Objectives: To market financial planning programs and provide financial counseling to ensure positive client relations

Employment

The employment section lists the jobs you've held. This information must be presented in reverse chronological order (your current job listed first, previous jobs listed next). This section must include the following:

- Your job title (if you have or had one)
- The name of the company you worked for
- The location of this company (either street address, city, and state or just the city and state)
- The time period during which you worked at this job
- Your job duties, responsibilities, and accomplishments

This last consideration is important. This is your chance to sell yourself. Merely stating where you worked and when you worked there will not get you a job. Instead, what did you achieve on the job? In this part of the resume, you should detail how you met deadlines, trained employees, cut expenses, exceeded sales expectations, decreased overage, and so forth. Plus, you want to quantify your accomplishments, as follows:

Note: Location, Location, Location!

Our discussion of work before education is arbitrary. You should present your most important section first. If education is your strength and will help you get the job, lead with education. If, in contrast, your work experience is stronger, begin your resume with work.

Assistant Manager
  McConnel Oil Change, Beauxdroit, LA
  2003 to present
  • Tracked and maintained over $25,000 inventory ◄——— Specific details highlight achievements.
  • Trained a minimum of four new employees quarterly
  • Achieved a 10 percent growth in customer car count for three consecutive years
  • Developed a written manual for hazardous waste disposal, earning a Citizen's Recognition Award from the Beauxdroit City Council

Education

In addition to work experience, you must include your education. Document your educational experiences in reverse chronological order (most recent education first; previous schools, colleges, universities, military courses, and training seminars next). When listing your education, provide the following information:

- Degree. If you have not yet received your degree, you can write "Anticipated date of graduation June 2005" or "Degree expected in 2005."
- Area of specialization.
- School attended. Do not abbreviate. Although you might assume that everyone knows what *UT* means, your readers won't understand this abbreviation. Is UT the University of Texas, the University of Tennessee, the University of Tulsa, or the University of Toledo?
- Location. Include the city and state.
- Year of graduation or years attended.

As you can see, this information is just the facts and nothing else. Many people might have the same educational history as you. For instance, just imagine how many of your current classmates will graduate from your school, in the same year, with the same degree. Why are you more hirable than they are? The only way you can differentiate yourself from other job candidates with similar degrees is by highlighting your unique educational accomplishments. These might include any or all of the following:

- Grade point average
- Academic club memberships and leadership offices held
- Unique coursework
- Special class projects
- Academic honors, scholarships, and awards
- Fraternity or sorority leadership offices held
- Percentage of your college education costs you paid for while attending school
- Technical equipment you can operate

Please note a key concern regarding your work experience and education. You should have no chronological gaps when all of your work and education are listed. You can't omit a year without a very good explanation. (A missing month or so is not a problem.)

Professional Skills

If you are changing professions or reentering the workforce after a long absence, you will want to write a functional resume. Therefore, rather than beginning with education or work experience, which won't necessarily help you get a job, focus your reader's attention on your unique skills. These could include

- Proficiency with computer hardware and software
- Procedures you can perform
- Special accomplishments and awards you have earned
- On-the-job training you have received
- Training you have provided
- New techniques you have invented or implemented
- Numbers of and types of people you have managed
- Publications you have created or worked on
- Certifications you have earned
- Languages you speak, read, and write

These professional skills are important because they help show how you are different from all other applicants. In addition, they show that although you have not been trained in the job for which you are applying, you can still be a valuable employee.

example

Professional Skills

- Proficient in Microsoft Word, Excel, PowerPoint, and FrontPage
- Knowledge of HTML, Java, Visual Basic, and C++
- Certified OSHA Hazardous Management Safety Trainer
- Fluent in Spanish and English

Military Experience

If you served for several years in the military, you might want to describe this service in a separate section. You would state your

- Rank
- Service branch
- Location (city, state, country, ship, etc.)
- Years in service
- Discharge status
- Special clearances
- Achievements and professional skills
- Training seminars attended and education received

Professional Affiliations

If you belong to regional, national, or international clubs or have professional affiliations, you might want to mention these. Such memberships might include the Rotarians, Lions Club, Big Brothers, Campfire Girls, or Junior League. Maybe you belong to the Society for Technical Communication, the Institute of Electrical and Electronic Engineers, the National Office Machine Dealers Association, or the American Helicopter Society. Listing such associations emphasizes your social consciousness and your professional sincerity. However, if you maintain many such affiliations, your prospective employer might question your ability to handle your job and your numerous memberships.

Optional Resume Components

Personal Data

Some job recruiters say that personal data rounds out the person. Others believe that personal data is just that, personal, and therefore irrelevant for job purposes. You be the judge. If you want to include personal data, be sure they relate to the job you are seeking.

Don't include any of the following information: birthdate, race, gender, religion, height, weight, religious affiliations, marital status, or pictures of yourself. Equal opportunity laws disallow employers from making decisions based on these factors.

References

If you choose to include references (three or four colleagues, supervisors, teachers, or community individuals who will recommend you for employment), list their names, titles, addresses, and phone numbers. Instead of listing names, you might prefer to follow a reference heading by stating, "Supplied on request," "Available on request," or "Furnished on request."

References are not mandatory. As with all elements of a resume, you should include only information that will benefit you in your job search. If you do not have anyone who will provide you an outstanding reference, omit this section. However, if you have great references, include these individuals—after asking their permission first. Excellent references, if checked before an interview, can give you an edge.

Another reason you might have for not listing several references by name, address, and phone number is that these references take up space on your resume. If your job experiences and education are limited and you need additional information to fill up a page, include a list of references. However, your priority should be to emphasize your skills, the ways you will benefit a prospective employer. Don't waste precious space by listing references if you can provide more valuable information about yourself.

Style

The preceding information suggests *what* you should include in your resume. Your next consideration is *how* this information should be presented. As mentioned throughout this textbook, format or page layout is essential for effective technical writing. The same holds true for your resume.

Avoid Sentences

Sentences create three problems in a resume. First, if you use sentences, the majority of them will begin with the first-person pronoun *I*. You'll write, "I have . . . ," "I graduated . . . ," or "I worked . . . ." Such sentences are repetitious and egocentric. Second, if you choose to use sentences you'll run the risk of committing grammatical errors: run-ons, dangling modifiers, agreement errors, and so forth. Third, sentences will take up room in your resume, making it longer than necessary.

Itemize Your Achievements

Instead of sentences, highlight your resume with easily accessible lists. Set apart your achievements by bulletizing your accomplishments, awards, unique skills, and so on.

Begin Your Lists with Verbs

To convey a positive, assertive tone, use verbs when describing your achievements. Following is a list of verbs you might use.

accomplished	initiated
achieved	installed
analyzed	led
awarded	made
built	maintained
completed	managed
conducted	manufactured
coordinated	negotiated
created	ordered
customized	organized
designed	planned
developed	prepared
diagnosed	presented
directed	programmed
earned	reduced
established	resolved
expanded	reviewed
fabricated	sold
gained	supervised
implemented	trained
improved	

Quantify

Your resume should not tell your readers how great you are; it should prove your worth. To do so, quantify, specify, precisely explain your achievements. For example, don't write, "Maintained positive customer relations with numerous clients." Instead write, "Maintained positive customer relations with 5,000 retail and 90 wholesale clients." Don't write, "Improved field representative efficiency through effective training." Instead write, "Improved field representative efficiency by writing corporate manuals for policies and procedures." Again, do not write, "Achieved production goals." Write, "Achieved 98 percent production, surpassing the company's desired goal of 90 percent."

Format Your Resume for Reader-Friendly Ease of Access

You can make your resume more accessible in many ways. For example, you can capitalize and boldface headings. You can indent subheadings to create white space. You can underline subheadings and italicize major achievements. As already mentioned, you should use bullets to create lists. You also might want to use different font sizes to highlight headings, subheadings, and achievements. However, don't use every highlighting technique at your disposal. Select only a few; otherwise, your resume will be overwhelmed with technique instead of substance.

Make It Perfect

You cannot afford to make an error in your resume. Remember, your resume is the first impression you'll make on your prospective employer. Errors in your resume will make a poor first impression.

METHODS OF DELIVERY

When writing either a reverse chronological or a functional resume, you can deliver your document in several ways:

Mail Version

The traditional way to deliver a resume is to insert it into an envelope and mail it. This resume can be highly designed, using bullets, boldface, horizontal rules, indentations, and different font sizes. Because this document will be a hard copy, what the reader sees will be exactly what you mail.

Do not be tempted to overdesign your resume, however. For example, avoid decorative fonts, like Bauhaus, *Lucida Handwriting,* or Tekton. "Refrain from using clip art, borders, photos and other images." Do not print your resume on unusual colors, like salmon, baby blue, tangerine, or yellow. "What you think may separate your resume from the rest of the pack may just only add unnecessary clutter" ("Powerful Resume" 2003). Instead, stick to heavy white paper and standard fonts, like Times New Roman and Arial.

Figures 7.1 and 7.2 on pages 231–232 are excellent examples of traditional resumes, ready to be mailed.

Web (HTML) Resume

In addition to your hard-copy resume, you might want to consider creating an online resume for transmittal over the Internet. Online resumes provide several benefits not reaped by the traditional resume printed on paper (Dikel 1999, 2). Your Web-based resume will

- Allow thousands of international employers to access it instantly
- Allow you to include portfolios of your work with lengthy documents, full-color graphics, animation, and sound
- Allow you to update it quickly and easily for global access
- Allow you to show your expertise with HTML (Hypertext Markup Language) and Web-ready graphics

If you create a Web site for your resume, using HTML coding, consider our criteria for effective Web sites discussed in Chapter 13:

- Use a navigational bar, allowing your readers to access the various parts of your online resume.
- Include a Home button, taking your reader back to the first screen of your Web site.
- Maintain optimum contrast between text and background for easy readability.

Resume Templates

Microsoft Word provides you a resume template if you need help getting started. On the menu bars, click on File, New, and then click on the General Templates hyperlink in the new Document task pane. On the Other Documents tab you will find optional resume layouts (Elegant Resume, Professional Resume, Contemporary Resume, and a Resume Wizard).

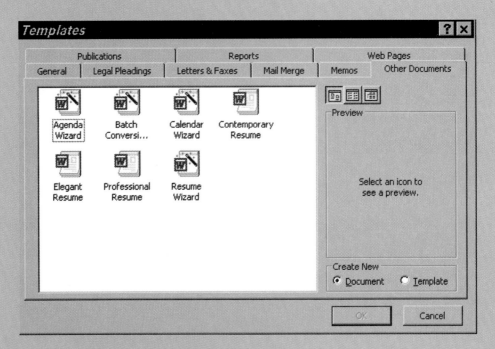

These resume templates provide benefits as well as create a few problems.

On the positive side, the templates are great reminders of what to include in your resume. They show you where your name goes and remind you to include objectives, work experience, education, and skills.

In contrast, the templates also limit you. They mandate font sizes and page layout. More important, if you use the same templates that everyone else does, then how will your resume stand out as unique?

A good compromise is to review the templates for ideas and then create your own resume with your unique layout.

- Include headings and subheadings to ensure your readers do not get lost in cyberspace.
- Limit line length to no more than approximately two-thirds of your viewable screen.
- Include highlighting techniques, such as graphics, white space, boldface headings, horizontal rules, and font sizes appropriate for various heading and subheading levels. However, do not use highlighting excessively and avoid blinking text, garish colors, or standard clip art.

We discourage using the word *resume* when you write a hard-copy resume. This fact will be obvious to your readers. However, the word *resume* should be included on your Web resume. Internet search engines usually categorize Web sites in one of two ways: either by your document title or by your URL (the Web site's address). Thus, to accommodate this fact, you should include the word *resume* in your document title and in your URL as the directory or file name (Skarzenski 1996, 18).

Sharon J. Barenblatt

787 Rainbow Avenue
Boston, MA 12987

Phone: 202-555-2121
E-mail: sharonjb@juno.com

Objective   Employment as a computer maintenance technician in ⟵ *State precise career objectives.*
the telecommunications industry

Experience   **Salesperson/Assistant Department Manager**
Electronic Warehouse   Boston, MA   2000–present
• Prepare nightly deposits, input daily receipts
• Open and close the store
• Troubleshoot computer systems   ⟵ *Use present tense verbs for current jobs.*
• Provide customer service
• Sell and support computer systems

Computer Lab Assistant
Boston Community College   Boston, MA   1999–2000
• Assisted students with computer application programs
• Administered and proctored tests   ⟵ *Use past tense verbs for prior employment.*
• Performed troubleshooting of computer hardware and
software problems

Salesperson
Online Computer Center   Boston, MA   1997–1999
• Demonstrated and sold computer systems

Market Research
PRM Research Company   Newcastle, MA   1996–1997
• Performed telephone surveys (20 hours per week while
going to high school)

Education   Boston Community College   Boston, MA
2005, Anticipated date of graduation
• AS, Computer Science
• 3.7 GPA while working 30 hours a week and going to
college full time

Expertise   Troubleshooting computer systems, telecommunications,
programming, and data entry

References   Available upon request

Figure 7.1   Reverse Chronological Resume

E-mail Resume

The U.S. mail—often referred to as "snail mail"—can take several days. E-mail is faster. Let's say that you have just found a job opening and you want to submit your resume *now!* The quickest way to get it to the prospective employer is as an e-mail attachment. Speed isn't the only issue. "Hiring managers and recruiters have become as addicted to e-mail as everyone else. More than one-third of human-resource professionals reported a preference for e-mailed resumes,

tech link

Read Kirsten Dixson's article titled "Crafting an E-mail Resume": *http://www. businessweek.com/careers/ content/nov2001/ ca20011113_7790.htm.*

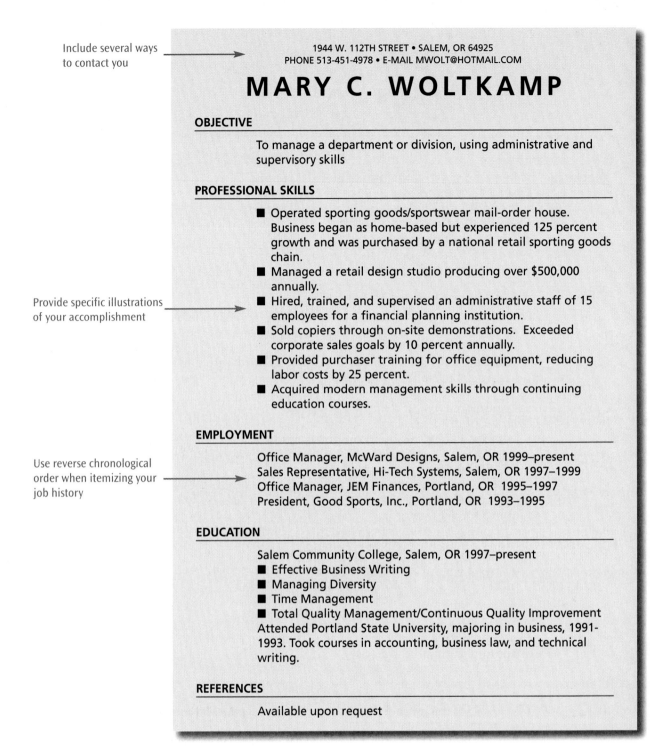

Include several ways to contact you

1944 W. 112TH STREET • SALEM, OR 64925
PHONE 513-451-4978 • E-MAIL MWOLT@HOTMAIL.COM

MARY C. WOLTKAMP

OBJECTIVE

To manage a department or division, using administrative and supervisory skills

PROFESSIONAL SKILLS

Provide specific illustrations of your accomplishment

- Operated sporting goods/sportswear mail-order house. Business began as home-based but experienced 125 percent growth and was purchased by a national retail sporting goods chain.
- Managed a retail design studio producing over $500,000 annually.
- Hired, trained, and supervised an administrative staff of 15 employees for a financial planning institution.
- Sold copiers through on-site demonstrations. Exceeded corporate sales goals by 10 percent annually.
- Provided purchaser training for office equipment, reducing labor costs by 25 percent.
- Acquired modern management skills through continuing education courses.

EMPLOYMENT

Use reverse chronological order when itemizing your job history

Office Manager, McWard Designs, Salem, OR 1999–present
Sales Representative, Hi-Tech Systems, Salem, OR 1997–1999
Office Manager, JEM Finances, Portland, OR 1995–1997
President, Good Sports, Inc., Portland, OR 1993–1995

EDUCATION

Salem Community College, Salem, OR 1997–present
- Effective Business Writing
- Managing Diversity
- Time Management
- Total Quality Management/Continuous Quality Improvement
Attended Portland State University, majoring in business, 1991–1993. Took courses in accounting, business law, and technical writing.

REFERENCES

Available upon request

Figure 7.2   Functional Resume

according to a [recent] survey by the Society for Human Resource Management" (Dixson 2001).

Your obvious choice, when using e-mail, is to send the resume as an attachment. Be careful, however! Before you send the resume as an attachment, "heed the directions of the employer you're e-mailing." Some employers may want an attachment, "while others may specify that you send no attachments" (Dixson

2001). Instead, they might want you to send the resume as the actual e-mail message. This is called an ASCII resume, which we discuss next in this chapter.

If you choose to submit the resume as an attachment, be sure to write a brief e-mail cover message (we discuss this later in the chapter as well, under letters of application). In addition, you must clarify what software you have used for this attachment—Microsoft Word, Works, or WordPerfect, for example.

ASCII Resume (Scannable)

An ASCII resume (pronounced *As Key*) can be e-mailed or sent through the U.S. mail. To create an ASCII resume, you type your text using Notepad for Windows, Simpletext for Macintosh, or Note Tab, which is available as freeware. You also could type your resume using Microsoft Word and save the document as a text file, with a *.txt* extension (Dikel 1999, 3).

tech link

Go to *http://www. quintcareers.com/ scannable_resumes.html* to read Randall Hansen's article "Scannable Resume Fundamentals: How to Write Text Resumes."

You want to create an ASCII resume for two reasons:

1.   *ASCII is compatible with all systems.* ASCII files can be read on any e-mail provider, "as a downloadable addition to a Web resume," in "online job boards," on a PC or Mac platform, and by any word processing software such as Microsoft Word, Works, WordPerfect, and so on (Dixson 2001). Here is why.

   All computers read standard ASCII. In contrast, all computers do not read extended ASCII. Standard ASCII contains the following characters:

- Letters *A* through *Z*, both upper- and lowercase
- Numbers 0 through 9
- The following punctuation marks: ! " " # $ & ' ( ) * + , − . / : ; <=> ? ' _ ~

   If you are trying to send your online resume to a computer that supports only standard ASCII but you have used extended ASCII, then the person receiving your online resume will not be able to read all that you have written. For instance, computers supporting only standard ASCII cannot display em dashes or "smart" quotation marks, italics, colored fonts, dingbats, reverse text, underlining, and shadowing.

   Use these techniques for creating your ASCII resume for transmittal by e-mail (Dikel 1999, 3; LeVie 2000, 11; "Polishing" 2003; Skarzenski 1996.).

- Avoid italics, underlining, colors, horizontal and vertical bars, and iconic bullets.
- White space is still important, but do not use your Tab key for spacing. Tabs will be interpreted differently in different computer environments. Use your space bar instead.
- Avoid organizing information in columns.
- Do not center text.
- Limit your screen length to approximately 60 characters per line.
- Hit a hard return at the end of each line instead of using word wrap.
- Do not worry about limiting yourself to the traditional one-page resume. You cannot fit as many words on an ASCII page, so it is acceptable to submit an ASCII resume of two pages or more. Furthermore, in e-mail, the reader will scroll.

Obviously, an ASCII resume is a "bare-bones document." It is missing many of the typical design elements that you use in other technical writing. That is important to ensure that your resume is readable on all electronic platforms. However, you still want the resume to be visually appealing, and you want your content to be accessible. Therefore, use these simple highlighting techniques:

- Use all-cap headings and place your text below the headings, spacing for visual appeal.
- Create bullets using an asterisk (*) or a hyphen (-).
- Create horizontal rules by using the equal sign (=) or underline (_).

Figure 7.3 on page 235 shows an excellent example of an ASCII resume.

2. *An ASCII resume is scannable.* Many companies no longer hire people to read resumes. Instead, they use computers to screen resumes with a modern technique called electronic applicant tracking. The company's computer program scans resumes as raster (or bitmap) images. Then optical character recognition (OCR) software translates the bitmap images into standard ASCII. Next, the software uses artificial intelligence to read the ASCII text, scanning for keywords. If your resume contains a sufficient number of these keywords, the resume will then be given to someone in the human resources department for follow-up. If your online resume is created using extended ASCII, OCR software will have trouble translating your text (McNair 1997). ASCII text is so plain that nothing will get in the way of the scanner.

To ensure that your scannable document is clear of distractions, achieve these goals:

- Use high-quality paper (send an original), optimum contrast (black print on white paper), and paper without wrinkles or folds (if you are mailing the resume).
- Use a Courier, Helvetica, or Arial typeface (10- to14-point type).
- Place your name at the top of the page. "Scanners assume that whatever is at the top is your name. If your resume has two pages, place your name and a 'page two' designation on the second page, and attach with a paper clip—no staples" (Kendall 2003).
- Include a summary of qualifications *full of keywords*.

Keywords are the most important feature of scannable resumes. OCR searches focus on keywords and phrases specifically related to the job opening. Therefore, this must be your primary focus as well (Hansen 2003).

The keywords include job titles, skills and responsibilities, corporate buzzwords, acronyms and abbreviations related to hardware and software, academic degrees, and certifications.

When using keywords, you want to be specific. For example, you do not want to write that you have knowledge of "various software products"—that is too vague. Be more precise: write that you can "create online help using RoboHelp" and that you can use "PageMaker and Quark." Similarly, don't say you are familiar with "computer technology." Instead, write that you are proficient in "multimedia, HTML, and Windows and Macintosh platforms" (McNair 1997, 14).

Other industry keywords are shown in Table 7.2.

In addition to hardware and software skills, OCR also scans for "soft skills," as shown in Table 7.3.

Rochelle J. Kroft
1101 Ave. L
Tuscaloosa, AL 89403
Home: 313-690-4530
Cell: 313-900-6767
E-mail: rkroft90@aol.com

OBJECTIVES
==

Show reader benefit.

To use HAZARDOUS WASTE MANAGEMENT experience and knowledge to ensure company compliance and employee safety.

PROFESSIONAL ACCOMPLISHMENTS
==

* HAZARDOUS WASTE MANAGEMENT with skills in teamwork, end-user support, quality assurance, problem solving, and written documentation.
* Five years' experience working with international and national businesses and regulatory agencies.
* Skilled in assessing environmental needs and implementing hazardous waste improvement projects.
* Able to communicate effectively with multinational, cross-cultural teams, consisting of clients, vendors, coworkers, and local and regional stakeholders.
* Excellent customer service, using strong problem-solving techniques.
* Effective project management skills, able to multitask.

COMPUTER PROFICIENCY
==
Microsoft Windows XP, FrontPage, PowerPoint, C++, Visual Basic, Java, CAD/CAM

EXPERIENCE
==

No designer highlighting techniques are used in a scannable resume. Instead use all capital letters, equal signs, double spaces, and asterisks for visual appeal.

Hazardous Waste Manager
Shallenberger Industries, Tuscaloosa, AL (2000–present)
* Assess client needs for root cause analysis and recommend strategic actions.
* Oversee waste management improvements, using project management skills.
* Conduct and document follow-up quality assurance testing on all newly developed applications to ensure compliance.
* Develop training manuals to ensure team and stakeholder safety. Shallenberger had had NO injuries throughout my management.
* Manage a staff of 25 employees.
* Achieved Citizen's Recognition Award from Tuscaloosa City Council for safety compliance record.

Hazardous Waste Technician
CleanAir, Montgomery, AL (1998–2000)
* Developed innovative solutions to improve community safety.
* Created new procedure manuals to ensure regulatory compliance.

EDUCATION
==
BS Biological Sciences, University of Alabama, Tuscaloosa, AL (1997)
* Biotechnology Honor Society president (1996)
* Golden Key National Honor Society

AFFILIATIONS
==
Member, Hazardous Waste Society International

Figure 7.3   ASCII Scannable Resume

TABLE 7.2 HARDWARE AND SOFTWARE KEYWORDS	
NT	FrontPage
DOS	AutoCAD
Microsoft Office	LAN/WAN
Corel Office Suite	IBM Client Access
Lotus Notes	CAD/CAM
Novell GroupWise	Windows XP
Internet Explorer	Visual Basic
Netscape Communicator	IBM AS/400
Flow Chart	C++
Excel	Dreamweaver
PowerPoint	Flash

TABLE 7.3 EMPLOYMENT SKILLS KEYWORDS	
Oral presentations	Customer service
Oral communication	Telemarketing
Effective writing skills	Marketing
Interpersonal communication	Product information
Teamwork	Self-motivated
Flexibility	Organization
Time management	Innovation
Management	Ethical
Web design	Quality assurance
Project management	Training
Supervision	Problem solving

CRITERIA FOR EFFECTIVE LETTERS OF APPLICATION

Your resume, whether hard copy or online, will be prefaced by a letter of application (or cover e-mail if you submit the resume as an e-mail attachment). These two components of your job package serve different purposes.

The resume is generic. You'll write one resume and use it over and over again when applying for numerous jobs. In contrast, the letter of application is specific. Each letter of application will be different, customized specifically for each job. Whereas the resume will always stay the same, except for periodic updating, your letters of application will change with each new job search.

Criteria for an effective letter of application include the following.

Letter Essentials

As discussed in Chapter 6, letters contain certain mandatory components: your address, the date, your reader's address, a salutation, the letter's body, a complimentary close, your signed name, your typed name, and an enclosure notation if applicable. If you are submitting an electronic resume along with an e-mail cover message, you will not need these letter essentials. We discuss the e-mail cover message later in this chapter.

Introduction

In your introductory paragraph, include the following:

- Tell where you discovered the job opening. You might write, "In response to your advertisement in the May 31, 2005, *Lubbock Avalanche Journal . . .*" or "Bob Ward, manager of human resources, informed me that . . ."

- State which specific job you are applying for. Often, the classified section of your newspaper will advertise several jobs at one company. You must clarify which of those jobs you're interested in. For example, you could write, "Your advertisement for a computer maintenance technician is just what I have been looking for."

- Sum up your best credentials. "My BS in chemistry and five years of experience working in a hazardous materials lab qualify me for the position."

Discussion

In the discussion paragraph(s), sell your skills. To do so, describe your work experience, your education, and your professional skills. This section of your letter of application, however, is not meant to be merely a replication of your resume. The resume is generic; the letter of application is specifically geared toward your reader's needs. Therefore, in the discussion, follow these guidelines:

- Focus on your assets uniquely applicable to the advertised position. Select only those skills from your resume that relate to the advertisement and which will benefit the prospective employer.

- Don't explain how the job will make you happy: "I will benefit from this job because it will teach me valuable skills." Instead, using the pronouns *you* and *your*, show reader benefit: "My work with governmental agencies has provided me a wide variety of skills from which your company will benefit."

- Quantify your abilities. Don't just say you're great ("I am always looking for ways to improve my job performance"). Instead, prove your assertions with quantifiable facts: "I won the 2005 award for new ideas saving the company money."

Conclusion

Your final paragraph should be a call to action. You could say, "I hope to hear from you soon" or "I am looking forward to discussing my application with you in greater detail." You could tell your reader how to get in touch with you: "I can be reached at 913- 469-8500." If you are more daring, you could write, "I will contact you within the next two weeks to make an appointment at your convenience. At that time I would be happy to discuss my credentials more thoroughly."

In addition to these suggestions, you should mention that you have enclosed a resume. You can do this either in the introduction, discussion, or conclusion. Select the place that best lends itself to doing so.

Style

Your letter of application should follow the guidelines for all successful technical writing.

- **Accuracy**—You cannot commit any errors (whether grammatical or autobiographical). Grammatical errors in your letter of application will suggest that you might make errors at work as well. Autobiographical errors constitute a lie. In either instance, you will have hurt your employment opportunities.
- **Conciseness**—A low fog index will help your readers focus their attention on your skills rather than on your elaborate sentence structure. Don't ramble on and on. Decide exactly what needs to be said, and then say it.
- **Audience recognition**—Show how you will benefit the employer. Don't tell the employer how his or her job will help you. Try to limit use of *I*; emphasize *you* and *your*.
- **Clarity**—Show clearly how your skills relate to the advertised position and quantify your assertions.

E-MAIL COVER MESSAGE

If you submit the resume as an attachment to e-mail, you should write a brief e-mail cover message. This e-mail message will serve the same purpose as a hard-copy letter of application. Therefore, you want to include an introductory paragraph, a body, and a conclusion.

- **Introduction**—Tell the reader which job you are applying for and where you learned of this position.
- **Body**—State that you have attached a resume. Tell the reader which software you have used to write your resume.

> **example**
>
> I have attached a resume for your review. To open this document, you will need Microsoft Word.
>
> I have attached a resume, saved as an RTF (rich text file).

Next, briefly explain why you are the right person for the job. You can do this in a short paragraph or by briefly listing three to five key assets. Remember, however, that your e-mail message will be more concise than a hard-copy letter. Whereas a letter can contain over 50 lines of text, the e-mail should not exceed 20–25 lines. You want to limit your reader's need to scroll (we discuss this point in detail in Chapter 5).

- **Conclusion**—Sum up your e-mail message pleasantly. Tell your reader that you would enjoy meeting him or her and that you look forward to an interview.

In addition to the standard three-paragraph text, your e-mail cover message requires the additional key concerns (Dixson 2001):

- **Do not use your current employer's e-mail system**—That clearly will tell your prospective employer that you misuse company equipment and company time.

- **Avoid unprofessional e-mail addresses**—Addresses such as Mustang65@aol.com, Hangglider@yahoo.com, or ILuvDaBears@juno.com are inappropriate for business use. When you use e-mail to apply for a job, it is time to dump your old e-mail address and become more professional. Use your initials or your name instead.
- **Send one e-mail at a time to one prospective employer**—Do not mass mail resumes. No employer wants to believe that he or she is just one of hundreds to whom you are writing.
- **Include a clear subject line**—Announce your intentions or the contents of the e-mail: "Resume—Vanessa Diaz" or "Response to Accountant Job Opening."

SAMPLE E-MAIL COVER MESSAGES AND LETTERS OF APPLICATION

Figure 7.4 shows an effective cover e-mail message prefacing an attached resume. Figures 7.5 and 7.6 are examples of effective letters of application.

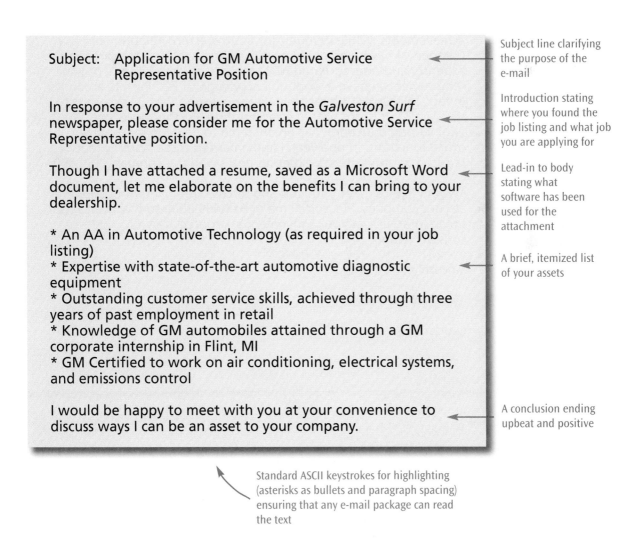

Subject line clarifying the purpose of the e-mail

Introduction stating where you found the job listing and what job you are applying for

Lead-in to body stating what software has been used for the attachment

A brief, itemized list of your assets

A conclusion ending upbeat and positive

Standard ASCII keystrokes for highlighting (asterisks as bullets and paragraph spacing) ensuring that any e-mail package can read the text

Figure 7.4   Cover E-mail Message

11944 West 112th Street
Denton, TX 77892

December 11, 2005

Mr. Oscar Holway
Valley Telephone Cooperative, Inc.
Raymonville, TX 78580

Dear Mr. Holway:

I am responding to your advertisement in the November 24, 2005, issue of *Telephony* for an accounting office supervisor. Because of my five years' experience in the telecommunications industry and my BS in electrical engineering, I believe I have the skills you require.

Although I have enclosed a resume, let me elaborate on my achievements. While working for Southwestern Telephone Company, I have managed the accounting activities for forecasting, expense, and nonregulated FCC accounts. As listed in one of your requirements, I have a working knowledge of FCC Part 32 accounts. In fact, my current project involves converting Part 31 accounts to the Part 32 uniform system of accounting. Furthermore, my engineering education has enabled me to develop mechanized office functions. In one year, I implemented three systems that reduced Southwestern's expenses and saved the company $50,000.

I would like to meet with you and discuss employment possibilities at your company. Please call me at 913-469-8500 so that we can set up an interview at your convenience. I appreciate your consideration.

Sincerely,

Leonard Liss

Leonard Liss

Enclosure

In the introduction, state where you learned about the job and which job you are applying for.

In the body, focus on ways you meet the requirements of the job.

Provide a contact number in the conclusion. End positively to create a good tone.

Figure 7.5   Letter of Application

4628 Little Avenue
Coral Gables, FL 33133

September 9, 2005

Ms. Karen Pechis
King Petroleum Company
Oakland Division
P.O. Box 189924
Oakland, CA 94610

Dear Ms. Pechis:

The September 4, 2005, *Oakland Voice* contained an advertisement from King Petroleum Company for an environmental protection specialist. After reading the required qualifications, I believe I can meet King's needs.

As my enclosed resume indicates, I have an MS in groundwater chemistry. I currently work for a county environmental agency where I have gained six years' experience with the various applications you mentioned in your advertisement. I have performed landfill groundwater monitoring at regional and national parks and supervised a team of 12 solid waste disposal monitoring specialists. In addition, I coordinated spill cleanups off the Atlantic and Gulf coasts and reported hazardous waste disposal procedures for local, state, and federal environmental agencies. In each of these instances, I successfully achieved governmental compliance. In fact, my supervisors cited my work as exemplary and entrusted me with training our new personnel.

Provide specific examples of your experience.

I welcome the challenge of working with King Petroleum as an environmental specialist. So that I can answer any of your questions, I would appreciate your calling me at 312-555-2121. I look forward to interviewing with you.

Include contact information.

Sincerely,

Glenn Pfeiffer

Glen Pfeiffer

Enclosure: resume

Figure 7.6   Letter of Application

Techniques for Interviewing Effectively

The goal of writing an effective resume and letter of application is to get an interview. The resume and letter of application open the door; only a successful interview will win you the job. In fact, some sources suggest that the interview is *the most important stage of your job search*. The Society for Human Resource Management states that 95 percent of respondents to a 2003 survey ranked "interview performance" as a "very influential" factor when deciding to hire an employee. "Interview performance [is] more influential than 17 other criteria, including years of relevant work experience, resume quality, education levels, test scores or references" (Stafford 2003, L1).

Here is how to interview successfully.

tech link

Go to *http://www.prenhall. com/gerson* for Web links, samples, and interactive activities.

1. *Dress for success.* Although this adage has become a cliché, it's still valid. The key to successful dressing is to wear clean, conservative clothing. No one expects you to spend money on high-fashion, stylish clothes, but everyone expects you to look neat and acceptable. Men and women's suits are still best.

2. *Be on time.* In fact, plan to arrive at your interview at least 20 to 30 minutes ahead of schedule. This will help you stay on time if you get lost or have difficulty finding a parking place. Also, being early will allow you to stop off in the restroom for a final hair brushing or tie straightening. Finally, being 5 or 10 minutes early makes a good impression on your prospective employer, who will recognize your enthusiasm.

3. *Don't chew gum, smoke, or drink too much caffeine (coffee, pop, etc.).* The gum will distort your speech, the cigarette could be offensive, and the caffeine might make you jittery.

4. *Watch your body language.* To make the best impression, don't slouch. Instead, sit straight in your chair, even leaning forward a little. Look your interviewer in the eye.

5. *Control your speaking.* Speak slowly, and don't ramble on and on. Once you've answered the questions satisfactorily, stop. Don't monopolize the interview.

6. *Come prepared.* This entails three things. First, you can bring to the interview a few supporting documents, including letters of recommendation, articles you've published or papers you've written, employer performance appraisals, transcripts, extra copies of your resume, laboratory results, schematics, and so on. Don't go overboard, however. Just two or three items will do.

Second, research the company so you can ask a few informed questions about its organization. These will show the interviewer that you are sincerely interested in the company. This is essential when preparing for your interview. Dr. Judith Evans, vice president of Right Management Consultants of New York, says that the most successful job candidates show interviewers that they "know the company inside and out" (Kallick 2003, D1). Researching a company differentiates you from other candidates. You can accomplish this goal easily by scanning a company's Internet site. These will give you an understanding of the company's culture, mission statement, product line, and services offered. Research will tell you if the company does business internationally, which companies represent the primary competition, where corporate offices are located, and who is on the board of directors. Then, with this knowledge, you can interview "like an insider," showing that you are thorough and have come prepared (Kallick 2003, D1).

Of equal importance, your research will help you decide if the company is right for you. After all, an interview answers two questions: do they want to hire you, and do you want to work for them?

Finally, know the types of questions you will be asked and be ready with answers. Some typical questions include the following:

- What are your strengths?
- What are your weaknesses?
- Why do you want to work for this company?
- Why are you leaving your present employment?
- What did you like most about your last job?
- What did you like least about your last job?
- How will you benefit this company?
- What computer hardware are you familiar with, and what computer languages do you know?
- What machines can you use?
- What special techniques do you know, or what special skills do you have?
- How do you get along with colleagues and with management?
- Can you travel?
- Will you relocate?
- What starting salary would you expect?
- What do you want to be doing in 5 years, 10 years?
- What about this job appealed to you?
- How would you handle this (hypothetical) situation?
- What was your biggest accomplishment in your last job or while in college?
- What questions do you have for us?

7. *When answering questions, focus on the company's specific needs.* For example, if the interviewer asks if you have experience using FrontPage, explain your expertise in that area, focusing on recent experiences or achievements. Be specific. In fact, you might want to tell a brief story to explain your knowledge. This is called "behavioral description interviewing" (Ralston et al. 2003, 9). It allows an interviewer to learn about your speaking abilities, organization, and relevant job skills. To respond to a behavioral description interview question, answer questions as follows (Ralston, et al. 2003, 11):

- Organize your story chronologically.
- Tell who did what, when, why and how.
- Explain what came of your actions (the result of the activity).
- Depict scenes, people, and actions.
- Make sure your story relates exactly to the interviewer's needs.
- Stop when you are through—do not ramble on. Get to the point, develop it, and conclude.

If, however, you do not have the knowledge required, then "explain how you can apply the experience you *do* have" (Hartman 2003, 24). You could say, "Although I've never used FrontPage, I have built Web sites using Netscape Composer and HTML coding. Plus, I'm a quick learner. I was able to learn RoboHelp well enough to create online help screens in only a week. Our customer was very happy with the result." This will show that you understand the job and can adapt to any task you might be given.

CRITERIA FOR EFFECTIVE FOLLOW-UP CORRESPONDENCE

Once you have interviewed, don't just sit back and wait, hoping that you will be offered the job. Write a follow-up letter or e-mail message. This follow-up accomplishes three primary things: it thanks your interviewers for their time, keeps your name fresh in their memories, and gives you an opportunity to introduce new reasons for hiring you.

A follow-up letter or e-mail message contains an introduction, discussion, and conclusion.

1. *Introduction.* Tell the readers how much you appreciated meeting them. Be sure to state the date on which you met and the job for which you applied.

2. *Discussion.* In this paragraph, emphasize or add important information concerning your suitability for the job. Add details that you forgot to mention during the interview, clarify details that you covered insufficiently, and highlight your skills that match the job requirements. In any case, sell yourself one last time.

3. *Conclusion.* Thank the readers for their consideration, or remind them how they can get in touch with you for further information. Don't, however, give them any deadlines for making a decision.

SAMPLE FOLLOW-UP LETTER

Figure 7.7 is an example of an effective follow-up letter. This letter succeeds for several reasons. First, it is short, merely reminding the reader of the writer's interest, instead of overwhelming him or her with too much new information. Second, the letter is positive, using words such as *enjoyed, skills, welcome, opportunity, exciting,* and *thank you.* Finally, the letter reemphasizes why the writer is perfect for the job.

CHECKLISTS

Seeking a job is not easy. However, it can be a manageable activity if you approach it as a series of separate but equal tasks. To get the job you want, you must search for an appropriate position, write an effective resume and letter of application, interview successfully, and write a follow-up letter.

4628 Little Avenue
Coral Gables, FL 33133

November 1, 2005

Ms. Karen Pechis
King Petroleum Company
Oakland Division
P.O. Box 189924
Oakland, CA 94610

Dear Ms. Pechis:

I enjoyed meeting you and your colleagues last Wednesday, October 27, 2005, to discuss the opening for an environmental protection specialist.

You stated in the interview that King Petroleum is planning to expand into offshore exploration. With my training skills and governmental compliance reporting abilities, I would welcome the opportunity to become involved in this exciting expansion.

Again, thank you for your time and consideration.

Sincerely,

Glenn Pfeiffer

Glen Pfeiffer

Remind the reader when you interviewed for a job.

In the body, include a follow-up to topics discussed in the interviews.

Figure 7.7   Follow-up Letter

Job Search Checklist

Job Openings

__ **1.** Did you visit your college or university job placement center?

__ **2.** Did you talk to your professors about job openings?

__ **3.** Have you networked with friends or past employers?

__ **4.** Have you checked with your professional affiliations or looked for job openings in trade journals?

__ **5.** Did you read the want ads in the newspapers?

__ **6.** Did you search the Internet for job openings?

Resume

__ **1.** Are your name, address, and phone number correct?

__ **2.** Is your job objective specific?

__ **3.** Are the dates within your education, work experience, and military experience sections accurate?

__ **4.** Have you included your degree, school, city, and state within the education section?

__ **5.** Have you included your job title, company name, city, and state within the work experience section?

__ **6.** Have you used verbs to introduce each of your professional skills?

__ **7.** Have you quantified each of your achievements?

__ **8.** Have you avoided using sentences and the word *I?*

__ **9.** Does your resume use highlighting techniques to make it reader-friendly?

__ **10.** Have you proofread your resume to find grammatical and mechanical errors?

__ **11.** Have you decided whether you should write a reverse chronological resume or a functional resume?

__ **12.** Have you decided to create an online resume? If so, have you considered the ways in which it differs from the traditional paper resume?

Letter of Application

__ **1.** Have you included all of the letter essentials?

__ **2.** Does your introductory paragraph state where you learned of the job, which job you are applying for, and your interest in the position?

__ **3.** Does your letter's discussion unit pinpoint the ways in which you will benefit the company?

__ **4.** Does your letter's concluding paragraph end cordially and explain what you will do next or what you hope your reader will do next?

__ **5.** Is your letter free of all errors?

Interview

__ **1.** Will you dress appropriately?

__ **2.** Will you arrive ahead of time?

__ **3.** Will you avoid gum, cigarettes, and caffeine?

__ **4.** Have you practiced answering potential questions?

__ **5.** Have you researched the company so you can ask informed questions?

__ **6.** Will you bring to the interview additional examples of your work or copies of your resume?

Follow-up Letter

__ **1.** Have you included all the letter essentials?

__ **2.** Does your introductory paragraph remind the readers when you interviewed and what position you interviewed for?

__ **3.** Does the discussion unit highlight additional ways in which you might benefit the company?

__ **4.** Does the concluding paragraph thank the readers for their time and consideration?

__ **5.** Does your letter avoid all errors?

CHAPTER HIGHLIGHTS

1. Use many different resources to locate possible jobs, such as college placement centers, instructors, friends, professional affiliations, want ads, and the Internet.

2. Use either a reverse chronological resume or a functional resume.

3. Write a traditional hard-copy resume or an online resume in ASCII or HTML format.

4. Indicate a specific career objective on your resume.

5. Write your letter of application so that it targets a specific job.

6. Prepare before your interview so you can anticipate possible questions.

7. Your follow-up letter or e-mail message will impress the interviewer and remind him or her of your strengths.

8. Over 60 percent of jobs are found through networking.

9. Relying only on the Internet for your job search is a mistake. Less than 4 percent of new jobs are obtained this way.

10. On your resume, place education first if that is your strongest asset, or begin with work experience if this will help you get the job.

CASE STUDY

Rewrite the following flawed resume. In doing so, revise the errors and create three different types of resumes for Macy G. Heart—a chronological resume, a functional resume, and an ASCII scannable resume.

1890 Arrowhead Dr.
Utica, MO 51246
710-235-9999

Resume of Macy G. Heart

Objective: Seeking a position in Computer Technology Customer Service where I can use my many technology and people-person skills. I want to help troubleshoot software and hardware problems and work in a progressive company that will give me an opportunity for advancement and personal growth.

Work Experience
Jan. 2004 to now   Aramco.net   St. Louis, Missouri   Tech Support Specialists
Primarily I provide customer support for customer problems with C++, Visual Basic, Java, Networking, and Databases (Access, Oracle, SQL). I provide solutions to software and hardware problems and respond to e-mail queries in a timely manner.

Oct. 2002 to Dec. 22, 2003   DocuHelp   Chesterfield, MO   Computer Consultant
Provide technical support for PCs and Macs. I also trained new PC and Mac users in hardware applications. When business was slow, I repaired computer problems, using my many technology skills.

May 2002 to Oct. 2002   Ram-on-the-Run   East St. Louis, Illinois   Computer Salesman
Sold laptops, PCs, printers, and other computer accessories to men and women. Answered customer questions. Won Salesman of the Month Award three months in a row due to exceeding sales quotas.

Jan. 2002 to May 2002   Carbondale High School   Carbondale, IL   Lab Tech
Worked in the school's computer lab, helping Mr. Jones with computer-related classwork. This included fixing computer problems and tutoring new students having trouble with assignments.

Education

Aug. 2004   Bachelor's Degree, Information Technology, Carbondale Institute of
Technology, Carbondale, IL
Concentration: Database/Programming Applications
Relevant classes: Business Information Systems, Hardware Maintenance, Database
Management, Visual Basic, Systems Analysis and Design, C++, Web Design

May 2000   Graduate Carbondale High School
Member of the Computer Technology Club
Member of FFA and DECA
Principal's Honor Roll, senior year

Computer Expertise

Cisco Certified, knowledge of C++, VB, Microsoft Office Suite XP, HTML, Java, SML

References

Mr. Oscar Jones, Computer Applications Teacher, Carbondale High School
Mr. Renaldo Gomez, Manager, Aramco.net
Mr. Ted Harriot, Technical Support Supervisor, DocuHelp

Additional Information

A good team player, who works well with others
Made all A's in my college major classes
Built my own computer from scratch in high school
Starting football player in high school, Junior Varsity tight end

EXERCISES

1. Practice a job search. To do so, find examples of job openings in newspapers, professional journals, at your college or university's job placement service, and online. Bring these job possibilities to class for group discussions. From this job search, you and your peers will get a better understanding of what employers want in new hires.

2. Make an appointment to visit a personnel manager at a company in your field of interest. Interview him or her, using the interviewing techniques discussed in Chapter 14. Discover what traits this manager looks for in a new employee and what he or she considers to be important in a resume and letter of application. Share your findings with your technical writing class through a brief oral presentation.

3. Write a resume, either online or traditional hard copy. To do so, follow the suggestions provided in this chapter. Once you have constructed this resume, bring it to class for peer review. In small groups, discuss each resume's successes and areas needing improvement. If you prefer, make transparencies of each resume and have an entire class review them for suggested improvements.

4. Using an Internet search engine, find job openings in your area of interest. Which companies are hiring, and what skills do they want from prospective employees? Make a list of companies advertising on the Internet, and share with your class your findings regarding their preferred skills.

5. Find examples of HTML resumes. You can do this by typing in the word *resume* on an Internet search engine. Then, using the criteria

provided in this chapter and in Chapter 13, determine which resumes are successful and which need improvement. Select one resume that can be improved and revise it.

6. Write a traditional hard-copy resume. Then, using our tips, rewrite it in either ASCII or HTML format.

7. Write a letter of application. To do so, find a job advertisement in your newspaper's classified section, in your school's career planning and placement office, at your work site's personnel office, or in a trade journal. Then construct the letter according to the suggestions provided in this chapter. Next, in small groups or on an overhead, review your letter of application for suggested improvements.

8. Practice a job interview in small groups, designating one student as the job applicant and other students as the interview committee. Ask the applicant the sample interview questions provided in this chapter or any others you consider valid. This will give you and your peers a feel for the interviewing process.

9. Write a follow-up letter. Again, use the suggestions provided in this chapter.

10. An *informational interview* can help you learn about the realities of a specific job or work environment. Interview a person currently working in your field of interest. You can find such employees as follows (Mulvaney 2003):

   - *Alumni Office*—ask your college or university for a list of alums who are willing to speak to students about careers.

   - *Career Center*—your school's career center might give you names of people to contact.

   - *Human Resources*—visit a company in your field of interest and ask the human resources staff for help.

   - *Networking*—do you have friends or family who know of employees in your field of interest?

   Once you find an employee willing to help, visit with him or her and ascertain the following:

   - What job opportunities exist in your field?

   - What does a job in your field require, in terms of writing, education, interpersonal communication skills, teamwork, and so on, as well as the primary job responsibilities?

   After gathering this information, write a thank-you letter to the employee who helped you. Then, write a report documenting your findings and give an oral presentation to your classmates.

WEB WORKSHOP

Access any Internet search engine to find information about the job search. You can go online to research resumes, cover letters, and follow-up thank-you notes. For example, check out Monster.com *(http://resume.monster.com/)* for resume guidelines, samples, resume makeovers, and do's and don'ts.

Research CareerJournal.com *(http://www.careerjournal.com/jobhunting/resumes/)* for up-to-date articles about resume writing (tips, samples, whether you should pay to have someone write a resume for you, and case studies about "red-flag" resumes).

CareerPerfect.com *(http://www.careerperfect.com/CareerPerfect/resumes.htm)* provides guidelines for electronic resume submission, techniques for creating scannable versions of your resume, and tips for creating ASCII resumes.

JobsMart.org *(http://jobsmart.org/)* and Monster.com *(http://resume.monster.com/archives/coverletter/)* offer you sample cover letters and cover letter do's and don'ts.

Check out JobSearchTech *(http://jobsearchtech.about.com/library/weekly/aa021003.htm)* to learn why and how to write follow-up thank-you letters, and to see samples.

Once you research any of these sites, analyze your findings. How do the letters and resumes compare to those discussed in this textbook? What new information have you learned from the online articles?

    a. Report your findings, either in an oral presentation or in writing (e-mail, memo, letter, or report).

    b. If you find sample letters or resumes that you dislike, rewrite them according to the criteria provided in this textbook.

QUIZ QUESTIONS

1. What are four places you can search for a job?

2. What is the primary goal of your resume?

3. What are two types of resumes?

4. When would you write a reverse chronological resume?

5. What is a career objectives line on a resume?

6. In what order should you list jobs you have held?

7. What should you include when you list your education?

8. What are some optional inclusions in your education section?

9. Why would you write a functional resume?

10. What types of personal data should you exclude?

11. Why should you avoid sentences in a resume?

12. Why should you bulletize on a resume?

13. How can you prove "you are great" on a resume?

14. How can you design a resume effectively?

15. What are the benefits of an online resume?

16. What is the main difference between a resume and a letter of application?

17. What are two ways to conclude a letter of application?

18. What are five things you can do to interview effectively?

19. What are three things you can accomplish in a follow-up letter?

20. How can networking help you find a job?

VISUAL APPEAL

3

Document Design

Explains the importance of document design and effective page layout; emphasizes organization, order, access, and variety to create visual appeal.

CHAPTER 9

Graphics

Focuses on tables and figures as visual communication in a written document; shows how visual aids achieve conciseness, clarity, and cosmetic appeal; provides criteria for various graphics, including line graphs, bar charts, topographical maps, CAD drawings, and icons.

Chapter Preview

Organization
Chunking—breaking a text into smaller units—allows you to create an overall organization in a technical document. You can use headings, white space, horizontal rules, or section dividers to create the chunks.

Order
Decide where to place material to prioritize the information in a document. Headings help you order the information.

Access
Visual ease of access is essential in a document. To help the reader quickly find information, use highlighting techniques such as bullets, white space, numbering, and boldface type.

Variety
Varying the look of a document enhances its readability. To accomplish this goal, print your document in landscape or portrait orientation, vary gutter widths and margins, and add graphics.

IMPORTANCE OF DOCUMENT DESIGN

In technical writing, are words your only concern? The answer is an emphatic "No!" What you say is important, but how the text looks on the page is equally important. If you give your readers excessively long paragraphs, pages full of wall-to-wall words, you have made a mistake. Ugly blocks of unreadable, unappealing text will get you and your company in trouble. Remember your purposes in technical communication.

Effective technical communication allows readers rapid access to the information, highlights important information, and graphically expresses your company's identity.

The Technical Writing Context

Why do people read correspondence? Although individuals read poetry, short stories, novels, and drama for enjoyment, few people read memos, letters, reports, or instructions for fun. They read these types of technical writing for information about a product or service. They read this correspondence while they talk on the telephone, while they commute to work, or while they walk to meetings.

Given these contexts, your readers want you to provide them information quickly, information they can understand at a glance. Reading word after word, paragraph after paragraph takes time and effort, which most readers cannot spare. Therefore, if your technical writing is visually unappealing, your audience might not even read your words. Readers will either give up before they have begun or be unable to remember what they have read. You cannot assume they will labor over your text to uncover its worth. Good technical writing allows readers rapid access to information.

Damages and Dangers

If your intended readers fail to read your text because it is visually inaccessible, imagine the possible repercussions. They could damage equipment by not recognizing important information which you have buried in dense blocks of text. The readers could give up on the text and call your company's toll-free hotline for assistance. This wastes your readers' and your coworkers' time and energy. Worse, your readers might hurt themselves and sue your company for failing to highlight potential dangers in the user manual.

Corporate Identity

Your document—whether it is a memo, letter, report, instruction, newsletter, or brochure—is a visual representation of your company, graphically expressing your company's identity. It might be the only way you meet your clients. If your text is unappealing, that is the corporate image your company conveys to the customer. If your text is not reader-friendly, that is how your company will appear to your client. Visually unappealing and inaccessible correspondence can negatively affect your company's sales and reputation. In today's competitive workplace, any leverage you can provide your company is a plus. Document design is one way to appeal to a client.

To clarify how important document design is, look at the inaccessible meeting minutes in Figure 8.1. These minutes are neither clear nor concise. You are given so much data in such an unappealing format that your first response upon seeing the correspondence probably is to say, "I'm not going to read that."

How can you make these minutes more inviting? How can you break up the wall-to-wall words and make key points jump off the page? To achieve effective document design, you will need to provide your readers visual

- Organization
- Order
- Access
- Variety

ORGANIZATION

The easiest way to organize your document's design is to break text into smaller chunks of information, a technique called *chunking*. When you use chunking to separate blocks of text, you help your readers understand the overall organization of your correspondence. They can see which topics go together and which are distinct (Keyes 1993, 640; Watzman 1987, ATA-49).

Chunking to organize your text is accomplished by using any of the following techniques:

- *Headings* (one to three words that summarize the content of a unit of information)
- *White space* (horizontal spacing between paragraphs, created by double- or triple-spacing)
- *Rules* (horizontal lines typed or drawn across the page to separate units of information)
- *Section dividers and tabs* (used in longer reports to create smaller units)

Notice how using some of these techniques improves the document design in Figure 8.2.

tech link

For a detailed article on chunking, go to *http://www. webstyleguide.com/site/ chunk.html.*

MINUTES

The meeting at the Carriage Club was attended by 30 members and guests. After the dinner, Roger Traver introduced the guest speaker, George Smith, university chancellor, and noted his accomplishments and experiences prior to education— U.S. Navy commander, Oak Ridge Laboratory researcher, and politician. Dr. Smith's talk, "Industry and Education Collaboration," was very interesting and included a history of special projects enjoyed by both academics and corporate heads. Dr. Smith suggested that we engineers could work with education to (1) provide training seminars, (2) help in urban development, and (3) provide intern opportunities. Recent industry–education collaborations include training seminars in computers, fiber optics, and human resource options. The chancellor's primary thrust was a request for $100,000 in financial aid for urban development. He said money had already been donated from three sources: a large realty firm, Capital Homes, had given $20,000; a philanthropic group, We Care, had donated a matching $20,000; Dr. Smith's university gave a matching $20,000. The remaining $40,000, Dr. Smith hoped, would come from industry donations. Finally, the chancellor noted that industry could help itself, as well as the community, by providing internships for university undergraduate majors. These internships could either be semester- or year-long arrangements, whereby students would work for minimum wage to learn more about the day-to-day aspects of their chosen fields. The chancellor said that these internships would not only increase the students' theoretical knowledge of engineering by giving them hands-on experience but also make them better future employees for the host engineering companies. Everyone would benefit. Dr. Smith noted that the students would receive a grade and credit for their work. After the speech, our VP introduced new business, calling for nominations for next year's officers; gave us the agenda for our next meeting; and adjourned the meeting.

Figure 8.1 Inaccessible Meeting Minutes

Order

Once a wall of unbroken words has been separated through chunking to help the reader understand the text's organization, the next thing a reader wants from your text is a sense of order. What's most important on the page? What's less important? What's least important? You can help your audience prioritize information by ordering—or *queuing*—ideas (Keyes 1993, 640–41; Watzman 1987, ATA-49). The primary way to accomplish this goal is through a hierarchy of headings set apart from each other through various techniques.

- **Typeface**—There are many different *typefaces* (or *fonts*), including Courier, Prestige, Helvetica, Arial, Bosanova, Future, LCD, Key, and

MINUTES ← First-level heading

The meeting at the Carriage Club was attended by 30 members and guests. After the dinner, Roger Traver introduced the guest speaker, George Smith, university chancellor, and noted his accomplishments and experiences prior to education—U.S. Navy commander, Oak Ridge Laboratory researcher, and politician. Dr. Smith's talk, "Industry and Education Collaboration" was very interesting and included a history of special projects enjoyed by both academics and corporate heads. Dr. Smith suggested that we engineers could work with education to accomplish three goals.

← Horizontal rule

Training Seminars
Recent industry–education collaborations include training seminars in computers, fiber optics, and human resource options.

Urban Development ← Second-level heading
The chancellor's primary thrust was a request for $100,000 in financial aid for urban development. He said money had already been donated from three sources: a large realty firm, Capital Homes, had given $20,000; a philanthropic group, We Care, had donated a matching $20,000; Dr. Smith's university gave a matching $20,000. The remaining $40,000, Dr. Smith hoped, would come from industry donations.

Internships
The chancellor noted that industry could help itself, as well as the community, by providing internships for university undergraduate majors. These internships could either be semester- or year-long arrangements, whereby students would work for minimum wage to learn more about the day-to-day aspects of their chosen fields. The chancellor said that these internships would not only increase the students' theoretical knowledge of engineering by giving them hands-on experience but also make them better future employees for the host engineering companies. Everyone would benefit. Dr. Smith noted that the students would receive a grade and credit for their work.

Conclusion
After the speech, our VP introduced new business, calling for nominations for next year's officers; gave us the agenda for our next meeting; and adjourned the meeting.

Figure 8.2 Document Design Using Chunking to Organize the Information

tech link

Go to *http://flamingtext.com/* to access hundreds of designer fonts.

Chicago. Whichever typeface you choose, it will either be a *serif* or *sans serif* typeface. Serif type has "feet" or decorative strokes at the edges of each letter. This typeface is commonly used in text because it is easy to read, allowing the reader's eyes to glide across the page (Benson 1985, 36–37).

SERIF ⟵ decorative feet

Sans serif (as seen in this parenthetical comment) is a block typeface that omits the feet or decorative lines. This typeface is best used for headings.

SANS SERIF ⟵ no decorative feet

tech link

For information on typography (links and news about font selection), go to *http://www.microsoft.com/ typography/default.mspx.*

- **Type size**—Another way of queuing for your readers is through the size of your type. A primary, first-level heading should be larger than subsequent, less important headings: second level, third level, and so forth. For example, a first-level heading could be in 18-point type. The second-level heading would then be set in 16-point type, the third-level heading in 14-point type, and the fourth-level heading in 12-point type (Keyes 1993, 641).

  Figure 8.3 shows examples of different typefaces and type sizes.

- **Density**—The *weight* of the type also prioritizes your text. Type *density* is created by boldfacing or double-striking words (Keyes 1993, 641; Benson 1985, 37).

- **Spacing**—Another queuing technique to help your readers order their thoughts is the *amount of horizontal space* used after each heading. The first-level heading should have more space following it than the second-level heading, and so forth.

- **Position**—Your headings can be centered, aligned with the left margin, indented, or outdented (*hung heads*). No one approach is more valuable or more correct than another. The key is consistency. If you center your first-level heading, for example, and then place subsequent heads at the left margin, this should be your model for all chapters or sections of that report (Keyes 1993, 641; Watzman 1987, VC-86).

Sans Serif Typefaces	Serif Typefaces
Avant Garde 12 point	Courier 12 point
Avant Garde 14 point	Courier 14 point
Avant Garde 18 point	Courier 18 point
Futura 12 point	Bookman 12 point
Futura 14 point	Bookman 14 point
Futura 18 point	Bookman 18 point
Helvetica 12 point	Goudy 12 point
Helvetica 14 point	Goudy 14 point
Helvetica 18 point	Goudy 18 point

Figure 8.3   Examples of Typefaces and Type Sizes

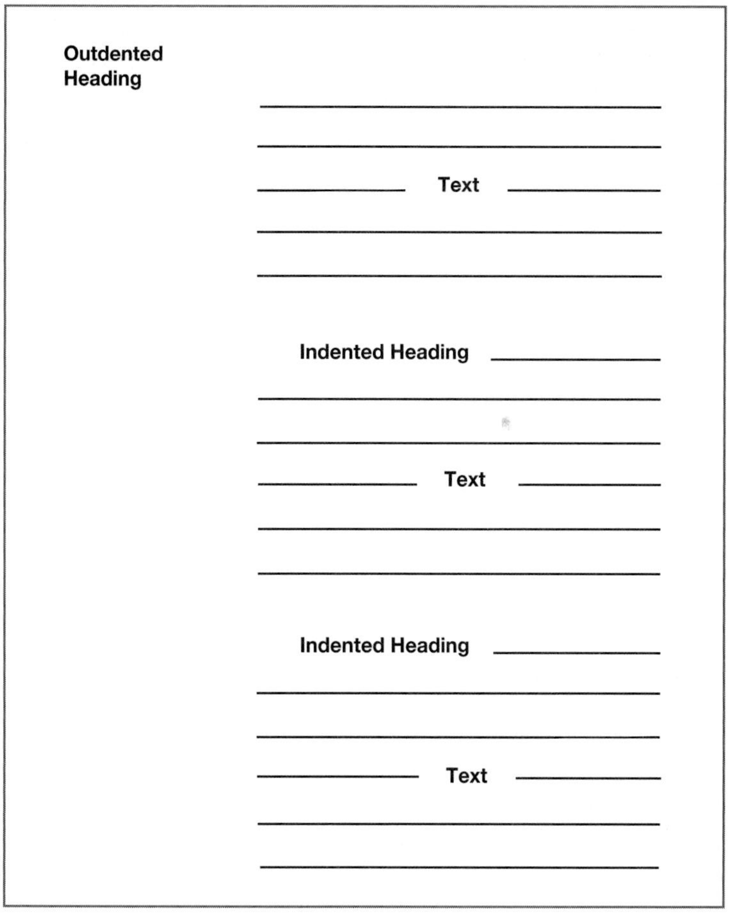

Figure 8.4   Outdented and Indented Headings

Figure 8.4 shows an *outdented* first-level heading with *indented* subsequent headings. Figure 8.5 shows a *centered* heading with subsequent headings aligned with the left margin.

Figure 8.6 reformats the meeting minutes seen in Figure 8.2 and uses queuing to order the hierarchy of ideas. The outdented first-level heading is set in a 12-point bold sans serif typeface, all caps. The second-level heading is set in a 10-point bold serif typeface and is separated from the preceding text by horizontal white space. The third-level heading is set in a 10-point bold serif typeface and is separated from the preceding text by double-spacing. It is also set on the same line as the following text.

Hierarchical heading levels allow the readers to visualize the order of information to see clearly how the writer has prioritized text.

ACCESS

Chunking helps the reader see which ideas go together, and a hierarchy of headings helps the reader understand the relative importance of each unit of information. Nonetheless, the document design in Figure 8.6 needs improvement. The reader still must read every word carefully to see the key points within each chunk of text. Readers are not that generous with their time. As writer, you should make your reader's task easier.

tech link

Go to *http://www.prenhall. com/gerson* for Web links, samples, and interactive activities.

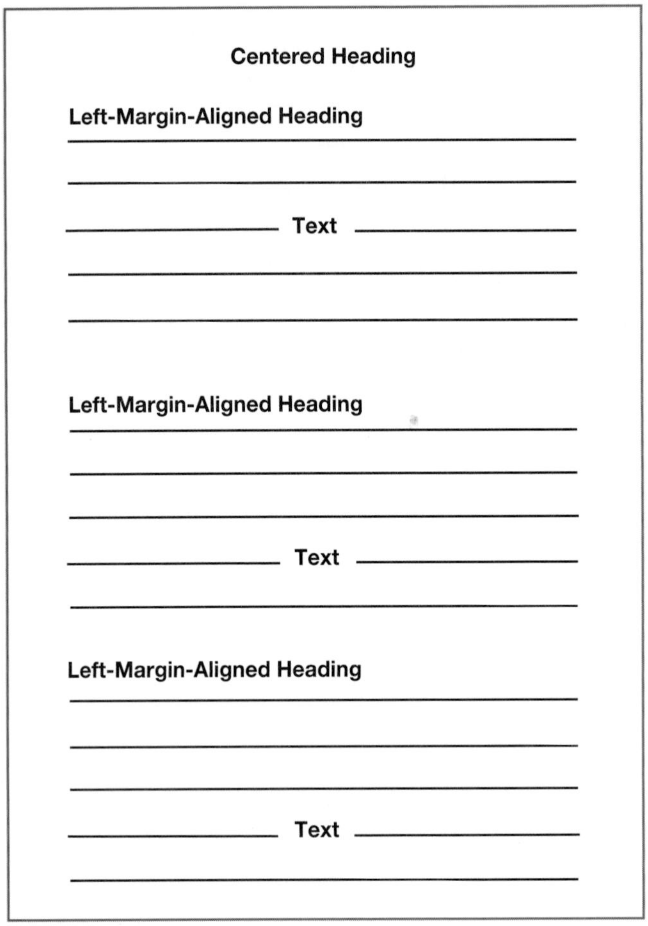

```
┌─────────────────────────────────────────┐
│                                         │
│            **Centered Heading**         │
│                                         │
│  **Left-Margin-Aligned Heading**        │
│  ─────────────────────────────────────  │
│                                         │
│  ─────────────────────────────────────  │
│                                         │
│  ──────────────── **Text** ───────────  │
│                                         │
│  ─────────────────────────────────────  │
│                                         │
│  ─────────────────────────────────────  │
│                                         │
│                                         │
│  **Left-Margin-Aligned Heading**        │
│  ─────────────────────────────────────  │
│                                         │
│  ─────────────────────────────────────  │
│                                         │
│  ──────────────── **Text** ───────────  │
│                                         │
│  ─────────────────────────────────────  │
│                                         │
│  **Left-Margin-Aligned Heading**        │
│  ─────────────────────────────────────  │
│                                         │
│  ─────────────────────────────────────  │
│                                         │
│  ─────────────────────────────────────  │
│                                         │
│  ──────────────── **Text** ───────────  │
│                                         │
│  ─────────────────────────────────────  │
│                                         │
└─────────────────────────────────────────┘
```

Figure 8.5   Centered and Left-Margin-Aligned Headings

A third way to assist your audience is by helping them *access* information rapidly—at a glance. You can use any of the following highlighting techniques to help the readers filter out extraneous or tangential information and focus on key ideas (Keyes 1993, 641; Watzman 1987, ATA-49).

- **White space**—In addition to horizontal space, created by double -or triple-spacing, you also can create *vertical space* by indenting. This vertical white space breaks up the monotony of wall-to-wall words and gives your readers breathing room. White space invites your readers into the text and helps the audience focus on the indented points you want to emphasize.

- **Bullets**—Bullets, used to emphasize items within an indented list, are created by using asterisks (*), hyphens (-), a lowercase *o*, degree signs (°), typographic symbols (■, ❏, ●, or ◆), or iconic dingbats (☞, ✆, or ✓).

- **Numbering**—*Enumeration* creates itemized lists that can show sequence or importance and allow for easy reference.

- **Boldface**—Boldface text emphasizes a key word or phrase.

- **All caps**—The technique of capitalizing text is an excellent way to highlight a WARNING, DANGER, CAUTION, or NOTE. However, capitalizing other types of information is not suggested because reading lowercase words is easier for your audience. All caps creates a block of letters in which individual letters aren't easily distinguished from each other.

MINUTES The meeting at the Carriage Club was attended by 30 members and guests. After the dinner, Roger Traver introduced the guest speaker, George Smith, university chancellor, and noted his accomplishments and experiences prior to education—U.S. Navy commander, Oak Ridge Laboratory researcher, and politician. Dr. Smith's talk, "Industry and Education Collaboration," was very interesting and included a history of special projects enjoyed by both academics and corporate heads. Dr. Smith suggested that we engineers could work with education to accomplish three goals.

This hanging head is set in all caps and a sans serif font.

Urban Development
The chancellor's primary thrust was a request for $100,000 in financial aid for urban development. He said money had already been donated from three sources: a large realty firm, Capital Homes, had given $20,000; a philanthropic group, We Care, had donated a matching $20,000; Dr. Smith's university gave a matching $20,000. The remaining $40,000, Dr. Smith hoped, would come from industry donations.

Create a hierarchy of headings by changing font size and style.

Internships
The chancellor noted that industry could help itself, as well as the community, by providing internships for university undergraduate majors. These internships could either be semester- or year-long arrangements, whereby students would work for minimum wage to learn more about the day-to-day aspects of their chosen fields. The chancellor said that these internships would not only increase the students' theoretical knowledge of engineering by giving them hands-on experience but also make them better future employees for the host engineering companies. Everyone would benefit. Dr. Smith noted that the students would receive a grade and credit for their work.

Training Seminars. Recent industry–education collaborations include training seminars in computers, fiber optics, and human resource options.

Headings create accessible content.

Conclusion
After the speech, our VP introduced new business, calling for nominations for next year's officers, gave us the agenda for our next meeting; and adjourned the meeting.

Figure 8.6   Document Design Using a Hierarchy of Heading Levels to Order the Information

- **Underlining**—Underlining should be used cautiously. If you underline too frequently, none of your information will be emphatic. One underlined word or phrase will call attention to itself and achieve reader access. Several underlined words or phrases will overwhelm your readers.

- **Italics**—Italics and underlining are used similarly as highlighting techniques.
- **Text boxes**—Place key points in a text box for emphasis. Here is an example.

> **NOTE:** Be sure to *hand-tighten* the nuts at this point. Once you have completed the installation, go back and securely tighten all nuts.

You can also italicize and use all caps within the text box, as we have in this example.

- **Fills**—You can further highlight text boxes through fills (lines, patterns, waves, bricks, gradients, and shadings).
- **Inverse type**—You can also help readers access information by using inverse type—printing white on black, versus the usual black on white.
- **Font type**—You could call attention to key ideas by changing your font type. If you have written the entire text in Times New Roman, for example, and you want to ensure that your reader clearly sees an important idea, you could type that word, phrase, or sentence in a different font. Shift to Arial, Tekton, or *Lucida Handwriting*, for example. However, some fonts—though interesting—are very unreadable. If used in excess, they can cause more trouble than they're worth. Imagine reading an entire letter in *Spring LP*, *Shelley Volante BT*, or AT Wedding. Such fonts cause eye strain and headaches; they don't aid communication.
- **Color**—Another way to make key words and phrases leap off the page is to color them. *Danger* would be red, for example, *Warning* orange, and *Caution* yellow. You can also use color to help a reader access the first-level heading, a header, or a footer. (*Headers* contain information placed along the top margin of text; *footers* contain information placed along the bottom margin of text.)

  For instance, if your text is typed in a black font color, then headings typed in a blue font would stand out more effectively. However, as with all highlighting techniques, a little bit goes a long way. Do not overuse color. Do not type several headings in different colors. Doing so could produce a very unprofessional impression.

  Example: Effective Use of Colored Headings

> **Committee Action**
> The Budget and Personnel Committee will vote to approve the audit report at the July meeting.
>
> **Recommendation**
> The committee will recommend that a proposal be submitted to improve roadway construction.
>
> **Staff Contacts**
> Mel Henderson
> Sean Thomson

Example: Ineffective Use of Colored Headings

Committee Action
The Budget and Personnel Committee will vote to approve the audit report at the July meeting.

Recommendation
The committee will recommend that a proposal be submitted to improve roadway construction.

Staff Contacts
Mel Henderson
Sean Thomson

All colors are not equal in visual value. Generally, dark colored fonts provide the most contrast against light colored backgrounds, or vice versa. For example, a black font on a white background (or a white font on a black background) creates optimum contrast. On the other hand, a light-colored font on a light background does not improve access.

Example: Good contrast helps access

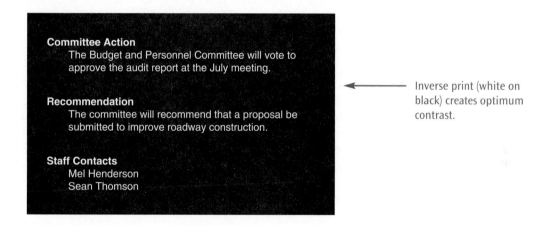

Inverse print (white on black) creates optimum contrast.

Example: Bad contrast hurts access

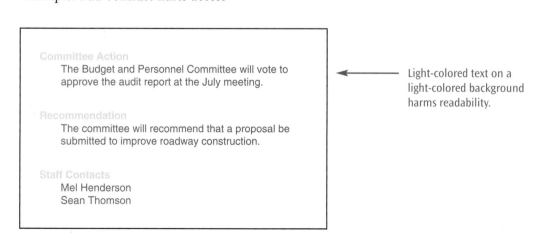

Light-colored text on a light-colored background harms readability.

You also should use colors tastefully. Avoid garish color combinations in graphics or backgrounds (red and orange, pink and green, or purple and yellow). Colors that clash will distract the reader more than aid access.

Example: Clashing Color Combinations that Hurt Access

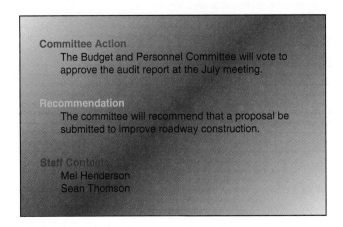

Here is a very important consideration: When it comes to using highlighting techniques, *more is not better.* A few highlighting techniques help your readers filter out background data and focus on key points. Too many highlighting techniques are a distraction and clutter the document design. Be careful not to overdo a good thing. Figure 8.7 (page 263) gives examples of several highlighting methods. Notice how Figure 8.8 (page 264) uses highlighting techniques to help the readers access information in the meeting minutes.

VARIETY

Each of the document designs in Figures 8.1, 8.2, 8.6, and 8.8 uses one column and is printed vertically on a traditional 8 1/2 × 11-inch page (a type of printing called portrait; see Figure 8.9 on page 265). This is not your only option. Your reader might profit from more variety (Watzman 1987, ATA-48). For example, you might want to use a smaller or larger paper; vary the weight of your paper (for example 10-pound, 12-pound, or heavier card stock), or even print your text on colored paper.

More important, you can vary the document design as follows:

- **Print horizontally**—Rather than print your text vertically—8 1/2 × 11-inch portrait—you could print *horizontally,* as an 11 × 8 1/2-inch landscape.

- **Use more columns**—Provide your reader two to five columns of text.

- **Vary gutter width**—Columns of text are separated by vertical white space called the gutter.

- **Use ragged-right margins**—Some text is fully justified (both right and left margins are aligned). Once this was considered professional, giving the text a clean look. Now, however, studies confirm that right-margin-justified text is harder for the audience to read. It's too rigid. In contrast, *ragged-right* type (the right margin is not justified) is easier to read and more pleasing to the eye. You can use this method to vary page layout (Everson 1990, 397).

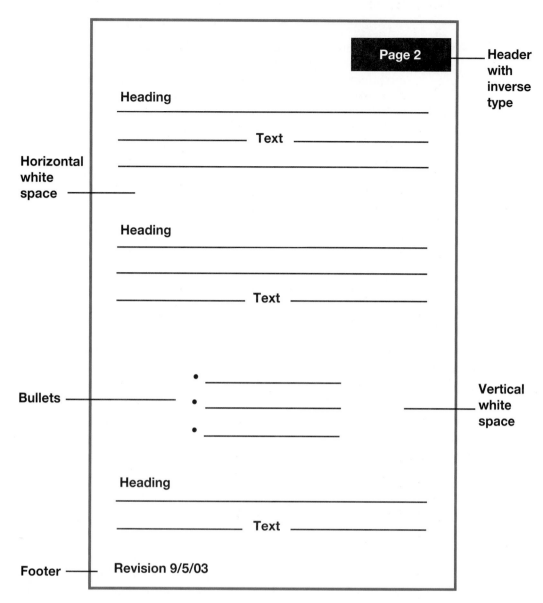

Figure 8.7 Highlighting Techniques for Access

Figure 8.10 shows how you can use columns, landscape orientation, and ragged-right margins to vary your document design.

Although you can vary your document design by printing horizontally and by using multiple columns, the audience is still confronted by words, words, and more words. The majority of readers do not want to wade through text. Luckily, words are not your only means of communication. You can reach a larger audience with different learning styles by varying your method of communication.

Graphics are an excellent alternative. Many people are more comfortable grasping information visually than verbally. Although it's a cliché, a picture is often worth a thousand words.

We discuss graphics (figures and tables) extensively in Chapter 9. However, to clarify our point about the value of variety, see Figure 8.11, which adds a graphic to the meeting minutes.

MINUTES The meeting at the Carriage Club was attended by 30 members and guests. After the dinner, Roger Traver introduced the guest speaker, George Smith, university chancellor, and noted his accomplishments and experiences prior to education:

- U.S. Navy commander
- Oak Ridge Laboratory researcher
- Politician

Dr. Smith's talk, "Industry and Education Collaboration," was very interesting and included a history of special projects enjoyed by both academics and corporate heads. Dr. Smith suggested that we engineers could work with education to accomplish three goals.

Urban Development

The chancellor's primary thrust was a request for $100,000 in financial aid for urban development. He said money had already been donated from three sources:

1. A large realty firm, Capital Homes, had given $20,000.
2. A philanthropic group, We Care, had donated a matching $20,000.
3. Dr. Smith's university also gave $20,000.

The remaining $40,000 would come from industry.

Internships

The chancellor noted that industry could help itself, as well as the community, by providing internships for university undergraduate majors:

1. Semester-long internships
2. Year-long internships

Students would work for minimum wage to learn more about the day-to-day aspects of their chosen fields. The chancellor said that these internships would not only increase the students' theoretical knowledge of engineering by giving them hands-on experience but also make them better future employees for the host engineering companies. <u>Everyone would benefit.</u> Dr. Smith noted that the students would receive a grade and credit for their work.

Training Seminars. Recent industry–education collaborations include training seminars in computers, fiber optics, and human resource options.

Conclusion

After the speech, our VP introduced new business, calling for nominations for next year's officers, gave us the agenda for our next meeting, and adjourned the meeting.

Labels pointing to the document:
- Bulletized list
- Numbered list
- Underlining

Figure 8.8   Document Design Using Highlighting Techniques to Help Readers Access Key Ideas

Figure 8.9   Portrait with One Column Fully Justified

Figure 8.10   Landscape with Two Columns and Ragged-Right Margins

MINUTES   The meeting at the Carriage Club was attended by 30 members and guests. After the dinner, Roger Traver introduced the guest speaker, George Smith, university chancellor, and noted his accomplishments and experiences prior to education:

- U.S. Navy commander
- Oak Ridge Laboratory researcher
- Politician

Bulleted list for accessibility

Dr. Smith's talk, "Industry and Education Collaboration," was very interesting and included a history of special projects enjoyed by both academics and corporate heads. Dr. Smith suggested that we engineers could work with education to accomplish three goals.

Urban Development
The chancellor's primary thrust was a request for $100,000 in financial aid for urban development. He said money had already been donated from three sources, but business and industry can still help significantly. The following pie chart clarifies what money has been encumbered and how industry donations are still needed.

Figure 8.11   Document Design Using a Pie Chart to Vary the Communication

Graphic for visual appeal and accessible data

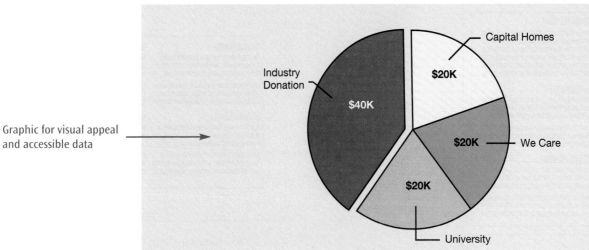

Figure 1 Donations

Internships
The chancellor noted that industry could help itself, as well as the community, by providing internships for university undergraduate majors:
1. Semester-long internships
2. Year-long internships
Students would work for minimum wage to learn more about the day-to-day aspects of their chosen fields. The chancellor said that these internships would not only increase the students' theoretical knowledge of engineering by giving them hands-on experience but also make them better future employees for the host engineering companies. <u>Everyone would benefit.</u>
Dr. Smith noted that the students would receive a grade and credit for their work.

Training Seminars. Recent industry–education collaborations include training seminars in computers, fiber optics, and human resource options.

Conclusion
After the speech, our VP introduced new business, calling for nominations for next year's officers, gave us the agenda for our next meeting, and adjourned the meeting.

Figure 8.11 Continued

SUMMARY

Creating effective document design is time consuming and creatively demanding. It's easier just to write the old-fashioned way—word after word, paragraph after paragraph, blocks of inaccessible text. Although it's easier, it's also less effective.

Your job is not just to dump data and hope your readers can figure it out. Instead, a successful technical writer works to involve the audience. Your goal is to help the readers understand the organization of your text, recognize its order, and access information at a glance. You also want to use varied types of communication, including graphics, for readers who absorb information better visually, rather than through words.

What's the payoff? Studies tell us that effective document design saves money and time.

One international customs department, after revising the design of its lost-baggage forms, reduced its error rate by over 50 percent. When a utilities company changed the look of its billing statements, customers asked fewer questions, saving the company approximately $250,000 per year. The U.S. Department of Commerce, Office of Consumer Affairs, reported that when several companies improved their documents' visual appeal, the companies increased business and reduced customer complaints (Schriver 1993, 250–51).

Document design isn't a costly frill. Effective document design is good for your company's business.

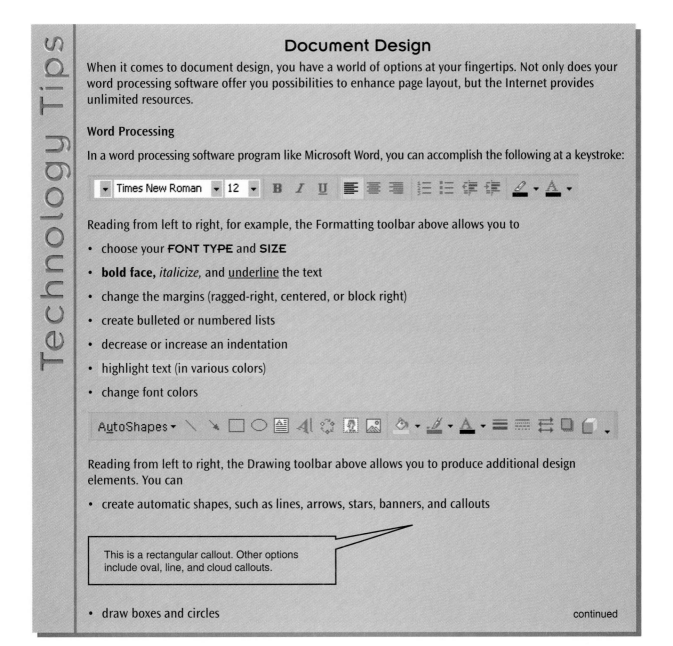

Technology Tips

Document Design

When it comes to document design, you have a world of options at your fingertips. Not only does your word processing software offer you possibilities to enhance page layout, but the Internet provides unlimited resources.

Word Processing

In a word processing software program like Microsoft Word, you can accomplish the following at a keystroke:

Reading from left to right, for example, the Formatting toolbar above allows you to

- choose your **FONT TYPE** and **SIZE**

- **bold face,** *italicize,* and underline the text

- change the margins (ragged-right, centered, or block right)

- create bulleted or numbered lists

- decrease or increase an indentation

- highlight text (in various colors)

- change font colors

Reading from left to right, the Drawing toolbar above allows you to produce additional design elements. You can

- create automatic shapes, such as lines, arrows, stars, banners, and callouts

This is a rectangular callout. Other options include oval, line, and cloud callouts.

- draw boxes and circles

continued

continued

- create text boxes

Text boxes can be enhanced further by changing the color of the lines or fill. You can also create a shadow for this box, as shown.

- insert WordArt

- insert clip art or pictures
- add horizontal rules (decorative lines in varying widths and designs)

In addition, by clicking on Format on your Menu bar, you can change many other features:

Format	Tools	Table	Window	Help

A Font...

≔ Bullets and Numbering...

Borders and Shading...

Change Case...

⬛ Styles and Formatting...

Object...

⌄⌄

- Font allows you to choose from dozens of font types and sizes, as well as to boldface and italicize, write in subscript or ^{superscript}, use the **shadow**, and **engrave** features.
- Bullets and Numbering allows you to customize bullets.

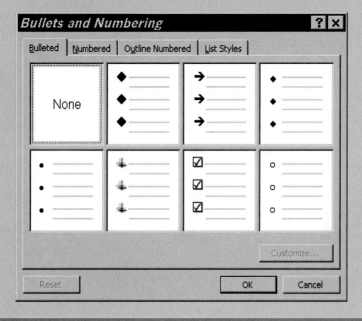

- Borders and Shading lets you alter the look of text boxes and tables. You can create shadows and 3-D effects for borders and create shadings to fill the boxes.

Finally, using your word processing software you can generate a hierarchy of headings:

Heading 1	¶
Heading 2	¶
Heading 3	¶

Internet

The Internet provides you even more resources for document design. Online, at sources like Google.com, you can find literally millions of images.

CHAPTER HIGHLIGHTS

1. Breaking your text into smaller chunks of information will help you create a more readable document.

2. When you create an order among the items in a document through use of typeface and type size, your audience can more easily prioritize the information.

3. Your reader should be able to glance at the document and easily pick out the key ideas. Highlighting techniques will help accomplish this goal.

4. Vary the appearance of your document by using columns, varying gutter widths, and printing in portrait or landscape orientation.

5. Your audience will access the content easily if you use white space, bullets, numbering, underlining, and text boxes.

EXERCISES

Team Projects

1. Bring samples of technical writing to class. These could include letters, brochures, newsletters, instructions, reports, or advertisements from magazines or newspapers.

   In groups of three to five class members, assess the document design of each sample. Determine which samples have successful document designs and which samples have poor document designs. Base your decisions on the criteria provided in this chapter: organization, order, access, and variety.

   Either orally or in writing, share your findings with other teams in the class.

2. In teams of three to five class members, take one of the less successful samples from team project 1 and reformat it to improve its document design. Focus on improving the sample's organization, order, access, and variety.

   Once your team has completed reformatting the text, make photocopies or transparencies of your work. Then review the various team projects, determining which team created the best document design. Select a winner, and explain why this text now has a successful document design.

3. In teams of three to five class members, reformat the following memo to improve its document design.

DATE: November 30, 2005
TO: Jan Hunt
FROM: Tom Langford
SUBJECT: CLEANING PROCEDURES FOR MANUFACTURING WALK-IN
          OVENS #98731, #98732, AND #98733

The above-mentioned ovens need extensive cleaning. To do so, vacuum and wipe all doors, walls, roofs, and floors. All vents/dampers need to be removed, and a tack cloth must be used to remove all loose dust and dirt. Also, all filters need to be replaced.

I am requesting this because loose particles of dust/dirt are blown onto wet parts when placed in the air-circulating ovens to dry. This causes extensive rework. Please perform this procedure twice per week to ensure clean production.

---

4. In teams of three to five class members, reformat the following summary to improve its document design.

SUMMARY

The City of Waluska wants to provide its community with a safe and reliable water treatment facility. The goal is to protect Waluska's environmental resources and to ensure community values.

To achieve these goals, the city has issued a request for proposal to update the Loon Lake Water Treatment Plant (LLWTP). The city recognizes that meeting its community's water treatment needs requires overcoming numerous challenges. These challenges include managing changing regulations and protection standards, developing financially responsible treatment services, planning land use for community expansion, and upholding community values.

For all of the above reasons, Hardtack and Sons (H & S) Engineering is your best choice. We understand the project scope and recognize your community's needs.

We have worked successfully with your community for a decade, creating feasibility studies for Loon Lake toxic control, developing odor-abatement procedures for your streams and creeks, and assessing your water treatment plant's ability to meet regulatory standards.

H & S personnel are not just engineering experts. We are members of your community. Our dynamic project team has a close working relationship with your community's regulatory agencies. Our Partner in Charge, Julie Schopper, has experience with similar projects worldwide, demonstrated leadership, and the ability to communicate effectively with clients.

H & S offers the City of Waluska an integrated program that addresses all your community's needs. We believe that H & S is your best choice to ensure that your community receives a water treatment plant ready to meet the challenges of the twenty-first century.

5. On the Internet, access 10 corporate Web sites. Study them to determine how these electronic documents are designed. Make a list of the techniques they use for visual appeal. Which Web sites are successful, and why? Which Web sites are unsuccessful, and why? How would you redesign these less successful Web sites to achieve better document design?

6. The following text is visually unappealing and inaccessible. Using the techniques discussed in this chapter (organization, order, access, and variety), improve the document's design.

In 1998, the Transportation Equity Act for the 21st Century (TEA-21) was passed and authorized Federal programs for roadway safety.

To limit its focus, the TEA-21 discussed safety as it was affected by facility planning, roadway design, and maintenance.

The TEA-21 was concerned about roadway safety for pedestrians, motorized vehicles, and non-motorized vehicles. Motorized vehicles included cars, vans, trucks, SUV's and motorcycles. Non-motorized vehicles include bicycles, roller skates, rollerblades, and scooters.

The TEA-21's planning guidelines consisted of many goals. These included communicating revised safety programs to pedestrians, non-motorized vehicle users, and motor vehicle users; organizing safety data for analysis by highway and city road departments; strategies for reducing fatalities for pedestrians and vehicle users; and developing city safety blueprints based on statewide prototypes.

One prototypical model that the TEA-21 recommends following is Operation Green Light. This model uses a traffic signal coordination system to maintain a steady flow of traffic. Steady traffic flow has been proven to reduce fatalities by 15 percent.

7. The following short report is poorly formatted. The text is so dense that readers would have difficulty understanding the content easily. Improve the document's design to aid access. Use highlighting techniques discussed in this chapter to revise the text.

Date   May 18, 2005
To: Martha Collins
From: Richard Davis
Subject: 2005 Switch Port Carriers

Attached are the supplemental 2005 Switch Port Carriers that are required to support this year's growth patterns. As we have discussed in previous phone conversations, the May numbers show a decrease in traffic, but forecasts still suggest increased traffic. Therefore, we are issuing plans for this contingency. If the June forecasts prove to be accurate, the ports being placed in the network via these plans will support our future growth except for areas where growth can not be predicted. Some areas, for example, are too densely populated for forecasting because the company did not hire enough survey personal to do a thorough job.

Following is an update of our suggested Port Additions. For Port 12ABR, add 16 Ports in Austin. For Port 13RgX, add 27 Ports in Houston. For Port 981D, add 35 Ports in San Antonio. For Port 720CT, add 18 Ports in Dallas. The total Port Additions will equal 96 and cost $3,590,625.

After working long hours on these suggestions, Port Additions should be considered mandatory. However, follow-up forecasts are probably needed due to the short time we were provided to do these studies. If you are going to perform follow-up forecasts, do so before September 1. The survey teams, if you want a successful forecast, need at least three months. Twenty-five team members should be sufficient.

8. Review the following PowerPoint screen. First, explain how the screen is visually flawed. Second, revise it for improved access using the techniques suggested in this chapter.

Recommendations

Table 1 outlines the investments recommended for 80 percent of the fiscal year 2005 funds.

The table focuses on 80 percent of the available funds. The remaining 20 percent are being withheld until fiscal year 2006.

Company	Costs
Harness	$2,300
MarTT3	$4,500
Notary Rex	$5,200
Mobile CRT	$2,750

9. Read the following headings and make them more accessible by creating a hierarchy using different font sizes and font types:

Meeting Minutes
Agenda
Discussion of Ongoing Projects
Recommendations
Pricing
Cost of Equipment
Cost of Facilities Update
Cost of Insurance Benefits
New Hiring Policies
Job Requirements
Employee Credentials
Licensing

10. Text boxes, like the one shown below, are an excellent way to call attention to key ideas, make hazards more emphatic, and add interest to a document. To practice creating text boxes, do the following:

a. Recreate the text box below.
b. Reformat the text box by changing the background color, the fill color, and the line color, style, and weight.
c. Add a shadow style to the exterior edges of the text box.

Warning

- Sharp blades.
- Cuts can occur.
- Wear protective gloves and keep your hands behind the blade guards.

QUIZ QUESTIONS

1. What are three reasons for designing a technical document effectively?

2. What is chunking?

3. What do you achieve by chunking?

4. In what four ways can you accomplish chunking?

5. What is the purpose of creating order or queuing your ideas?

6. How can you achieve a hierarchy of headings in your technical document?

7. What do you accomplish through a hierarchy of headings?

8. How can you help your readers access your information, filtering out extraneous or tangential information?

9. What can happen if you overuse highlighting techniques?

10. What are four ways in which you can vary your document design?

Graphics

BENEFITS OF VISUAL AIDS

Although your writing may have no grammatical or mechanical errors and you may present valuable information, you won't communicate effectively if your information is inaccessible. Consider the following paragraph:

example

> In January 2005, the actual rainfall was 1.50", but the average for that month was 2.00". In February 2005, the actual rainfall was 1.50", but the annual average had been 2.50". In March 2005, the actual rainfall was 1.00", but the yearly average was 2.50". In April 2005, the actual rainfall was 1.00", but annual averages were 2.50". The May 2005 actual rainfall was 0.50", whereas the annual average had been 1.50". No rainfall was recorded in June 2005. Annually, the average had been 0.50". In July 2005, only 0.25" rain fell. Usually, July had 0.50" rain. In August 2005, again no rain fell, whereas the annual August rainfall measured 0.25". In September and October 2005, the actual rainfall (0.50") matched the annual average. Similarly, the November actual rainfall matched the annual average of 1.50". Finally, in December 2005, 2.00" rain fell, compared to the annual average of 1.50".

If you read the preceding paragraph in its entirety, you are an unusually dedicated reader. Such wall-to-wall words mixed with statistics do not create easily readable writing.

The goal of effective technical writing is to communicate information easily. The example paragraph fails to meet this goal. No reader can digest the data easily or see clearly the comparative changes from one month's precipitation to the next.

To present large blocks of data or reveal comparisons, you can supplement, if not replace, your text with graphics. In technical writing, visual aids accomplish several goals. Graphics (whether hand drawn, photographed, or computer generated) will help you achieve conciseness, clarity, and cosmetic appeal.

Conciseness

Visual aids allow you to provide large amounts of information in a small space. Words used to convey data (such as in the example paragraph) double, triple, or even quadruple the space needed to report information. By using graphics, you can also delete many dead words and phrases.

Clarity

Visual aids can clarify complex information. Graphics help readers see the following:

tech link

For information about interactive three-dimensional graphics and animation, go to *http://research.microsoft.com/graphics/*.

- **Trends**—Certain trends, such as increasing or decreasing sales figures, are most evident in line graphs.
- **Comparisons between like components**—Comparisons such as actual monthly versus average rainfalls can be seen in grouped bar charts.
- **Percentages**—Pie charts help readers discern these.
- **Facts and figures**—A table states statistics more clearly than a wordy paragraph.

Cosmetic Appeal

Visual aids help you break up the monotony of wall-to-wall words. If you only give unbroken text, your reader will tire, lose interest, and overlook key concerns. Graphics help you sustain your reader's interest. Let's face it; readers like to look at pictures.

The two types of graphics important for technical writing are tables and figures. This chapter helps you correctly use both.

COLOR

All graphics look best in color, don't they? Not necessarily. Without a doubt, a graphic depicted in vivid colors will attract your reader's attention. However, the colors might not aid communication. For example, colored graphics could have these drawbacks (Reynolds and Marchetta 1998, 5–7):

1. The colors might be distracting (glaring orange, red, and yellow combinations on a bar chart would do more harm than good).
2. Colors that look good today might go out of style in time.
3. Colored graphics increase production costs.
4. Colored graphics consume more disk space and computer memory than black-and-white graphics.
5. The colors you use might not look the same to all readers.

tech link

Go to *http://www.prenhall.com/gerson* for Web links, samples, and interactive activities.

Let's expand on this last point. Just because you see the colors one way does not mean your readers will see them the same. We are not talking about people with vision problems. Instead, we are talking about what happens to your color graphics when someone reproduces them as black-and-white copies. We are also talking about computer monitor variations.

We discuss computer monitors in great detail in Chapter 13. The color on a computer monitor depends on its resolution (the number of pixels displayed) and the monitor's RGB values (how much red, green, and blue light is displayed). Because all monitors do not display these same values, what you see on your monitor will not necessarily be the same as what your reader sees. To solve this problem, test your graphics on several monitors. Also, limit your choices to primary colors instead of the infinite array of other color possibilities. More important, use patterns to distinguish your information so that the color becomes secondary to the design.

THREE-DIMENSIONAL GRAPHICS

Many people are attracted to three-dimensional graphics. After all, they have obvious appeal. Three-dimensional graphics are more interesting and vivid than flat, one-dimensional graphics. However, 3-D graphics have drawbacks. A 3-D graphic is visually appealing, but it does not convey information quantifiably. A word of caution: Use 3-D graphics sparingly. Better yet, use the 3-D graphic to create an impression; then include a table to quantify your data.

CRITERIA FOR EFFECTIVE GRAPHICS

Figure 9.1 is an example of a cosmetically appealing, clear, and concise graphic. At a glance, the reader can pinpoint the comparative prices per barrel of crude oil between 2001 and 2005. Thus, the line graph is clear and concise. In addition, the writer has included an interesting artistic touch. The oil gushing out of the tower shades just the parts of the graph that emphasize the dollar amounts. Envision this graph without the shading. Only the line would exist. The shading provides the right touch of artistry to enhance the information communicated.

The graph shown in Figure 9.1, although successful, does not include all the traits common to effective visual aids. Whether hand drawn or computer generated, successful tables and figures have these characteristics:

1. Are integrated with the text (i.e., the graphic complements the text; the text explains the graphic).
2. Are appropriately located (preferably immediately following the text referring to the graphic and not a page or pages later).
3. Add to the material explained in the text (without being redundant).
4. Communicate important information that could not be conveyed easily in a paragraph or longer text.
5. Do not contain details that detract from rather than enhance the information.

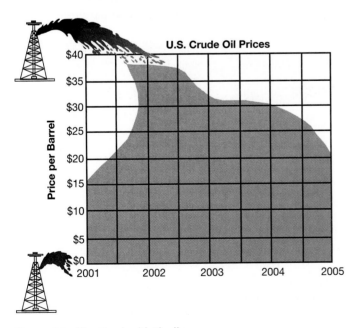

Figure 9.1   Line Graph with Shading

6. Are an effective size (not too small or too large).

7. Are neatly printed to be readable.

8. Are correctly labeled (with legends, headings, and titles).

9. Follow the style of other figures or tables in the text.

10. Are well conceived and carefully executed.

TYPES OF GRAPHICS

Graphics can be broken down into two basic types: tables and figures. Tables provide columns and rows of information. You should use a table to make factual information, such as numbers, percentages, and monetary amounts, easily accessible and understandable. Figures, in contrast, are varied and include bar charts, line graphs, photographs, pie charts, schematics, line drawings, and many more.

Tables

Let's tabulate the information about rainfall in 2005 presented earlier. Because effective technical writing integrates text and graphic, you will want to provide an introductory sentence prefacing Table 9.1, as follows:

> **example**
>
> Table 9.1 reveals the actual amount of rainfall for each month in 2005 versus the average documented rainfall for those same months.

This table has advantages for both the writer and the reader. First, the headings eliminate needless repetition of words, thereby making the text more readable. Second, the audience can see easily the comparison between the actual amount of rainfall and the monthly averages. Thus, the table highlights the content's significant differences. Third, the table allows for easy future reference. Tables could be created for each year. Then the reader could compare quickly the

TABLE 9.1 2005 MONTHLY RAINFALL VERSUS AVERAGE RAINFALL (ALL FIGURES IN INCHES)		
Month	2005 Rainfall	Average Rainfall
January	1.50	2.00
February	1.50	2.50
March	1.00	2.50
April	1.00	2.50
May	0.50	1.50
June	0.00	0.50
July	0.25	0.50
August	0.00	0.25
September	0.50	0.50
October	0.50	0.50
November	1.50	1.50
December	2.00	1.50

changes in annual precipitation. Finally, if this information is included in a report, the writer will reference the table in the Table of Contents' List of Illustrations. This creates ease of access for the reader.

Criteria for Effective Tables

To construct tables correctly, do the following:

1. Number tables in order of presentation (i.e., Table 1, Table 2, Table 3, etc.).

2. Title every table. In your writing, refer to the table by its number, not its title. Simply say, "Table 1 shows . . . ," "As seen in Table 1," or "The information in Table 1 reveals . . . "

3. Present the table as soon as possible after you have mentioned it in your text. Preferably, place the table on the same page as the appropriate text, not on a subsequent, unrelated page or in an appendix.

4. Don't present the table until you have mentioned it.

5. Use an introductory sentence or two to lead into the table.

6. After you have presented the table, explain its significance. You might write, "Thus, the average rainfall in both March and April exceeded the actual rainfall by 1.50 inches, reminding us of how dry the spring has been."

7. Write headings for each column. Choose terms that summarize the information in the columns. For example, you could write "% of Error," "Length in Ft.," or "Amount in $."

8. Because the size of columns is determined by the width of the data or headings, you may want to abbreviate terms (as shown in item 7). If you use abbreviations, however, be sure your audience understands your terminology.

9. Center tables between right and left margins. Don't crowd them on the page.

10. Separate columns with ample white space, vertical lines, or dashes.

11. Show that you have omitted information by printing two or three periods or a hyphen or dash in an empty column.

12. Be consistent when using numbers. Use either decimals or numerators and denominators for fractions. You could write 3 1/4 and 3 3/4 or 3.25 and 3.75. If you use decimal points for some numbers but other numbers are whole, include zeroes. For example, write 9.00 for 9.

13. If you do not conclude a table on one page, on the second page write *Continued* in parentheses after the number of the table and the table's title.

Table 9.2 is an excellent example of a correctly prepared table.

Figures

Another way to enhance your technical writing is to use figures. Whereas tables eliminate needless repetition of words, figures highlight and supplement important points in your writing. Like tables, figures help you communicate with your reader.

Types of figures include the following:

- Bar charts
  - Grouped bar charts
  - 3-D (tower) bar charts

TABLE 9.2	STUDENT HEADCOUNT ENROLLMENT BY AGE GROUP AND STUDENT STATUS, FALL 2003				
Age Group	New Students	Continuing Students	Readmitted	Other	Total
15–17	453	33	2	2	490
18–20	1,404	1,125	132	—	2,661
21–23	339	819	269	—	1,427
24–26	263	596	213	—	1,072
27–29	250	436	134	—	820
30–39	524	1,168	372	—	2,064
40–49	271	510	186	—	967
50–59	76	121	54	—	251
60+	19	48	16	—	83
Unknown	109	92	27	2	230
Total	3,708	4,948	1,405	4	10,065

Creating Graphics in Microsoft Word
(Pie Charts, Bar Charts, Line Graphs, etc.)

You can create customized graphics in Microsoft Word as follows:

1. Click on Insert on the Menu bar.
2. Point to Picture.
3. Click on Chart.

Word will open a datasheet and bar chart template that you can customize by inserting your own text and numbers.

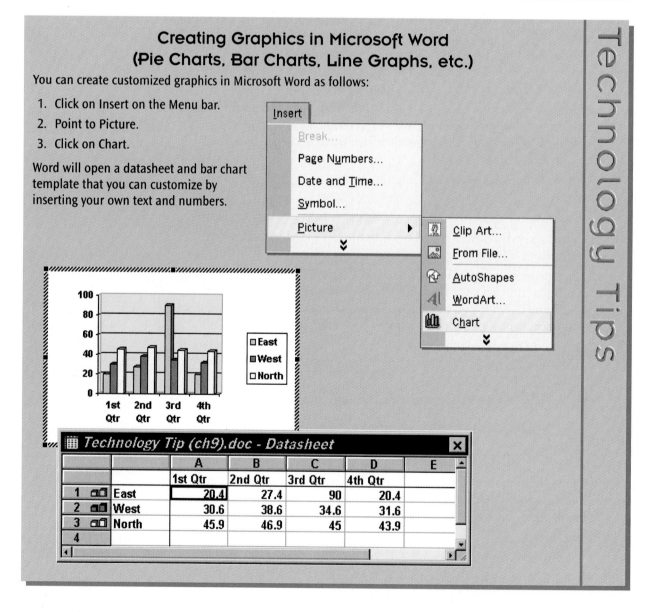

Once you have opened Word's graphic's datasheet and template, you can customize the graphic further as follows:

1. Choose the type of graphic you want by clicking on Chart on the Menu bar and scrolling to and selecting Chart Type. The Chart Type dialog box displays, allowing you to select a chart type.

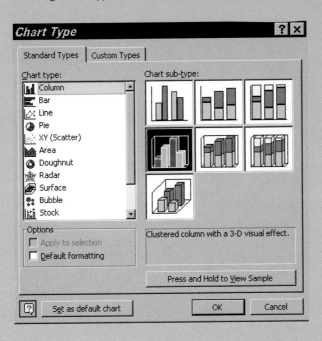

2. To add figure numbers, figure titles, legends, gridlines, data labels, and data tables, click on Chart and scroll to and select Chart Options.

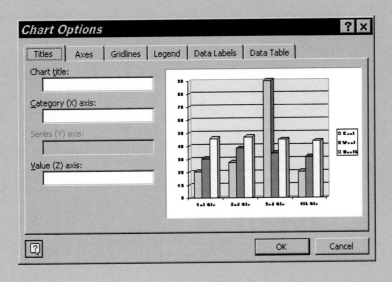

- Pictographs
- Gantt charts
- 3-D topographical charts
- Pie charts

- Line charts
  - Broken line charts
  - Curved line charts
- Combination charts
- Flowcharts
- Organizational charts
- Schematics
- Geologic maps
- Line drawings
  - Exploded views
  - Cutaway views
  - Super comic book look
  - Renderings
  - Virtual reality drawings
- CAD drawings
- Photographs
- Icons
- Internet downloadable graphics

All of these types of figures can be computer generated using an assortment of computer programs. The program you use depends on your preference and hardware.

Criteria for Effective Figures

To construct figures correctly, do the following:

1. Number figures in order of presentation (i.e., Figure 1, Figure 2, Figure 3, etc.).
2. Title each figure. When you refer to the figure, use its number rather than its title: for example, "Figure 1 shows the relation between the average price for houses and the actual sales prices."
3. Preface each figure with an introductory sentence.
4. Don't use a figure until you have mentioned it in the text.
5. Present the figure as soon as possible after mentioning it instead of several paragraphs or pages later.
6. After you have presented the figure, explain its significance. Don't let the figure speak for itself. Remind the reader of the important facts you want to highlight.
7. Label the figure's components. For example, if you are using a bar or line chart, label the x- and y-axes clearly. If you're using line drawings, pie charts, or photographs, use clear *call-outs* (names or numbers that indicate particular parts) to label each component.
8. When necessary, provide a legend or key at the bottom of the figure to explain information. For example, a key in a bar or line chart will explain what each differently colored line or bar means. In line drawings and photographs, you can use numbered call-outs in place of names. If you do so, you will need a legend at the bottom of the figure explaining what each number means.

9. If you abbreviate any labels, define these in a footnote. Place an asterisk (*) or a superscript number ([1], [2], [3]) after the term and then at the bottom of the figure where you explain your terminology.

10. If you have drawn information from another source, note this at the bottom of the figure.

11. Frame the figure. Center it between the left and right margins or place it in a text box.

12. Size figures appropriately. Don't make them too small or too large.

13. Try the super comic book look (figures drawn in cartoonlike characters to highlight parts of the graphic and to interest readers).

Bar Charts

Bar charts show either vertical bars (as in Figure 9.2) or horizontal bars (as in Figure 9.3). These bars are scaled to reveal quantities and comparative values. You can shade, color, or crosshatch the bars to emphasize the contrasts. If you do so, include a key explaining what each symbolizes, as in Figure 9.4. *Pictographs* (as in Figure 9.5) use picture symbols instead of bars to show quantities. To create effective pictographs, do the following:

1. The picture should be representative of the topic discussed.

2. Each symbol equals a unit of measurement. The size of the units depends on your value selection as noted in the key or on the *x*- and *y*-axes.

3. Use more symbols of the same size to indicate a higher quantity; do not use larger symbols.

Gantt Charts

Gantt charts, or *schedule charts* (as in Figure 9.6, page 285), use bars to show chronological activities. For example, your goal might be to show a client phases of a project. This could include planned start dates, planned reporting milestones, planned completion dates, actual progress made toward completing

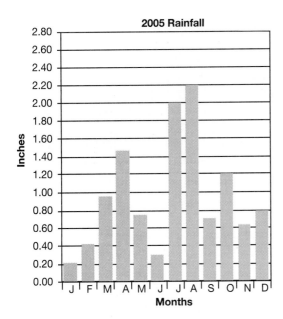

Figure 9.2   Vertical Bar Chart

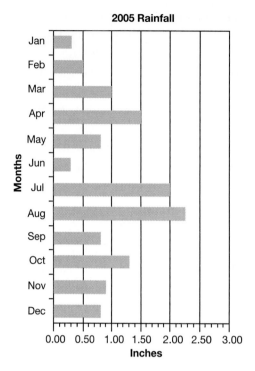

2005 Rainfall

Figure 9.3   Horizontal Bar Chart

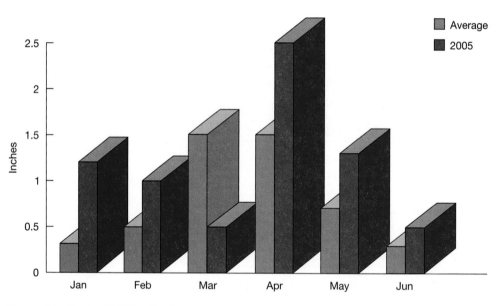

Comparison of 2005 Rainfall to Average

Figure 9.4   Grouped 3-D Bar Chart

the project, and work remaining. Gantt charts are an excellent way to represent these activities visually. They are often included in proposals to project schedules or in reports to show work completed. To create successful Gantt charts, do the following:

1. Label your *x*- and *y*-axes. For example, the *y*-axis represents the various activities scheduled, then the *x*-axis represents time (either days, weeks, months, or years).

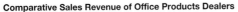

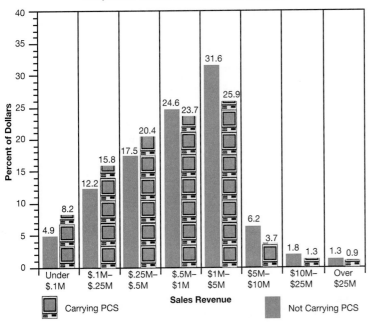

Figure 9.5   Pictograph

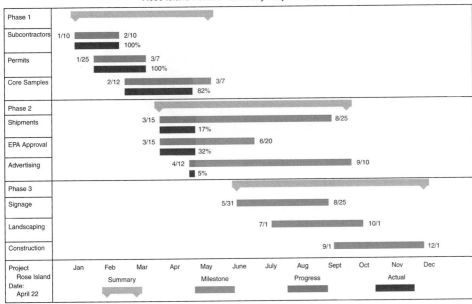

Figure 9.6   Gantt Chart

2. Provide gridlines (either horizontal or vertical) to help your readers pinpoint the time accurately.

3. Label your bars with exact dates for start or completion.

4. Quantify the percentages of work accomplished and work remaining.

5. Provide a legend or key to differentiate between planned activities and actual progress.

3-D Topographical Charts

Three-dimensional contour representations are not limited to land elevations. A three-dimensional surface chart could be used to represent many different forms of data. These 3-D "topos" (as shown in Figure 9.7 below) are used in industries as varied as aerospace, defense, education, research, oil, gas, and water. Applications include CAD/CAM, statistical analysis, and architectural design.

Pie Charts

Use pie charts (as in Figure 9.8) to illustrate portions of a whole. The pie chart represents information as pie-shaped parts of a circle. The entire circle equals 100 percent or 360 degrees. The pie pieces (the wedges) show the various divisions of the whole.

To create effective pie charts, do the following:

1. Be sure that the complete circle equals 100 percent or 360 degrees.
2. Begin spacing wedges at the twelve o'clock position.
3. Use shading, color, or crosshatching to emphasize wedge distributions.

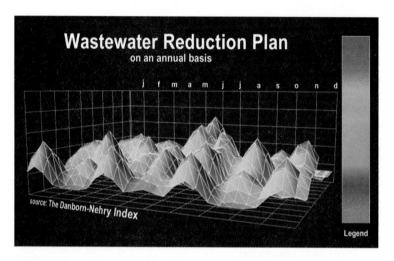

Figure 9.7   3-D Topographical Chart
Courtesy of Brandon Henry.

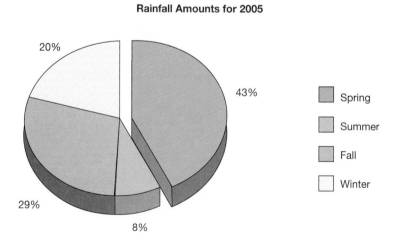

Figure 9.8   Pie Chart

4. Use horizontal writing to label wedges.

5. If you don't have enough room for a label within each wedge, provide a key defining what each shade, color, or crosshatching symbolizes.

6. Provide percentages for wedges when possible.

7. Do not use too many wedges—this would crowd the chart and confuse readers.

8. Make sure that different sizes of wedges are fairly large and dramatic.

Line Charts

Line charts reveal relationships between sets of figures. To make a line chart, plot sets of numbers and connect the sets with lines. These lines create a picture showing the upward and downward movement of quantities. Line charts of more than one line (see Figure 9.9) are useful in showing comparisons between two sets of values. However, avoid creating line charts with too many lines, which will confuse your readers.

Combination Charts

A combination chart reveals relationships between two sets of figures. To do so, it uses a combination of figure styles, such as a bar chart and an area chart (as shown in Figure 9.10, page 288) or a bar chart and a line chart. The value of a combination chart is that it adds interest and distinguishes the two sets of figures by depicting them differently.

Flowcharts

You can show chronological sequence of activities using a flowchart. Flowcharts are especially useful for writing technical instructions (see Chapter 12). When using a flowchart, remember that ovals represent starts and stops, rectangles represent steps, and diamonds equal decisions (see Figure 9.11, page 288).

Organizational Charts

These charts (as in Figure 9.12, page 289) show the chain of command in an organization. You can use boxes around the information or use white space to

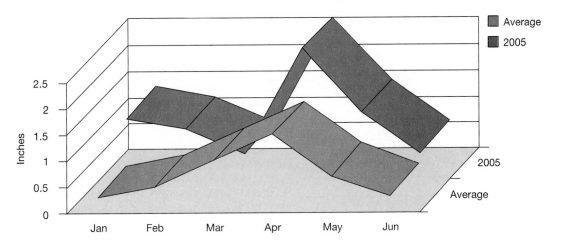

Figure 9.9   3-D Line Chart

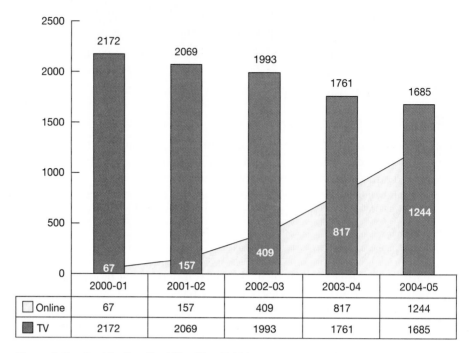

Figure 9.10   Combination Chart (Bar, Line, Table)

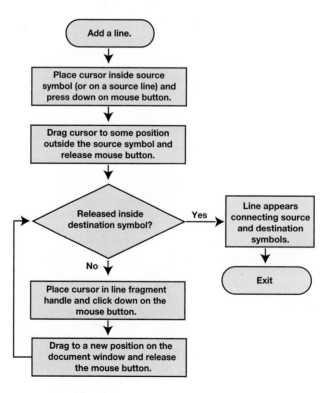

Figure 9.11   Flowchart

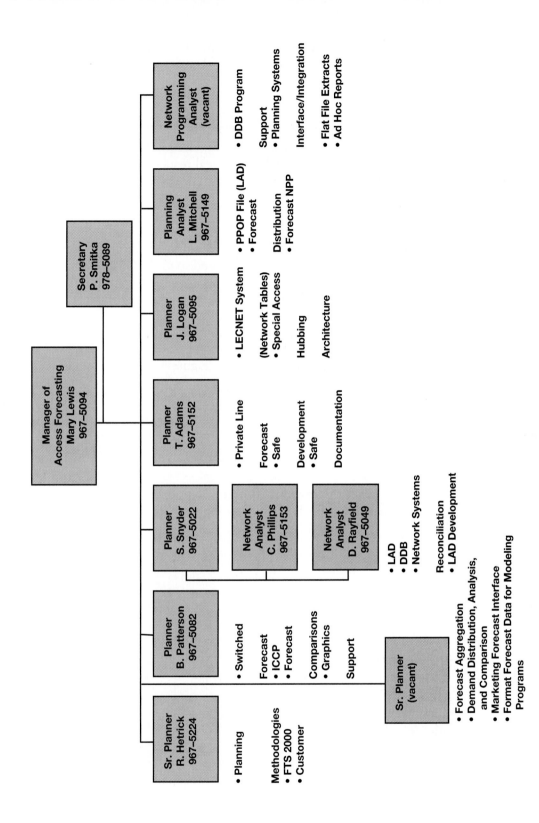

Figure 9.12 Organizational Chart

289

distinguish among levels in the chart. An organizational chart helps your readers see where individuals work within a business and their relation to other workers.

Schematics

Schematics are useful for presenting abstract information in technical fields such as electronics and engineering. A schematic diagrams the relationships among the parts of something such as an electrical circuit. The diagram uses symbols and abbreviations familiar to highly technical readers.

The schematic in Figure 9.13 shows various electronic parts (resistors, diodes, condensors) in a radio.

Geologic Maps

Maps of any sort help us understand locations. Usually these show cities, streets, roads, highways, rivers, lakes, mountains, and so forth. Geologic maps do more. They also show terrain, contours, heat ranges, the surface features of a place or object, or an analysis of an area. Often, to help readers orient themselves, these maps are printed on top of a regular map (called a base map). The base map is black and white. The geologic map, in contrast, uses colors, contact and fold lines, and special symbols to reveal the geology of an area. These features are then defined on a map legend or key (as in Figure 9.14).

Line Drawings

Use line drawings to show the important parts of a mechanism or to enhance your text cosmetically. To create line drawings, do the following:

1. Maintain correct proportions in relation to each part of the object drawn.

2. If a sequence of drawings illustrates steps in a process, place the drawings in left-to-right or top-to-bottom order.

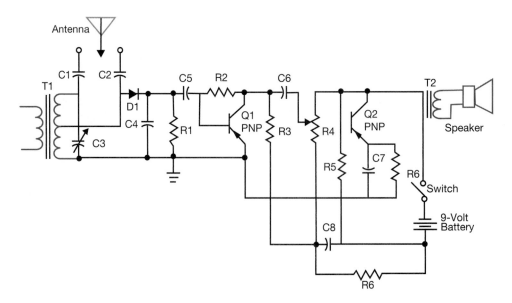

Figure 9.13   Schematic of a Radio

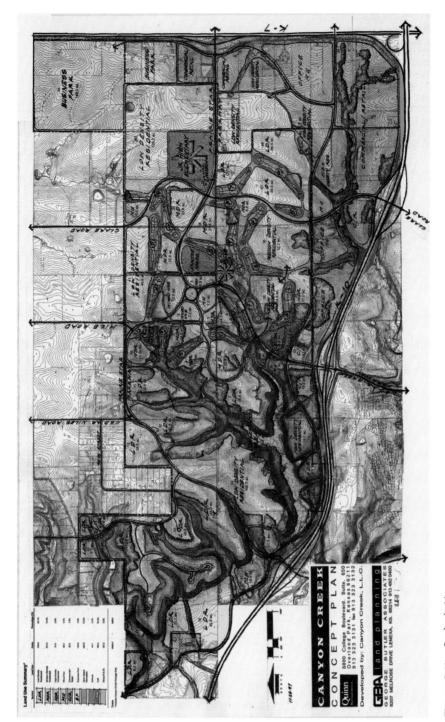

Figure 9.14   Geologic Map
Courtesy of George Butler Associates, Inc.

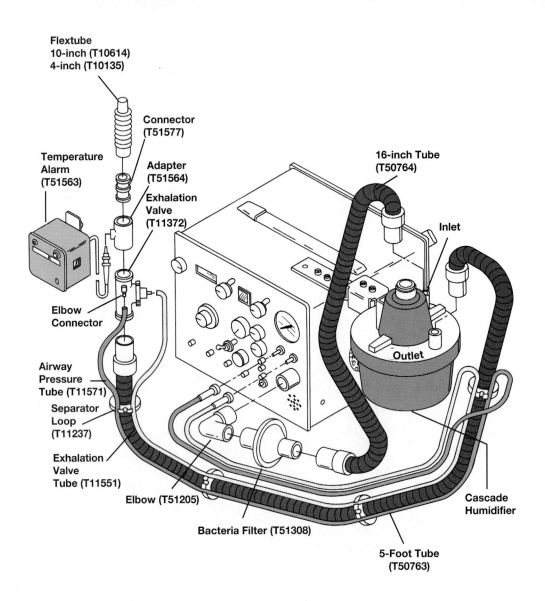

Flextube
10-inch (T10614)
4-inch (T10135)

Connector
(T51577)

Temperature
Alarm
(T51563)

Adapter
(T51564)

Exhalation
Valve
(T11372)

16-inch Tube
(T50764)

Inlet

Elbow
Connector

Outlet

Airway
Pressure
Tube (T11571)

Separator
Loop
(T11237)

Exhalation
Valve
Tube (T11551)

Elbow (T51205)

Cascade
Humidifier

Bacteria Filter (T51308)

5-Foot Tube
(T50763)

Figure 9.15   Line Drawing of Ventilator (Exploded View with Call-outs)
Courtesy of Nellcor Puritan Bennett Corp.

3. Using call-outs to name parts, label the components of the object drawn (see Figure 9.15).

4. If there are numerous components, use a letter or number to refer to each part. Then reference this letter or number in a key (see Figure 9.16).

5. Use exploded views (Figure 9.16) or cutaways (Figures 9.17, 9.18, and 9.19) to highlight a particular part of the drawing.

Renderings and Virtual Reality Drawings

Two different types of line drawings are renderings and virtual reality views. Both offer 3-D representations of buildings, sites, or things. Often used in the

Exhalation Valve Parts List		
Item	Part Number	Description
1	000723	Nut
2	003248	Cap
3	T50924	Diaphragm
4	Reference	Valve Body
5	Reference	Elbow Connector
—	T11372	Exhalation Valve

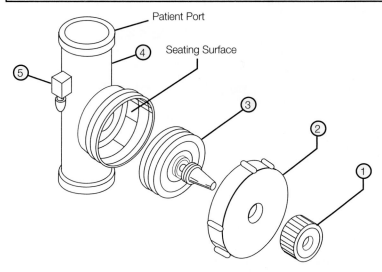

Figure 9.16   Line Drawing of Exhalation Valve (Exploded View with Key)
Courtesy of Nellcor Puritan Bennett Corp.

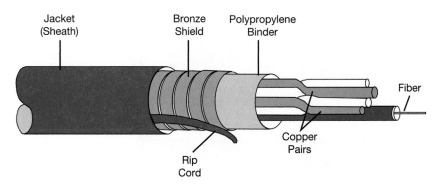

Figure 9.17   Line Drawing of Cable (Cutaway View)
Courtesy of Nellcor Puritan Bennett Corp.

architectural/engineering industry, these 3-D drawings (as shown in Figures 9.20 and 9.21) help clients get a visual idea of what services your company can provide. Renderings and virtual reality drawings add lighting, materials, and shadow and reflection mapping to mimic the real world and allow customers to see what a building or site will look like in a photorealistic setting.

CAD Drawings

Computer-aided design (CAD) drawings, such as the floor plan in Figure 9.22, use geometric shapes and symbols to provide a customer with a graphic view of

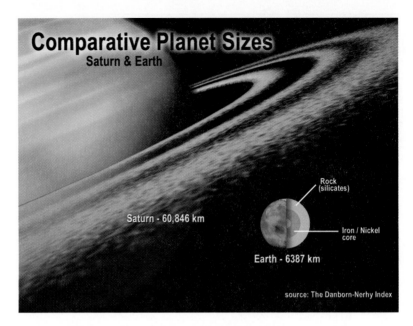

Figure 9.18 Comparison of Planets with Cutaway View
Courtesy of Brandon Henry.

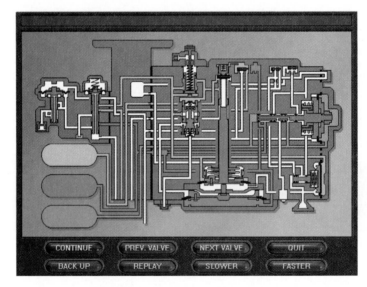

Figure 9.19 Cutaway View of a Railcar Braking System
Courtesy of Burlington Northern Santa Fe Railroad.

a setting drawn to a particular scale. CAD drawings include *notations* to define scale and a *title block*. The title block gives the date of completion, name of the draftsperson, company, client, and project name.

Photographs

A photograph can illustrate your text effectively. Like a line drawing, a photograph can show the components of a mechanism. If you use a photo for this purpose, you will need to label (name), number, or letter parts and provide a key. Photographs are excellent visual aids because they emphasize all parts equally. Their primary advantage is that they show something as it truly is.

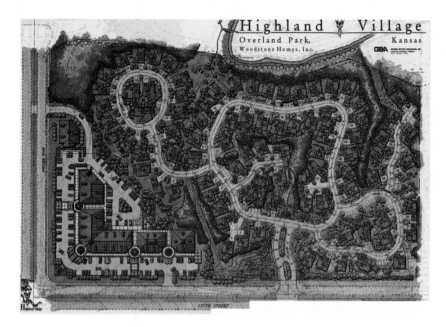

Figure 9.20   Architectural Rendering
Courtesy of George Butler Associates, Inc.

Figure 9.21   3-D Drawing
Courtesy of Johnson County Community College.

Photographs have one disadvantage, however. They are difficult to reproduce. Whereas line drawings photocopy well, photographs do not. See Figure 9.23.

Icons

Approximately 23 percent of America's population is functionally illiterate. In today's global economy, consumers speak diverse languages. Given these two facts, how can technical writers communicate to people who cannot read and to people who speak different languages? Icons offer one solution. Icons (as in

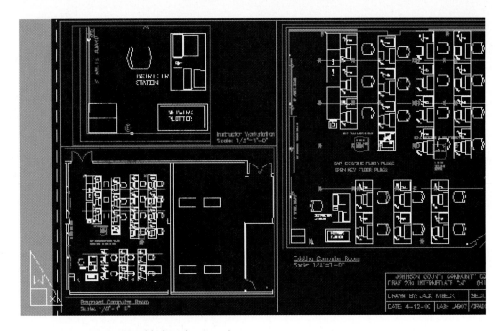

Figure 9.22   Computer-Aided Design Drawing

Figure 9.23   Photograph of Mechanical Piping
Courtesy of George Butler Associates, Inc.

Figure 9.24   An Icon of Explosives

Figure 9.25   An Icon of Dangerous Machinery

Figure 9.26   An Icon of Electric Shock

Figure 9.27   An Icon of Corrosive Material

Figures 9.24, 9.25, 9.26, and 9.27 shown above) are visual representations of a capability, a danger, a direction, an acceptable behavior, or an unacceptable behavior.

For example, the computer industry uses icons—open manila folders to represent computer files. In manuals, a jagged lightning stroke iconically represents the danger of electrocution. On streets, an arrow represents the direction we should travel; on computers, the arrow shows us which direction to scroll. Universally depicted stick figures of men and women greet us on restroom doors to show us which rooms we can enter and which rooms we must avoid.

When used correctly, icons can save space, communicate rapidly, and help readers with language problems understand the writer's intent.

To create effective icons, follow these suggestions:

1. *Keep it simple*—You should try to communicate a single idea. Icons are not appropriate for long discourse.

2. *Create a realistic image*—This could be accomplished by representing the idea as a photograph, drawing, caricature, outline, or silhouette.

3. *Make the image recognizable*—A top view of a telephone or computer terminal is confusing. A side view of a playing card is completely unrecognizable. Select the view of the object that best communicates your intent.

4. *Avoid cultural and gender stereotyping*—For example, if you are drawing a hand, you should avoid showing any skin color, and you should stylize the hand so it is neither clearly male nor female.

5. *Strive for universality*—Stick figures of men and women are recognizable worldwide. In contrast, letters—such as *P* for *parking*—will mean very

little in China, Africa, or Europe. Even colors can cause trouble. In North America, red represents danger, but red is a joyous color in China. Yellow calls for caution in North America, but this color equals happiness and prosperity in the Arab culture (Horton 1993, 682–93).

Internet Downloadable Graphics

More and more, you will be writing online as the Internet, intranets, and extranets become prominent in technical writing. (We discuss the Internet in detail in Chapter 13.) You can create graphics for your Web site in three ways:

Download Existing Online Graphics

The Internet contains thousands of Web sites that provide online clip art. These graphics include photographs, line drawings, cartoons, icons, animated images, arrows, buttons, horizontal lines, balls, letters, bullets, and hazard signs. In fact, you can download any image from any Web site. Many of these images are freeware, which you can download without cost and without infringing on copyright laws.

To download these images, just place your cursor on the graphic you want, then right-click on the mouse. A pop-up menu will appear. Click on Save Picture As. Once you have done this, a new menu will appear. You can save your image in the file of your choice, either on the hard drive or on your disk. The images from the Internet will already be GIF (graphics interchange format) or JPEG (joint photographic experts group) files. Thus, you will not have to convert them for use in your Web site.

Modify and Customize Existing Online Graphics

If you plan to use an existing online graphic as your company's logo, for example, you will need to modify or customize the graphic. You will want to do this for at least two reasons: to avoid infringing on copyright laws and to make the graphic uniquely yours.

To modify and customize graphics, you can download them in two ways. First, you can print the screen by pressing the Print Screen key (usually found on the upper right of your keyboard). This captures the entire screen image in a clipboard. Then you can open a graphics program and paste the captured image. Second, you can save the image in a file (as discussed) and then open the graphic in a graphics package. Most graphics programs will allow you to customize a graphic. Popular programs include Paint, Paint Shop Pro, PhotoShop, Corel Draw, Adobe Illustrator, Freehand, and Lview Pro. In these graphics programs, you can manipulate the images by changing colors, adding text, reversing the images, cropping, resizing, redimensioning, rotating, retouching, deleting or erasing parts of the images, overlaying multiple images, joining multiple images, and so forth. After you make substantial changes, the new image becomes your property.

You could also take any existing graphic from hard-copy text (magazines, journals, books, newsletters, brochures, manuals, reports, etc.), scan the image, crop and retouch it, save it, and then reopen this saved file in one of the graphics programs for further manipulation.

Some graphics programs, such as Paint, save an image only as a BMP, a bitmap image. Once the image has been altered, you will need to convert your bitmap file to a GIF or JPEG format for use in your Web site. Doing so is important because the Internet will not read BMP images.

A final option is to create your own graphic. If you are artistic, draw your graphic in a graphics program, save the image as a GIF or JPEG file, and then load the image into your Web site. This option might be more challenging and time consuming. However, creating your own graphic gives you more control over the finished product, provides a graphic precisely suited to your company's needs, and helps avoid infringement of copyright laws.

Technology Tips

Creating Screen Captures

Screen captures are an outstanding graphic aid for user manuals, Web sites, proposals, brochures, newsletters, and fliers.

1. To create a screen capture, find any graphic online and press the Print Screen button on the top right of your computer's keyboard. This captures the entire desktop image in a buffer.

2. To download and crop this image using any graphics software, open your graphics software program, click on Edit, and scroll to and select Paste.

Once you have downloaded your screen capture, you might need to format it to fit in your document. For example, you would format the graphic below as follows:

1. Right-click and select Format Picture.

2. In the Format Picture dialog box, click on the Layout tab and select your graphic's preferred orientation (in front of the text, behind, etc.).

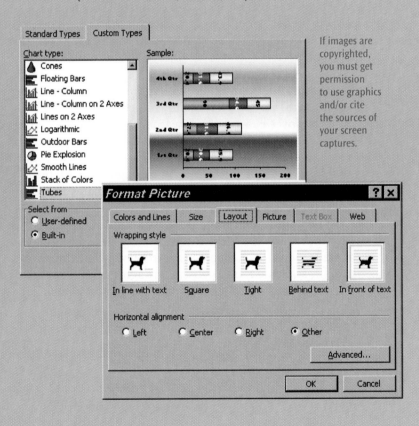

If images are copyrighted, you must get permission to use graphics and/or cite the sources of your screen captures.

Chapter Highlights

1. Using graphics can allow you to create a more concise document.

2. Graphics add variety to your text, breaking up wall-to-wall words.

3. Color and 3-D graphics can be effective. However, these two design elements also can cause problems. Your color choices might not be reproduced exactly as you planned, and a 3-D graphic could be misleading rather than informative.

4. Tables are effective for presenting numbers, dates, and columns of figures.

5. You can often communicate more easily with your audience when you use figures, such as bar charts, pie charts, line charts, flowcharts, and organizational charts, to highlight and supplement important parts of your text.

ACTIVITIES

CASE STUDY

Your company is submitting a proposal to a client. The proposal is about the creation of a company newsletter geared to the client's employees. Part of this proposal will include a tentative project schedule. You plan on informing the client that you will begin the project May 1 and conclude September 7.

Project activities include the following: (1) meetings between the newsletter staff and the client's marketing department, scheduled for May 1 through May 5, to determine objectives; (2) employee interviews regarding company recreational events and employee celebrations (birthdays, births, awards, etc.), scheduled for May 8 through May 19; (3) meetings with the accounting department, scheduled for May 22 through May 31, to ascertain stock information, annual corporate earnings, and employee benefits; (4) research regarding newsletter layout options (paper size, weight, color, and number of pages suggested per edition), scheduled for June 5 through June 16; (5) cost estimates for newsletter production based on the above findings, scheduled for June 19 through June 30; (6) follow-up meetings with client management to approve the cost estimates and tentative page layout, scheduled for July 3 through July 5; (7) writing the first draft of a newsletter, scheduled for July 10 through July 21; (8) submission of the draft to client management for approval, scheduled for July 24; (9) follow-up interviews of employees and the accounting department to update information for the final newsletter draft, scheduled for August 1 through 12; (10) newsletter production, scheduled for August 15 through 26; and (11) project completion—newsletters delivered to each client employee—scheduled for September 7.

Create a Gantt chart conveying the above information.

EXERCISES CREATING FIGURES AND TABLES

1. Present the following information in a pie chart, a bar chart, and a table.

   In 2004, the Interstate Telephone Company bought and installed 100,000 relays. It used these for long-range testing programs that assessed failure rates. It purchased 40,000 Nestor 221s; 20,000 VanCourt 1200s; 20,000 Macro R40s; 10,000 Camrose Series 8s; and 10,000 Hardy SP6s.

2. Using the information presented in activity 1 and the following revised data, show the comparison between 2004 and 2005 purchases through two pie charts, a grouped bar chart, and a table.

   In 2005, after assessing the success and failure of the relays, the Interstate Telephone Company made new purchases of 200,000 relays. It bought 90,000 VanCourt 1200s; 50,000 Macro R40s; 30,000 Camrose Series 8s; and 30,000 Hardy SP6s. No Nestors were purchased.

3. Create a table for the following information.

   When the voltage out is 13 V, the frequency out is 926 Hz. When the voltage out is 12.5 V, the frequency out is 1.14 K. When the voltage out is 12 V, the frequency out is 1.4 K. When the voltage out is 11 V, the frequency out is 1.8 K. When the voltage out is 10 V, the frequency out is 2.3 K. When the voltage out is 9 V, the frequency out is 2.8 K. When the voltage out is 8 V, the frequency out is 3 K. When the voltage out is 7 V, the frequency out is 0 Hz.

4. Create a line chart. To do so, select any topic you like. The subject matter, however, must include varying values. For example, present a line chart of your grades in one class, your salary increases (or decreases) at work, the week's temperature ranges, your weight gain or loss throughout the year, the miles you've run during the week or month, amounts of money you've spent on junk food during the week, and so forth.

5. Create a flowchart. To do so, select a topic about which you can write an instruction—for example, the steps for changing oil, winterizing your house, planting trees, seeding your yard, refinishing furniture, making a cake, changing a tire, interviewing for a job, wallpapering a room, building a deck, or installing a fence. Flowchart the sequential steps for any of these procedures.

6. Create three bar charts, using the information from activity 1. Make one of the bar charts in color, one as a 3-D bar chart, and the other as a one-dimensional bar chart. Print them and give them to others in class. Determine which of the three is most effective and why.

7. Find effective and ineffective photographs in business documents. Explain why the effective photographs succeed. Explain why the ineffective photographs fail to convey accurate information.

8. Go online to any of the following Web sites and find what you consider to be effective or ineffective examples of hazard icons. Explain why the effective icons aid clarity. Explain why the ineffective ones fail to communicate warnings clearly.

   • Pesticide Toxicity Hazard and Risk: *http://www1.agric.gov.ab.ca/ $department/deptdocs.nsf/all/prm2375?opendocument*

   • The Safety Sign Shop: *http://hh181.hiphip.com/merchant/frames/*

   • International Safety and Health Information Centre: *http://www.ilo. org/public/english/protection/safework/cis/products/icsc/dtasht/ symbols/index.htm*

   • Instant Art: *http://www.instant-art.com/homeset.html*

9. Analyze the following graphics and explain which ones succeed and which ones fail.

a. Vertical Bar Chart

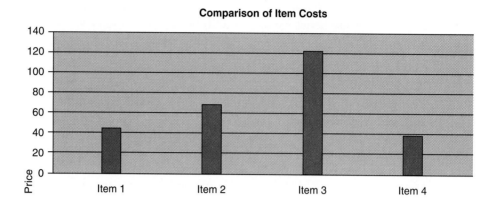

b. Pie Chart

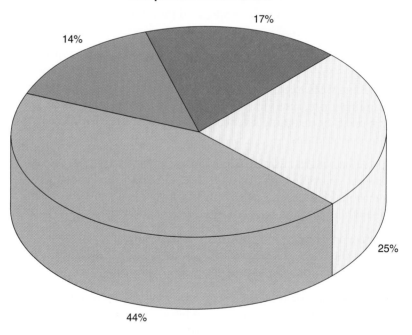

c. Vertical Bar Chart

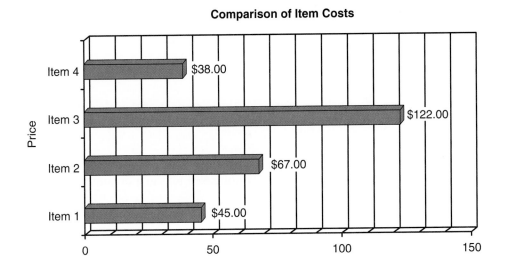

Comparison of Item Costs

d. Table

TABLE 1 COMPARISON OF ITEM COSTS		
	Price	Percentage
Item 1	$ 45.00	17%
Item 2	$ 67.00	25%
Item 3	$122.00	44%
Item 4	$ 38.00	14%
Total	$272.00	100%

10. Based on the criteria provided in this chapter, revise any of the poor graphics from the preceding assignment (number 9). Add any additional information necessary to correct the graphics.

TEAM PROJECTS

1. In small groups, create new icons for any or all of the following common computer functions: open, retrieve, close, save, password, print, undo, search, cut, and paste. Don't use any icons you've seen elsewhere; invent new ones. To do so, abide by the criteria provided in this chapter. Once the icons have been created, share them with other groups and select the best icons. Be prepared to justify your selections either in a group oral presentation or in a group-written report.

2. Find examples of good and bad tables and figures in a popular magazine, a technical journal, and the local newspaper. Bring these to class and, in small groups, discuss their successes and failures. Be prepared to justify your opinions either in a group oral presentation or in a group-written report.

ACTIVITIES

3. Select any of the unsuccessful graphics from team project 2. In small groups, decide how the flawed graphics could be improved and redraw or alter them by computer accordingly. Once these revisions are made, share your results with the class and select the best revisions.

QUIZ QUESTIONS

1. How can you effectively present large blocks of data or reveal comparisons?

2. How do graphics allow you to be concise?

3. What are four things graphics help readers to see?

4. What can happen when you use only unbroken text in a document?

5. When should you avoid using color in a graphic?

6. What are drawbacks to 3-D graphics?

7. When you use graphics, what are five criteria to follow to achieve successful graphics?

8. What are two things you achieve by using a table?

9. After a table, what should you provide in your text?

10. In a table, should you use decimals or numerators and denominators for fractions?

11. Figures accomplish what two things in a document?

12. What are five types of figures?

13. What are two types of bar charts?

14. What is a pictograph?

15. What are three ways to achieve an effective pictograph?

16. When would you use a Gantt chart?

17. What do pie charts illustrate?

18. What type sequence can you reveal in a flowchart?

19. How are schematics useful?

20. What are two reasons for using icons?

TECHNICAL APPLICATIONS

Fliers, Brochures, and Newsletters

Explains the criteria you should follow to write effective fliers, brochures, and newsletters; differentiates among these types of corporate communication; provides usability checklists to gauge the success of these technical applications; gives a case study to help you create these types of documents.

CHAPTER 11

Technical Descriptions

Discusses how to write technical descriptions including specifications for mechanisms, tools, or pieces of equipment; distinguishes between photographic and impressionistic word usage; provides brainstorming/listing as a prewriting technique.

CHAPTER 12

Instructions and User Manuals

Explains how to write the instructions and manuals that tell customers how to use or operate the equipment or software they have purchased; shows the importance of organization, audience recognition, graphics, and style to create instructions; uses flowcharts to help you prewrite; provides a usability checklist to help you test your instruction.

10 Fliers, Brochures, and Newsletters

Chapter Preview

Why Write Fliers?
Fliers are a cost-effective way to advertise your company.

Criteria for Writing Fliers
Well-written fliers will have reader appeal and sell your company.

Why Write Brochures?
A brochure will give you more space for presenting information.

Criteria for Writing Brochures
You will place specific information in the various panels of the brochure.

Effective Brochure Usability Checklist
Apply our checklist to the draft of your brochure.

Why Write Newsletters?
Newsletters are an excellent way to share news about your company.

Criteria for Writing Newsletters
Adhering to the specific parts of a typical newsletter will allow you to create effectively.

Process
Use the writing process to create a flier, a brochure, or a newsletter.

Effective Newsletter Usability Checklist
Apply our checklist to the draft of your newsletter.

Writing at Work

Tech Toolshop

TechToolshop is a full-service automobile business. It sells parts, provides car maintenance and repair, and offers corporate auto and truck fleet services.

Located in Big Springs, Iowa, this business of 1,200 employees began as a storefront auto parts store in 1995. But, through hard work, innovation, vision, and a willingness to change with the times, the Oleander brothers have expanded their business significantly over the years.

Despite the company's extraordinary growth, James and Harold Oleander have never forgotten their commitment to customer service. TechToolshop maintains a connection to its customers in several ways. For example, each month, the company connects with clients and vendors through a corporate newsletter—*ToolHelp*. In this newsletter, the company provides the following:

- Answers to automobile FAQs—frequently asked questions concerning repair and maintenance
- Updates on employees (promotions, birthdays, weddings, awards, newly achieved certifications, etc.)
- Community service projects sponsored by TechToolshop. These include gathering food for Big Spring's community food pantry, painting and upkeep for homes owned by Big Spring's senior citizens, visits to the local hospice, and cleanup at the city's arboretum and zoo
- New clients served by TechToolshop's auto and truck fleet program
- Job opportunities

TechToolshop has expanded its corporate coverage by writing online newsletters as well. These e-mail newsletters, which can be updated more frequently than their hard-copy text, have three distinct audiences: corporate clients with large fleets of vehicles; local families who use TechToolshop for parts and auto maintenance; and governmental legislators and automobile lobbyists.

Whereas the newsletters provide corporate information, TechToolshop uses brochures and fliers for promotional sales purposes. TechToolshop has three, full-color brochures that focus on its primary functions: auto parts, repair and maintenance, and fleet service. The company has many fliers, however. The full-color brochures are costly to write and publish, but the single-sided, postcard-sized fliers are more cost effective. In addition, these fliers, which can be produced quickly, focus on individual topics where quick turnaround is important: what's on sale this month, recent auto industry news that affects the car owner, innovative vehicle maintenance information, repair and replacement steps for auto recalls, best prices for used fleet vehicles, and so on.

One of TechToolshop's major goals is maintaining excellent customer relations. Good work and client referrals obviously help achieve this goal. However, the Oleanders have found that continuous communication is another way to accomplish client rapport. Though TechToolshop's business is good, it wants to keep it that way. Outstanding technical writing through brochures, fliers, and newsletters equals good business for TechToolshop.

OBJECTIVES

Imagine these scenarios. A business has just opened in your city. How will you learn about the company's products or services? *Or* you're visiting a city, a museum, an amusement park, a zoo, a college campus, or a historical site. How do you learn what these locations offer? *Or* how do you receive news from your bank, your child's school district, your brokerage firm, or your healthcare organization? You don't receive information by way of an essay, that's for sure. Instead, you'll either read a flier, a newsletter, or a brochure.

Maybe you won't even be physically at a site to pick up the brochure. Maybe you won't even be mailed the newsletter or flier. You might be online researching real estate, automobiles, travel, health concerns, or insurance. To find the information you seek, you'll access an *electronic* flier, brochure, or newsletter. Yahoo, for example, an Internet browser, lists online newsletters about the following diverse topics:

- Business
- Health
- Electronic Commerce
- Computers
- Law
- News and Media
- Education
- Real Estate
- Science
- Psychology

Corporate communication is accomplished in more ways than by mailing letters or sending e-mail messages. This chapter will introduce you to three other valuable tools that will help you communicate with your clients and colleagues—fliers, brochures, and newsletters.

Table 10.1 compares the various features of these tools.

TABLE 10.1	COMPARING FLIERS, BROCHURES, AND NEWSLETTERS		
	Fliers	Brochures	Newsletters
Size	The size of a postcard, one side of an 8 1/2 × 11-inch sheet of paper, or 2 pages maximum.	Usually six panels of text, on the front and back of an 8 1/2 × 11-inch sheet of paper.	Anywhere from 1 to 12 pages.
Topics	Limited—due to their size, fliers must focus on one, key point of a specific topic.	More—with increased length, brochures can focus on 5–10 different aspects of a topic.	Most—many topics can be discussed in great detail.
Audience	Dependent on the topic, fliers focus on a very specific high-tech or low-tech audience.	With expanded coverage, brochures can be geared toward lay, low, and high-tech audiences. However, most brochures are geared toward potential clients, readers external to a company.	Newsletters are sent to high and low-tech company employees as well as lay corporate stakeholders.
Cost	Least expensive due to size.	More expensive due to size and additional graphics.	Most expensive due to size, additional graphics, and needed software.
Technology	Fliers can be produced using standard word processing tools.	Brochures can be produced using standard word processing tools.	Standard word processing tools can be used to produce newsletters. However, more effective newsletters require additional software (Quark or FrameMaker) and hardware (printers capable of printing on custom-sized paper).
Longevity	Due to size and limited coverage, fliers have a short shelf life. They can be and should be updated quickly.	Due to cost and coverage, brochures can have a long shelf life. Revising them is not easy or cost effective.	Newsletters are generally published on a weekly, monthly, or quarterly basis. Thus, their longevity, regardless of cost and coverage, is limited.

WHY WRITE FLIERS?

Here's a story to show the value of writing fliers. The president of a technical writing firm told us that he contracted with an advertising company to create a super-slick brochure for his corporation. The brochure cost approximately $10,000. The company owner and his colleagues were thrilled with the brochure's quality, but they were a bit overwhelmed at the price. How could they generate enough income from such a costly communication vehicle to justify its expense?

Next, this technical writing company tried to create its own newsletter—to advertise as well as to share information with its clients regarding innovative, emerging technologies. The newsletter was professional looking and admirably depicted the company as a legitimate player in the technical writing field. In addition, the newsletter was far more cost effective. It could be produced in-house for approximately $350 an issue. Furthermore, in contrast to the brochure, which was a one-shot deal highlighting limited aspects of the company, the newsletter could cover many topics and be updated periodically. Still, at $350, produced quarterly, costs could mount.

Finally, the company took a different tack. It decided to try one-page fliers. The fliers would cost almost nothing to produce in terms of labor, they could be mass printed on the company's own desktop publishing equipment, and then the fliers could be mailed to clients for the cost of a stamp. Thus, this company concluded that fliers were

- **Cost effective**—Each flier would cost far less than an expensive advertising campaign and could be produced in-house.
- **Time efficient**—Creating a flier could entail only a few hours of labor by the company's employees.
- **Responsive to immediate needs**—Different fliers could be created for different audiences and purposes, at a moment's notice, to meet unique, emerging needs.
- **Personalized**—Fliers could be created with a specific market or client in mind. Then these fliers could either be mailed to that client or hand-delivered for a more personal touch (something not achievable with a Web site, for instance).

CRITERIA FOR WRITING FLIERS

When writing your flier, consider these points:

Keep the Flier Short

Though one page might be preferable, you could create a two-page flier, using the front and back of an 8 1/2 × 11-inch sheet of paper. However, if you keep your flier to one page (front only), then you can save money by folding the flier in thirds, stapling it, and using the blank side for mailing purposes (addresses and stamp). A flier could be even smaller, the size of a postcard, for example.

tech link
Go to *http://www.prenhall.com/gerson* for Web links, samples, and interactive activities.

Focus on One Idea, Topic, or Theme per Flier

Brochures and newsletters can cover topics more expansively. In contrast, a flier should distill an idea to its essence. It should make *one key point*. This is how you give the flier punch, pizzazz, and impact. This is how you make the flier's content relevant to your audience—by fulfilling that audience's unique need. For example, if your company's focus is automotive parts, don't write a flier covering every car accessory. Instead, one flier might focus on stereo systems, another flier on windshield wiper blades, a third flier on batteries, and so on.

Use a Title to Identify the Theme

The title can be one or two words long, you could use a phrase, or you might want to write an entire sentence at the top of the flier. Perhaps a more effective approach would be to begin your flier with a question to immediately arouse reader interest: "Software giving you a headache?" or "How usable is your Web site?"

Limit Your Text

This goes along with distilling an idea. A flier should be read at one sitting—or even at a glance. You can help your reader accomplish this goal by limiting your words. As already noted, if you want to expand on an idea, use a different communication medium, such as a brochure or newsletter. The goal of a flier should be more precise. You should limit the text to fifty or so words per page. If you can get by with fewer words, that's even better. You want your ideas to hit the reader hard. Fewer words can accomplish a lot if you select those words carefully, with each word actively driving home a key point.

Increase the Font Size

With so few words, you can increase your font size. Usually in a letter or memo, you will use a 12-point font. For a flier, in contrast, you could use a 16-point font and up for text, and a 20-point font and up for titles. This will make the text more readable and more dramatic.

Use Graphics

One graphic, at least, should emphasize your theme and visually make your point memorable. Another graphic could include your company logo (for corporate identity and namesake recognition). The logo could be placed anywhere you choose on the flier, but it would be a good idea to always have this logo in the same place for all fliers. Again, the goal is to constantly, even repetitiously, remind your readers of your company's identity. In addition, the logo should be accompanied by an address or phone number so clients can contact you or visit your site.

Use Color

You should use color either with your font or the graphics for audience appeal. However, as noted in Chapter 8 on document design, don't overuse color. Excess will distract your reader. Pick one dominant color to emphasize key points, maybe a color in your company logo, to habitually remind your reader of your company's identity.

Use Highlighting Techniques

Highlight text using techniques such as bullets, white space, tables, boldface, italics, headings, subheadings, or columns. Again, don't overdo it. A little highlighting goes a long way, especially on a one-page flier. Too much makes a jumble of your text and distracts from the message.

Find the Right Phrase

Select a catchy phrase, which you can use continually in all your fliers, to personify your company's primary focus. For example, as of 2004, McDonald's repeats the line "We love to make you smile"; Pepsi uses "It's the Cola"; Ford writes "100 Years of Automotive Achievement"; Harley-Davidson's "It's time to ride"; and Pizza Hut's "Gather 'Round the Good Stuff." What phrase captures your company's personality?

Recognize Your Audience

You want to show the readers how your product or service will benefit them. Thus, you must understand your audience's needs and direct the flier to meet those concerns. In addition, you want to engage the reader. To do this, use *pronouns* that speak to the reader on a personal level and *positive words* that motivate the reader to action. Finally, remember to speak at the reader's level of understanding, defining terms as necessary.

Avoid Grammatical Errors

The flier might be all your reader ever learns about your company. If the flier is attractive, informative, and grammatically correct, that depicts your company positively. If, in contrast, your flier has errors in it—for everyone to see—then what have you accomplished? You have destroyed your professionalism. Surely a client can find a vendor who can do a job without mistakes (and grammar errors in your flier show that you are prone to making mistakes).

See Figure 10.1 for a sample flier.

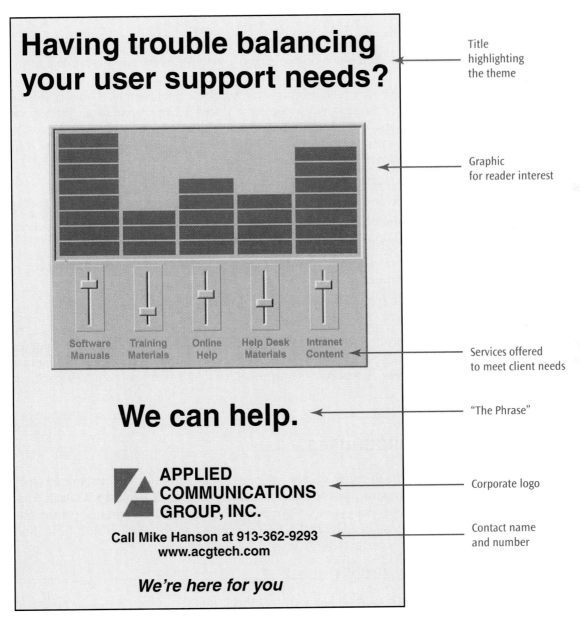

Figure 10.1   Flier
Courtesy of Applied Communications Group, Inc.

Because fliers are very short by definition, you can use a very simple *prewriting* technique to gather data. Try listing. This could be as easy as listing *who* your intended audience is, *why* you're creating this flier, *what* limited idea you hope to focus on, and *how* your company can meet your reader's needs. Then in the *writing* stage of the process, draft the flier by sketching the overall layout (complete with graphic and text). Finally, in the *rewriting* phase, make sure that you have met the criteria for successful fliers. You especially want to double-check that you have included your company's name, phone number, street address, e-mail address, perhaps a contact person, "the phrase" that hooks your reader, appropriate highlighting techniques, and correct grammar.

WHY WRITE BROCHURES?

tech link

For Hewlett Packard's guide to creating brochures, go to *http://www.hp.com/sbso/ productivity/howto/ marketing_main/marketing_ brochure/*.

A flier *must* be short—one or two pages. Obviously, then, not a great deal of information will be conveyed in a flier. If you have more information to convey than can fit on a one-page flier, a brochure might be a good option. Think of all the brochures you have seen when you have visited a bank, gone to a museum, or toured a new city. These brochures offer you a detailed overview of products, services, options, and opportunities, complete with photographs, maps, or charts. You might want to write a brochure for any of the following reasons (Le Vie 2000, 21):

- Create awareness of your company, product, or service
- Increase understanding of a product, service, or your company's mission
- Advertise new aspects about your company, product, or service
- Change negative attitudes
- Show ways in which your company, product, or service surpasses your competition
- Increase frequency of use, visit, or purchase
- Increase market share

Whatever your motives are, however, the brochures must be concise enough to put in your back pocket while on a trip, easy enough to read while standing in a crowded subway. A brochure is more detailed than a flier but less developed than a newsletter.

CRITERIA FOR WRITING BROCHURES

Brochures come in many shapes and sizes. They can range from a simple front and back, four-panel, 8½ × 5½-inch brochure (one landscape 8½ × 11-inch page folded in half vertically) to 6-, 8-, or even 12-panel brochures printed on any size paper you choose (see Figure 10.2). Your topic and the amount of information will determine your brochure's size.

Title Page (Front Panel)

Usually, the title page includes at least three things:

- **Topic**—In the top one-third of the panel, name the topic about which you are writing. This could include a product name, a service, a location, or the subject of your brochure.

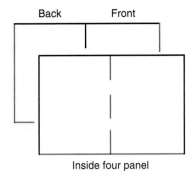
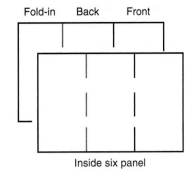

Figure 10.2 Four- and Six-Panel Brochures

- **Graphic**—In the middle third of the panel, you will want to include a graphic to appeal to your reader's need for a visual representation of your topic. The graphic will sell the value of your subject (its beauty, its usefulness, its location, it significance, etc.) or visually represent the focus of your brochure.
- **Contact Information**—In addition, place identifying and contact information on the bottom third of this panel. You could include your name, your company's name, street address, city, state, zip code, telephone number, fax number, and e-mail address.

Back Panel

The back panel can accomplish any one of at least four goals:

- **Conclusion**—You could use this panel to summarize your brochure's content, using this back page as you would a concluding paragraph to any form of communication. Thus, you could restate the highlights of your topic or suggest a next step for your readers to pursue.
- **Mailing**—The back panel could be used like the face of an envelope. On this panel, when left blank, you could provide your address, a place for a stamp or paid postage, and your reader's address.
- **Coupons**—As a tear-out, this panel could be an incentive for your readers to visit your site or use your service. Here you could provide discounts or complimentary tickets.
- **Location**—A final consideration would be to provide your reader with your address, hours of operation, phone numbers, e-mail address, and a map to help them locate you.

Body Panels (Fold-in and Inside)

The number of body panels and the text you include depend entirely on you, your topic, and your budget. Though we can't tell you exactly what to include, here are some suggestions for creating the brochure's text:

- **Provide headings and subheadings.** These act as navigational tools to guide your readers, direct their attention, and help them find the information they need. The headings and subheadings should follow a consistent pattern of font type and size, as discussed in Chapter 8. That is, your first-level headings should be larger and more emphatic than your second-level subheadings. Also, the headings must be parallel to

TABLE 10.2	DEVELOPING IDEAS			
Location(s)	Maps	Technical Specifications	Bonding/ Insurance	Activities
Prices	Credentials	Company History	Unique Characteristics	Employment Opportunities
Options	Directions	Delivery (Dates/ Methods)	Personnel Biographies	Payment Plans

each other grammatically. For example, if your first heading is titled "Introduction," a noun, all subsequent headings must be nouns, like "Location," "Times," "Payment Options," and "Technical Specifications." If, instead, your first heading is a complete sentence, like "This is where it all began," then your subsequent headings must also be complete sentences: "It's still beautiful," "Here's how to find us," "Prices are affordable," etc.

- **Use graphics.** Graphics such as photographs, maps, line drawings, tables, figures, and so on help vary the page layout, add visual appeal, and enhance your text.
- **Develop your ideas.** Answer reporter's questions (*who, what, when, where, why,* and *how*). Consider including the information listed in Table 10.2, dependent upon your topic.

Document Design

Try using only three or so different highlighting techniques for your document design. Even though you want to make your brochure interesting to look at, don't overdo the use of highlighting techniques—too much of anything is always a problem.

- Limit sentence length to 10–12 words and paragraph length to 4–6 lines. When you divide the paper into panels, text can become cramped very easily. Long sentences and long paragraphs then become difficult to read. By limiting the length of your text, you will help your readers access the information.
- Use white space instead of wall-to-wall words. Indent and itemize information so readers won't have to wade through too much detail.
- Use color for interest, variety, and emphasis. For example, you can use a consistent color for your headings and subheadings. But, as already noted, don't overuse color or use too many different colors, which will compete with your content.
- Bullet key points.
- Boldface or underline key ideas.
- Do not trap yourself within one panel. For variety and visual appeal, let text and graphics overlap two or more of the panels.
- Place graphics at angles (occasionally) or alternate their placement at either the center, right, or left margin of a panel. Panels can become very rigid if all text and graphics are square. Find creative ways to achieve variety.

Audience Recognition and Involvement

Whom are you writing to? Is your audience high tech, low tech, lay, or a combination of the three? Are you writing to customers, vendors, colleagues in other companies, coworkers, supervisors, or subordinates? To avoid problems, define any terminology, abbreviations, or acronyms that could cause confusion. Also, brochures are very friendly methods of communication. You want to personalize your text through pronoun usage and contractions.

Accuracy

As in all technical writing, your brochure represents your company and yourself. What the reader sees in the brochure is a reflection of your professionalism—or lack thereof. You don't want to embarrass yourself or your colleagues. Proofread to catch grammatical or contextual errors.

See Figure 10.3 to get a better idea of what a typical brochure looks like.

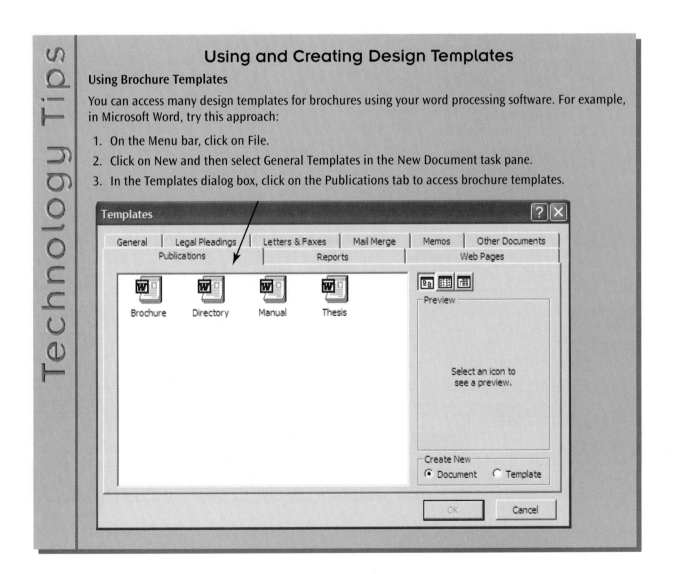

Once you double-click on the Brochure icon, you will find a six-panel template, which you can expand as needed. The template will not tell you what to write in each panel. Instead, Word just provides you a design layout, a shell for you to fill with your content.

Although the template can help you design your brochure, it also can be limiting. You will be provided font type and sizes, as well as predetermined horizontal rules.

To avoid these limitations, you can create your own brochure layout.

Creating Your Own Brochure Layout

Instead of using a predetermined template, create your own as follows:

1. Change the page orientation from portrait to landscape by clicking on File on the Menu bar and then Page Setup.

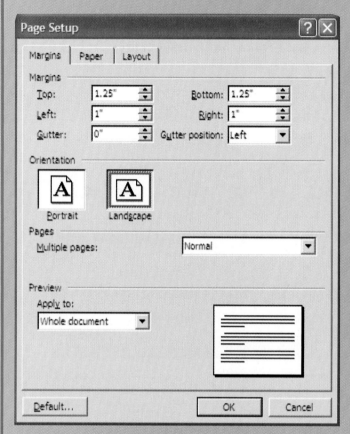

2. Create three panels for your brochure by clicking on the Columns icon on the Standard toolbar at the top of your screen.

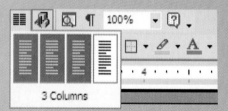

3. Then, on the Formatting toolbar choose the font type and size to create your brochure.

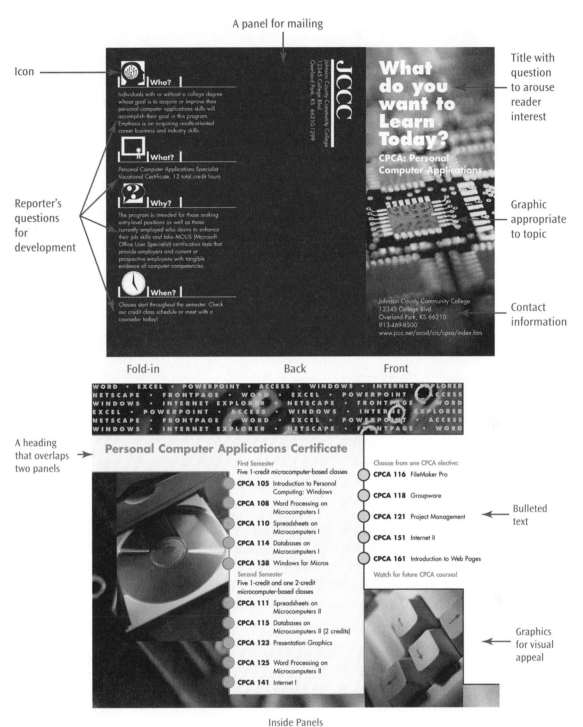

Figure 10.3   Brochure
Courtesy of Johnson County Community College.

The following labels appear around the figure:

- A panel for mailing
- Icon
- Reporter's questions for development
- Title with question to arouse reader interest
- Graphic appropriate to topic
- Contact information
- A heading that overlaps two panels
- Bulleted text
- Graphics for visual appeal

Fold-in   Back   Front

Inside Panels

The Writing Process		
Prewriting	Writing	Rewriting

• Examine your purposes
• Determine your goals
• Consider your audience
• Gather your data
• Determine how the content will be provided
	• Organize the draft according to some logical sequence that your readers can follow easily
	• Format the content to allow for ease of access
		• Revise
		  ° Add missing details
		  ° Delete wordiness
		  ° Simplify word usage
		  ° Enhance the tone of your communication
		  ° Reformat your text for ease of access
		  ° Practice the speech or review the text
		• Proofread
		  ° Correct errors

PROCESS

Now that you have an overview of what goes into a brochure, it's time to write. How do you start working on your brochure? As shown throughout this book, the best way to approach any writing activity is as a process. First, prewrite to gather data, then write a rough draft, and finally rewrite by revising your text. Remember that the writing process is dynamic, with the three steps frequently overlapping.

Prewriting

We have suggested different types of prewriting techniques to help you gather ideas and organize your information. Some techniques work best for specific types of correspondence. For example, reporter's questions are an excellent way to prewrite for letters, whereas clustering works well for memos, and organizational charts for Web sites. A new prewriting technique especially effective for brochures is *cubing*.

Cubing

Visualize a three-dimensional cube, with six sides (see Figure 10.4). Then try looking at a topic from the following six angles or perspectives:

1. *Describe your topic*—What does it look like? Consider focusing on all five senses. We realize that sight, sound, smell, taste, and touch might not be appropriate for every topic. Just use whichever senses apply to your subject matter and disregard the rest.

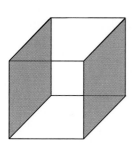

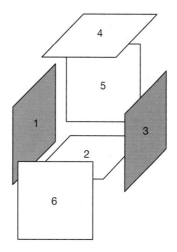

Figure 10.4   Cubing

2. *Compare/contrast* the subject to other topics to discover the similarities and dissimilarities. Make a list of 5 to 10 similarities and 5 to 10 differences. Your topic should compare favorably, of course, to your competition.

3. *Associate your topic*—What does your subject remind you of? Create analogies, similes, and metaphors. People often get a better understanding of a topic when you help them envision what it's similar to. For example, one could say that Las Vegas is *an oasis* in the Nevada desert. That's "sort of" true because the literal definition of "oasis" is a small watering hole surrounded by palm trees and dunes. To call Las Vegas an oasis is more figuratively metaphorical than actual.

4. *Analyze it*—Tell how it's made. What are its components, its various facets or aspects? Decide how it will benefit the reader or meet the readers' needs.

5. *Apply it*—What can you do with your topic? How can it be used on the job, in the home, for educational purposes, for pleasure, and so on?

6. *Argue the pros and cons*—What's good about your topic? Conversely, what's bad? Again, answer these questions with a specific audience in mind. Who is your client, for example? How will your topic help him or her (that represents the pros)? How might your topic not help the audience (that would represent the cons)? The reason for focusing on negatives, as well as positives, is that by knowing one's complaints beforehand, you can refute potential criticisms.

Cubing allows you to analyze your topic from multiple perspectives and thus gather a great deal of information. Some information you will use; some you won't. The key to prewriting is to delve into a topic deeply and to gather data—perhaps more than you will need. The time for deleting information and for sharpening your focus comes later, during the rewriting phase. In addition to helping you develop your ideas thoroughly, cubing helps you organize your thoughts. Each side of the cube might represent a panel in your brochure.

Writing

Now that you have gathered information, the next step is to start constructing your brochure. This is the drafting stage, when words are placed on the individual panels. Follow this procedure:

Create the Panels

Technical writing differs from other forms of communication because it is limited by space. For example, a novel can be 100, 500, or 1,000 pages in length. Technical writing, in contrast, must "fit in the box." A car's user manual has to fit in the glove compartment; a brochure's content must fit inside its panels. You might as well set the parameters of your brochure now. If you envision a four-panel brochure, set your page layout to "landscape" and create two columns. If you envision a six-panel brochure, set the page layout to "landscape" and create three columns (or use an existing template, present in your word processing software). Doing so allows you to work within the boundaries mandated by a brochure format.

Title Your Brochure

In the top third of the right-hand panel of your brochure, write a title. For example, you could name your company or product, plus you could give an accompanying descriptive phrase: "Moody Gardens . . . a World of Wonders"; "17 Historic Sites: Place Yourself in Kansas History." The title should be relatively short, should attract your reader's attention, and should clearly inform your reader of your brochure's primary focus.

Select a Graphic

On the title page, include a graphic that pictorially represents your topic and entices the reader to open the brochure for more information.

Subdivide Your Topic

What are you planning to discuss about your topic? Benefits, prices, locations, uses, specifications, options, warranties, and so forth? Make this decision based on your prewriting. These components—subdivisions of your topic—become the content for your panels.

Use Headings for Your Panels

Once you have topics for discussion, give each panel a heading. The heading serves two purposes. It keeps you on track as the writer, helping you to maintain your focus, ensuring that you do not wander off the topic. Of equal importance, the heading acts as a signpost for your reader, like street signs along a roadway. The heading gives your reader direction, guiding your audience through the brochure.

Write the Text

Writing the text, of course, is your most important job. Your ultimate goal is to communicate information about your topic. To do so, develop your ideas by answering reporter's questions (*who, what, when, where, why,* and *how*), drawing from your prewriting. What have you discovered while cubing that will help your reader better understand your topic? Add this information to your brochure.

Rewriting

To help you revise your rough draft, consider the following effective brochure usability checklist. What is "usability"? Usability helps you determine whether your reader can "use" the brochure effectively and whether your brochure meets your user's needs and expectations. That is, does your brochure work? The reader not only wants to find information that helps him or her better understand the topic, but the audience also wants the text to be readable, accurate, up-to-date, and easy to access. Thus, usability focuses on four key factors (Dorazio 2000).

1. *Retrievability*—Can the user find specific information quickly and easily?

2. *Readability*—Can the user read and comprehend information quickly and easily?

3. *Accuracy*—Is the information complete and correct?

4. *User satisfaction*—Does the brochure present information in a way that is easy to learn and remember?

You can revise your brochure to achieve usability by following this checklist.

Effective Brochure Usability Checklist

First Panel

___ **1.** Does the first panel name the product, company, or service?

___ **2.** Does the first panel provide a graphic to attract the reader's attention and pictorially represent the topic?

___ **3.** Does the first panel provide contact information, such as address, phone number, e-mail address, fax number, and so on?

Body Panels

___ **1.** Are headings presented similarly, maintaining parallelism?

___ **2.** Are graphics effectively used for interest and to clarify ideas?

___ **3.** Does the brochure vary font sizes and type to emphasize key points and add visual appeal?

___ **4.** Are bullets and numbers used to itemize ideas for better access?

___ **5.** Has ample white space been used to help the reader access information and to make reading easier?

___ **6.** Has color been used effectively for visual appeal?

Content

___ **1.** Are all unfamiliar terms defined?

___ **2.** Is the technical content correct, verified by peer review?

___ **3.** Is the brochure grammatically correct?

___ **4.** Is the text clear, answering reporter's questions (*who, what, when, where, why*, and *how*)?

___ **5.** Is the text concise, using short words, short sentences, and short paragraphs?

___ **6.** Does the brochure meet the writer's goals: to create awareness of the company, product, or service; to increase understanding; to advertise new aspects; to change negative attitudes; to show ways in which the topic surpasses the competition?

___ **7.** Are maps used to help the reader find the writer's location?

___ **8.** Are discounts or promotional incentives offered to ensure reader participation?

Audience

___ **1.** Is the level of writing appropriate for the audience (high tech, low tech, lay, multiple readers)?

___ **2.** Has the appropriate tone been achieved through positive words and personalized pronouns?

Once you have answered these usability questions, revise accordingly.

- Add missing information for clarity, such as a missing e-mail address, phone number, building address, maps, and so on.
- Delete unnecessary words, phrases, and content for conciseness.
- Simplify words and phrases, or define acronyms and abbreviations to better communicate with your audience.
- Move information within the brochure. What is the best place to locate a map? Where should you list your credentials? What should be on the first panel your reader sees? You will have to assess these issues individually. However, remember that the most important information should probably be the first thing your reader sees, not something buried on the last panel of the brochure.
- Reformat for access. Now is the time to add the bullets, italics, or boldfacing. Maybe you need to reconsider your color choice. Is the background you've used too dark for readability? Is your font selection too hard to read? If your paragraphs are too long, cut them in half, and add more white space. Add more graphics.
- Enhance the tone of your brochure. Add more pronouns and contractions to make the brochure more friendly and casual.
- Correct errors. Nothing will destroy your professionalism more quickly than typographical errors or errors in content. Imagine how your business will be hurt if you have typed the wrong e-mail address or phone number.

WHY WRITE NEWSLETTERS?

When we open our mailboxes, in addition to letters and bills, we often find newsletters—from our bank, our children's schools, our stock brokerage firm, our home's association, the professional organizations to which we belong, and our church. Each of these entities is hoping to share news with us. We learn about

- New teachers, parent–teacher meetings, and sports' scores in the school newsletters
- New opportunities for investment at the brokerage firm
- Rules and regulations for snow removal, fence building, and roof repair in our neighborhood newsletter
- Revised earning rates for certificates of deposit, bank hours, or ATM machine locations from our bank
- Conferences and membership benefits from our professional organizations
- Hours for services, a president's message from our church's executive board, and news of the upcoming chili cook-off

Companies use newsletters at work to tell their employees when and where the company picnic will be held, who's newly hired at the firm, how stock prices are faring, if the rumored merger is true, how building remodeling is going, and so on. For these issues, and many more, a flier won't work because it is too short. Brochures won't work. Not only are they often too sales oriented, but the panels don't allow enough space to develop lengthy stories. The best option, then, for sharing detailed information in a rather personal and friendly manner might be newsletters.

CRITERIA FOR WRITING NEWSLETTERS

Though newsletters can vary in size, typically they are 8 1/2 × 11- inches with a "portrait" orientation, divided into two or three columns (see Figure 10.5). A newsletter has the following features:

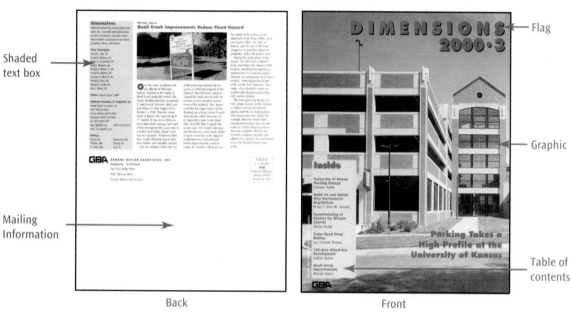

Shaded text box

Mailing Information

Back

Flag

Graphic

Table of contents

Front

Initial cap

Pop-up quotes

Graphics

Inside pages

Figure 10.5   Newsletter
Courtesy of George Butler Associates, Inc.

The Flag

Often mistaken as the "masthead" (which actually names the newsletter's editors and authors), the flag gives the title of your newsletter, the date of its publication, and perhaps the publication volume and issue number. The flag will be placed at the top of your newsletter in a large, bold font, clearly identifying your newsletter.

Columns

You will want to divide your newsletter into two to three columns. You can vary this, however. Occasionally, you might want some pages of your newsletter to have two columns whereas others have three. Even within a page, you could strive for variety; the top half of the page could have two columns and the bottom half would use three. The key is visual appeal, as shown in the storyboard examples of Figure 10.6. In the storyboarding, note how the writer varies the layout with a shaded sidebar, photographs, pop-up quotes, and two or three columns.

Headlines and Subheadings

To help your readers navigate the text, provide a headline for the major article on page one, first-level headings for all other articles, as well as second-level subheadings. We discuss the importance of font size and style, as well as parallelism, for first- and second-level headings in Chapter 8. When writing your headlines and headings, consider these suggestions:

- Use present tense instead of past.
- Avoid excessive punctuation.
- Omit titles such as *Dr., Mr., Mrs., Ms.,* and so on.
- Avoid abbreviations and acronyms.
- Use strong verbs.
- Use short words and phrases.

Pop-up Quotes

One way to gain your reader's interest is through pop-up quotes. These "talk-bubbles" or "pull-out quotes" are usually

- printed in a larger font than the articles
- centered on a page or within a column

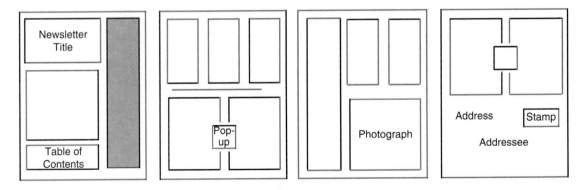

Figure 10.6   Sample Storyboard for a Four-Page Newsletter

- printed inside a border
- printed in a different color

Even though the pop-ups repeat text within an article, these are the quotes that the writer considers essential. Thus, they call attention to key concerns.

Sidebars

Another way to add variety to your page layout is through sidebars. These are often shaded or gradient-colored text boxes that present late-breaking, additional, or contrasting short news items.

Table of Contents

To help your readers find information within the newsletter, use a table of contents. To add visual appeal, place this table within a box, use a different font size and type, add color, and number your articles. You should list the headings within this table, possibly the page numbers on which your reader will find the articles, and you might want to include abstracts (brief overviews of what each story entails).

Initial Caps

Many newsletters begin each article with an initial cap for visual appeal. This is a larger, dropped capital letter that prefaces the first word of the first sentence of an article—as shown in this paragraph.

Newsletter Style

As in all effective technical writing, you should limit your word length (1–2 syllables), sentence length (10–15 words), and paragraph length. The length of paragraphs is especially important when writing a newsletter because columns make your writing more cramped. You want to limit your paragraphs to 4–6 sentences.

Highlighting

Other ways to add interest to your newsletter include typical highlighting techniques, such as tables and figures, font changes, bullets, numbering, boldface, and italics. As always, you don't want to overuse any highlighting techniques. Limit yourself to about three methods per page.

PROCESS

Now it is time to write your newsletter. How should you proceed? As always, we suggest that you follow a simple, three-step process: prewrite, write, and rewrite.

Prewriting

What will you discuss in each of your articles? What should the lead story be? *Brainstorming* would be a good way to gather data for your text. Make a random list of topics that your readers will be interested in or bounce ideas off a colleague. Then, to develop information about these topics, you will probably need to do research, interview the relevant sources, and answer reporter's questions (*who*, *what*, *when*, *where*, *why*, and *how*).

Once you have decided on your content, an especially valuable prewriting technique for newsletters is storyboarding. Draw a preliminary sketch of how you envision your newsletter will look. When doing so, consider these questions:

- Where do you want to place the flag? Should you center it across the entire top of the first page, or perhaps it could occupy only the left or right third?
- Do you want the flag to be printed in a different color or even within a colored text box?
- Where will you place the table of contents—across the bottom of the first page or along either the left or right margins?
- What color scheme do you have in mind for the entire newsletter?
- Will you use colored, shaded, or gradient sidebars?
- How many graphics do you envision using? Will you include photographs as well as line art? Where will you place these? And do you have permission to use the art?
- How many columns will you use—two, three, more?
- How long will your newsletter be? How many articles do you plan to include? If you envision a four-page newsletter, that's one 8½ × 17-inch sheet of paper printed front and back and folded. If you plan a longer newsletter, then you'll have to factor in some kind of binding method.

Drawing a storyboard sketch of your newsletter will help you plan your layout, length, and graphical needs (see Figure 10.6 on page 324).

Writing

Once you have gathered data for your correspondence and determined how your newsletter will look, start writing. To do so, follow this procedure:

Select the Appropriate Software

To write your newsletter, you could use any word processing package and divide the page into columns. For better results, however, you'll want to use a software package geared toward newsletter writing, such as Adobe FrameMaker or QuarkXpress. These packages allow you to maneuver text and graphics easily and shift from two to three columns without trouble.

Check the Criteria for Writing Newsletters

Make sure that you have included a flag, a headline, headings, subheadings, and a table of contents.

Draft the Text

Remember that in the writing stage of the process, all you want to do is create a rough draft of the text. Don't worry about every grammatical consideration, the placement of graphics, color, or your style. You can revise the length of your words, sentences, and paragraphs later. Now is the time to just get your information on the page.

Rewriting

The final stage of the process is rewriting, the phase where you make sure that you've accomplished what you intended and that your work is professional.

To help you revise your rough draft, consider the following effective newsletter usability checklist. As noted earlier in this chapter, "usability" is a way of determining whether your reader can "use" the newsletter effectively, whether your newsletter meets your user's needs and expectations. The reader not only wants to find information that helps him or her better understand the subject matter, but the audience also wants the text to be readable, accurate, up-to-date, and easy to access. Usability focuses on

1. *Retrievability*—Can the user find specific information quickly and easily?
2. *Readability*—Can the user read and comprehend information quickly and easily?
3. *Accuracy*—Is the information complete and correct?
4. *User satisfaction*—Does the newsletter provide information that the reader can benefit from ? (Dorazio 2000)

Effective Newsletter Usability Checklist

First Page

___ 1. Does the first page provide a flag that names the newsletter and gives the date, volume, and issue number?

___ 2. Does the first page provide a table of contents to help the reader find information throughout the newsletter?

___ 3. Is the headline story the most important article in the newsletter?

Text

___ 1. Are headings used effectively to clarify the article's content as well as to arouse the reader's interest?

___ 2. Do the headings
 • use present tense instead of past?
 • avoid excessive punctuation?
 • omit titles, such as *Dr., Mr., Mrs.*, and *Ms.*?
 • avoid abbreviations and acronyms?
 • use strong verbs?
 • use short words and phrases?

___ 3. Are sidebars used to introduce short, additional, contrasting, or late-breaking information?

___ 4. Is the text clear, answering reporter's questions (*who, what, when, where, why,* and *how*)?

___ 5. Does the newsletter meet the writer's goals: to create awareness of the company, product, or service; to increase understanding; to advertise new aspects; to change negative attitudes; to show ways in which the topic surpasses the competition; to inform the reader of new information?

Access

___ 1. Are subheadings used to break up blocks of paragraphing and to help the readers navigate the text?

___ 2. Do the headings and subheadings vary font sizes and type for emphasis and visual appeal?

___ 3. Are bullets and numbers used to itemize ideas for better access?

___ 4. Has ample white space been used to help the reader access information and to make reading easier?

___ 5. Has color been used effectively for visual appeal?

___ 6. Have photographs, figures, and tables been used to add interest as well as to make information more clear?

___ 7. Have pop-up quotes been used for interest and clarity?

___ 8. Has the newsletter varied its use of two and three columns?

___ 9. Have color, shading, or gradients been used for interest?

___ 10. Have initial drop caps been used at the beginning of an article or new paragraph for interest, visual appeal, and to call attention to a new idea?

Style

___ 1. Is the text concise, using
 • short words (1–2 syllables)?
 • short sentences (10–15 words long)?
 • short paragraphs (preferably no longer than 6 sentences)?

continued

continued

Audience

__ **1.** Are all unfamiliar terms defined?

__ **2.** Is the level of writing appropriate for the audience (high tech, low tech, lay, multiple readers)?

__ **3.** Has the appropriate tone been achieved through positive words and personalized pronouns?

Accuracy

__ **1.** Is the technical content correct, verified by peer review?

__ **2.** Is the newsletter grammatically correct?

Once you have answered these usability questions, revise your newsletter accordingly.

- Add missing information for clarity. To do so, ask yourself which of the reporter's questions you have omitted. Be sure that you've included the date, volume number, and issue on your flag.

- Delete unnecessary words, phrases, and content for conciseness.

- Simplify words and phrases or define acronyms and abbreviations to better communicate with your audience.

- Move information within the newsletter. What should be in the headline story? This should be the most important article in your newsletter. Is it? Where should you place the other articles within the newsletter? You will have to answer this question individually. Information at the end of the newsletter won't be read with the same level of interest as earlier topics.

- Reformat for access. Now is the time to add the bullets, italics, or boldfacing. Maybe you need to reconsider your color choice for the shaded boxes or for the flag. Is your font selection too hard to read? If your paragraphs are too long, cut them in half, and add more white space. Add more graphics. Add pop-up quotes. Change your use of columns. Perhaps you will want to add a third column to a page, delete a column, split the page in half with three columns on top and two below, and so on.

- Enhance the tone of your newsletter. Add more pronouns and contractions to make the text more friendly and casual.

- Correct errors. Nothing will destroy your professionalism more quickly than typographical errors or errors in content. Think how your business will be hurt or how your employees will be miffed if you have typed incorrect information about a child's birth date, the wrong stock prices, the wrong dates for the company picnic, the wrong Web site link for purchasing products, and so forth.

CHAPTER HIGHLIGHTS

1. Fliers are cost effective, time efficient, responsive to immediate needs, and personalized.

2. Fliers should be short, focused on one idea, titled, have limited text, and be visually appealing.

3. Recognize your audience and their needs when you write a flier, brochure, or newsletter.

4. Use the writing process when you create fliers, brochures, or newsletters to ensure successful writing.

5. Brochures can have anywhere from 4 to 12 panels.

6. Software templates can be used to help you create a brochure.

7. The front panel of a brochure contains a title and a graphic that depicts the topic.

8. The interior panels of a brochure contain information about the topic.

9. The back panel of a brochure can be used for mailing purposes.

10. Document design is important when you create a flier, brochure, or newsletter.

11. Cubing is a good prewriting technique for brochures.

12. The usability checklists will help you meet your readers' needs for a brochure and newsletter.

13. Newsletters include a flag, columns, headlines, headings, subheadings, pop-up quotes, sidebars, and a table of contents.

14. Storyboarding is a good prewriting technique for newsletters.

15. Use headings and subheadings to design the content in fliers, newsletters, and brochures.

Case Study

After reading the case study, do any or all of the following:

- create a flier about *one* of the company's most unique offerings
- write a brochure
- write a newsletter

You can create any information you'd like (such as a map, upcoming events, or personal testimonials from employees) to flesh out the details of this case study.

A company has just expanded into your city. Having been in business in the Midwest since 1995, TechToolshop provides sales, service, maintenance, and installation for all automobile makes and models.

Through an online catalog and storefront site, it sells automobile parts.

TechToolshop has proudly worked with many large companies, including McDonald's, Pepsi, Ford, Texaco, Sprint, Transamerica, GE, JCPenney, and Chase Manhattan Bank, providing automobile fleet service.

TechToolshop's home office is in Big Springs, Iowa, at 11324 Elm, where over 1,200 employees work. The phone number at this site is 212–345–6666, and the e-mail address is *ToolHelp@TechTools.com.* The company's Web site can be found at *http://www.TechTools.com.* TechToolshop's new local address in your city is 5110 Nueces Avenue. The phone number is 345–782–8776.

TechToolshop is most proud of its Product Support and Performance Guarantees. It offers free answers to auto repair questions, 24 hours a day, at 1–800–TechHelp. By moving to your city, it also guarantees arrival at your site within 2 hours of any service emergency call. The company's greatest innovation is the installation of auto FAQ's kiosks in every mall, library, and bank in your city. By keying in your Personal Identification Number (obtained by paying a

Fliers, Brochures, and Newsletters

329

small monthly fee), you can have answers to technical questions at your convenience. Of course, you can get help via TechToolshop's Web site. In addition, the company warrants all products and services—money back—for 90 days, covering defects in material and installation.

The company is owned by James Wilcox Oleander (president) and his brother Harold Robert Oleander (CEO). They started this company after graduating from Midwestern State University, with degrees in auto technology. Their first store had only three employees, but through hard work, their business grew 450 percent in the first two years of operation. Expert recruiting of the best State U graduates increased their workforce, as did the Oleanders' philosophy of "employee ownership." TechToolshop's workers take pride in their work, and their success increases their stock options (TechToolshop's stock at Nasdaq opened in 1996 at $18 a share but has listed as high as $45). The Oleanders plan to open at least 12 new stores each year throughout the United States, as well as to pursue franchise options. They have high hopes for their future success.

EXERCISES

1. *Fliers*, due to their restricted length, will focus on a limited topic—one specific aspect of a subject. With this in mind, write a flier about any topic, such as a

   - new computer software application
   - computer hardware innovation
   - recent automotive technology breakthrough
   - Web site addition
   - method for solving a company's problem with billing, accounting, shipping, delivery, and so on
   - new telecommunications technology
   - solution to HVAC or plumbing needs
   - unique landscaping option
   - drafting skill to help meet architectural or engineering needs
   - new electrical engineering design
   - fire protection technique
   - environmental service
   - home improvement ability

2. *Brochures* will develop ideas more fully than fliers, but with less detail than newsletters. Write a brochure about a topic, such as

   - a school organization or club
   - a church activity or club
   - a city organization, agency, or company (the local computer store, animal shelter, nursing home, community center, Rotary club, or city government)
   - your business
   - your place of employment
   - a vacation spot
   - a site or location (park, museum, zoo, amusement park, historical site, athletic field or stadium, etc.)

- a product or line of products (tools, automotive parts, computer accessories, software, etc.)
- a service (home improvement, hobbies, consulting, animal services, lawn and garden care, installation, automotive care, etc.)
- an author, painter, sculptor, inventor, teacher, sports figure, or historical figure
- a vehicle

3. *Newsletters* offer the most detailed information of the three types of communication discussed in this chapter. Write a newsletter about

- your business
- your place of employment
- your church
- your school
- your school club or organization
- your sports team
- your professional organization or affiliation
- your club or group of hobbyists (dinner club, bridge club, model train club, square dancing club, etc.)
- your favorite singer, musical group, actor, sports star
- your city or region or state's unique ethnic culture (Hispanic, African American, Japanese, Swedish, Native American, etc.)

4. Find examples of professional fliers, brochures, or newsletters. Bring these to class for discussion. Determine which of these types of communication are successful and why (based on the criteria provided in this chapter). Rewrite the flawed examples to improve them.

5. Reformat the following poorly designed flier. To do so, follow the criteria provided in this chapter. Provide any additional information you need.

Come to the 2005 Classic Corvette Show

9:00 AM to 5:00 PM

$10 per carload

Food and Prizes!

**Oak Park Convention Center
May 30, 2005**

WEB WORKSHOP

You can find thousands of examples of online brochures, fliers, and newsletters by searching for these terms in any Internet search engine. However, what you find may surprise you.

Hard-copy brochures, fliers, and newsletters abide by very precise criteria. Paper brochures are always multipaneled documents, printed front and back. Fliers are always very short, limited to a page or so. Newsletters have columns and headings, for example. In contrast, online brochures, fliers, and newsletters differ from their hard-copy namesakes.

Access an Internet browser and search for online brochures, fliers, or newsletters. Once you find these examples, analyze your findings. How do the online documents compare to the hard-copy versions discussed in this textbook?

    a. Report your findings, either in an oral presentation or in writing (e-mail message, memo, letter, or report). What are the differences? Why are the online versions so different? Are the online differences valid, or should online versions be more similar to hard-copy documents?

    b. Print out online samples and rewrite them as hard-copy documents, according to the criteria provided in this textbook.

QUIZ QUESTIONS

1. What type information do you include in a flier?

2. How can you access an electronic flier, brochure, or newsletter?

3. List four benefits of using a one-page flier.

4. How many key points should you employ in a flier?

5. List three ways to highlight your flier.

6. What can a "catchy phrase" accomplish in your flier?

7. When would a brochure be more effective than a flier?

8. List five reasons why you would write a brochure.

9. What should you include in the title page of a brochure?

10. How do headings and subheadings assist the reader in a brochure?

11. List four ways document design can be used effectively in a brochure.

12. When considering the audience for a brochure, how do you avoid problems?

13. Why is cubing an effective prewriting technique?

14. List three topics you could discuss in a brochure.

15. Explain usability.

16. List four key factors on which usability focuses.

17. What can you write about in a newsletter?

18. List the key parts of a newsletter.

19. What highlighting techniques can you use in a newsletter?

20. Why would you use a storyboard to help create a newsletter?

Technical Descriptions

Chapter Preview

Criteria for Writing Technical Descriptions
When you write a technical description, follow our criteria for a title, overall organization, internal organization, development, word usage, and visual aids.

Process
The writing process allows you to prewrite, draft your text, and revise your technical description.

Process Log
Following a student-written process log of a technical description will help you with your own writing.

TYPES OF TECHNICAL DESCRIPTIONS

A *technical description* is a part-by-part depiction of the components of a mechanism, tool, or piece of equipment. Technical descriptions are important features in several types of correspondence.

Operations Manuals

Manufacturers often include an *operations manual* in the packaging of a mechanism, tool, or piece of equipment. This manual helps the end user construct, install, operate, and service the equipment. Operations manuals often include technical descriptions.

Technical descriptions provide the end user with information about the mechanism's features or capabilities. For example, this information may tell the user which components are enclosed in the shipping package, clarify the quality of these components, specify what function these components serve in the mechanism, or allow the user to reorder any missing or flawed components. Here is a brief technical description found in an operations manual:

example

> The Modern Electronics Tone Test Tracer, Model 77A, is housed in a yellow, high-impact plastic case that measures 1¼ inch × 2 inch × 2¼ inch, weighs 4 ounces, and is powered by a 1604 battery. Red and black test leads are provided. The 77A has a standard four-conductor cord, a three-position toggle switch, and an LED for line polarity testing. A tone selector switch located inside the test set provides either solid tone or dual alternating tone. The Tracer is compatible with the EXX, SetUp, and Crossbow models.

Product Demand Specifications

Sometimes a company needs a piece of equipment that does not currently exist. To acquire this equipment, the company writes a *product demand* specifying its exact needs:

Subject: Requested Pricing for EDM Microdrills

Please provide us with pricing information for the construction of 50 EDM Microdrills capable of meeting the following specifications:

- Designed for high-speed, deep-hole drilling
- Capable of drilling to depths of 100 times the diameter using 0.012-inch to 0.030-inch diameter electrodes
- Able to produce a hole through a 1.000-inch-thick piece of AISI D2 or A6 tool steel in 1.5 minutes, using a 0.020-inch diameter electrode

We need your response by January 13, 2005.

Study Reports Provided by Consulting Firms

Companies hire a consulting engineering firm to study a problem and provide a descriptive analysis. The resulting *study report* is used as the basis for a product demand specification requesting a solution to the problem.

One firm, when asked to study crumbling cement walkways, provided the following technical description in its study report:

example

The slab construction consists of a wearing slab over a ½ inch-thick waterproofing membrane. The wearing slab ranges in thickness from 3½ inches to 8½ inches and several sections have been patched and replaced repeatedly in the past. The structural slab varies in thickness from 5½ inches to 9 inches with as little as 2 inches over the top of the steel beams. The removable slab section, which has been replaced since original construction, is badly deteriorated and should be replaced. Refer to Appendix A, Photo 9, and Appendix C for shoring installed to support the framing prior to replacement.

Sales Literature

Companies want to make money. One way to market equipment or services is to describe the product. Such descriptions are common in sales letters, brochures, proposals, and on the Internet. The following technical description is from Hewlett-Packard's Web site:

Key Specifications	Hewlett-Packard DeskJet 930C
Speed, Black and White	9 ppm (pages per minute)
Speed, Color	7.5 ppm
Resolution, Black and White	600 dpi (dots per inch)
Resolution, Color	2,400 dpi
Memory	4 MB (megabytes)
Paper Handling Support	Cards, Envelopes, Labels, Standard paper, Transparencies, Banners, Iron-on, Photo paper
Sheet Capacity	100 sheets
Weight	12.6 lb
Dimension	17.32 in. × 7.72 in. × 15.74 in. (W × D × H)

Not all technical writing is photographic and quantifiable information. Some descriptions are impressionistic. Note that the following sales piece uses impressionistic adjectives and adverbs (italicized), allowing your imagination to depict the product's components:

Inside the *elegant* Zephyr you'll find *comfortably designed* seats done in *rich, tailored* cloths. The *special* suspension system surrounds you with a driving experience quieter than Zephyr's past. Outside you'll find a *new bright* grille, sport mirrors, and a *smart* dual pinstripe. Finally, underneath all this comfort and luxury beats the *road-hugging* heart of an American classic.

Whatever your purpose in writing a technical description, the following will help you write more effectively:

- Criteria
- Process (with sample process log)
- Technical description examples

CRITERIA FOR WRITING TECHNICAL DESCRIPTIONS

As with any type of technical writing, there are certain criteria for writing technical descriptions.

Title

Preface your text with a title precisely stating the topic of your description. This could be the name of the mechanism, tool, or piece of equipment you're writing about.

Overall Organization

In the *introduction* you specify and define what topic you are describing, explain the mechanism's *functions* or *capabilities*, and list its *major components*.

The Apex Latch (#12004), a mechanism used to secure core sample containers, is composed of three parts: the hinge, the swing arm, and the fastener.

The DX 56 DME (Distance Measuring Equipment) is a vital piece of aeronautical equipment. Designed for use at altitudes up to 30,000 feet, the DX 56 electronically converts elapsed time to distance by measuring the length of time between your transmission and the reply signal. The DX 56 DME contains the transmitter, receiver, power supply, and range and speed circuitry.

In the *discussion,* you use highlighting techniques (itemization, headings, underlining, white space) to describe each of the mechanism's components.

Your *conclusion* depends on your purpose in describing the topic. Some options are as follows:

- **Sales**—"Implementation of this product will provide you and your company . . . "
- **Uses**—"After implementation, you will be able to use this XYZ to . . . "
- **Guarantees**—"The XYZ carries a 15-year warranty against parts and labor."
- **Testimony**—"Parishioners swear by the XYZ. Our satisfied customers include . . . "
- **Comparison/contrast**—"Compared to our largest competitor, the XYZ has sold three times more . . . "
- **Reiteration of introductory comments**—"Thus, the XYZ is composed of the aforementioned interchangeable parts."

Use *graphics* in your technical descriptions. Today, many companies use the *super comic book look*—large, easy-to-follow graphics that complement the text. You can use line drawings, photographs, clip art, exploded views, or sectional cutaway views of your topic, each accompanied by *call-outs* (labels identifying key components of the mechanism). These types of graphics are discussed in Chapter 9. When using graphics, try one of the techniques shown in Figures 11.1, 11.2, and 11.3.

Internal Organization

When describing your topic in the discussion portion of the technical description, itemize the topic's components in some logical sequence. Components of a piece of equipment, tool, or product can be organized by importance (as discussed in Chapter 3).

However, *spatial* organization is better for technical descriptions. When a topic is spatially organized, you literally lay out the components as they are seen, as they exist in space. You describe the components as they are seen either from left to right, from right to left, from top to bottom, from bottom to top, from inside to outside, or from outside to inside.

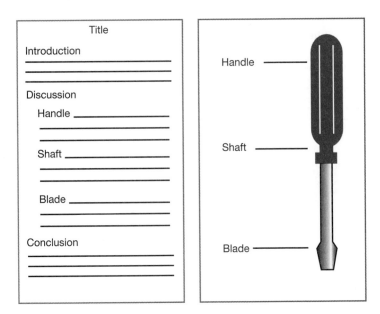

Figure 11.1   Text and Graphics Separate

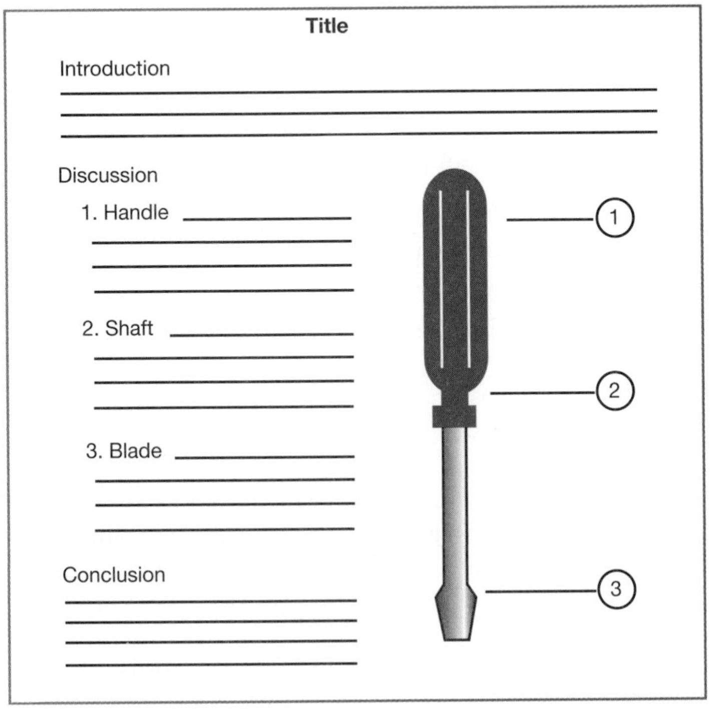

Figure 11.2   Text and Graphics Merged

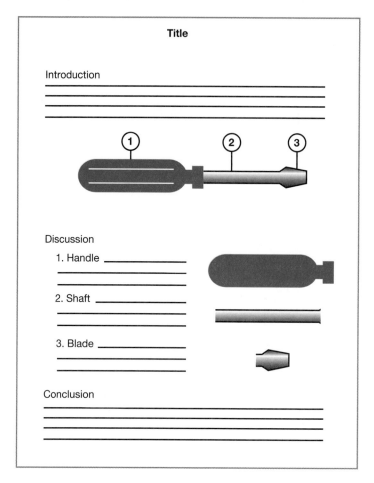

Figure 11.3 Text and Graphics Merged with Exploded View

Development

To describe your topic clearly and accurately, detail the following:

Weight	Materials (composition)
Size (dimensions)	Identifying numbers
Color	Make/model
Shape	Texture
Density	Capacity

Word Usage

Your word usage, either photographic or impressionistic, depends on your purpose. For factual, objective technical descriptions, use photographic words. For subjective, sales-oriented descriptions, use impressionistic words. Photographic words are denotative, quantifiable, and specific. Impressionistic words are vague and connotative. Table 11.1 shows the difference.

Another way to note the difference between impressionistic words and photographic detail is through the ladder of abstraction, shown in Figure 11.4.

TABLE 11.1 PHOTOGRAPHIC VERSUS IMPRESSIONISTIC WORD USAGE	
Photographic	Impressionistic
6′ 9″	tall
350 lb	heavy
gold	precious metal
6,000 shares of United Can	major holdings
700 lumens	bright
0.030 mm	thin
1966 XKE Jaguar	impressive car

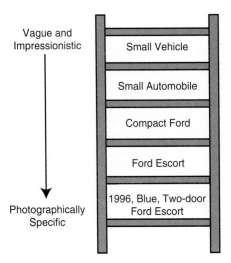

Figure 11.4   Ladder of Abstraction

PROCESS

Now that you know what should be included in a technical description, your next question is, "How do I go about writing this type of document?" As always, effective writing follows a process. To write your technical description,

- *Prewrite* to gather data and determine objectives
- *Write* a draft of the description
- *Rewrite* by revising the draft, thereby making it as perfect as possible

Remember that the parts of the writing process frequently overlap.

Prewriting

We've discussed reporter's questions and clustering/mind mapping as techniques for gathering data and determining objectives. Either of these two prewriting tools can be used for your technical description. However, another prewriting technique can be helpful: *brainstorming/listing.*

The Writing Process		
Prewriting	Writing	Rewriting

• Examine your purposes
• Determine your goals
• Consider your audience
• Gather your data
• Determine how the content will be provided
	• Organize the draft according to some logical sequence that your readers can follow easily
	• Format the content to allow for ease of access
		• Revise
		  ° Add missing details
		  ° Delete wordiness
		  ° Simplify word usage
		  ° Enhance the tone of your communication
		  ° Reformat your text for ease of access
		  ° Practice the speech or review the text
		• Proofread
		  ° Correct errors

Brainstorming/listing, either individually or as a group activity, is a good, quick way to sketch out ideas for your description. To brainstorm/list, do the following:

Title Your Activity

Write the title of your activity at the top of the page, to help you maintain your focus (for example, *Description: XYZ model 2267 Light Pen*).

List Any and All Ideas

Jot down any ideas or thoughts you might have on the topic. Don't editorialize. Avoid criticizing ideas at this point. Just randomly jot down as many aspects of the topic as possible, without making value judgments.

Edit Your List

Reread the list and evaluate it. To do so, (a) select the most promising features, (b) cross out any that don't fit, and (c) add any obvious omissions.

Writing

Once you've gathered your data and determined your objectives, the next step is writing your description.

Review Your Prewriting

After a brief gestation period (an hour, a half-day, or a day) in which you wait to acquire a more objective view of your prewriting, go back to your brainstorming/

listing and reread it. Have you omitted all unneeded items and added any important omissions? If not, do so.

Organize Your List

Make sure the items in the list are organized effectively. To communicate information to your readers, organize the data so readers can follow your train of thought. This is especially important in technical descriptions. You want your readers not only to understand the information but also to visualize the mechanism or component. This visualization can be achieved by organizing your data spatially.

Title Your Draft

At the top of the page, write the name of your piece of equipment, tool, or mechanism.

Write a Focus Statement

In one sentence, write the

- name of your topic (plus any identifying numbers)
- possible functions of your topic or your reason for writing the description
- number of parts composing your topic

This will eventually act as the introduction. For now, however, it can help you organize your draft and maintain your focus.

Draft the Text

Use the sufficing technique discussed in earlier chapters. Write quickly without too much concern for grammatical or textual accuracy, and let what you write suffice for now. Just get the information on the page, focusing on overall organization and a few highlighting techniques (headings, perhaps). The time to edit is later.

Sketch a Rough Graphic of Your Equipment, Component, or Mechanism

You can refine this rough graphic later during the rewriting stage.

Rewriting

Rewriting is the most important part of the process. This is the stage during which you fine-tune and polish your technical description, making it as perfect as it can be. A perfected description ensures your credibility. An imperfect description makes you and your company look bad.
   Revise your draft as follows:

tech link
Go to *http://www.prenhall. com/gerson* for Web links, samples, and interactive activities.

Add Any Detail Required to Communicate More Effectively

Add information such as brand names, model numbers, and specificity of detail (size, material, density, dimensions, color, weight, etc.).

Delete Dead Words and Phrases for Conciseness

Refer to our discussion of effective technical writing style in Chapter 3.

Simplify Long-Winded, Old-Fashioned Words and Phrases for Easier Understanding

Chapter 3 will help you with this important aspect of your writing.

Move (Cut and Paste) Information to Ensure Spatial Organization

You don't want to give your reader a distorted view of your mechanism. To avoid doing so, the technical description must maintain a spatial order—left to right, right to left, and so on. For example, let's say you are describing a pencil, graphically depicted horizontally so the eraser end is to the left and the graphite tip is to the right. This pencil consists of the eraser, metal ferrule, wooden body, and sharpened point (that's the correct spatial order). If you describe the eraser first, then the tip, then the body, and then the ferrule, you've distorted the spatial order.

Reformat Your Text for Ease of Access Through Highlighting Techniques

Use highlighting techniques such as headings, boldface type, font size and style, itemization, and so forth. Rewriting is also the time to perfect your graphics. When revising your description, make sure that your graphics are effective. To do so,

- Add a figure number and title. Place these beneath the graphic (for example, "Figure 11.5. Exhalation Valve with Labeled Call-outs").
- Draw or reproduce graphics neatly.
- Place the graphics in an appropriate location, near where you first mention them in the text.
- Make sure that the graphics are an effective size. You don't want the graphics to be so small that your reader must squint to read them, nor do you want them to be so large that they overwhelm the text.
- Label the components. Use call-outs to name each part, as in Figure 11.5.

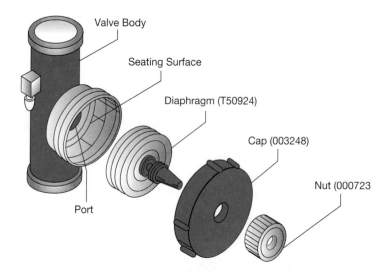

Figure 11.5   Exhalation Valve with Labeled Call-Outs
Courtesy of Nellcor Puritan Bennett Corp.

Enhance the Style of Your Text

Make sure that you have avoided impressionistic words. Instead, use photographic words, which are precise and specific. In addition, strive for a personalized tone. It is difficult to incorporate pronoun usage in technical descriptions. However, even when writing about a piece of equipment, remember that "people write to people." Although you are describing an inanimate object, you're writing to another human being. Therefore, use pronouns to personalize the text. (Examples later in this chapter clarify how you can accomplish this.)

Correct Your Draft

Proofread to correct grammar errors as well as errors in content: measurements, shapes, textures, materials, colors, and so on.

Avoid Biased Language

Remember that your text will be read by multiculturally diverse audiences as well as by readers of different ages and genders.

The following checklist will help you in writing technical descriptions.

Technical Description Checklist

___ **1.** Does the technical description have a title noting your topic's name (and any identifying numbers)?

___ **2.** Does the technical description's introduction (a) state the topic, (b) mention its functions or the purpose of the mechanism, and (c) list the components?

___ **3.** Does the technical description's discussion use headings to itemize the components for reader-friendly ease of access?

___ **4.** Do you need to define the mechanism and its main parts?

___ **5.** Is the detail within the technical description's discussion photographically precise? That is, does the discussion portion of the description specify the following?

Colors	Capacities
Sizes	Textures
Materials	Identifying numbers
Shapes	Weight
Density	Make/model

___ **6.** Are all of the calculations and measurements correct?

___ **7.** Do you sum up your discussion using any of the optional conclusions discussed in this chapter?

___ **8.** Does your technical description provide graphics that are correctly labeled, appropriately placed, neatly drawn or reproduced, and appropriately sized?

___ **9.** Do you write using an effective technical style (low fog index) and a personalized tone?

___ **10.** Have you avoided biased language?

___ **11.** Have you avoided grammatical and mechanical errors?

PROCESS LOG

The following student-written process log clarifies the way process is used in writing technical descriptions.

Prewriting

We asked students to describe a piece of equipment, tool, or mechanism of their choice. One student chose to describe a cash register pole display. To do so, he first had to gather data using brainstorming/listing. He provided the following list:

1. Three parts
2. Pole printed circuit board (PCB)
3. Case assembly
4. Filter
5. Plastic materials

Next, we asked students to go back to their initial lists and edit them. Because this student's list was so sketchy, he did not need to omit any information. Instead, he had to add omissions and missing detail, as in Figure 11.6.

Writing

After the prewriting activity, we asked students to draft a technical description. Focusing on overall organization, highlighting, detail, and a sketchy graphic, this student wrote a rough draft (Figure 11.7).

Pole PCB
- length—15 mm
- width—5.1 mm
- tube length—10.8 mm
- face plate width—2.3 mm
- thick—1.7 mm
- PCB thickness—0.2 mm
- 10 inch stranded wire with female connectors
- fiberglass and copper construction

Pole Case Assembly
- long—15.5 mm
- bottom width—2.5 mm
- top width—0.9 mm
- mounting pole—5 mm high × 3.2 mm diameter
- tounge for mounting—3.1 mm
- lower mounting tounge—1.5 mm
- side mounting tounge—0.8 mm high
- high—6.1 mm
- almond-colored plastic

Filter
- long—15.6 mm
- high—6.2 mm
- thick—0.7 mm
- plastic, blue

Figure 11.6   Student's Brainstorming/Listing

The QL169 Customer Pole Display provides the viewing of all transaction data for the customer. The display consist of a printed circuit board, a case assembly, and a filter display.

Display Circuit Board

- length—15 mm
- width—5.1 mm
- tube length—10.8 mm
- face plate width—2.3 mm
- thick—1.7 mm
- PCB thickness—0.2 mm
- 10 inch stranded wire with female connectors
- fiberglass and copper construction

Display Case Assembly

- long—15.5 mm
- bottom width—2.5 mm
- top width—0.9 mm
- mounting pole—
  5 mm high × 3.2 mm diameter
- tounge for mounting—3.1 mm
- lower mounting tounge—1.5 mm wide
- side mounting tounge—0.8 mm high
- high—6.1 mm
- almond-colored plastic

Display Filter

- long—15.6 mm
- high—6.2 mm
- thick—0.7 mm
- plastic, blue

Figure 11.7   Student's Rough Draft

Rewriting

Once students completed their rough drafts, we used peer review groups to help each student revise his or her paper. In Figure 11.8, students

- Added new detail for clarity
- Corrected any errors
- Perfected graphics

The student incorporated these suggestions and prepared the finished copy (Figure 11.9).

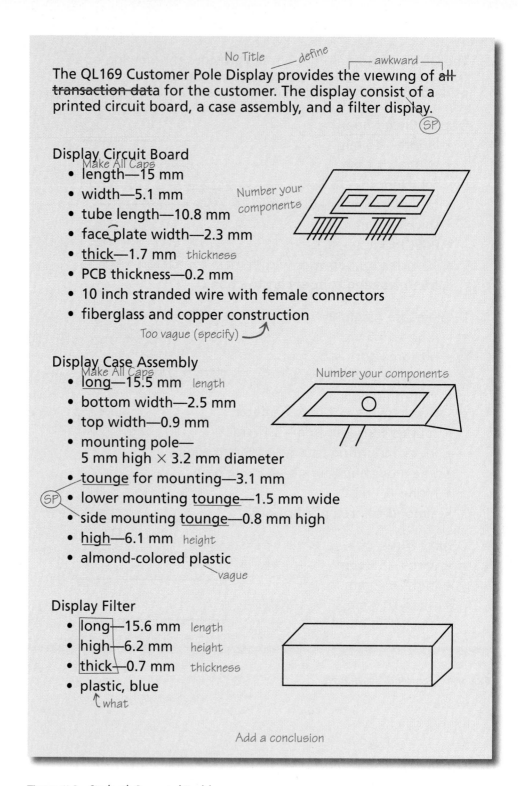

The QL169 Customer Pole Display provides the viewing of ~~all transaction data~~ for the customer. The display consist of a printed circuit board, a case assembly, and a filter display.

No Title ___define___ ___awkward___ (SP)

Display Circuit Board
Make All Caps

- length—15 mm
- width—5.1 mm
- tube length—10.8 mm
- face plate width—2.3 mm
- thick—1.7 mm *thickness*
- PCB thickness—0.2 mm
- 10 inch stranded wire with female connectors
- fiberglass and copper construction

Too vague (specify)

Number your components

Display Case Assembly
Make All Caps

- long—15.5 mm *length*
- bottom width—2.5 mm
- top width—0.9 mm
- mounting pole—
  5 mm high × 3.2 mm diameter
- tounge for mounting—3.1 mm
- lower mounting tounge—1.5 mm wide (SP)
- side mounting tounge—0.8 mm high
- high—6.1 mm *height*
- almond-colored plastic

vague

Number your components

Display Filter

- long—15.6 mm *length*
- high—6.2 mm *height*
- thick—0.7 mm *thickness*
- plastic, blue

what

Add a conclusion

Figure 11.8 Student's Suggested Revisions

QL169 Pole Display

The QL169 pole is an electronic mechanism that provides an alphanumeric display for customer viewing of cash register sales. The display consists of a printed circuit board (PCB) assembly, a case assembly, and a display filter.

Item Description

1. PCB ASSEMBLY
2. CASE ASSEMBLY
3. DISPLAY FILTER

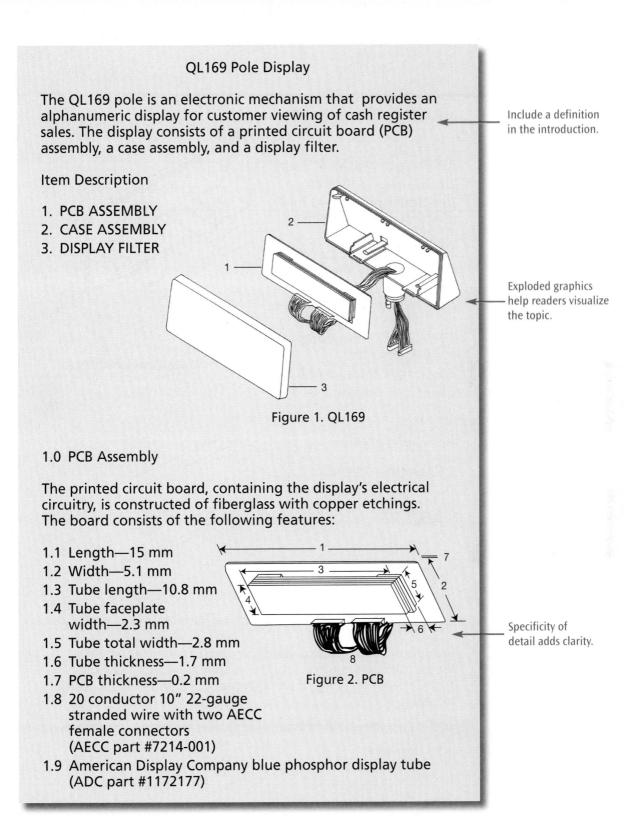

Figure 1. QL169

1.0 PCB Assembly

The printed circuit board, containing the display's electrical circuitry, is constructed of fiberglass with copper etchings. The board consists of the following features:

1.1 Length—15 mm
1.2 Width—5.1 mm
1.3 Tube length—10.8 mm
1.4 Tube faceplate width—2.3 mm
1.5 Tube total width—2.8 mm
1.6 Tube thickness—1.7 mm
1.7 PCB thickness—0.2 mm
1.8 20 conductor 10" 22-gauge stranded wire with two AECC female connectors (AECC part #7214-001)
1.9 American Display Company blue phosphor display tube (ADC part #1172177)

Figure 2. PCB

Include a definition in the introduction.

Exploded graphics help readers visualize the topic.

Specificity of detail adds clarity.

Figure 11.9   Student's Finished Copy

2.0 Case Assembly

Almond-colored ABS plastic, used to construct the case assembly, protects the PCB.

2.1 Length—15.5 mm
2.2 Bottom width—2.5 mm
2.3 Top width—0.9 mm
2.4 Mounting pole—5 mm high and 3.2 mm in diameter
2.5 Mounting tongue inside width—3.1 mm from side of assembly
2.6 Lower mounting tongue—1.5 mm wide
2.7 Side mounting tongue—0.8 mm high
2.8 Tongue thickness—0.2 mm
2.9 Height—6.1 mm

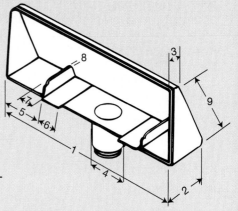

Figure 3. Case Assembly

3.0 Display Filter

3.1 Length—15.6 mm
3.2 Height—6.2 mm
3.3 Thickness—0.7 mm

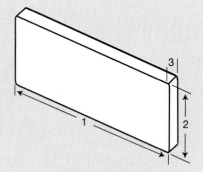

Figure 4. Display Filter

Transparent blue plastic, used to construct the display filter, allows the customer to view the readings. The QL169 pole provides easy viewing of clerk transactions and ensures cashier accuracy.

Conclusion focuses on end-user benefits.

Figure 11.9   continued

SAMPLE TECHNICAL DESCRIPTION

On page 349 is a professionally written technical description merging graphic and text (Figure 11.10). This is a highly effective technical description with these characteristics:

- A title
- An excellent introduction that lists topic, function, and components

A1 Feed Switch

The A1 Feed Switch is an electronic device that provides an automatic solid-state closure (normally open type) after sensing the lack of presence of food. The feed switch can detect dry concentrate, meal, rolled barley, and fine and coarse feeds, as well as special feeds such as soybean meal and high-moisture corn. Figure 1 shows the three principal parts of the feed switch.

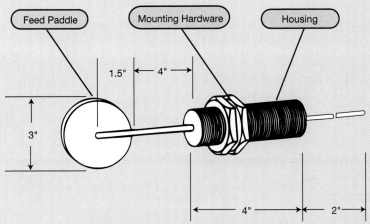

Figure 1. Electronic A1 Feed Switch

- HOUSING: The housing consists of a blue molded polycarbonate cylinder, 4" in length and 1" in diameter. Encircling the complete length of the housing are 1"–18 threads. At the middle of one end of the housing is a 2" Belden 6906 cable. Attached to the middle of the other end of the housing is a feed paddle.

- MOUNTING HARDWARE: The mounting hardware consists of two white, flat, hexagonal, plastic Delryn locking nuts, Grade 2, 1" ID, 1 1/4" OD, 1/8" thick, and 1"–18 EF threads per inch.

- FEED PADDLE: The feed paddle consists of a 3" diameter, 16-gauge (0.060") No. 304 galvanized steel, Finish 2D or better, welded to a No. 101 Finish 3D or better galvanized steel shaft 5 1/2" in length and 1/4" in diameter. Exposed threads 1/4"–20 and 1/2" in length protrude from the free end of the shaft. Welded on three sides to the middle of the 3" diameter steel is the 1 1/2" of the unthreaded shaft end. The feed paddle is white Teflon coated to a thickness of 1/64".

The Electronic A1 Feed Switch, with its five-year warranty, is reliable in dusty and dirty environments. The easy-to-install, short-circuit-protected switch is internally protected for transient voltage peaks to a maximum of 5 KV, for up to 10 msec duration.

Figure 11.10   Professionally Written Technical Description

- An effectively drawn and placed graphic
- A reader-friendly discussion with headings exactly corresponding to the call-outs in the drawing
- A precisely detailed discussion focusing on materials, color, dimensions, and so on
- An effective conclusion

CHAPTER HIGHLIGHTS

1. Technical descriptions are often used in user manuals, product specifications, reports, and sales literature.
2. The introduction of a technical description explains the mechanism's function and major components.
3. Graphics make a technical description useful to the reader.
4. Follow a logical pattern of organization in a technical description.
5. Word usage can be either photographic or impressionistic.

EXERCISES

1. Write a technical description, either individually or as a team. To do so, first select a topic. You can describe any tool, mechanism, or piece of equipment. However, don't choose a topic too large to describe accurately. To provide a thorough and precise description, you will need to be exact and minutely detailed. A large topic, such as a computer, an oscilloscope, a respirator, or a Boeing airliner, would be too demanding for a two- to four-page description. On the other hand, do not choose a topic that is too small, such as a paper clip, a nail, or a shoestring. Choose a topic that provides you with a challenge but that is manageable. You might write about any of the following topics:

Hammer	Computer disk	VCR remote control
Wrench	Computer mouse	X-acto knife
Screwdriver	Light bulb	Ballpoint pen
Pliers	Calculator	Pencil
Wall outlet	Automobile tire	Credit card

Once you or your team have chosen a topic, prewrite (listing the topic's components), write a draft (abiding by the criteria provided in this chapter), and rewrite (revising your draft).

2. Find examples of professionally written technical descriptions in technical books and textbooks, professional magazines and journals, or on the Internet. Bring these examples to class and discuss whether they are successful according to the criteria presented in this chapter. If the descriptions are good, specify how and why. If they are flawed, state where and suggest ways to improve them. You might even want to rewrite the flawed descriptions.

3. Select a simple topic for description, such as a pencil, coffee cup, toothbrush, or textbook (you can use brainstorming/listing to come up

ACTIVITIES

with additional topics). Describe this item without mentioning what it is or providing any graphics. Then, read your description to a group of students/peers and ask them to draw what you have described. If your verbal description is good, their drawings will resemble your topic. If their drawings are off base, you haven't succeeded in providing an effective description. This is a good test of your writing abilities.

Web Workshop

You are ready to purchase a product. This could include printers, monitors, digital cameras, scanners, PCs, laptops, speakers, cables, adapters, automotive engine hoists, generators, battery chargers, jacks, power tools, truck boxes, screws, bolts, nuts, rivets, hand tools, and more. A great place to shop is online. By going to an online search engine, you can find not only prices for your products but also technical descriptions or technical specifications. These will help you determine if the product has the size, shape, materials, and capacity you are looking for.

Go online to search for a product of your choice and review the technical description or specifications provided. Using the criteria in this chapter and your knowledge of effective technical writing techniques, analyze your findings.

- How do the online technical descriptions compare to those discussed in this textbook?
- Are graphics used to help you visualize the product?
- Are call-outs used to help you identify parts of the product?
- Are high-tech terms defined?
- Is the use of the product explained?
  a. Report your findings, either in an oral presentation or in writing (e-mail message, memo, letter, or report).
  b. Rewrite any of the technical descriptions that need improvement according to the criteria provided in this textbook.

Quiz Questions

1. What is a technical description?
2. Why would you use a technical description?
3. What are the components of a technical description?
4. What is the super comic book look?
5. What is a call-out?
6. In what three places can you position graphics in a technical description?
7. What is the best way to organize a technical description?
8. What information is included in the development of a technical description?
9. What is the difference between photographic and impressionistic words?
10. What are three ways you can rewrite your technical description?

Chapter Preview

Criteria for Writing Short Instructions
A successful instruction requires a title, effective organization, audience recognition, graphics, and correct technical writing style.

Process
Follow the writing process to write an effective instruction.

Process Log
A collaboratively written student log takes you through the writing process. To write their instruction, the students gathered data, determined objectives, sequenced their information, drafted a text, and revised.

Criteria for Writing a User Manual
User manuals typically contain more than a simple instruction. They often include a cover page, hazard alerts, a table of contents, an introduction, definitions, descriptions, and corporate contacts in addition to step-by-step instructions.

Writing at Work

PhlebotomyDR

As the baby-boom generation ages, medical needs are expanding—sometimes faster than medical care facilities and medical professionals can manage. One area in which this has been felt most acutely is in medical laboratories. Thousands of medical technicians have needed to be trained to accommodate increased demand.

PhlebotomyDR is a medical consulting firm seeking to solve this problem. Its primary area of concern is training newly hired technicians responsible for performing blood collection. PhlebotomyDR facilitates training workshops to teach venipuncture standards and venipuncture procedures.

PhlebotomyDR focuses on the following venipuncture instructions:

- Proper patient identification procedures
- Proper equipment selection, sterilization, use, and cleaning
- Proper labeling procedures
- Order of phlebotomy draw
- Patient care before, during, and following venipuncture
- Safety and infection control procedures
- Procedures to follow when meeting quality assurance regulations

Each of the above instructions requires numerous steps, complete with visual aids.

PhlebotomyDR offers its audience various communication channels. Hospitals, labs, and treatment centers can access PhlebotomyDR's instructions as follows:

- Hard-copy instructional manuals
- Online instructions, accessible at *http://www. phlebotomydr.com*

- Videos showing step-by-step performances of blood collection, complete with case studies enacted by technicians, patients, and supervisors
- Computer-aided instruction (CAI) for individual tutorials
- One-on-one tutorials with trained phlebotomists
- Instructional workshops geared toward groups of seminar participants

PhlebotomyDR's outstanding staff realizes that trained technicians make an enormous difference. Training, achieved through instructional manuals, electronic aids, and individual facilitation, ensures the health and safety of patients. Of equal importance, excellent training also benefits many stakeholders. Untrained technicians make errors that cost us all. Medical errors create insurance problems, the need to redo procedures, increased medical bills, the potential involvement of regulators and legislators, and dangerous repercussions for patients.

In contrast, effective communication, achieved through successful instructions, saves lives, time, and money.

Instructions and user manuals are important in technical writing. Every manufactured product comes complete with either a short (one- to four-page) instruction booklet or with a more detailed user manual (user manuals can be anywhere from four pages to a book-length text).

tech link

Go to *http://www.prenhall. com/gerson* for Web links, samples, and interactive activities.

You can receive short instructions for baking brownies; making pancakes; putting together children's toys; or changing a tire, the oil in your car, or the coolant in your engine. Longer user manuals help people set up stereo systems, construct electronic equipment, maintain computers, and operate fighter planes.

There are even instructions for making chicken noodle soup. One major soup manufacturer recently stopped printing instructions for soup preparation on its cans of soup. Executives assumed that such information was unnecessary. They were mistaken. The company received thousands of calls from consumers asking how to prepare the soup. Responding to befuddled customers, the company reverted to printing these instructions. "Pour soup in pan. Slowly stir in one can of water. Heat to simmer, stirring occasionally."

Even pop-top cans of soda include instructions: "1. Lift up. 2. Pull back. 3. Push down." When such simple items as soup and soda need instructions, you can imagine how necessary instructions are for complex products and procedures.

Today, we frequently see short instructions about computer-related problems. If you are at home or in the office, for example, and you can't figure out how to operate one of your computer's applications, you will dial a 1–800 hotline and speak to a computer technician. Following is an e-mail message we received in such an instance. Notice how this e-mail includes an introduction, a numbered body, a cautionary note, and a conclusion:

From: pscsupport@earthlink.net
Sent: Saturday, April 28, 2005, 7:23 P.M.
To: Steve Gerson
Subject: Technical Support

Thank you for contacting us about your computer problem. Recently, we have encountered cases where an online Earthlink update modified existing profiles. This change can affect your e-mail. To solve this problem, create a new profile in Earthlink 5.0. [Note: This will remove old e-mail and your address book.]

1. Click on the icon to open up the Earthlink 5.0 sign-on screen.
2. Click on Configure.
3. Click on New profile.
4. Click on Manage profiles.
5. Click on Add an existing account to this computer.
6. Click in the user name field and type in your user name in lowercase letters.
7. Click in the password field and type in your password in lowercase letters.
8. Click Next.
9. When asked, "Do you want to configure Internet Explorer for Earthlink," click Yes.
10. When asked, "Do you want to configure Netscape Communicator for Earthlink," click Yes.
11. Print out your information.
12. Click Finish.

If you need further help, please let us know. You can always access our customer support site at http://support.earthlink.net or click on the Support tab in your personal start page.

Include instructions or user manuals whenever your audience needs to know how to

Operate a mechanism	Restore a product
Install equipment	Correct a problem
Manufacture a product	Service equipment
Package a product	Troubleshoot a system
Unpack equipment	Care for a plant
Test components	Use software
Maintain equipment	Set up a product
Clean a product	Implement a procedure
Monitor a system	Construct anything
Repair a system	Assemble a product

Whatever your objectives are, this chapter will help you write your short instructions or help you construct a user manual.

CRITERIA FOR WRITING SHORT INSTRUCTIONS

Follow these criteria for writing an effective instruction.

tech link

Go to *http://www.diynet.com* (the Do It Yourself Network) for excellent, step-by-step examples of instructions, complete with graphics.

Title

Preface your text with a title that explains two things: *what* you are writing about (the name of your product or service) and *why* you are writing (the purpose of the instruction). For example, to merely title your instruction "Overhead Projector" would be uninformative. This title names the product, but it does not explain why the instruction is being written. Will the text discuss maintenance, setup, packing, or operating instructions? A better title would be "Operating Instructions for the XYZ Overhead Projector."

Organization

As the writer of the instruction, you know what your objectives are. You know why you are writing and what you want to accomplish. However, your readers, upon picking up the instruction, are unaware of your intentions. Whereas you know where you are going, they must patiently await further information regarding the destination. To help your readers follow your train of thought, provide them with a road map. One way to do this is through organization. A neatly organized instruction can orient your readers, thereby helping them follow your directions. To organize your instruction effectively, include an *introduction*, a *discussion*, and a *conclusion*.

Introduction

In the first paragraph of your instruction, tell your readers three things: what *topic* you will be discussing, your *reasons* for writing, and the *number of steps* involved in the instruction. The topic names the product or service. Your reason for writing either explains the purpose of the instruction ("maintaining the machine will increase its longevity") or comments about the product's capabilities or ease of use. Stating the number of steps allows your readers to plan ahead and organize their thoughts and their time. Look at the following examples:

> **example**
>
> Good
>
> The following seven steps will help you operate the Udell PQ 4454 Overhead Projector.

> **example**
>
> Better
>
> The Udell PQ 4454 is an overhead projector for business and school. It's easy to use and requires little maintenance. Follow these eight steps for operating the machine.

> **example**
>
> Best
>
> The Udell PQ 4454 Overhead Projector is used to project written or graphic material onto a screen or wall. Because of its capability to enlarge and project, it is ideal for use in schools or businesses. Six simple steps will help you operate the projector.

The first of the preceding examples includes product plus steps; the second adds ease of use. The third example, however, is best because it mentions product, steps, ease of use, plus capabilities.

Required Tools or Equipment   In addition to this introductory overview focusing on the topic and steps to be performed, you might want to tell your readers what tools or equipment they will need to perform the procedures. You can provide this information through a simple list, or perhaps you'll want to add graphics (the type of tool needed plus a picture of that tool). This depends on whether your readers are high tech, low tech, or lay.

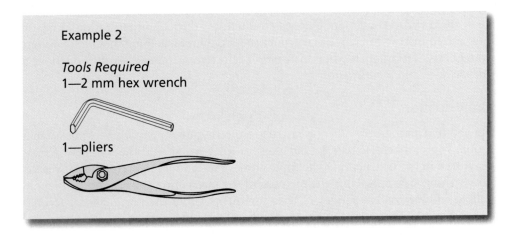

Example 1

Tools Required
1—2 mm hex wrench
1—pliers

Example 2

Tools Required
1—2 mm hex wrench

1—pliers

Hazard Notations   Finally, decide whether you should preface your instructions with dangers, warnings, cautions, or notes. This is an essential consideration to avoid costly lawsuits and to avoid potentially harming an individual or damaging equipment.

When including hazard alert messages, you need to consider the following:

- **Placement.** Where should the caution or warning notice go? Do you place the information before or after the appropriate text? If you place the information after the text, *after the reader has already performed the task*, the damage is done. Your readers already will have hurt themselves or harmed the equipment. The hazard alert is placed before the step. In fact, you might want to state your warnings or cautions in the introduction, before the audience even begins to perform a task.

- **Access.** If your hazard alert is printed before the appropriate step, but you haven't used any highlighting techniques to make it obvious, your readers won't see the danger. You need to make the warning or caution notice obvious. To do so, vary your typeface and type size, use white space to separate the warning or caution from surrounding text, or box the warning or caution.

- **Definitions.** What does *caution* mean? How does it differ from *warning, danger*, or *note*? Unfortunately, these terms are not defined consistently.

    Three primary institutions that seek to provide a standardized definition of terms are American National Standards Institute (ANSI), the U.S. military (MILSPEC), and the Occupational Safety and Health Administration (OSHA). However, these institutions often define terms differently. For example, both ANSI and MILSPEC define *note* similarly as essential operating information. OSHA defines *notice* as general instructions relative to safety measures. ANSI defines *caution* as a hazardous situation that might cause minor or moderate injury, whereas

MILSPEC defines *caution* as something that could result in damage or destruction to equipment. OSHA uses *caution* to designate potential hazards regarding unsafe practices. Personal injury, damage to equipment, and hazardous unsafe practices should be distinguished more clearly. ANSI and MILSPEC define both *warning* and *danger* as an activity possessing the potential for serious injury or death. OSHA defines *danger* as immediate danger. The words overlap and don't distinguish between death and serious harm.

To avoid confusion, we suggest the following hierarchy of definitions, which clarifies the degree of hazard:

1. *Note.* Important information, necessary to perform a task effectively or to avoid loss of data or inconvenience.

2. *Caution.* The potential for damage or destruction of equipment.

3. *Warning.* The potential for serious personal injury.

4. *Danger.* The potential for death.

Our goal is to avoid confusion, which can occur if one writes, "Cautionary Warning Notice!"

- **Colors.** Another way to emphasize your hazard message is through a colored window or text box around the word. Usually, *Note* is printed in blue or black, *Caution* in yellow, *Warning* in orange, and *Danger* in red.

- **Text.** To further clarify your terminology, provide the readers text to accompany your hazard alert. Your text should have the following three parts:

  1. *A one- or two-word identification alerting the reader.* Words such as *High Voltage, Hot Equipment, Sharp Objects,* or *Magnetic Parts,* for example, will warn your reader of potential dangers, warnings, or cautions.

  2. *The consequences of the hazards, in three to five words.* Phrases like "Electrocution can kill," "Can cause burns," "Cuts can occur," or "Can lead to data loss," for example, will tell your readers the results stemming from the dangers, warnings, or cautions.

  3. *Avoidance steps.* In three to five words, tell the readers how to avoid the consequences noted: "Wear rubber shoes," "Don't touch until cool," "Wear protective gloves," or "Keep disks away."

- **Icons.** Equipment is manufactured and sold globally; people speak different languages. Your hazard alert should contain an icon—a picture of the potential consequence—to help everyone understand the caution, warning, or danger. (Icons are discussed in Chapter 9.)

Figure 12.1 shows an effective page layout and the necessary information to communicate hazard alerts.

Discussion

Itemize and thoroughly discuss the steps in your instruction. Organize them *chronologically*—as a step-by-step sequence. Obviously, you cannot tell your readers to do step 6, then go back to step 2, then accomplish step 12, then do step 4. Such a distorted sequence would fail to accomplish your goals. To operate machinery, monitor a system, or construct equipment, your readers must follow a chronological sequence. Be sure that your instruction is chronologically accurate.

tech link

For a detailed article on the use of art (icons and color) to communicate safety and hazards, go to *http://www.osha.gov/SLTC/ hazardcommunications/ hc2inf2.html#*.1.3.1.*

tech link

To access information about hazard icons, go to the following Web sites:
- *http://www.hse.gov.uk/ pubns/safesign.htm*
- *http://www.speedysigns. com/signs/*
- *http://www.atsdr.cdc. gov/NFPA/nfpa_label.html*

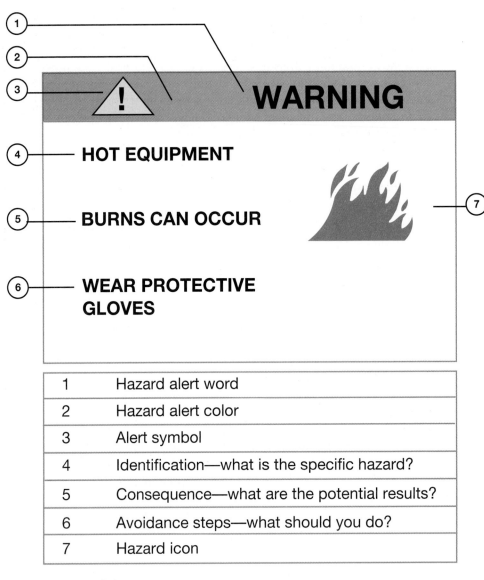

1	Hazard alert word
2	Hazard alert color
3	Alert symbol
4	Identification—what is the specific hazard?
5	Consequence—what are the potential results?
6	Avoidance steps—what should you do?
7	Hazard icon

Figure 12.1   Hazard Alert

Conclusion

As with a technical description, you can conclude your instruction in various ways. You can end your instruction with a (a) comment about warranties; (b) sales pitch highlighting the product's ease of use; (c) reiteration of the product's applications; (d) summary of the company's credentials; (e) troubleshooting guide; (f) frequently asked questions (FAQs); and (g) corporate contact information. (For examples, see the discussion on conclusions in Chapter 11.)

Another valid way to conclude, however, is to focus on *disclaimers*, as in the following example:

example

This operating guide to the Udell PQ 4454 Overhead Projector is included primarily for ease of customer use. Service and application for anything other than normal use or replacement parts must be performed by a trained and qualified technician.

Audience Recognition

How many times have you read an instruction and been left totally confused? You were told to "place the belt on the motor pulley," but you didn't know how. Or you were told to "discard the used liquid in a safe container," but you didn't know what was safe for this specific type of liquid. You were told to "size the cutting according to regular use," but you had never regularly performed this activity.

Here is a typical instruction written without considering the audience:

example

> To overhaul the manual starter, proceed as follows: Remove the engine's top cover. Untie the starter rope at the anchor and allow the starter rope to slowly wind onto the pulley. Tie a knot on the end of the starter rope to prevent it from being pulled into the housing. Remove the pivot bolt and lift the manual starter assembly from the power head.

Although many high-tech readers might be able to follow these instructions, many more readers will be confused. How do you remove the engine's top cover? Where is the anchor? Where is the pivot bolt and how do you remove it? What is the power head?

The problem is caused by writers who assume that their readers have high-tech knowledge. This is a mistake for several reasons. First, even high-tech readers often need detailed information because technology changes daily. You cannot assume that every high-tech reader is up-to-date on these technical changes. Thus, you must clarify. Second, low-tech and lay readers—and that's most of us—carefully read each and every step, desperate for clear and thorough assistance.

As the writer, you should provide your readers with the clarity and thoroughness they require. To do this, recognize accurately who your readers are and give them what they want, whether that amounts to technical updates for high-tech readers or precise, even simple, information for low-tech or lay readers. The key to success as a writer of instructions is, "Don't assume anything. Spell it all out—clearly and thoroughly!"

Graphics

As with technical descriptions, clarify your points graphically using the *super comic book look*. Use drawings or photographs that are big, simple, clear, keyed to the text, and labeled accurately. Not only do these graphics make your instructions more visually appealing, but also they help your readers and you. What the reader has difficulty understanding, or you have difficulty writing clearly, your graphic can help explain pictorially.

A company's use of graphics often depends on the audience. With a high-tech reader, the graphic might not be needed. However, with a low-tech reader, the graphic is used to help clarify. Look at the following examples of instructions for the same procedure. Figure 12.2 uses text with the super comic book look; Table 12.1 uses text without graphics.

Note that Figure 12.2 uses large, bold graphics to supplement the text, clearly depicting what action must be taken. This is appropriate for a low-tech end user. On the other hand, Table 12.1, written for a manufacturer's technician (high-tech), assumes greater knowledge and, therefore, deletes the graphic.

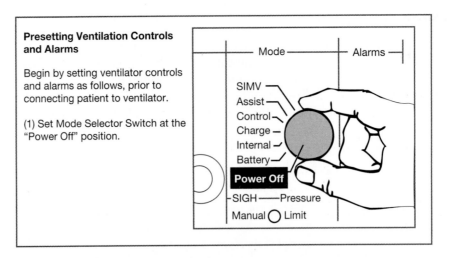

Presetting Ventilation Controls and Alarms

Begin by setting ventilator controls and alarms as follows, prior to connecting patient to ventilator.

(1) Set Mode Selector Switch at the "Power Off" position.

Figure 12.2   Text with Super Comic Book Graphics
Courtesy of Puritan Bennett Corp.

TABLE 12.1 TEXT WITHOUT GRAPHICS FOR HIGH-TECH READERS		
Location Item	Action	Remarks
	Note For additional operating instructions, refer to Companion 2800 operator's manual.	
1. MODE switch	Set to POWER OFF.	
2. Rear panel	Ensure that instructions on all labels are observed.	
3. Patient tubing circuit	Assemble and connect to unit.	See Table 2, step 3.
4. Power cord	Connect to 120/220 V ac grounded outlet.	If power source is external battery, connect external battery cable to unit per Table 2, step 4.
5. Cascade I humidifier	Connect to unit.	See Table 2, step 2.

Courtesy of Puritan Bennett Corp.

Style

1. *Number your steps*—Do not use bullets or the alphabet. Numbers, which you can never run out of, help your readers refer to the correct step. In contrast, if you used bullets, your readers would have to count to locate steps—seven bullets for step 7, and so on. If you used the alphabet, you'd be in trouble when you reached step 27.

2. *Use highlighting techniques*—Boldface, different font sizes and styles, emphatic warning words, color, italics, and so on call attention to special concerns. A danger, caution, warning, or specially required technique must be evident to your reader. If this special concern is buried in a block of

unappealing text, it will not be read. This could be dangerous to your reader or costly to you and your company. To avoid lawsuits or to help your readers see what is important, call it out through formatting.

3. *Limit the information within each step*—Don't overload your reader by writing lengthy steps:

OVERLOADED STEPS

1. Start the engine and run it to idling speed while opening the radiator cap and inserting the measuring gauge until the red ball within the glass tube floats either to the acceptable green range or to the dangerous red line.

Before

Instead, separate the distinct steps:

SEPARATED STEPS

1. Start the engine and run it to idling speed.

2. Open the radiator cap and insert the measuring gauge.

3. Note whether the red ball within the glass tube floats to the acceptable green range or up to the dangerous red line.

After

4. *Develop your points thoroughly*—Don't say, "After rotating the discs correctly, grease each with an approved lubricant." Instead, clarify what you mean by correct rotating and specify the approved lubricant, as in the following example. Unless you tell your readers what you mean, they won't know.

WELL-DEVELOPED STEPS

1. Rotate the disks clockwise so that the tabs on the outside edges align.

2. Lubricate the discs with 2 oz of XYZ grease.

After

5. *Use short words and phrases*—Remember the fog index. (See Chapter 3.)

6. *Begin your steps with verbs—the imperative mood*. Note that each of the numbered steps in the following example begins with a verb:

VERBS BEGIN STEPS

1. *Number* your steps.

2. *Use* highlighting techniques.

After

3. *Limit* the information within each step.

4. *Develop* your points thoroughly.

5. *Use* short words and phrases.

6. *Begin* your steps with verbs.

7. *Personalize your text*—Remember, people write to people. Involve your readers in the instruction by using pronouns. (See Chapter 4.)

8. *Do not omit articles*—Articles—*a, an,* and *the*—are part of our language. Although you might see instructions that omit these articles, please don't do so yourself. Articles do not take up much room in your text, but they do make your sentences read more fluidly.

PROCESS

Now that you know what should be included in an instruction, it is time to write. But how do you start? Where do you begin?

As in other technical writing activities, follow a process: prewriting to gather data and determine objectives, writing to draft your correspondence, and rewriting to revise and perfect.

The Writing Process		
Prewriting	Writing	Rewriting

• Examine your purposes
• Determine your goals
• Consider your audience
• Gather your data
• Determine how the content will be provided
	• Organize the draft according to some logical sequence that your readers can follow easily
	• Format the content to allow for ease of access
		• Revise
		  ° Add missing details
		  ° Delete wordiness
		  ° Simplify word usage
		  ° Enhance the tone of your communication
		  ° Reformat your text for ease of access
		  ° Practice the speech or review the text
		• Proofread
		  ° Correct errors

Prewriting

For writing an instruction, reporter's questions would be a good place to start gathering data and determining objectives. Ask yourself:

- *Who* is my reader?
  - High tech?
  - Low tech?
  - Lay?
- *What detailed* information is needed for my audience? For example, a low-tech or lay reader will be content with "inflate sufficiently" whereas an instruction geared toward high-tech readers will require "inflate to 25 psi."
- *How* must this instruction be organized?
  - Chronologically
- *When* should the procedure occur?
  - Daily, weekly, monthly, quarterly, biannually, or annually?
  - As needed?
- *Why* should the instruction be carried out?
  - To repair, maintain, install, operate, and so on. (See the list of reasons for using an instruction at the beginning of this chapter.)

Once you have gathered this information, you can use brainstorming to sketch out a rough list of the steps required in the instruction and the detail needed for clarity. After that, your major responsibility is to sequence your steps chronologically. To accomplish this goal, use a prewriting technique called *flowcharting*.

Flowcharting

Flowcharts chronologically trace the stages of an instruction, visually revealing the flow of action, decision, authority, responsibility, input/output, preparation, and termination of process. Flowcharting is not just a graphic way to help you gather data and sequence your instruction, however. It is two-dimensional writing. It provides your reader content as well as a panoramic view of an entire sequence. Rather than just reading an instruction step by step, having little idea what's around the corner, with a flowchart your reader can figuratively stand above the instruction and see where it's going. With a flowchart, readers can anticipate cautions, dangers, and warnings, since they can see at a glance where the steps lead.

To create a flowchart, use the International Organization for Standardization (ISO) flowchart symbols shown in Figure 12.3.

Figure 12.4 shows an excellent example of a flowchart, visually depicting an instruction's sequence.

Writing

Once you've graphically depicted your instruction sequence using a flowchart, the next step is to write a rough draft of the instruction.

Review Your Prewriting

In reviewing your prewriting, you will accomplish the following:

- **Clarify your audience.** Now is a good time to review your reporter's questions and to determine whether your initial assumptions regarding

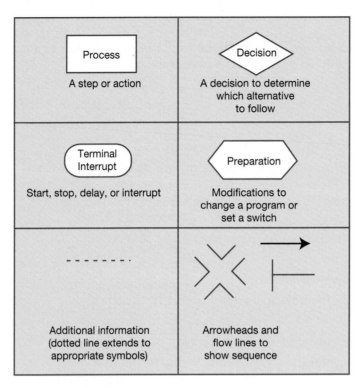

Figure 12.3   Flowcharting Symbols

audience are still correct. If your attitude toward your audience has changed, redefine your readers and their needs.

- **Determine your focus.** In your prewriting, you answered the question *why*. Are you still certain of your objectives for writing the instruction? If so, write a one-sentence thesis or statement of objectives to summarize this purpose. For example, you could write, "This instruction should be followed so that glitches in the system can be discovered and corrected." This thesis or topic sentence will help you maintain your focus throughout the instruction. If you are uncertain about your purpose for writing the instruction, however, rethink the situation. You will not be able to write an effective instruction until your objectives are clearly stated.
- **Organize your steps.** Review your flowchart to determine whether any steps in the instruction have been omitted or misplaced. If the chronological sequence is incorrect, reorganize your detail. If the chronology is correct, you're ready to move on to the next step in drafting.

Review the Instruction Criteria

Before you write the rough draft, remind yourself what the instruction will consist of. You'll need an introduction, a discussion with itemized steps, graphics to help your reader perform the requested procedures, and a conclusion.

Write the Draft

You've gathered your data and determined your objectives. You've correctly sequenced the steps and reminded yourself of the instruction criteria. At this point, you should feel comfortable with the prospect of writing. Follow the sufficing

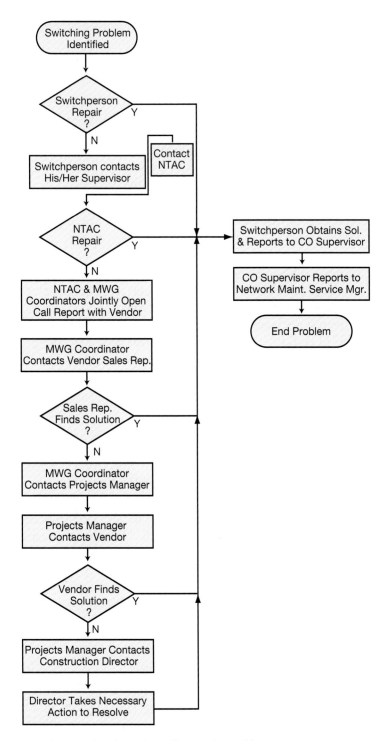

Figure 12.4 Flowchart of Handling Service Problems

technique again, roughly drafting your instruction. Don't worry about perfect grammar or graphics. Perfecting the text comes in the next step, rewriting.

Rewriting

Rewriting, the last stage in the process, is the time to perfect your instruction. For manuals, this stage calls for *usability testing*. You've written the instruction, but is it usable? If the instruction doesn't help your audience complete the task, what have you accomplished? To determine the success of your manual, test the instruction.

Usability Testing

To test the usability of an instruction, follow this procedure:

1. *Select a test audience*—The best audience for usability testing would include a representative sampling of individuals with differing levels of expertise. A review team composed only of high-tech readers would skew your findings. In contrast, a review team consisting of high-tech, low-tech, and lay readers would give you more reliable feedback.

2. *Ask the audience to test the instructions*—The audience members would attempt to complete the instructions, following the procedure step by step.

3. *Monitor the audience*—What challenges do the instructions seem to present? For example, while observing the audience, has the audience completed all of the steps easily? Are the correct tools and equipment listed? Are terms defined as needed and presented where they are necessary? Has the test audience abided by the hazard notations? Are any steps overloaded and, thus, too complicated? Has each step been explained specifically?

4. *Time the team members*—How long does it take each member to complete the procedure? More important, why has it taken some team members longer to complete the task?

5. *Quantify the audience's responses*—Once your test audience has completed the procedure, you must debrief these individuals to determine what problems they encountered. Use the Effective Instruction Checklist to help gather quantifiable information about the instruction's usability.

Your peer evaluators do not have to suggest ways to fix any problems encountered. Their goal is just to give you feedback. If they say, "I don't understand this word," "Step 3 seems too long," "I could use a graphic here to clarify your meaning," or "More white space would keep me from feeling overwhelmed," you can revise the text accordingly. To do so, reread the instruction, consider the suggestions made during usability testing, and improve the draft by using the eight-step rewriting procedure explained here.

Add Detail for Clarity

Don't hope you've said enough to clarify your steps. Tell your readers *exactly* how to do it or why to do it. Part of detail involves warnings and cautions, for example. Make sure that you've added all necessary dangers or concerns. In addition, assess each step. Have you added sufficient detail to clarify your intent? For instance, if you are writing to a lay reader, you don't want to say merely "Insert the guide posts." You must say, "Before inserting the guide post, dig a 2-ft-deep by 2-ft-wide hole. Pour the ready-mixed concrete into the hole. Insert the post into the hole and hold it steady until the concrete starts to set (approximately one minute)."

Delete

Omit unnecessary words and phrases to achieve conciseness. In addition, delete extraneous information that a high-tech reader might not find useful. This, however, could be dangerous. What you might see as extraneous, your reader might see as necessary. If you delete any information, do so carefully.

Effective Instruction Checklist

___ 1. Does the instruction have an effective title?
   - Is the topic mentioned?
   - Is the reason for performing the instruction mentioned?

___ 2. Does the instruction have an effective introduction?
   - Is the topic mentioned?
   - Is the reason for performing the instruction mentioned?
   - Is the ease of use or capabilities mentioned?
   - Have the number of steps been listed?
   - Has a list of required tools or equipment been provided?

___ 3. Are hazard alert messages used effectively?
   - Are the hazard alert messages placed correctly?
   - Is the correct term used (*Danger, Warning, Caution,* or *Note*)?
   - Is the correct color used with the appropriate term?
   - Does the hazard alert text identify the hazard, provide the consequences, and suggest avoidance steps?
   - Is an effective icon used to depict the hazard?

___ 4. Is the instruction's discussion effective?
   - Is the instruction organized chronologically?
   - Does each step avoid overloading by presenting one clearly defined action?
   - Does each step begin with a verb?

   - Does the instruction's discussion contain the same number of steps mentioned in the introduction?

___ 5. Is the instruction well developed?
   - Does the instruction avoid assuming that readers will understand and, instead, clearly explain each point?

___ 6. Does the instruction recognize audience effectively?
   - Is the instruction written for a high-tech, low-tech, or lay audience?
   - Has sexist language been avoided?
   - Does the text include language appropriate for a multicultural audience?
   - Is the text personalized using pronouns?

___ 7. Is the conclusion effective?
   - Have warranties been mentioned?
   - Is a sales pitch provided showing ease of use?
   - Has the conclusion reiterated a reason for performing these steps?
   - Are disclaimers provided?

___ 8. Is the instruction's document design effective?
   - Are highlighting techniques effectively used? These include graphics (such as comic book look photographs or line drawings), varied typefaces and type sizes, color, and white space.

___ 9. Is correct technical writing style used, abiding by the fog index?

___ 10. Have grammatical and textual errors been avoided?

Simplify

Don't overload your steps. If you have too much information in any one stage, divide this detail into smaller units, adding new steps. Doing so will help your readers follow your instructions. Simplify your word usage; refer to the fog index.

Move Information

Make sure that the correct chronological sequence has been followed.

Reformat Using Highlighting Techniques

If you include dangers, cautions, warnings, or additional information for clarity, set these off (make them emphatic) through formatting. Use color, text boxes, font size and style, emphatic words, graphics, and so on.

Enhance Your Text

Use the imperative mood for each step in your instruction. Whenever an action is required, begin with a verb to emphasize the action. In addition, you can enhance your text through personalization. Remember, people write to people. To achieve this person-to-person touch, add pronouns to your instruction.

Correct Your Instruction for Accuracy

Proof your text for grammatical correctness as well as for contextual accuracy.

Avoid Biased Language

Your instructions could be read by an international audience. Avoid cultural biases and write for readers diverse in age and gender.

PROCESS LOG

Following is an example of an instruction collaboratively written by a group of students. To write this instruction, the students gathered data, determined objectives and sequenced their instruction through prewriting, drafted an instruction in writing, and finally revised their draft in rewriting.

Their assignment was to write an instruction on how to use an overhead projector, an item readily available in most classrooms.

Prewriting

Figure 12.5 illustrates reporter's questions and students' flowcharting.

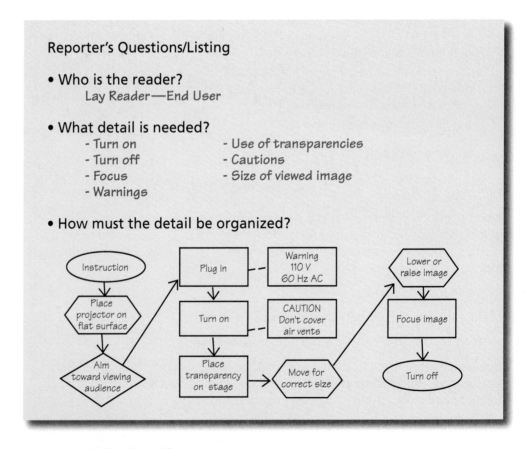

Figure 12.5   Students' Prewriting

Writing

After completing the prewriting, the students composed a draft, complete with sketchy graphics (Figure 12.6).

Rewriting

The students revised the rough draft, considering all the rewriting techniques discussed in this text. The revision is shown in Figure 12.7 on page 370. The students then prepared their finished copy (Figure 12.8 on pages 371–373).

The Udell Overhead Projector projescts material on a flat surface. Follow these five steps to do so.

1. Place the machine on a flat surface within viewing range of the screen.
2. Plug power plug into a wall outlet and turn the machine on.
3. Put the transparency on the projectors horizontal viewing surface.
4. Move the image up or down for better viewing and focus image by adjusting the focus knob.
5. Turn machine off when completed and umplug.

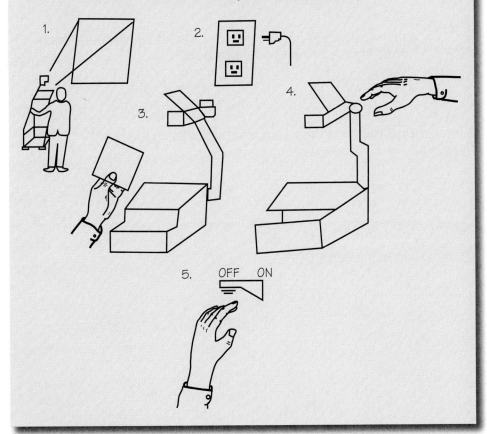

Figure 12.6   Students' Rough Draft

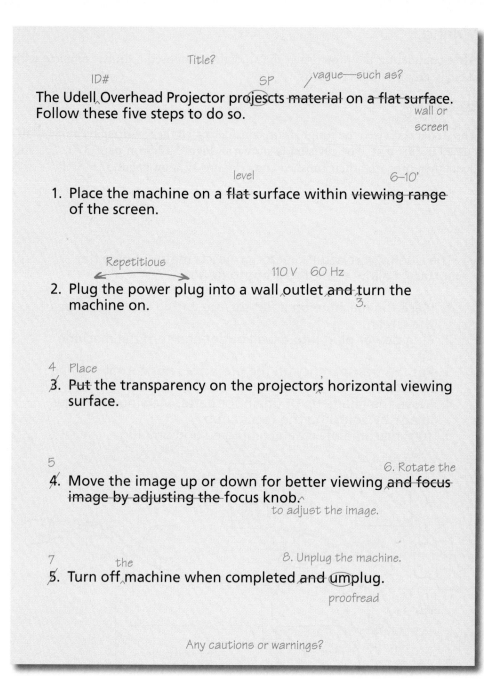

The following text appears within the student revision image:

Title?

ID#   *SP*   *vague—such as?*

The Udell Overhead Projector projescts material on a flat surface. Follow these five steps to do so.   *wall or screen*

level   *6–10'*

1. Place the machine on a flat surface within viewing range of the screen.

Repetitious   *110 V 60 Hz*

2. Plug the power plug into a wall outlet and turn the machine on.   *3.*

4 Place

3. Put the transparency on the projectors horizontal viewing surface.

5   *6. Rotate the*

4. Move the image up or down for better viewing and focus image by adjusting the focus knob.   *to adjust the image.*

7   *the*   *8. Unplug the machine.*

5. Turn off machine when completed and umplug.   *proofread*

Any cautions or warnings?

Figure 12.7   Students' Suggested Revisions

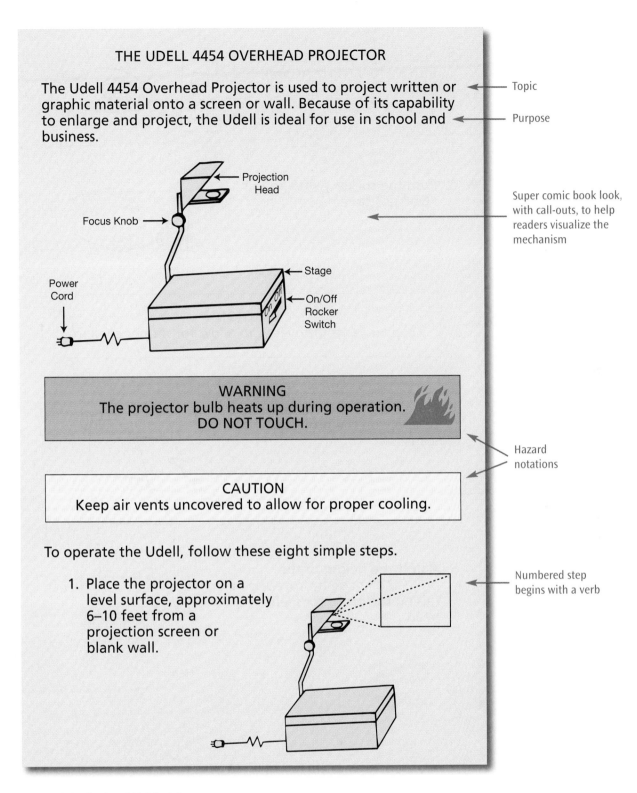

THE UDELL 4454 OVERHEAD PROJECTOR

The Udell 4454 Overhead Projector is used to project written or graphic material onto a screen or wall. Because of its capability to enlarge and project, the Udell is ideal for use in school and business.

Topic

Purpose

Projection Head

Focus Knob

Super comic book look, with call-outs, to help readers visualize the mechanism

Stage

Power Cord

On/Off Rocker Switch

WARNING
The projector bulb heats up during operation.
DO NOT TOUCH.

Hazard notations

CAUTION
Keep air vents uncovered to allow for proper cooling.

To operate the Udell, follow these eight simple steps.

1. Place the projector on a level surface, approximately 6–10 feet from a projection screen or blank wall.

Numbered step begins with a verb

Figure 12.8   Students' Finished Copy

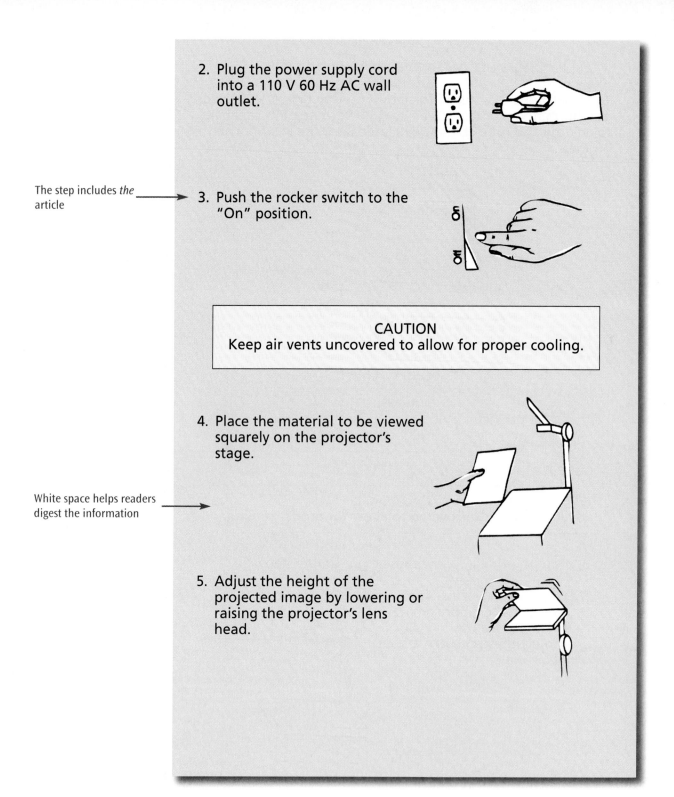

2. Plug the power supply cord into a 110 V 60 Hz AC wall outlet.

The step includes *the* article

3. Push the rocker switch to the "On" position.

On
Off

CAUTION
Keep air vents uncovered to allow for proper cooling.

4. Place the material to be viewed squarely on the projector's stage.

White space helps readers digest the information

5. Adjust the height of the projected image by lowering or raising the projector's lens head.

Figure 12.8   Students' Finished Copy (continued)

6. Rotate the focus knob for clear viewing of the projected image.

7. Push the rocker switch to the "Off" position when you are through viewing your material.

8. Unplug the unit's power cord.

Following these eight simple steps will help you use the Udell overhead projector.

Figure 12.8 Students' Finished Copy (continued)

Figure 12.9 is an excellent student-written instruction.

Title clarifying intent ⟶ **INSTRUCTIONS FOR DRY MOUNTING PHOTOGRAPHS**

Dry mounting photographs keeps them safe for your family and generations to come. To successfully dry mount your photographs, follow the seven steps we've provided.

Dry mounting requires the following materials:

Materials

Required tools and equipment listed ⟶
1. Press
2. Tacking iron
3. Mat knife
4. Mat board
5. Dry-mount tissue
6. Ruler
7. Print (photographed image)

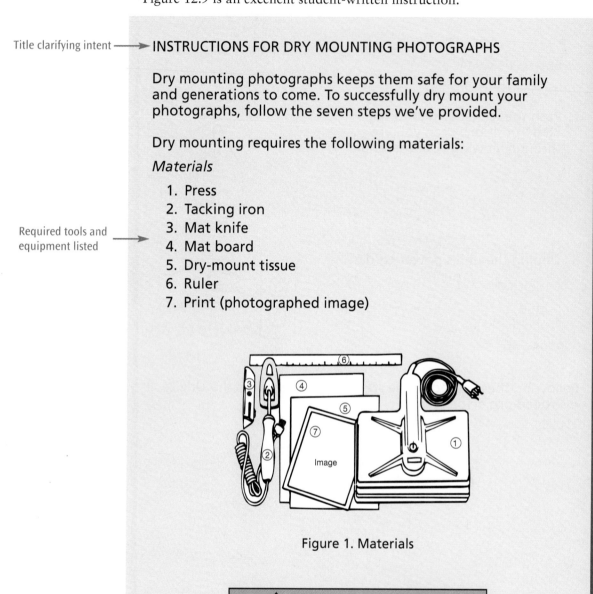

Figure 1. Materials

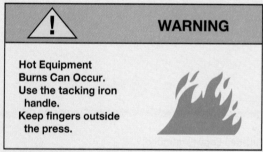

⚠ WARNING

Hot Equipment
Burns Can Occur.
Use the tacking iron
  handle.
Keep fingers outside
  the press.

Figure 12.9   Dry Mounting Instruction

1.0 DRYING THE MATERIAL

1.1 Plug in the press and turn it on. Wait to see the red light.
1.2 Heat the press to 250 degrees by adjusting the dial on the lid.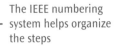

The IEEE numbering system helps organize the steps

1.3 Predry the mount board and the print.
1.4 Place them in the press and hold down the lid for 30 seconds.

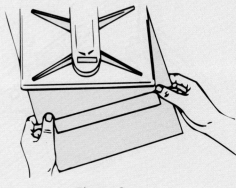

Figure 2

2.0 WIPING THE MATERIALS CLEAN

2.1 After the materials have been baked, let them cool for two to three minutes.
2.2 Wipe away any loose dust or dirt with a clean cloth.

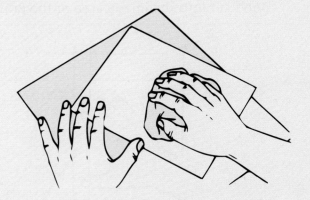

Visual aids enhance the instructions

Figure 3

Figure 12.9   Dry Mounting Instruction (continued)

3.0 TACKING THE MOUNTING TISSUE TO THE PRINT

3.1 Plug in the tacking iron and turn it on.
3.2 With the print face down and the mounting tissue on top, tack the tissue to the center of the print.

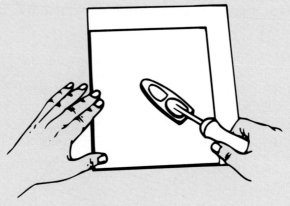

Figure 4

4.0 TRIMMING THE MOUNTING TISSUE

4.1 Place a piece of smooth cardboard under the print so that you won't cut your table.
4.2 Trim off excess mounting tissue using the ruler.

> **CAUTION**
> Press firmly on the ruler to prevent slipping!
> Don't cut into the image area of the print!

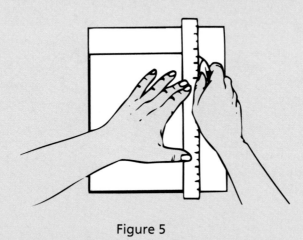

Figure 5

Figure 12.9   Dry Mounting Instruction (continued)

5.0 ATTACHING THE MOUNTING TISSUE AND PRINT TO THE BOARD

> **CAUTION**
> The tissue must lie completely flat on the board
> to prevent wrinkles under the print!

5.1 Position the print and the tissue, face up, on the top of the mount board.

5.2 Slightly raise one corner of the print and touch the tacking iron to the tissue on top of the board.

5.3 Press down the corner lightly, taking care to avoid wrinkles.

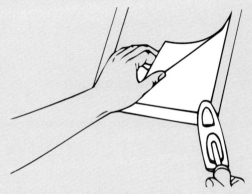

Figure 6

6.0 MOUNTING THE PRINT

6.1 Put the board, tissue, and print (with cover sheet on top) into the press.

6.2 Hold down the press lid firmly for 30 seconds.

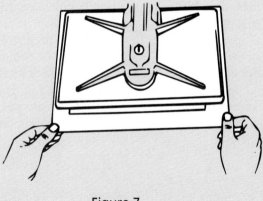

Figure 7

Figure 12.9 Dry Mounting Instruction (continued)

7.0 TRIMMING THE MOUNTED PRINT

7.1 Pressing down firmly on the ruler to avoid slipping, trim the edges of the mounted print with a sharp mat knife.

Image

Figure 8

Conclusion to sum up the instruction → By carefully dry mounting your photographs, they will look beautiful and last longer. Enjoy your artistry.

Figure 12.9   Dry Mounting Instruction (continued)

CRITERIA FOR WRITING A USER MANUAL

In business, many topics require more detail than can be discussed in a short instruction. For instance, if you work for General Motors and your job is to write the instructions for operating a new vehicle, you won't be able to do so if you are limited to four pages. An automobile, with thousands of parts, will require a more extensive document. That's when you need to write a user manual.

Your user manual might focus on installation, operation, maintenance, or troubleshooting. It might be packed with the product or produced separately to be sold at a later date. The user manual might be sized to fit in your car's glove compartment, or it might be as large as an encyclopedia. Whatever the size, shape, or purpose of your user manual, it will include an instruction or several instructions, abiding by the criteria already provided in this chapter.

However, in contrast to a short instruction, user manuals contain additional information, including *any* or *all* of the following components (the order of these components will vary):

Cover Page

The cover page accomplishes at least two goals. First, the cover page will name the product or service being discussed *and* explain the purpose of the manual.

> **example**
>
> All CALLS Cordless Telephone
> Instruction Manual

AutoCall Cordless Telephone
Use and Care Guide

Second, in addition to naming the product or service and stating the purpose of the manual, a cover page might also graphically depict the product or service. This could be accomplished either with a photograph or a line drawing. Doing so allows the readers to identify the product or service more readily.

Hazard Alerts

Usually, when readers turn the cover page to open the manual, they see hazard alerts or safety guidelines. Companies want to protect their customers from harm, minimize potential damage to equipment, and avoid costly lawsuits. Hazard alerts could be a list of safety tips, such as the following guidelines from a roofing user manual:

- Please read all instructions.
- Use care with ladders.
  - Rest ladders on steady ground.
  - Lean ladders against a wall.
  - Allow only one person on a ladder.
- Do not touch electric lines.
- Stay off your roof during strong winds.
- Use a scaffolding system for steep roofs.

Better than a list, however, would be a hazard alert like those already discussed in this chapter. A more effective hazard alert would include the appropriate word (*Danger*, *Warning*, *Caution*, or *Note*), color, hazard identification, consequence, avoidance steps, and icon depicting the hazard.

Table of Contents

Your user manual will have several sections. As already noted, you'll include numerous instructions (setup, installation, maintenance, operation, troubleshooting, for example). In addition, you might include specifications, warranties, guarantees, and customer service contact numbers.

However, readers might want to access only one of these sections. An effective table of contents will allow your readers to find the one or two items they specifically want. (We discuss successful tables of content in Chapter 16.)

Introduction

Companies need customers. A user manual might be the only contact a company has with its customer. Therefore, user manuals often are reader-friendly and seek to achieve audience recognition and audience involvement. The manuals try to reach out and touch the customer.

Look at these introductions from two user manuals:

Thank you for your purchase. Installation and operating procedures are contained in this booklet about your new color TV. The minutes you spend reading this book will contribute to hours of viewing pleasure.

Welcome to the world of modern electronics. The cordless phone you have purchased is the product of advanced technology. This manual will help you install your new phone easily.

Each of these introductions uses pronouns (*you*, *your*, and *our*) to personalize the manual. The introductions also use positive words, such as *Welcome*, *Thank you*, and *pleasure* to achieve a positive customer contact. This is important because companies know that without customers, they're out of business. An effective introduction promotes good customer-company relationships.

Definitions of Terminology

If your manual uses the abbreviations *BDC*, *CCW*, or *CPR*, will the readers know that you are referring to "bottom dead center," "counterclockwise," or "continuing property records"? Probably not.

If the readers are not familiar with your terminology, they might miss important information and perform an operation incorrectly.

To avoid this problem, define your abbreviations, acronyms, and symbols. Because *Danger*, *Warning*, *Caution*, and *Note* are rarely defined the same way in different manuals, an effective manual defines what it means by these hazard alerts. You can give your definitions early in the user manual, throughout the manual, or in a glossary located at the end of your manual (we discuss glossaries in Chapters 15 and 17). The following example defines terms alphabetically:

⚠	This symbol designates an important operating or maintenance step.
BDC	Bottom dead center
CCW	Counterclockwise
Danger	This hazard alert designates the possibility of death. Be extremely careful when performing an operation.
RMS	Root mean square

Technical Descriptions

In addition to instructions, many user manuals contain technical descriptions of the product or system. A description could be a part-by-part explanation or labeling of a product or system's components. Such a description helps readers recognize parts when they are referred to in the instruction. For example, if the user manual tells the reader to lay shingles with the tabs pointing up, but the reader doesn't know what a "tab" is, the step cannot be performed. In contrast, if a de-

TABLE 12.2	CORDLESS TELEPHONE SPECIFICATIONS
Base Unit	Specifications
Transmit Frequency	46.6 to 46.9 MHz
Receive Frequency	49.6 to 49.9 MHz
Power Requirements	117 V, 60 Hz, 6 W
Size	84 mm (W) × 60 m (H) × 246 mm (D)
Weight	755 g
Handset	Specifications
Transmit Frequency	49.6 to 49.9 MHz
Receive Frequency	46.6 to 46.9 MHz
Power Requirements	Rechargeable nickel cadmium batteries
Size	60 mm (W) × 210 mm (H) × 44 mm (D)
Weight	345 g

scription with appropriate call-outs is provided, then the reader's job has been simplified.

Perhaps the user manual will contain a list of the product's specifications, such as its size, shape, capacity, capability, and materials of construction. Table 12.2 is an example of a specification from a cordless telephone user manual. Such a specification allows the user to decide whether the product meets his or her needs. Does it state the desired frequency? Is it the preferred weight and size? The specification answers these questions.

Finally, a user manual might include a schematic, depicting the product or system's electrical layout. This would help the readers troubleshoot the mechanism. (Chapter 9 provides a sample schematic.)

Warranties

Warranties protect the customer and the manufacturer. Many warranties tell the customer, "This warranty gives you specific legal rights, and you may also have other rights that vary from state to state." A warranty protects the customer if a product malfunctions sooner than the manufacturer suggests it might: "This warrants your product against defects due to faulty material or installation." In such a case, the customer usually has a right to free repairs or a replacement of the product.

However, the warranty also protects the manufacturer. No product lasts forever, under all conditions. If the product malfunctions after a period of time designated by the manufacturer, then the customer is responsible for the cost of repairs or a replacement product. The designated period of time differs from product to product. Furthermore, many warranties tell the customers, "This warranty does not include damage to the product resulting from normal wear, accident, or misuse."

Disclaimers are another common part of warranties that protect the manufacturer. For example, some warranties include the following disclaimers:

- Note: Any changes or modifications to this system not expressly approved in this manual could void your warranty.
- Proof of purchase by way of a receipt clearly noting that this unit is under warranty must be presented.

- The warranty is only valid if the serial number appears on the product.
- The manufacturer of this product will not be liable for damages caused by failure to comply with the installation instructions enclosed within this manual.

Accessories

A company always tries to increase its income. One way to do so is by selling the customer additional equipment. A user manual promotes such equipment in an accessories list. This equipment isn't mandatory, required for the product's operation. Instead, accessories lists offer customers equipment such as extra-long cables or cords, carrying cases, long-life rechargeable batteries, extendable antennas, a modem, CD-ROM capabilities, and video instruction packages. Often the specifications are also provided for these accessories.

Frequently Asked Questions

Why take up your customer support employees' valuable time by having them answer the same questions over and over? By including a frequently asked questions (FAQs) page in the user manual, common consumer concerns can be addressed immediately. This will save your company time and money while improving customer relations.

Corporate Contact Information

The user manual helps customers purchase accessories, answers customer questions, and solves customer complaints. Therefore, most user manuals conclude by giving telephone numbers and addresses for local and regional service locations, 24-hour service hotlines, and consumer information bureaus.

Usability Evaluation Checklist

Audience Recognition

___ 1. Are technical terms defined?

___ 2. Are examples used to explain difficult steps at the reader's level of understanding?

___ 3. Do the graphics depict correct completion of difficult steps at the reader's level of understanding?

___ 4. Are the tone and word usage appropriate for the intended audience?

___ 5. Does a cover page explain for the audience your manual's purpose?

___ 6. Does the introduction involve the audience and clarify how the reader will benefit?

Development

___ 1. Are steps precisely developed?

___ 2. Is all required information provided, including hazards, technical descriptions, warranties, accessories, and required equipment or tools?

___ 3. Is irrelevant or rarely needed information omitted?

Conciseness

___ 1. Are words, sentences, and paragraphs concise and to the point?

___ 2. Are the steps self-contained so the reader doesn't have to remember important information from the previous step?

___ 3. Are the steps cross-referenced to help the reader refer to information provided elsewhere?

Consistency

___ 1. Is a consistent hierarchy of headings used?

___ 2. Are graphics presented consistently (same location, same use of figure titles and numbers, similar sizes, etc.)?

___ 3. Does wording mean the same throughout (technical terms, cautions, warnings, notes, etc.)?

___ 4. Is the same system of numbering used throughout?

continued

continued

Ease of Use

__ 1. Can readers easily find what they want because instructions include
  - Table of contents
  - Glossary
  - Hierarchical headings
  - Headers and footers
  - Index
  - Cross-referencing or hypertext links
  - Frequently asked questions (FAQs)

Document Design

__ 1. Do graphics depict how to perform steps?

__ 2. Is white space used to make information accessible?

__ 3. Does color emphasize hazards, key terms, or important parts of a step?

__ 4. Do numbers or bullets divide steps into manageable chunks?

__ 5. Are boldface or italics used to emphasize important information?

Figure 12.10 (pages 384–391) is a sample user manual with screen captures to highlight procedures.

Figure 12.11 (pages 392–402) is a second sample user manual.

How to Place Graphics in Your Correspondence

Figure 12.10   User Manual with Screen Captures
Screen shots reprinted by permission from Microsoft Corporation and George Butler Associates, Inc.

Hazards

⚠	**CAUTION**

Internet graphics are protected by copyright laws.

You can be assessed actual or statutory damages if you infringe upon a Web site's copyrights.

Alter the image or seek approval before using an existing graphic.

Introduction

All correspondence benefits from graphics (pie charts, bar charts, photographs, line drawings, tables, etc.).

Graphics enhance your communication, making it more visually appealing. The best source for graphics might be the Internet. Here you can find literally millions of images.

This manual will help you
- *Download* graphics from the Internet
- *Insert* graphics into your correspondence
- *Move* graphics within your correspondence

Figure 12.10   User Manual with Screen Captures (continued)

Table of Contents

Glossary

BMP	Bitmap image
pop-up menu	Menu options that "pop up" when you right-click
GIF	Graphics Interchange Format
JPEG	Joint Photographic Experts Group

ii

Figure 12.10   User Manual with Screen Captures (continued)

1.0 Downloading Graphics from the Internet

1.1 **Right-click** on the graphic.
1.2 **Scroll** down to Save Picture As and **left-click**.

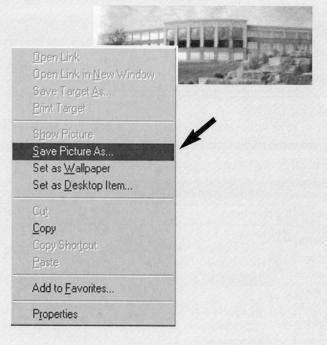

Figure 12.10   User Manual with Screen Captures (continued)

1.3 Left-click on the Save in down arrow to select the file of your choice (A: drive, C: drive, etc.).

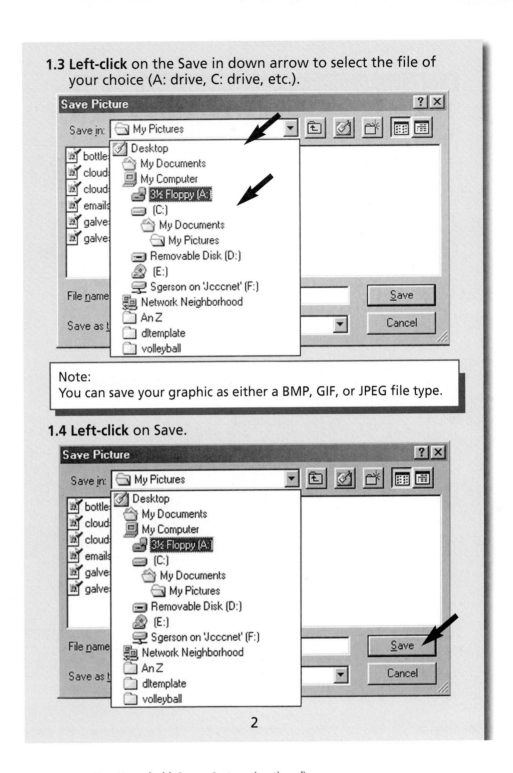

Note:
You can save your graphic as either a BMP, GIF, or JPEG file type.

1.4 Left-click on Save.

Figure 12.10   User Manual with Screen Captures (continued)

2.0 Inserting Graphics in Your Correspondence

2.1 Place your cursor where you want the graphic to appear.

2.2 Left-click on Insert, scroll to Picture, and choose From File.

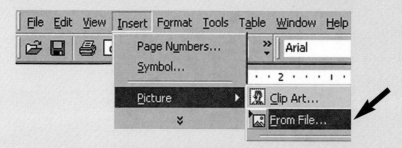

2.3 Left-click on the Look in down arrow to select the file of your choice (A: drive, C: drive, etc.).

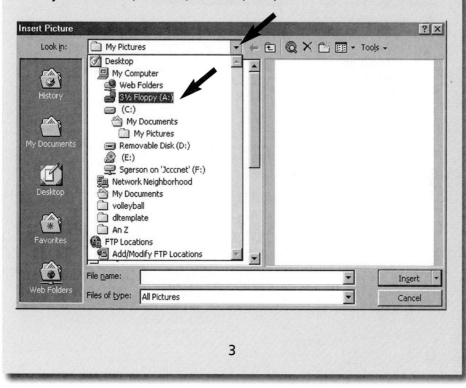

3

Figure 12.10   User Manual with Screen Captures (continued)

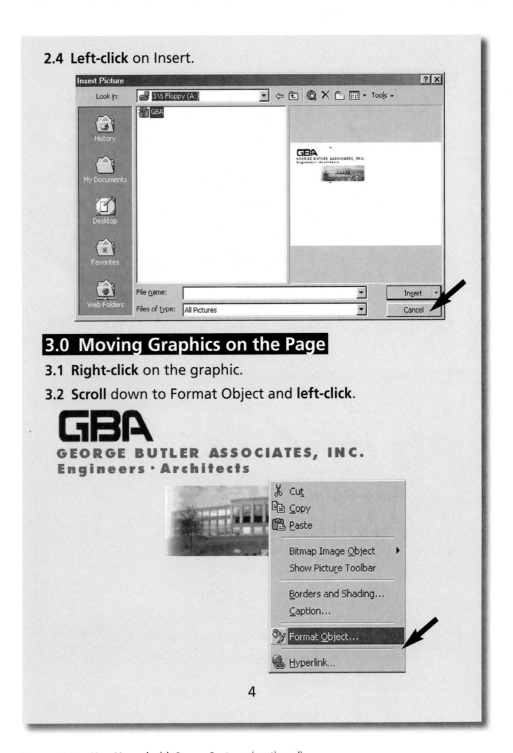

2.4 Left-click on Insert.

3.0 Moving Graphics on the Page

3.1 Right-click on the graphic.

3.2 Scroll down to Format Object and **left-click**.

4

Figure 12.10   User Manual with Screen Captures (continued)

3.3 Left-click on Layout, Behind text, and OK.

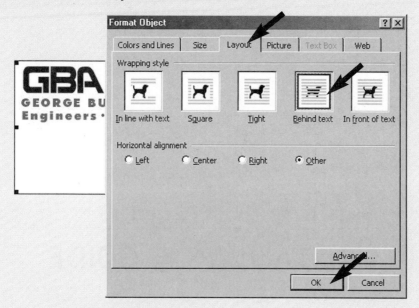

Troubleshooting Guide

Problem	Solution
I can't find the graphic that I've saved.	Be sure to look in the correct file. Check the desktop, your A: drive, or your C: drive.
I can't get a pop-up menu to appear when I click on the graphic.	Be sure that you **right-clicked** on the graphic instead of left-clicking.
My graphic is too large to fit on the page.	To learn how to resize your graphic, call *1-800-graphichelp*.

5

Figure 12.10   User Manual with Screen Captures (continued)

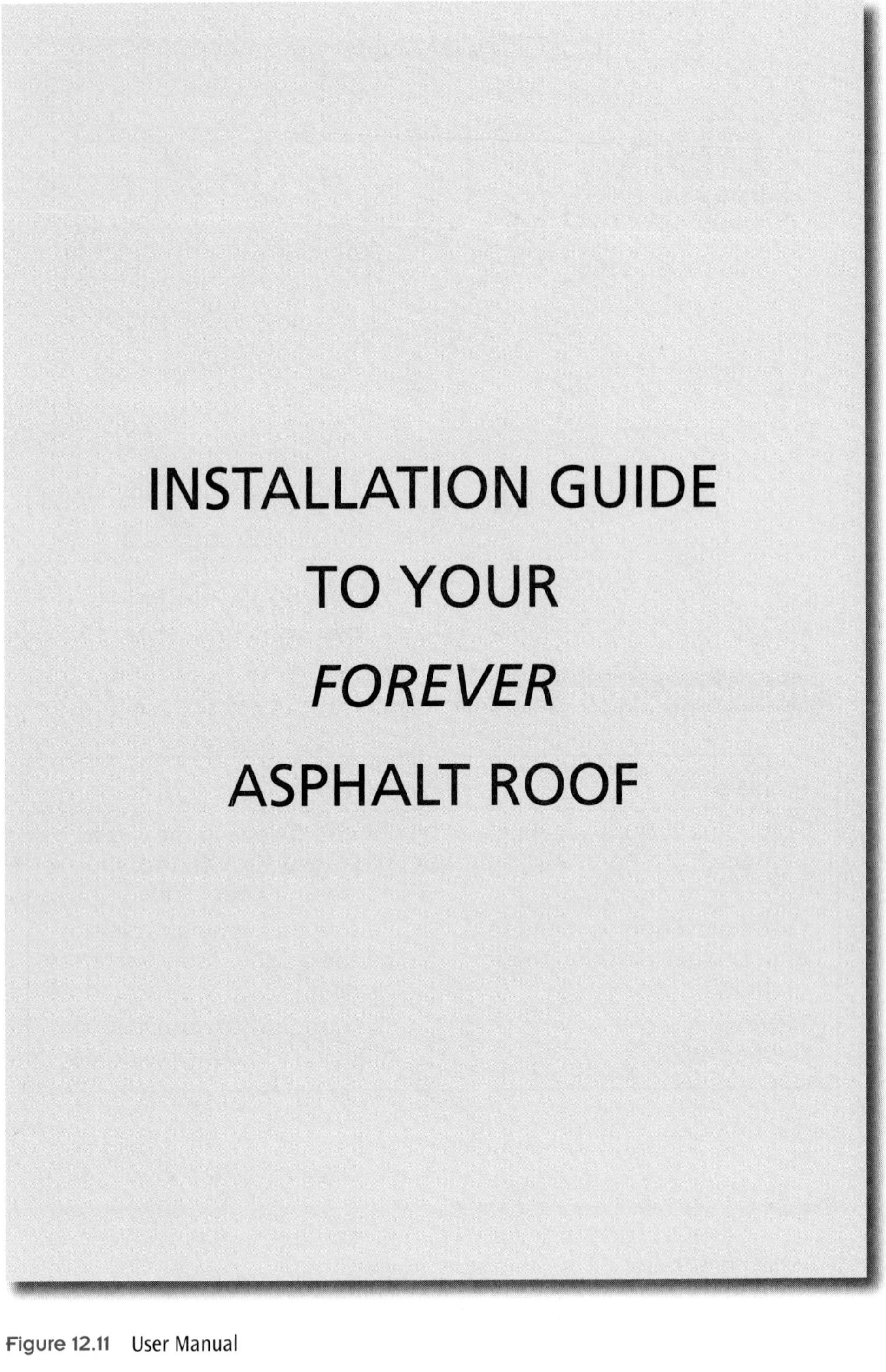

INSTALLATION GUIDE
TO YOUR
FOREVER
ASPHALT ROOF

Figure 12.11   User Manual

IMPORTANT SAFETY TIPS

 Please read the entire manual before starting installation of *FOREVER* Asphalt Shingles.

1. DO NOT touch electric lines above your house.

2. Use care in climbing your ladder.
   • Rest ladders on steady ground.
   • Lean ladders against a wall.
   • Allow only one person on a ladder.

3. On steep roofs, use a scaffolding system.

4. Stay off your roof during strong winds or thunderstorms.

2

Figure 12.11   User Manual (continued)

TABLE OF CONTENTS

3

Figure 12.11   User Manual (continued)

TOOLS REQUIRED FOR INSTALLATION AND SPECIFICATIONS

TOOLS	SPECIFICATIONS
Ladder	Appropriate size
Hammer	Claw-type, smooth face
Utility Knife	6" retractable safety blade
Tape Measure	10' minimum
Chalk Line	Light-colored chalk
Shovel	Flat-end, long-handled tear-off shovel
Underlayment	#15 asphalt-saturated roofing felt One roll covers 200 sq ft of roof.
Three-tab *FOREVER* Asphalt Shingles	One square of shingles covers 100 sq ft of roof. Add 10% for capping, borders, and waste.
Nails	2 1/2 lb, 11- or 12-gauge, 3/8" diameter hot-galvanized nails will cover 1 square of shingles.

4

Figure 12.11 User Manual (continued)

DEFINITION OF TERMS

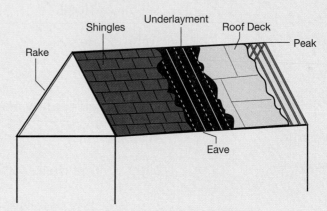

Figure 1. Parts of the Roof

- Rake—The vertical edge or side of the roof
- Shingles—Asphalt coverings for your roof
- Underlayment—Saturated felt to protect your roof deck from rot and mildew
- Peak—The upper horizontal edge of the roof
- Roof deck—The plywood base that covers your rafters
- Eave—The lower horizontal edge of the roof
- Tab—The cutout side of the shingle

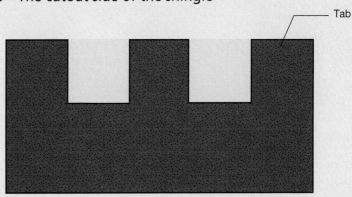

Figure 2. Shingle

5

Figure 12.11   User Manual (continued)

INTRODUCTION:
Benefits of Choosing
FOREVER
Asphalt Shingles

Congratulations! You have made an outstanding decision by choosing to install *FOREVER* Asphalt Shingles. Our shingles are superior to other types of roofing materials for the following reasons:

TABLE 1. BENEFITS OF ASPHALT SHINGLES

Item	Asphalt	Wood	Slate	Ceramic
Inexpensive	X	X		
Beautiful	X	X	X	X
Durable	X		X	
Easy to install	X			

6

Figure 12.11   User Manual (continued)

PREPARING YOUR ROOF DECK

Before you can install your new *FOREVER* Asphalt Shingles, you need to prepare your roof deck. To do so, follow these steps:

TOOLS REQUIRED
 Hammer
 Shovel

1. Use the hammer to pry up old shingles, the old underlayment, and old nails.
2. Scoop up and remove the roof debris with the shovel.
3. After removing the old roofing material, check the roof deck for sagging, rotting wood.

> NOTE: If you find any sagging or rotting wood, you will need to replace those sections with new plywood. Please call 1-800-FOREVER for information about our quality plywood products.

7

Figure 12.11   User Manual (continued)

LAYING AND MARKING THE UNDERLAYMENT

Once you have prepared the roof deck, next you need to install the underlayment. To do so, follow these steps:

TOOLS/EQUIPMENT REQUIRED
  Underlayment
  Chalk

> NOTE: Underlayment roofing felt comes in rolls 4' wide X 50' long. Each roll covers 200 sq ft. You will need half as many rolls as squares of shingles.

1. Install the underlayment by rolling it out perpendicular to the peak and the eave.

> NOTE: Be sure to overlap each piece of underlayment by 2" to provide weatherproofing coverage.

2. Use the chalk to put horizontal guidelines on the underlayment.
3. Make the first line 12" from the eave and parallel to it.
4. Mark subsequent lines every 5" to the roof peak (see Figure 3, "Marking the Underlayment").

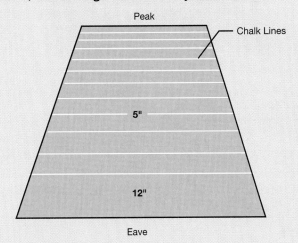

Figure 3. Marking the Underlayment

8

Figure 12.11   User Manual (continued)

INSTALLING YOUR *FOREVER* ASPHALT SHINGLES

Starter Course
The starter course is the first row of shingles installed. This row is closest to the eave. To lay this starter course, follow these steps:

EQUIPMENT/TOOLS REQUIRED
  Hammer
  Nails
  Shingles

1. Starting at the left rake, lay the shingles side by side on the roof with the tabs pointing up to the roof peak.
2. Nail the shingles to the roof using three nails per shingle (see Figure 4, "Starter Course," for nail placement).

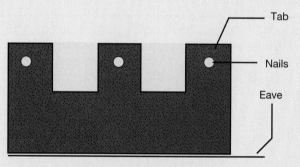

Figure 4. Starter Course

First Course

> NOTE: This course and all subsequent courses will be laid with the <u>tabs pointing down</u> to the eave, not up to the peak.

1. Beginning at the left rake eave, lay a <u>full</u> shingle directly over the starter course shingle.
2. Nail this shingle to the starter course using three nails per shingle.
3. Continue to the right two to three full shingles.

9

Figure 12.11   User Manual (continued)

Second and Other Courses

1. Cut off a tab on the left side of a shingle.
2. Overlapping the starter course halfway, place this shingle over the first shingle in the first course.
3. Nail the shingle to the first course using two nails per shingle.
4. Alternating full shingles with shingles missing a tab, install all subsequent courses as noted above. (See Figure 5, "Shingle Placement for Second and Other Courses.")

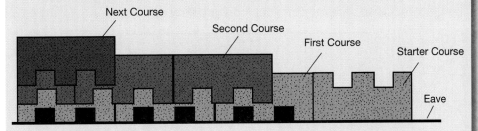

Figure 5. Shingle Placement for Second and Other Courses

10

Figure 12.11   User Manual (continued)

<u>WARRANTY</u>

FOREVER Asphalt Shingles are warranted to give normal service for 20 years or as long as you own your home. We will replace any shingle that cracks, chips, flakes, or peels under normal conditions. As with all roofing material, some nonwarranted damage may occur in severe hail or windstorms or if the roof is not properly installed. To receive warranty service, save your dated sales receipt and proofs of purchase from each square of shingles you purchase.

For tips on installation, call our 24-hour service hotline at
1-800-FOREVER

FOREVER Asphalt Shingles
8942 Hancock Center
Clearwater, FL 40043

11

Figure 12.11   User Manual (continued)

Chapter Highlights

1. To organize your instruction effectively, include an introduction, a discussion of sequenced steps, and a conclusion.

2. Be sure to number your steps and start each step with a verb.

3. Hazard alerts are important to protect your reader and your company. Place these alerts early in your instruction or before the appropriate step.

4. In the instruction, follow a chronological sequence.

5. Your readers do not know how to perform the procedure; that's why they are reading the instruction. Therefore, be precise.

6. Graphics will help your reader see how to perform each step.

7. Highlighting techniques emphasize important points, thus minimizing damage to equipment or injury to its users.

8. Don't compress several steps into one excessively long and demanding step.

9. The writing process, complete with usability testing, will help you construct an effective instruction.

10. In the introduction to a user manual, you can promote good customer–company relations by using personal pronouns and positive words.

Exercises

1. Write an instruction. To do so, first select a topic. You can write an instruction telling how to monitor, repair, test, package, plant, clean, operate, manage, open, shut, set up, maintain, troubleshoot, install, use software, and so on. Choose a topic from within your field of expertise or one that interests you. Follow the writing process techniques to complete your instruction. Prewrite (using a flowchart), write a draft (abiding by the criteria for instructions presented in this chapter), and rewrite to perfect your text.

2. Write a user manual. To do so, first select a topic. For example, you could write a user manual that explains how to operate a portable CD player, a cassette tape player, or a combination radio-alarm clock. The user manual could focus on the installation and operation of any one of the following computer accessories: printer, modem, CD-ROM unit, or software package. In addition, TV sets, VCRs, ceiling fans, wall telephone units, or classroom overhead projectors could serve as the topic of a user manual that explains installation and operation. Finally, a user manual could explain the installation, operation, and maintenance for a coffeemaker, a kitchen blender, a stereo system with CD player, a tape cassette, or a VCR. Search your home, garage, or workplace for items about which you could write a user manual.

3. Find examples of instructions or user manuals for consumer products. These can include instructions for assembling children's toys, refinishing furniture, insulating attics or windows, setting up stereo systems, flushing out a radiator, installing ceiling fans, and so on. Try rummaging around in your kitchen pantry for examples of

ACTIVITIES

instructions—look at boxes and cans of food that tell you how to bake brownies or cook spaghetti. Instructions are everywhere.

Once you find some examples, bring them to class. Then, applying the criteria for good instructions presented in this chapter, determine the success of the examples. If they are successful, explain why and how. If they fail, show where the problems are and rewrite the instructions to improve them.

4. Find examples of instructions or user manuals written in your work environment. Bring these to class. Using the criteria for instructions presented in this chapter, decide whether the instructions are successful or unsuccessful. If the instructions are good, show how and why. If they are flawed, explain the problem(s). Then rewrite the instructions to improve them.

5. Find an example of a successful instruction or user manual, either from your work environment or home (a consumer product). To make this instruction even more effective, construct a flowchart that will graphically enhance the writing and make the instruction easier to follow.

6. Find an example of a successful instruction or user manual, either from your work environment or home (a consumer product). To make this instruction even more effective, draw super comic book pictures to accompany the text. These drawings will graphically enhance the writing, making it easier for your reader to follow the instructions.

7. Good writing demands revision. Following is a flawed instruction. To improve it, rewrite the text, abiding by the criteria for instructions and the rewriting techniques included in this chapter.

> DATE: November 30, 2005
> TO: Maintenance Technicians
> FROM: Second Shift Supervisor
> SUBJECT: OVEN CLEANING
>
> The convection ovens in kiln room 33 need extensive cleaning. This would consist of vacuuming and wiping all walls, door, roofs, and floors. All vents and dampers need to be removed and a tack cloth used to remove loose dust and dirt. Also, all filters need replacing. I am requesting this because when wet parts are placed in the ovens to cure the paint, loose particles of dust and dirt are blown onto the parts, which causes extensive rework. I would like this done twice a week to ensure cleanliness of product.

8. One of the keys to success in writing instructions is audience recognition. One reason flawed instructions fail is lack of audience recognition. For example, an instruction uses high-tech terms but is intended for low-tech readers.

Find an example of an instruction or user manual geared to a high-tech audience that correctly uses high-tech terms. Then find an example of an instruction geared to low-tech readers that incorrectly uses high-tech words. Next, find an instruction geared to low-tech or lay readers that has correctly defined its terms and explained its technology. Finally, find an example of an instruction that talks down to the reader, using low-tech or lay terms when high-tech words or phrases would have been better.

9. Find examples of instructions or user manuals that do not provide effective hazard alerts. Using the criteria for effective hazard alerts provided in this chapter, decide where the alerts are flawed. Then improve the instructions or user manuals by creating effective hazard alerts.

10. Microsoft Word's Help program provides instructional steps for hundreds of word processing operations. Access the Word Help Index to find instructions. Open 3–5 of these instructions and accomplish the following:
    a. discuss how they are similar to or different from the instructions discussed in this chapter.
    b. select one of the instructions and rewrite it, adding graphics (screen captures), cautions or notes, and a glossary of terms.

WEB WORKSHOP

Review any of the following Web site's online instructions. Based on the criteria provided in this chapter, are the instructions successful or not?

- If the answer is yes, explain why and how the instructions succeed.
- If the answer is no, explain why the instructions fail.
- Rewrite any of the flawed instructions to improve them.

Web Sites	Topics
http://www.hometips.com/diy.html	Electrical systems, plumbing, kitchen appliances, walls, windows, roofing, and siding
http://www.hammerzone.com	Kitchen projects, tubs, sinks, toilets, showers, and water heaters
http://dmoz.org/Home/Home_Improvement/	Links to step-by-step procedures for painting, welding, soldering, plumbing, walls, windows, and door repair and installation
http://directory.google.com/Top/Home/Home_Improvement/	Links to sites for instructions on decorating, electrical, flooring, furniture, lighting, painting, plumbing, welding, windows, and doors
http://www.quakerstate.com/pages/carcare/oilchange.asp	Instructions for changing oil
http://www.csaa.com/yourcar/takingcareofyourcar	AAA instructions for car care, including general maintenance and checking fluids, hoses, drive belts, electrical systems, and tires
http://www.gateway.com/index.shtml	Access the FAQs for upgrading systems or correcting problems with printers, drivers, monitors, memory, and more

QUIZ QUESTIONS

1. What are five reasons you would include instructions for your audience?

2. What are the three parts of a well-organized instruction?

3. What should you include in the introduction?

4. Why would you use a hazard alert?

5. Where are hazard alerts placed?

6. How can you create an accessible hazard alert?

7. What are the differences among warning, danger, and caution hazard alerts?

8. When is a note used?

9. What colors highlight a warning, a danger, and a caution?

10. What elements are included in a hazard alert?

11. How do you organize instructions?

12. What is a disclaimer?

13. Why should you consider the audience when you write an instruction?

14. What do graphics achieve in instructions?

15. Why should you be concerned with length of sentences in your instructions?

16. In what way should you begin each step in an instruction?

17. How can you approximate "usability testing" of your instructions in a classroom?

18. What are some features of usability testing?

19. Where would you place definitions in a user manual?

20. How can you promote good customer–company relations in the introduction to a user manual?

ELECTRONIC COMMUNICATION

13 Online Help and Web Sites

Writing at Work

Future Promise

Future Promise is a not-for-profit organization geared toward helping at-risk high school students. This agency realizes that to reach its target audience (teens aged 15–18), it needs an Internet presence.

Future Promise's CEO, Brent Searing, has decided to form a cross-functional team to create the agency's Web site. Brent will encourage the team to work collaboratively to determine the Web site's content, its level of interactivity, and its design features. Brent wants the Web site to include

- College scholarship opportunities
- After-school intramural sports programs
- Job-training skills (resume building and interviewing)
- Service learning programs to encourage civic responsibility
- An FAQ page
- Future Promise's 800-hotline (for suicide prevention, STD information, depression, substance abuse, and peer counseling)
- Additional links (for donors, sponsors, educational options, job opportunities, etc.)

To accommodate these Web components, the Future Promise Web team will consist of the agency's accountant, sports and recreation director, public relations manager, counselor, training facilitator, graphic artist, and computer and information systems director. In addition to these Future Promise employees, Brent also has asked two local high school principals, two local high school students, and a representative from the mayor's office to serve on the committee. Future Promise's public relations manager Jeannie Kort will chair the committee.

Jeannie has a big job ahead of her. First, she must coordinate everyone's schedules. The two principals and high school students, for example, can attend meetings only after school hours. Next, she must meet Brent's deadline; he wants the Web site up and running within three months. Jeannie also must manage this diverse team (with varying ages, levels of responsibility, and levels of knowledge).

Another challenge involves hardware and software. Jeannie realizes that not all of her Web readers will have state-of-the-art computers. Therefore, she must ask her Web design team to create a Web site that is accessible on many platforms, even low-end hardware. Thus, though Jeannie wants the site to be colorful, interactive, and entertaining, the site also must load quickly, avoid frames, accommodate many different monitor resolutions, and be both PC-and Mac-friendly.

The task is daunting, but the end product will be invaluable for the city and the city's youth. Jeannie and Brent know that by conveying information about jobs, training, scholarships, and counseling to their end users (at-risk teens), Future Promise can improve the quality of many people's lives. Jeannie's Web design team has an exciting project on its hands.

During the last decade, the written word underwent a quantum metamorphosis, leaping from the printed page into cyberspace. This change affected technical writing significantly. Correspondence, once limited to hard-copy letters, reports, and memos, is now often online as e-mail. Corporate brochures and newsletters, once paperbound, now are online. Product and service manuals, once paperbound, now are online. Resumes are online. Research is online. Technical writing is increasingly electronic. To keep up with this communication revolution, you must go online also. By teaching you how to write successful online help screens and create an effective Web site, this chapter will help you learn how to work in cyberspace.

The Internet—the "Information Superhighway"

Your online help screens and Web sites will be transmitted via the Internet, an intranet, or an extranet. But what do these words mean?

The Internet is a decentralized medium, with very few external controls. To use it, you can choose which Internet service provider (ISP) you wish, and you can access any Internet site you desire. Thus, as the synonym "information superhighway" suggests, the Internet allows the user to access data from various locations almost without restriction. The Internet lets you travel from your computer to information sites worldwide. You can speak to and hear from educators, engineers, lawyers, doctors, businesspeople, hobbyists, or anyone linked to this international communication web, day or night.

How do you find the information you need on the Internet? If you were on a highway looking for a specific site, you might consult a road atlas. To help you find information on the Internet, you can use something resembling a road atlas—the World Wide Web (WWW or Web), along with Web browsers such as Netscape Navigator and Microsoft's Internet Explorer. The World Wide Web supports documents (Web sites) formatted in a special computer language called Hypertext Markup Language (HTML). (See Table 13.1 for Web abbreviations and acronyms.) HTML allows the webmaster (the creator of the Web site) to include graphics, audio, video, and animation as well as hypertext links to other sites and other Web screens.

TABLE 13.1	COMMON WEB SITE/INTERNET ABBREVIATIONS AND ACRONYMS
<A HREF>	Anchor hypertext reference
<BG>	Background
CGI	Common Gateway Interface
.com	commercial business—domain name
.edu	educational institution—domain name
FTP	File Transfer Protocol
.gif	Graphics Interchange Format file extension
.gov	governmental—domain name
HTML	Hypertext Markup Language
HTTP	Hypertext Transfer Protocol
<IMG SRC>	image source
.jpeg, .jpg	Joint Photographic Experts Group file extensions
<LI>	listed item
.mil	military—domain name
MOO	MUD, Object Oriented
MUD	multiuser dungeon
.net	network—domain name
<OL>	ordered list
.org	organization—domain name
SGML	Standard Generalized Markup Language
TCP/IP	Transmission Control Protocol/Internet Protocol
.tif, .tiff	Tagged Image File extensions
<UL>	unordered list
URL	Uniform Resource Locator

THE INTRANET—A COMPANY'S INTERNAL WEB

Whereas the Internet opens the entire world of information to us, an intranet is more limited. It is a contained network. An intranet might include several, linked local area networks (LANs), and it might include connections to the Internet. Like the Internet, an intranet uses Transmission Control Protocol/Internet Protocol (TCP/IP), Hypertext Transfer Protocol (HTTP), and other Internet protocols. The difference, however, is that an intranet belongs to a company. Its purpose is "to share company information and computing resources among employees" ("Intranet" 1998). Individuals outside the company cannot access the intranet unless they have corporate authorization; a "firewall" prevents access to an intranet's private network ("Firewall" 1998, 1).

How does a company benefit by using an intranet? Rather than printing hundreds of copies of a document for employees who might be officed in buildings located around a city or around the world, a company can make this information accessible to its employees through an intranet. Companies can place on their corporate intranets

- Web-based discussion forums. These allow employees to post, read, respond to, and then archive corporate e-mail messages.
- Web-based multimedia and kiosks for instructional or informational purposes.

- Online polls. Employees can respond to corporate polls. Then the results can be tabulated, and employees around the company can view the results.
- Company forms. These can be filled out and submitted electronically.
- Policy and procedure manuals. Whereas paper documents take up space, get dusty, and get lost, online intranet manuals save space and are available instantly.
- Employee phone directories. Company employees come and go. Online, additions and deletions can be updated easily.
- Organizational charts. Company hierarchies change rapidly. Online, they can be updated and conveyed to employees immediately.

The benefits of an intranet are numerous. An intranet is paper free. Companies can lower mailing, binding, and printing costs. Information can be updated easily. Data can be archived without taking up space. Procedures, polls, directories, forums, and forms are at an employee's fingertips rather than buried under paper or on the top shelves of inaccessible bookshelves. Most importantly, every corporate employee throughout a company—whether in the office or off-site—can access corporate information online at the touch of a key. Speed, efficiency, and cost savings—these are the benefits of a corporate intranet ("Intranet Applications" 1998).

THE EXTRANET—A WEB WITHIN A WEB

We have seen that the Internet gives us worldwide access to online information, and we have explained that an intranet gives employees within a company access to corporate information. What's an extranet? An extranet is "a collaborative network" that uses Internet protocols to "link businesses with their suppliers, customers, or other businesses" that have common goals. An extranet thus allows several companies to share information but also to keep this information within the confines of their collaborative unit. Extranets can include Web-based training programs of mutual benefit to the participating companies, product catalogs, private newsgroups relating to a specific industry, and management approaches for companies working on a shared project ("Extranet" 1998).

ONLINE VERSUS PAPER—WHAT'S THE DIFFERENCE?

Online communication (online help screens and Web sites) represents an entirely different mode of communication. Not only do online help screens and Web sites differ from paper text, but also the "e-reader" approaches online communication differently than he or she will tackle hard-copy text.

What are the differences between paper text and online communication?

Characteristics of the Online E-reader

Topic Specific

In libraries or book stores, we wander up and down aisles, looking for any book that interests us. In contrast, e-readers tend to access the Web or online help screens with specific goals in mind. You go online to search for specific information found in specific Web sites: CD prices, automobile loan rates, hotel room availability, restaurant menus, technical specifications for laser printers, the start date for your college's spring semester, and so forth. Similarly, when you open an online help screen, you want to know exactly how to perform a specific task: how to fill out a form, file taxes, reconfigure your e-mail options, change margins on your word processing page, and so on (Moore 2003, 16–17). E-readers are goal oriented online.

Speed

Because e-readers have a specific goal when they go online, they want that information quickly. Whereas those who read novels happily will linger over a long work of fiction for days, weeks, or months, e-readers often scan, skim, and skip over text, looking just for the information they want and ignoring the rest of the text.

In fact, readers want to find information in "ten seconds. Your web site visitors love skim-readable pages. They're not lazy. They're just in a hurry. They don't want to waste time and money reading the wrong web page, when there are millions of other pages to choose from" (McAlpine 2002).

As Roger Black says, "Faster beats fancier. If visitors must wait a minute to download art, they'll leave your site and never come back" ("Web Sites That Work" 2004).

Finally, e-readers rarely revisit Web sites. "In most cases, the e-reader gets the information all in one shot or not at all" (Moore 2003, 16).

Note:
E-readers want to find information in ten seconds.
Source: From McAlpine 2002.

Electronic Platforms

Another difference between an e-reader and the same person who reads a hard-copy novel is the way in which he or she accesses the document. Most hard-copy text is printed on either book-sized pages or on $8\frac{1}{2} \times 11$-inch sheets of paper. In contrast, e-readers can access online help screens and Web sites "using cell phones, PDAs, and other wireless devices" (Moore 2003, 17). Thus, screen resolution and the size of type font are key elements for online readability. Furthermore, e-readers will be using Netscape, Internet Explorer, AOL, Macs and PCs, and a host of other electronic platforms to view your text. Each of these electronic platforms differs in subtle ways. To construct effective online help screens and Web sites, you must consider how your reader will access your document.

Languages

Remember that *WWW* stands for "World Wide Web." The Internet is international, and global access has grown tremendously. Look at China, for example. Individuals with Internet access grew from approximately 2 million users in 1999 to over 25 million by 2003 (St. Amant 2003, 15). Though English is "commonly used throughout the world, not all online users speak or read English" (St. Amant 2003). Thus, your e-reader is multilingual and multicultural.

Characteristics of Online Communication

Because e-readers are unique, you must alter the way you write online communication. This means changing your mindset as a writer. When you write an online help screen or Web site, you must reconsider

- **The Screen Versus the Page**—Forget $8\frac{1}{2} \times 11$. Online help screens are smaller, and good Web screens rarely require scrolling.

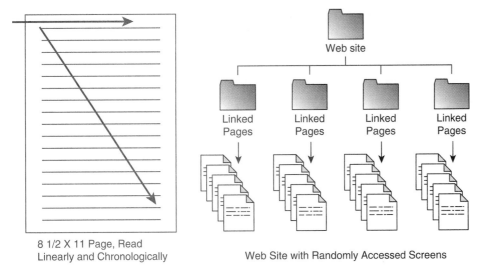

8 1/2 X 11 Page, Read
Linearly and Chronologically

Web Site with Randomly Accessed Screens

Figure 13.1   How We Read Books Compared to How We Read Web Sites

- **Skimming Versus Linear Reading**—We read books "linearly," line by line. In contrast, online help screens and Web screens are skimmed and scanned.
- **Hypertext Links Versus Chronological Reading**—We read books from beginning to end, sequentially. Web sites, however, allow us, even encourage us, to leap randomly from screen to screen or from Web site to Web site (see Figure 13.1).

Page Layout

Most hard-copy text is $8\frac{1}{2} \times 11$ inches, with 1-inch margins (top, sides, and bottom). When you get to the bottom of the page, you turn to the next page. In contrast, the size of online help screens and Web sites varies.

Page Length   On-screen, less is best. When writing online, "you must be aware that computer screens display smaller amounts of information than a printed page" (Hemmi 2002, 11). A successful Web page should limit pertinent information to one screen, approximately 6 to 7 inches long, without requiring the reader to scroll. This equates to approximately 20–22 lines of text, versus the 50–55 lines that can fit on an $8\frac{1}{2} \times 11$-inch sheet of paper.

Online help screens are even smaller, rarely filling an entire page. They are more likely the size of 3×5-inch index cards.

Thus, the "material that fits on one printed page might require three to six screens online" (Hemmi 2002, 11).

Margins   Hard-copy text, when read from margin to margin, is approximately 80 characters long (a character is every letter, punctuation mark, and space). The size and type of your font, of course, affects the number of characters you will type per line. Online, horizontal margins vary from monitor to monitor (21 inches, 19 inches, 17 inches, 15 inches, etc.). On a 17-inch monitor, for example, text runs approximately 12 inches, left margin to right margin. On a 15-inch monitor, text runs approximately 10 inches.

On a Web page, your audience is forced to read more words per line if you use the entire screen. That's difficult. Readers have trouble maintaining their focus (visually and mentally) when expected to read long lines of horizontal text.

Vertically, the problem increases. Regardless of the monitor size, cybertext can run forever. The reader can scroll, and scroll, and scroll. Whereas paper sizes are controlled by convention and printing companies, you must control the size of a Web page.

A successful Web page should limit lines of text to perhaps two-thirds of the screen, with a graphic, white space, or hypertext links placed in the remaining third. This breaks up the monotony of long lines into manageable chunks. Thus, the reader can maintain focus more easily (Eddings et al. 1998).

Font   Your computer's word processing software probably defaults to Times New Roman, 12-point. That is considered the best type and size font for readable hard-copy text. You can enhance your hard-copy text with designer fonts (stylized and cursive), boldface, or underlining for visual effect. In contrast, sans serif fonts, like Arial, seem to work best for online reading. "Stylized, cursive, italicized, and decorative typefaces are difficult to read online" (Hemmi 2002, 11).

Underlining is especially troublesome online, because online, hypertext links are shown as underlined text. To avoid confusing your readers online, you must avoid using underlining as a highlighting technique.

Structure

Paper text is read sequentially. We read from page 1 through the end of the text. Think of a book: you would never consider reading page 102 before you have read page 18 or Chapter 7 before you have read Chapter 4. In fact, you are expected to read sequentially—page after page after page. After all, the book (or newsletter or manual) is bound; each page is physically stitched, glued, or stapled to the next. In contrast, online help screens and Web sites are composed of randomly "stacked" screens. For online help screens, you access a search mechanism, find the topic you want, and click on a link to access it. For a Web site, you scan the home page, and then you decide what you want to read and when you want to read it. Once you make this decision, you click on hypertext buttons, access any screen, and read content in any order. Thus, you receive "just in time" or "as needed" information. That is, if you want to access the linked text, you can. If you don't want to read this screen, you don't have to. Again, on-screen, less is best for the random reader who plans to skim and scan (Goldenbaum and Calvert 1998). Hard-copy documentation follows a page-by-page format, whereas online text is nonlinear. The structure of online text is modular and flexible (Hemmi 2002, 11).

Noise

Though hard-copy text is rigidly linear, it can be easier to read than online text. Paper is dull and absorbs light. Most hard-copy text is colorless and motionless; a book is composed of black-and-white sentences. For online text, however, the screen is composed of glass, which reflects light, creating visual glare. Many of our screens are smudged where we have used our fingertips to point out interesting images.

To help e-readers, you need to consider a key challenge of reading online: computer "noise"—sound and visual distractions. Extended viewing of a computer screen is more demanding than continued reading of paper text. Web sites often contain lots of color, blinking text, animated graphics, frames of layered text, and sound and video. Noise—multiple distractions—inundates the screen. In such a busy communication vehicle as a Web site, the less we give the reader on one screen, the better.

ONLINE HELP

Imagine this possibility. Your boss calls you into a meeting and says, "As of today, we are no longer producing paper documents" (McGowan 2000, 23). It's not as farfetched as it sounds. Online Web help is becoming a major part of the technical writer's everyday job for several reasons, including

- The increased use of computers in business, industry, education, and the home
- The reduced dependence on hard-copy manuals by consumers
- The need for readily available online assistance
- Proof that people learn more effectively from online tutorials than from printed manuals (Pratt 1998, 33)

What is an online help system? Online help systems, which employ computer software to help users complete a task, include procedures, reference information, wizards, and indexes (see Figure 13.2). Typical online help navigation provides readers hypertext links, tables of contents, and full-text search mechanisms (Zubak 2001, 9).

Help menus on your computer are excellent examples of online help systems. As the computer user, you pull down a help menu, search a help list, and click on the topic of your choice, revealed as a *hot button* (discussed under "Web Sites"). When you select your topic, you could get a *pop-up* (a small window superimposed on your text), or your computer might *link* to another full-sized screen layered over your text. In either instance, the pop-up or hyperlink gives more information about your topic (Henselmann 1997, 475; Pratt 1998, 34–36):

- **Overviews**—explanations of why a procedure is required and what outcomes are expected.
- **Processes**—discussions of how something works.
- **Definitions**—an online glossary of terms.
- **Procedures**—step-by-step instructions for completing a task.
- **Examples**—feedback verifying the completion of a task or graphic depictions of a completed task. These could include a screen capture or a description with call-outs.
- **Cross-references**—hypertext links to additional information.
- **Tutorials**—opportunities to practice online.

Online help systems allow the technical writer to create interactive training tools and informational booths within a document. These online systems can be created using a wide variety of authoring tools. Some popular ones include *RoboHELP, HelpBreeze, ForeHelp, Help Magician, Visual Help,* and *WYSI–Help* (Zubak 1996, 11). The value of online help is immense. Because online help offers "just in time" learning or "as needed" information, readers can progress at their own pace while learning a program or performing a task.

Techniques for Writing Effective Online Help

To create effective online help screens, consider these suggestions:

Organize Your Information for Easy Navigation

Poorly organized screens lead to readers who are "lost in cyberspace." Either they cannot find the information they need, or they have accessed so many hypertext links that they are five or six screens deep into text.

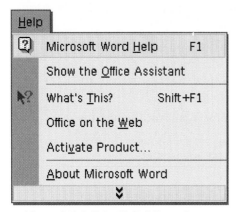

Microsoft Help Tab with Pull-Down Screen

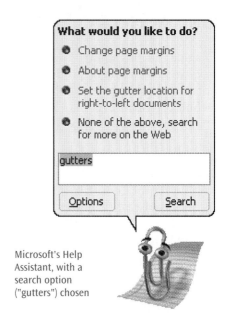

Microsoft's Help
Assistant, with a
search option
("gutters") chosen

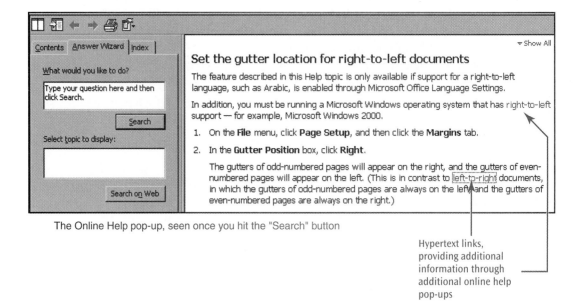

The Online Help pop-up, seen once you hit the "Search" button

Hypertext links,
providing additional
information through
additional online help
pop-ups

Figure 13.2   Online Help Systems

You can avoid such problems and help your readers access information in various ways:

- Allow users to record a history of the screens they've accessed through a bookmark or a Help Topics pull-down menu (Stevens 1997, 411).

- Provide an online Contents menu, allowing users to access other, cross-referenced help screens within the system (Stevens 1997, 411).

- Provide a Back button or a Home button to allow the readers to return to a previous screen (Goldenbaum and Calvert 1998).

A good test is the "three clicks rule." Readers should not have to access more than three screens to find the answer they need. Similarly, readers should not have to backtrack more than three screens to return to their place in the original text (Timpone 1997).

Recognize Your Audience

Online help must be user oriented. After all, the only goal of online help is to help the user complete a task. Thus, a successful online help system must be "designed at the same level of detail as the user's knowledge and experience" (Wagner 1997). This means that technical writers have to determine a user's level of knowledge.

If the system is transmitted by way of an intranet or extranet, you might be tempted to write at a high- or low-tech level. Your readers, you assume, will work within a defined industry and possess a certain level of knowledge. That, of course, is probably a false assumption. Even within a specific industry or within a specific company, you will have coworkers with widely diverse backgrounds: accountants, engineers, data processors, salespeople, human resource employees, management, technicians, and so on.

Don't assume. Find out what information your readers need. You can accomplish this goal through usability testing (discussed in Chapter 2), focus groups, brainstorming sessions, surveys (discussed in Chapter 17), and your company's hotline help desk logs (Timpone 1997). Then don't scrimp on the information you provide. Don't just provide the basic or the obvious. Provide more detailed information and numerous pop-ups or links. Pop-up definitions are an especially effective tool for helping a diverse audience (see Chapter 4 for help with definitions). Remember, with online help screens, readers who don't need the information can skip it. Those readers who do need the additional information will appreciate your efforts. They will more successfully complete the tasks, and your help desk will receive fewer calls.

Achieve a Positive, Personalized Tone

Users want to be encouraged, especially if they are trying to accomplish a difficult task. Thus, your help screens should be constructive, not critical. Your text should be "written in the affirmative," a concept supported by human-computer interaction (HCI) concerns (Wagner 1997). HCI-driven online help systems "coach" the users rather than "command" them ("Human Computer Interaction" 1998, 2). The messages also should be personalized, including pronouns to involve the reader.

Excellent examples of this positive, personalized tone can be seen in Microsoft's Office Assistants, found in Microsoft Word, Excel, and Outlook. This online help system allows you to choose your help wizard from several options. For example, you could choose an Einstein caricature called the Genius, who says, "Hello. Can I assist you with your work in electronic space?" Another wizard, Mother Nature, "provides gentle help and guidance." A third option, the Dot, introduces itself by stating, "When you need help of any kind, just give me a click."

Design Your Document

How will your help screens look? Document design is important because it helps your readers access the information they need. (See Figure 13.3.) To achieve an effective document design, consider these points:

- **Use color sparingly**—Color causes several problems in online documents. Bright colors and too many colors strain your reader's eyes. Furthermore, a color that looks good on a high-resolution monitor

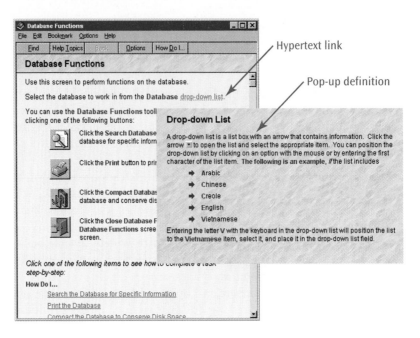

Hypertext link

Pop-up definition

Figure 13.3   Help Screen
Courtesy of Earl Eddings, technical communications specialist.

might be difficult to read on a monitor with poor resolution. Your primary goal is contrast. To help your audience read your text, you want to maximize the contrast between the text color and the background color (Goldenbaum and Calvert 1998). Black text on a white background offers the optimum contrast. Many help screens, such as those provided by Microsoft Office and our sample, use black text on a pale yellow background. This combination provides contrast and differentiates the help screens from the running text. WordPerfect's help system also uses black text on a pale yellow background. Once you select a topic for assistance, however, the pop-up is shown with black text on a white background; headings are blue, and hot buttons are green. In all instances, contrast is achieved.

- **Be consistent**—Pick a color scheme and stick with it. Your headings should be consistent, along with your word usage, tone, placement of help screen links and pop-ups, graphics, wizards, and icons (Henselmann 1997, 475). Readers expect to find things in the same place each time they look. If your help screens are inconsistent, readers will be confused.

- **Use a 10-point sans serif font**—A 12-point type size is standard for most printed documents, but 10-point type will save you valuable space online. Serif fonts are the standard for most technical writing. (See our discussion of typeface and type size in Chapter 8.) A serif font, with small, horizontal "feet" at the bottom of each letter, helps guide the eye while reading printed text. However, on lower resolution monitors, serif text is more difficult to read.

- **Use white space**—Don't clutter your help screens. "Avoid excessive emphasis techniques" (Stevens 1997, 410). Minimize your reader's overload by adding ample horizontal and vertical white space. Online, less is best.

Be Concise

"It takes 20 to 30 percent longer to read on-screen than in print, so you must minimize text" (Timpone 1997). Furthermore, online space is limited. Chapter 3 provides you many techniques for achieving conciseness. In addition to limiting word and sentence length, a help screen should avoid horizontal and vertical scrolling (Stevens 1997, 410). Each screen should include one "self-contained" message (Timpone 1997).

Be Clear

Your audience reads the help screen only to learn how to perform a task. Thus, your only job is to meet the reader's needs—clearly. To accomplish this goal, you need to be specific. (See Chapter 3 for suggestions.) In addition, clarity online could include the following:

- Tutorials to guide the reader through a task. Microsoft gives its readers a "How?" button to meet this need. WordPerfect's help system offers "How Do I?" links to show step-by-step procedures.

- Graphics that depict the end result. Microsoft gives its readers a "Show me" button followed by screen captures. WordPerfect's help system provides an "Examples" link with pictures of what you can do and how to do it.

- Cross-references. WordPerfect provides an "Additional Help" link, complete with a glossary, an index, customer support, a Contents button, and a speech recognition system (available on CD-ROM). Microsoft programs also provide a help index and a table of contents.

- Pop-up definitions.

Provide Access on Multiple Platforms

You will want every reader to be able to access your online help, not just those using a PC or a Mac or UNIX. After all, if the reader can't access your online help, how have you helped him or her? (The only exception would be if you were creating online help for an intranet or extranet over which your company had complete control.) To ensure that all readers can access your system,

- Do not use platform-specific technologies, such as ActiveX, which only runs on 32-bit Windows.

- Create content that can be viewed on multiple browsers, including Internet-Exchange and Navigator.

- Create your online help for the oldest version that you will support— the lowest common denominator. Remember, all of your users won't have the latest version browser, software, or hardware.

- Test your online help on at least three different browsers, versions, and platforms (Zubak 1996, 10).

Correct Your Grammar

As in all technical writing, incorrect grammar online leads to two negative results: a lack of clarity and a lack of professionalism. Don't embarrass your company or confuse your reader with grammatical errors. Proofread.

WEB SITES

Online help screens are short and have limited focuses. Web sites, in contrast, can convey any amount of information. Web sites are created by companies, organizations, schools, and government agencies, not to mention individuals. That's why a Web site's URL (Uniform Resource Locator—the Web site's address) reads *.com* (commercial), *.org* (organization), *.edu* (education), *.gov* (government), and so on.

The Democratic National Party has a Web site. So do McDonald's, Ford Motor Company, the city of Las Vegas, the University of Texas, and Greenpeace. There are hundreds of Web sites dedicated to Elvis Presley. Whatever topic you are interested in researching, you can find it online in someone's Web site. In December 1996, the Internet search engine Yahoo listed 161,068 company Web sites in its Business and Economy link. By March 1999, this number had almost tripled to 431,034 companies with Web sites listed in Yahoo's Business and Economy page. As of January 2004, Yahoo listed 729,142 business and economy Web sites. And that number increases every day.

Web sites range in size from one online screen—called a *home page*—to numerous screens linked by hypertext hot buttons. These hypertext buttons can be highlighted words or phrases that are boldfaced, underlined, or colored to distinguish them from other text. Or the hypertext links can be graphical user interfaces (GUIs), icons that are "hot." When you point your cursor at the highlighted hot button, the pointer turns into a small hand. Then you click on the hot button to jump to another screen.

Criteria for a Successful Web Site

Follow this criteria when creating your site.

Home Page

The home page on a Web site is your welcome mat (Figure 13.4). It is the first thing your reader sees; thus, the home page sets the tone for your site. A successful home page should consist of the following components:

Identification Information   Who are you? What is the name of your company, product, or service? How can viewers get in touch with you if they want to purchase your product? A good home page should clearly name the company, service, or product. Further, a successful home page should provide the reader access to a corporate phone number, an e-mail address, a fax number, an address, and customer service contacts (Berst 1998).

Graphic   Don't just tell the reader who or what you are. Show the reader. An informative, attractive, and appealing graphic depicting your product or service will convey more about your company than mere words will. Check out various Web sites to prove this point. Notice how inviting and family-friendly the McDonald's Web site is, complete with a sparkling golden arch. Pepsi's Web site is also inviting but avant-garde, symbolizing "Generation Next." Your home page's graphic, like a corporate logo, should represent your company's ideals.

Lead-in Introduction   In addition to a graphic, provide a phrase or sentence that tells your reader who and what you do. What's your company's focus? Applied Communications Group uses the phrase "A Software Consulting Firm" on its home page to define itself. At this textbook's writing, the home page of Black & Veatch (an international engineering company) introduced the company to the reader through this heading and follow-up sentence:

Building a World of Difference

Through our engineering, construction, and related activities around the globe, we are challenging the frontiers of knowledge to provide energy, information, water, and infrastructure for a better world. (Black & Veatch)

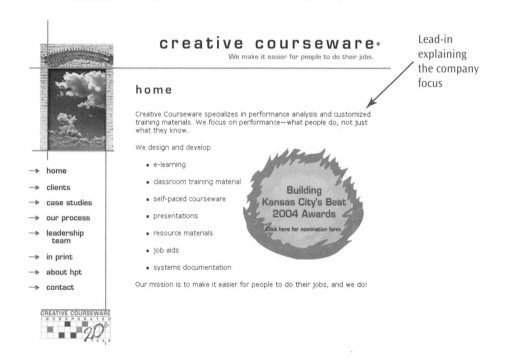

Figure 13.4 Web Site Home Page with Lead-in Introduction
Courtesy of Applied Communications Group, Inc.

Navigation Bar By providing the reader with either hypertext links or graphical user interfaces (GUIs), the home page's navigation bar acts as an interactive table of contents or index. The reader selects the topic he wishes to pursue, clicks on that link, and jumps to a new screen.

There is no one way to present these hypertext link buttons or GUIs on the home page. You could provide a vertical or horizontal navigation bar listing your hypertext links. Graphic links can be presented on the home page in any manner you choose.

Linked Pages

Once your reader clicks on the hypertext links from the home page, he or she will jump to the designated linked pages. These linked pages should contain the following:

Headings and Subheadings The reader clicks on the home page link and arrives at a new screen. Will the reader know where he or she is now? Or will the reader be lost in cyberspace? To ensure that readers know where they are in the context of the Web site, you need to use headings. These give the readers visual reminders of their location. Successful headings on linked pages are in the same location, font size, and font type.

Development As in all correspondence, you need to develop your ideas thoroughly. In Chapter 3, we suggest responding to reporter's questions and specifying. The same applies online. Each linked page will develop a new idea. Prove your points precisely.

tech link
Go to *http://flamingtext.com/* to make customized headings, using hundreds of designer fonts.

Navigation

As mentioned, readers can easily get lost in cyberspace. If readers are viewing one screen but want to return to a screen they looked at before, how do they get there? There are no pages to turn. You need to help the readers navigate through cyberspace. You can do this in two ways:

- **Home buttons**—The reader needs to be able to return to the home page easily from any page of a Web site. Remember, the home page acts as a table of contents or index for all the pages within the site. By returning to the home page, the reader can access any of the other pages. To ensure this easy navigation, you need to provide a hypertext-linked Home button on each page.

- **Links between Web pages**—Why make the reader return to the home page each time he or she wants to access other pages within a site? If each page has a navigation bar with a hypertext link to all pages within the site, then the reader can access any page, in any order of discovery.

Document Design

tech link

For information on Web typography and headings, go to *http://www.microsoft.com/typography/web/designer/css01.htm.*

You can do amazing things on a Web site. You can add distinctive backgrounds, colored fonts, different font faces and sizes, animated graphics, frames, and highlighting techniques (such as lines, icons, and bullets). However, just because you can doesn't mean you should! Document design should enhance your text and promote your product or service, not distract from your message. As mentioned before, on a Web site, less is best.

Background   On Web sites, you can add backgrounds, running the gamut from plain white, to various colors, to an array of patterns: fabrics, marbled textures, simulated paper, wood grains, grass and stone, psychedelic patterns, water and cloud images, and waves. They all look exciting. However, they are not all effective. When choosing a background, you want to consider your corporate image. If you are an engineering company, do you want a pink background? Perhaps a subtle stone or "brick" background would be more effective in portraying your corporate identity. If you are a child-care center, do you want a dark background? Maybe this will be too dour. A white background with toys as a watermark might more successfully convey your center's mission statement.

tech link

Go to *http://www.ender-design.com/rg/* for Web backgrounds, bullets, and horizontal lines.

In addition, remember that someone will attempt to read your Web site's text. Very few font colors or styles are legible against a psychedelic background—or against most patterned backgrounds, for that matter. Instead, to achieve readability, you want the best contrast between text and background. Despite the vast selection of backgrounds at your disposal, the best contrast is still black text on a white background.

Font   How hard can this be? You just use any font type and size you want, right? No. As already noted, it's hard to read text online. Text tends to "flicker," glare impedes, and we get eye strain. Thus, font selection is very important. Consider these suggestions (Williams 2000, 389):

- Use sans serif fonts, like Arial, or one specifically designed for the Web—Verdana. Conventional wisdom has always suggested that serif fonts, like Times New Roman, work best for correspondence. However, online, a serif (the "foot" at the bottom of text) displays irregularly on low-resolution screens.

- Use 12- to 14-point type, the size that most readers with normal vision find easiest to read. A smaller font will cause problems for readers. Then use a larger font for headings.

- Avoid overusing bold and italics. First, these should only be used for emphasis, and then used sparingly. Boldface is often poorly formed on-screen because it must add pixels to a letterform. Italics, due to the oblique orientation, does not always work well with the horizontal and vertical orientation of the pixel grid.

- Avoid using underlining, which readers will confuse with a hypertext link.

- Avoid using all caps. As with e-mail, text typed in all caps makes you look as if you are shouting. In addition, studies show that we read lowercase text about 13 percent faster than we read text in all caps. Uppercasing, thus, will add additional reading problems for your audience, already stressed by the challenges of reading online.

Color   When you create a Web site, you can use any color in the spectrum, but do you want to? What corporate image do you have in mind? The font color should be suitable for your Web site, as well as readable. A yellow font on a blue background will cause headaches for your reader. A red font on a black background is nightmarish. The issue, however, is not just esthetics. A primary concern is contrast. Red, blue, and black font colors on a white background are very legible because their contrast is optimum. Other combinations of color don't offer this contrast; thus, the reader will have trouble deciphering your words (Goldenbaum and Calvert 1998).

In addition, when you use color on your Web site, the color you see on your monitor will *not* necessarily be the color your reader sees. As a webmaster, you cannot solve this problem because you cannot control other people's monitors. All you can do is try to reduce the problem. First, use restraint with color. Stick with black, red, and blue font color for text. Variations such as salmon, lime green, cornflower blue, medium aquamarine, and goldenrod won't be true. Next, test your use of color on several different monitors. You may well be surprised at what you see, but now you can change your font color if needed for clarity (Eddings et al. 1998).

Graphics   Graphics affect download time. For example, a "small" graphic (approximately 90 × 80 pixels, 1½ × 1½ inches) consumes about 2.5 kilobytes. A "medium" graphic (approximately 300 × 200 pixels, 5 × 4 inches) consumes about 25 kilobytes. A "large" graphic (approximately 640 × 480 pixels, 10 × 8 inches) consumes about 900 kilobytes. When you use graphics with varying amounts of color depth, your file size increases. The larger your file, the longer it will take for the images to load. The longer it takes for your file to load, the less interested your reader will be in your Web site. As mentioned already, less is best. Limit the size and number of your graphics.

Furthermore, use suitable graphics. If you work for an accounting firm, do you want graphics depicting a smiling calculator operated by an oversized flamingo? Create and insert your graphics carefully. (We discuss ways in which you can create and download Internet graphics in Chapter 9.)

Highlighting Techniques   Font sizes and styles, lines, icons, bullets, frames, java applets, animation, video, audio—you can do it all on your Web page, but should you? Nothing is more distracting than a Web site full of highlighting techniques. Each design element is like spice. Your Web site is the stew. If you throw in too many different spices, the stew becomes inedible. A little spice goes a long way.

Effective Online Highlighting

1. *Lines* (horizontal rules) can separate headings and subheadings from the text.
2. *Bullets and icons* enliven your text and break up the monotony of wall-to-wall words.
3. *First- and second-level headings,* achieved by changing your font size and style, separate key ideas.
4. *Boldface* also emphasizes important points.
5. *White space,* created by indenting, makes text more readable (Spyridakis 2000, 367; Yeo 1996, 12–14).

Ineffective Online Highlighting

1. *Frames* are considered to be one of the worst highlighting techniques. Jesse Berst, editorial director of ZDNet, considers frames to be one of the "Seven Deadly Web Site Sins." He says, "Too many frames . . . produce a miserable patchwork effect." If you want to achieve the frame look, but without the hassle, use tables instead. They are easier to create and revise, they load faster, and they are less distracting to your readers.
2. *Italics* are not easy to read online.
3. *Underlining* creates problems for e-readers. Remember, hypertext links will usually appear as underlined words or phrases. Thus, underlining for emphasis could mislead your reader.
4. *Java applets* take a long time to load.
5. *Video* requires an add-on users must download before they can enjoy your creation.
6. *Animation* can be very sophomoric and distracting.

Style

1. *Conciseness*—Conciseness is important in all technical correspondence. (We discuss this in great detail in Chapter 3.) Conciseness is even more important in a Web site. As noted earlier in this chapter, a successful Web page should be limited to one viewable screen, and a line of text should rarely exceed two-thirds of the screen.

2. *Personalized tone*—The Internet is a very friendly medium of communication, in contrast to many books and journals. On a Web site, you are allowed, and in fact encouraged, to use pronouns, contractions, and positive words. Create a personalized tone that engages your readers.

Grammar

Perhaps because of the friendly, casual nature of the Internet, many Web sites are poorly proofread. That's a problem. Casualness is one thing; lack of professionalism is another. If your Web site represents your company, you don't want prospective clients to perceive your company as lacking in quality control. In fact, that's what poor grammar online denotes—poor attention to detail.

It's easy to see why many sites are grammatically flawed. First, anyone can go online almost instantaneously. All you do is code in your HTML and transfer the file to your Web server, and you are online. That's great for business, but it's bad for proofreading. Good proofreading takes time, a concept almost antithetical to the Internet. Next, though most word processing packages now have spell checks and grammar checks that tell you immediately when you have made an error, many Web conversion programs do not contain these amenities. Thus, when constructing your Web site, you must be vigilant. The integrity of your Web site demands correct grammar.

Samples

Figures 13.5–13.9 are examples of well-designed Web pages. Notice how the Missouri Department of Transportation's (MoDOT) home page (Figure 13.5) is divided approximately into thirds. The left third contains the navigation bar. The middle third contains a visual to catch the reader's attention. The final third provides a pull-down menu for additional information about individual counties.

In Figure 13.6, MoDOT uses color to aid accessibility. When the cursor hovers over the "Business" link, the color change helps orient the e-reader to that Web page's content. In Figure13.7, the same color is used for the Web page's heading, ensuring that the e-reader does not get lost in "cyberspace." Also note how the "Business" page's text is placed in the right third of the screen. This helps the reader access the text without having to read from margin to margin. MoDOT's state icon in the top left screen location acts as a "Home" button.

Figure 13.8 shows how color is used again on the navigation bar to differentiate one Web page from the others. Finally, Figure 13.9 highlights the value of graphics for visual appeal.

Figure 13.5   Missouri Department of Transportation (MoDOT) Web Site Home Page
Reprinted with the Permission of the Missouri Department of Transportation.

Pull-down
screen
with
color
change

Figure 13.6 MoDOT Home Page with Pull-down Screen and Cursor Pointing at Business Hyperlink
Reprinted with the Permission of the Missouri Department of Transportation.

MoDOT logo
used as a
"Home" button

Banner color
coded to match
heading

Heading
for screen

Graphic
for
visual
appeal

Figure 13.7 MoDOT Business Page
Reprinted with the Permission of the Missouri Department of Transportation.

Figure 13.8 MoDOT Home Page with Pull-down Screen and Cursor Pointing at Other Transportation Hyperlink
Reprinted with the Permission of the Missouri Department of Transportation.

Pull-down
screen with
color change

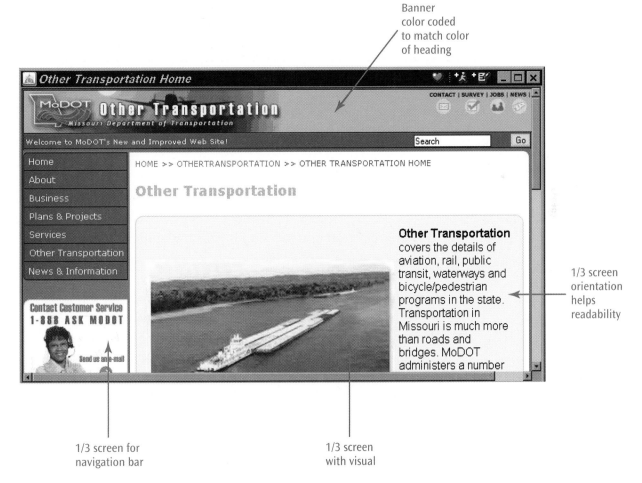

Banner
color coded
to match color
of heading

1/3 screen
orientation
helps
readability

1/3 screen for
navigation bar

1/3 screen
with visual

Figure 13.9 MoDOT Other Transportation Page
Reprinted with the Permission of the Missouri Department of Transportation.

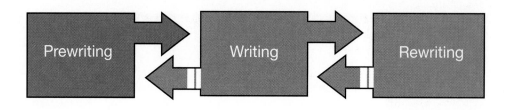

The Writing Process		
Prewriting	Writing	Rewriting

• Examine your purposes
• Determine your goals
• Consider your audience
• Gather your data
• Determine how the content will be provided
	• Organize the draft according to some logical sequence that your readers can follow easily
	• Format the content to allow for ease of access
		• Revise
		  ° Add missing details
		  ° Delete wordiness
		  ° Simplify word usage
		  ° Enhance the tone of your communication
		  ° Reformat your text for ease of access
		  ° Practice the speech or review the text
		• Proofread
		  ° Correct errors

Process

Throughout this textbook, we suggest that you write according to a process: a step-by-step approach to writing that is geared toward your success. The same process applies to Web site construction.

Prewriting

Before you begin to construct your Web site, gather data. Answer these reporter's questions.

1. *Why* are you creating a Web site? Is your corporation creating the Web site to

- Inform the public about products and services?
- Sell products via online order forms?
- Encourage new franchisees to begin their own businesses?
- Advertise new job openings within the company?
- Provide a company profile, including annual reports and employee resumes?
- List satisfied clients?
- Provide contact information?
- All of the above?

2. The answers to several questions tell you *who* your audience is (Wilkinson 1997, 15). Are you writing to low-tech clients, prospective employees, the general lay public, stockholders, or high-tech associates in similar businesses? Will your audience be

- Customers looking for product information, such as specifications, pricing information, guarantees, warranties, instructions, and facility locations?
- Current users of a product or service looking for upgrades, enhancements, changes in pricing, guarantees, warranties, schedules, personnel, facilities locations, and so on?
- New customers wanting to purchase the product or service by way of an online order form?
- Customers looking for online help with instructions or procedures?
- Customers hoping to contact customer service?
- Prospective new hires seeking employment?

3.   Answering these questions will help you decide *what* to include in your site. You could include

- Order forms
- Forms for customer comments
- Frequently asked questions (FAQs)
- Online resumes of your personnel
- Product descriptions or specifications
- Product or service support
- Job openings
- Customer contacts
- Company history, qualifications, affiliations
- Costs

4.   *How* will you construct the site? Will you use HTML coding or HTML converters and editors? Where will you get the graphics you will need? Will you download existing graphics, modified to meet legal and ethical standards, or will you create new graphics with the help of your in-house technical artists?

5.   *Where* will your Web site reside? Will it be an Internet site open to the World Wide Web, or will you focus on an intranet or extranet site?

Once you have gathered this information, use storyboarding for a preliminary sketch of your Web site's layout.

Sketch the Home Page   How do you want the home page to look? What color scheme do you have in mind? How many graphics do you envision? Where will you place your graphics and text—margin left, centered, margin right? Using a storyboard to sketch the graphic or graphics, lead-in, and key hypertext buttons will save you time. If you decide up front what your home page will look like, you will be able to research your graphics, determine how many linked pages you will need, and begin to draft the text.

Sketch the Linked Page(s)   How many linked pages do you envision? Will they have graphics, online forms, e-mail links, audio or video, headings and sub-headings, or links to other Internet, intranet, or extranet sites? Will you use frames or tables? What will your color scheme for these pages be? Sketch your design for these screens by storyboarding. Doing so will allow you to preview the entire site. This will help you estimate the time required to complete the project, as well as give you a better understanding of what resources (personnel,

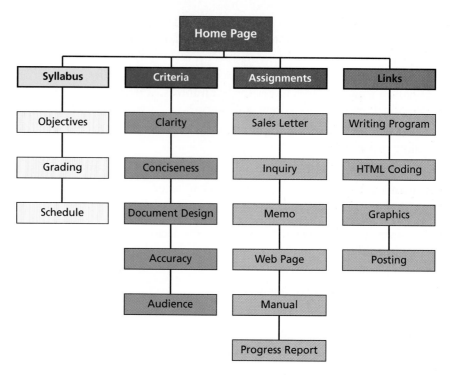

Figure 13.10   Organizational Chart for Technical Writing Online

hardware, software, etc.) you will need to construct your Web site. Then you can meet deadlines or bill accordingly.

Another prewriting technique that works well for Web sites is an organizational chart. As with a storyboard, an "org chart" gives you an overview of your Web site: the home page, key links, and additional screens (see Figure 13.10).

Writing

The second stage of the process encourages you to create a rough draft of your Web site. To do so, follow this procedure:

Study the Successful Web Site Checklist   This will remind you what to include in your rough draft and how best to structure the Web site.

Review Your Prewriting   Have you answered all the reporter's questions? How does your Web site storyboarding sketch look? Do you see any glaring omissions or unnecessary items? If so, add what's missing and omit what's unneeded.

Draft the Web Site   You can create your rough draft in several ways.

- Write a rough draft in any word processing program. Then, after revision, you can save your text as a Web page.

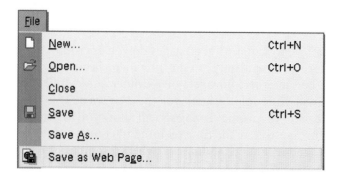

- Write a rough draft in any word processing program. Then, after revision, you can copy and paste the text into an HTML editor. For Windows users, these include Ant HTML, FrontPage, HotDog Web Editor, Netscape Navigator 4.0 and above, Microsoft Internet Assistant for Word 6.0 for Windows and above, Corel WordPerfect 8.0, and others. Macintosh users could try Ant HTML, Alpha-Text, WebDoor, and others (Bryne 1996, 9).
- Draft your text directly in any of the above HTML editors. In this chapter, we provide Technology Tips for using FrontPage.
- Write your draft using HTML coding. Knowing how to use Hypertext Markup Language is a valuable skill. First, when you code your own text, you control the page. In contrast, depending on an HTML editor can limit your options (Scott 1998,16–17). Second, if you use an HTML editor and you do not like the result, you will be unable to change it unless you know HTML coding (Bryne 1996, 9).

Figures 13.11a and 13.11b show a home page in FrontPage's Normal view and then with its HTML code revealed.

HTML Coding

We know that HTML coding looks daunting. However, it's not as hard as it looks. Actually, HTML is a fairly easy-to-learn language. Table 13.2 provides you with the simple codes you will need to draft your own text. These HTML codes are not case-sensitive.

tech link

Go to *http://www. htmlgoodies.com/* for tutorials on HTML and Java coding.

Rewriting

The final stage of the writing process is rewriting. Once you've completed drafting your Web site, is it what you had in mind? If not, revise the site. To do so, consider the web site usability checklist. What's "usability"? As discussed in Chapter 3, usability is a way to determine whether your reader can "use" the Web site effectively, whether your site meets your user's needs and expectations. Not only does the reader want to find information that helps him or her better understand the topic, but also the audience wants the Web site to be readable,

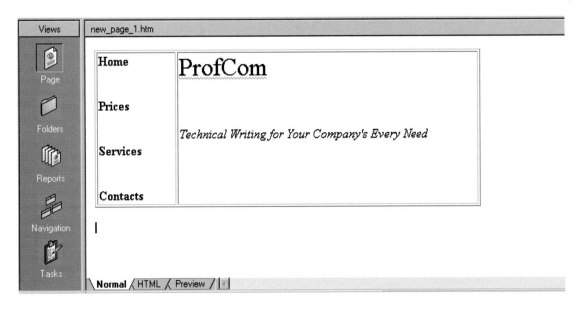

Figure 13.11a   FrontPage's Normal View

new_page_1.htm

```
<html>

<head>
<meta http-equiv="Content-Language" content="en-us">
<meta http-equiv="Content-Type" content="text/html; charset=windows-1252">
<meta name="GENERATOR" content="Microsoft FrontPage 4.0">
<meta name="ProgId" content="FrontPage.Editor.Document">
<title>New Page 1</title>
</head>

<body>

<div align="left">
  <table border="1" width="505" height="64">
    <tr>
      <td width="101" height="64" valign="top"><b>Home</b>
        <p><b><br>
        Prices</b></p>
        <p><b><br>
        Services</b></p>
        <p><b><br>
        Contacts</b></p>
      </td>
      <td width="404" height="64" valign="top"><font size="6">ProfCom</font>
```

Page   Folders   Reports   Navigation   Hyperlinks   Tasks

Normal   HTML   Preview

Figure 13.11b   FrontPage's HTML View

TABLE 13.2	**HTML Codes**	
	On	Off
Document setup	<html>	</html>
	Creates a Web page.	The forward slash mark (/) turns off all the codes.
Title	<title>	</title>
	Title does not appear on the Web page.	
Body	<body>	</body>
	Begins the text.	
Emphasis techniques		
Boldface	<b>	</b>
Italics	<i>	</i>
Blinking	<blink>	</blink>
	Text blinks on and off.	
Font sizes		
Largest	<H1>	</H1>
	<H2>	</H2>
	<H3>	</H3>
	<H4>	</H4>
	<H5>	</H5>
Smallest	<H6>	</H6>

Dividers		
Double return	\<P\> Creates a double-space.	\</P\>
Line break	\<BR\> Acts like a hard return.	\</BR\>
Horizontal rule	\<HR\> Places a thin line across the page.	\</HR\>
Center	\<center\> Centers text or graphics.	\</center\>
Itemized lists		
Ordered list	\<OL\> \<LI\>apples \<LI\>oranges \<LI\>peaches \</OL\>	Looks like this: 1. apples 2. oranges 3. peaches
Unnumbered list	\<UL\> \<LI\>apples \<LI\>oranges \<LI\>peaches \</UL\>	Looks like this: • apples • oranges • peaches
Backgrounds		
Colors	\<body bgcolor="red"\> or any color you choose. The "bg" stands for "background." \<font color="red"\> or any color you choose. This changes the font colorfrom the default black. These colors can also be designated by their numerical hex codes. Black, for example, is numerically written "#000000." Red is "#FF0000,"Pink is "#BC8F8F," and Sky Blue is "#3299CC." There are literally thousands of such numerical delineations for different colors.	\</font color\>
Patterns	\<body background="waves.gif"\> We've used "waves.gif" as an example. You can find thousands of patterns on the Internet.	
Images/graphics	The Internet is perhaps the largest garage sale on Earth. If you want it, you can find it on the Internet. The same holds true for graphics (pictures, icons, lines, bullets, animations, backgrounds, buttons, etc.). Most of it is free. You merely code it in your HTML text as follows: \<img src="filename.gif"\> or \<img src="filename.jpg"\> We've used "filename" as an example. Each graphic will have its own name.	
Hypertext links	Links are what allow you to jump from linked page to linked page. The code looks like this: \<a href="filename.html"\> We've used "filename" as an example. You would put within the less than/greater than symbols (called "tags") the name of the file you plan to link to.	\</a\>

accurate, up-to-date, and easy to access. Thus, usability focuses on three key factors (Dorazio 2000):

1. *Retrievability*— the user wants to find specific information quickly and be able to navigate easily between the screens.

2. *Readability*— the user wants to be able to read and comprehend information quickly and easily.

3. *Accuracy*— the user wants complete, correct, and up-to-date information.

The goal of usability testing is to solve problems that might make a Web site hard for people to use. By successfully testing a site's usability, a company can reduce customer and colleague complaints; increase employee productivity (reducing the time it takes to get work done); increase sales volume because customers can purchase products or services online; and decrease help desk calls and their costs. When these goals are achieved, you have provided user satisfaction. You can revise your Web site to achieve usability by following this checklist.

Web Site Usability Checklist

Audience Recognition and Involvement

___ 1. Does the Web site meet your reader's needs?

___ 2. Does the Web site give your audience a reason to return (tutorials, tips, comics, links to other interesting sites, regular updates, etc.)?

___ 3. Does the Web site solicit user feedback for interactivity and display customer comments?

___ 4. Does the Web site make it easy for your audience to purchase online through online forms?

___ 5. Does the Web site use pronouns to engage the reader?

Home Page

___ 1. Does the home page provide identification information (name of service or product, company name, e-mail, fax, city, state, street address, etc.)?

___ 2. Does the home page provide an informative and appealing graphic that represents the product, service, or company?

___ 3. Does the home page provide a welcoming and informative introductory phrase, sentence, or paragraph?

___ 4. Does the home page provide hypertext links connecting the reader to subsequent screens?

Linked Pages

___ 1. Do the linked pages provide headings clearly indicating to the reader which screen he or she is viewing?

___ 2. Do the linked pages develop ideas thoroughly (appropriate amount of detail, specificity, valuable information)?

___ 3. Are the linked pages limited to one or two primary topics?

Navigation

___ 1. Does the Web site allow for easy return from linked pages to the home page?

___ 2. Does the Web site allow for easy movement between linked pages?

___ 3. Does the Web site provide access to other Web sites for additional information?

Document Design

___ 1. Does the Web site provide an effective background, suitable to the content and creating effective contrast for readability?

___ 2. Does the Web site use color effectively, in a way suitable to the content and creating effective contrast for readability?

___ 3. Does the Web site use a consistent document design (colors, background, graphics, font), carried throughout the entire site?

___ 4. Does the Web site use headings and subheadings for easy navigation?

___ 5. Does the Web site vary font size and type to create a hierarchy of headings?

___ 6. Does the Web site use graphics effectively— suitable to the content, not distracting, and load quickly?

continued

continued

___ **7.** Does the Web site use highlighting techniques effectively (white space, lines, bullets, icons, audio, video, frames, font size and type, etc.) in a way that is suitable to the content and not distracting?

- Line length limited to approximately two-thirds of the screen (40–60 characters per line)
- Text per Web page limited to one viewable screen, minimizing the need to scroll

Style

___ **1.** Is the Web site concise? (Remember that reading online is a challenge to end users.)
- Short words (1–2 syllables)
- Short sentences (10–12 words per sentence)
- Short paragraphs (4 typed lines maximum)

Accuracy

___ **1.** Does the Web site avoid grammatical errors?

___ **2.** Does the Web site ensure that information (phone numbers, e-mail, fax numbers, addresses, content) is current and correct?

___ **3.** Do the internal and external links work?

If your Web site does not meet the criteria provided in the usability checklist, revise the site as follows:

1. *Add new detail for clarity*—Have you said all you need to say? Have you provided contact information, such as phone numbers, e-mail addresses, fax numbers, and a mailing address? Do you need to include a form to help your customers order goods online? Do you want to advertise job openings? Would your Web site profit by including an online catalog in which you describe your products?

Text looks different online than when you code it in. Once you see the first draft of your Web site, you might change your mind about its components. If you do so and want to add new information, the rewriting stage is the time to make those changes.

2. *Delete anything unnecessary for conciseness*—Do your text and graphics all fit on one viewable screen? Or do you make your readers scroll endlessly to gather all their data? Have you loaded your Web site with too many bells and whistles? These could include java scripting, animation, or audio and video plug-ins. If your readers have to wait too long for your site to load, you could lose their interest—and their business. Review and revise your Web site for conciseness. Delete the unnecessary add-ons. Delete dead words and phrases. Reduce the size of your graphics. More is not always better.

3. *Simplify words and phrases*—Web sites are very friendly types of correspondence. Multisyllabic legalese is rarely appropriate for this modern medium. Think simple. (See Chapter 3 for help on this point.)

4. *Move information for emphasis*—Look at several Web sites. Where do other companies place information? You'll notice that most companies list their webmasters (the people or companies who create their Web sites) at the bottom of a screen. Why? Because it would place too much emphasis on the writer and too little emphasis on the product or service to list webmasters at the top of a screen. Similarly, notice where companies mention their product or service: usually at the top of a screen so they will receive immediate recognition. As in real estate, location, location, location sells. Decide what is most important on your Web site and then place this information where it can be seen clearly.

5. *Reformat for access*—No communication medium depends more on layout than a Web site. In fact, rarely do you hear people say, "That Web site has great content." What you hear is, "That's a great-looking Web site" or "That's a terrible-looking Web site." Your Web site's design will make or break it.

Once you load your site, determine how it looks. Is there ample white space? Have you indented text and used bullets or icons to break up the monotony of wall-to-wall words? Where have you placed your graphics? Do the graphics integrate well with the text, and are they pleasing to the eye? What about your headings and subheadings? Are your first-level headings larger than your second-level headings? Should you center them? Would your site profit from a horizontal rule or line separating text from headings? Have you used color effectively? Is there enough contrast between your background and your font color? Where do you want to place your navigational links? Early in the history of the Internet, most companies placed the links at the bottom of the page. Now, however, we see the navigational links placed either along the left or the right margin. These links are viewable instantly, rather than forcing the readers to scroll down to find them. (For further suggestions regarding document design, see Chapter 8.)

6. *Enhance the tone and style of your Web site*—As noted, Web sites tend to be rather informal. Don't write a Web site using the same stuffy tone that you would use when writing a report. Enhance the tone of your site by using contractions, positive words, and lots of pronouns. (See Chapter 4 for help.)

7. *Correct errors*—How many people might read your Web site? Remember, this is the World Wide Web we're talking about. In contrast to a memo or letter, which perhaps one or two people might read, a successful Web site could be seen by millions—worldwide. Now imagine how bad grammatical errors will look when viewed through a magnifying glass the size of the entire earth. People in dozens of countries and on numerous continents will learn that you can't use commas correctly. Don't let this happen to you or to your company. Correct your errors. Proofread your Web site diligently.

8. *Make sure it works*—Test your Web site. How long does it take to load? How do your graphics and colors look on "multiple browsers, platforms, . . . monitors, and with different transmission speeds" (Yeo 1996, 14)? A good idea is to test your site on at least three different computer systems (Eddings et al. 1998). Do your hypertext links work? First, determine whether your readers can easily navigate from one screen to the next within your Web site. Next, if you have created hypertext links to external sites, test these frequently. External sites change constantly. You might need to change your hypertext codes to match these external changes (Yeo 1996, 14). Does your e-mail link work? How about the online order form? Can someone actually submit a request for goods on this form? Do your audio and video plug-ins work? No matter how beautiful your Web site looks on your monitor, the site will not be successful unless all the components work.

Creating Your Web Site

Consider using Microsoft's FrontPage to create your Web site. To do so, follow these steps:

1. On a FrontPage blank page, create space for a navigation bar and the home page text by clicking on Table and then Draw Table.

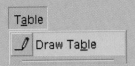

2. Use your cursor to draw a rectangle as shown.

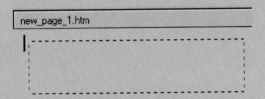

3. Split the screen into two columns.

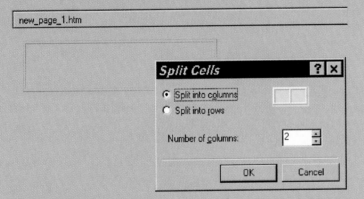

4. Adjust the center cell line and the size of the table by clicking and dragging on the borders.

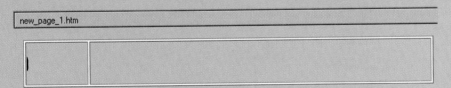

5. Insert any text and graphics you want (just as you would in word processing).

continued

continued

6. Change the look as you see fit (color of text, color of navigation bar, etc.). You can change the navigation bar column as follows:

   a. Right-click within the left cell and scroll to and click on Cell Properties.

   b. In the Cell Properties dialog box, click on the Background Color pull-down arrow and select any color you like.

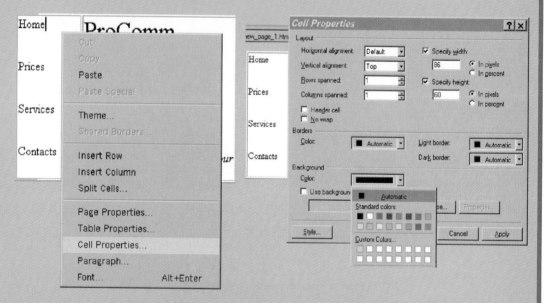

7. Create hypertext links by highlighting any word or graphic and clicking on the Hyperlink icon.

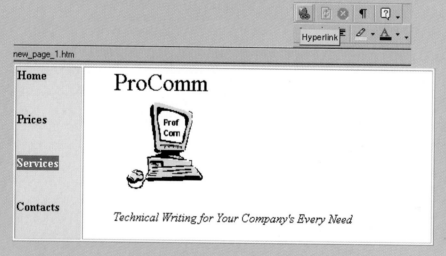

FrontPage offers this and many more options for creating and customizing your Web site.

CHAPTER HIGHLIGHTS

1. To be an effective communicator in business and industry today, you have to communicate electronically.

2. The Internet, the information superhighway, connects computers internationally.

3. The information accessible on the Internet is virtually unrestricted.

4. Web browsers, such as Netscape Navigator and Microsoft Internet Explorer, allow you to access information on the Web.

5. Companies use intranets to communicate internally.

6. An extranet allows several companies to share information confidentially.

7. Interactive online help systems provide "just in time" and "as needed" information to help your audience complete a task.

8. Web sites are created by companies, the government, schools, organizations, and individuals.

9. Use our criteria when creating your Web sites.

10. The process approach to writing applies to Web site construction.

CASE STUDIES

Future Promise

Future Promise is a not-for-profit organization geared toward helping at-risk high school students. This agency realizes that to reach its target audience (teens aged 15–18), it needs an Internet presence.

To do so, it has formed a 12-person team, consisting of the agency's accountant, sports and recreation director, public relations manager, counselor, technical writer, graphic artist, computer and information systems director, two local high school principals, two local high school students, and a representative from the mayor's office. Jeannie Kort, the PR manager, is acting as team leader.

The team needs to determine the Web site's content, design, and levels of interactivity. Jeannie's boss, Brent Searing, has given the team a deadline and a few components that must be included in the site:

- College scholarship opportunities
- After-school intramural sports programs
- Job-training skills (resume building and interviewing)
- Service learning programs to encourage civic responsibility
- Future Promise's 800-hotline (for suicide prevention, STD information, depression, substance abuse, and peer counseling)
- Additional links (for donors, sponsors, educational options, job opportunities, etc.)

Jeannie has a big job ahead of her.

Assignment

Form a team (or become a member of a team as assigned by your instructor) and design Future Promise's Web site. To do so, follow the criteria for Web design provided in this chapter. Then, prewrite, write, and rewrite as follows:

- **Prewrite:**
  - ◆ Research the topics listed above (either in a library, online, or through interviews) to gather details for your Web site.
  - ◆ Consider your various audiences and their respective needs and interests.
  - ◆ Focus on your Web site's purpose—are you writing to inform, persuade, instruct, or build rapport?
  - ◆ Draw a storyboard of how you would like the Web site to look or use an organizational chart to lay out the Web design.
  - ◆ Divide your labors among the team members (who will research, write, create graphics, etc.?).

- **Write:**
  - ◆ Draft your text either in a word processing program (you can save as an HTML file) or directly in FrontPage (or any other Web design program of your choice).

- **Rewrite:**
  - ◆ Review the Web criteria in this chapter.
  - ◆ Add, delete, simplify, move, enhance, and correct your Web site.
  - ◆ Test the site to make sure all links work and that it meets your audience's needs.

Proposal for Web Site

The following short proposal recommends that upper management approve the construction of a corporate Web site. The company's management has agreed with the proposal. You and your team have received a memo directing you to build the Web site. Using the proposal, build your company's Web site.

DATE: November 11, 2005
TO: Distribution
FROM: Shannon Conner
SUBJECT: RECOMMENDATION REPORT FOR NEW CORPORATE
          WEB SITE

Introduction
In response to your request, I have researched the impact of the Internet on corporate earnings. Companies are still striving to position themselves on the World Wide Web and maximize their earnings potential. I concluded that from 1999 to 2001, only 16 percent of companies with Internet sites were making a profit from their online services. This fact, however, has changed. From 2003 to 2005, many more corporate Web sites earned a profit for their companies. The time is right for our company, Java Lava, to go online.

Discussion

Why should we go online? The answer is simple: We can maximize our profits by making our local product a global product. How can we accomplish this goal?

1. *International bean sales.* Currently, coffee bean sales account for only 27 percent of our company's overall profit. These coffee bean sales depend solely on walk-in trade. The remaining 73 percent stems from over-the-counter beverage sales. If we go global via the Internet, we can expand our coffee bean sales dramatically. Potential clients from every continent will be able to order our coffee beans online.
2. *International promotional product sales.* Mugs, T-shirts, boxer shorts, jean jackets, leather jackets, key chains, paperweights, and calendars could be marketed. Currently we give away items imprinted with our company's logo. By selling these items online, we could make money while marketing our company.
3. *International franchises.* We now have three coffeehouses, located at 1200 San Jacinto, 3897 Pecan Street, and 1801 West Paloma Avenue. Let's franchise. On the Internet, we could offer franchise options internationally.
4. *Online employment opportunities.* Once we begin to franchise, we'll want to control hiring practices to ensure that Java Lava standards are met. Through a Web site, we could post job openings internationally and list the job requirements. Then potential employees could submit their resumes online.

In addition to this information, used to increase our income, we could provide the following:

- A map showing our three current sites.
- Our company's history—founded in 1948 by Hiram and Miriam Coenenburg, with money earned from their import/export bean business. "Hi" and "Mam," as they were affectionately called, handed their first coffeehouse over to their sons (Robert, John, and William) who expanded to our three stores. Now a third generation, consisting of Rob, John, and Bill's six children (Henry, Susan, Andrew, Marty, Blake, and Stefani) could take us into the next millennium.
- Sources of our coffee beans—Guatemala, Costa Rica, Columbia, Brazil, Sumatra, France, and the Ivory Coast.
- Freshness guarantees—posted ship dates and ground-on dates; 100 percent money-back guarantee.
- Corporate contacts (addresses, phone numbers, e-mail, fax numbers, etc.).

Conclusion

Coffee is a "hot" commodity now. The international market is ours for the taking. We can maximize our profits and open new venues for expansion. The Web is our tarmac for takeoff. I'm awaiting your approval.

EXERCISES

Online Help

1. Study the help menus in your word processing package. Distinguish between the pop-up windows and the hyperlinks. What are the differences? What are the similarities?

2. In a small group, have each individual ask a software-related question regarding a word processing application. These could include questions such as "How do I print?" and "How do I set margins?" and so on. Then, using a help menu, find the answers to these questions. Are the answers provided by way of pop-up windows or hyperlinks? Rewrite any of these examples to improve their conciseness or visual layout.

3. Take an existing document (one you have already written in your technical writing class or writing from your work environment) and rewrite it as online help. To do so, find "hot" topics in the text and expand on them as either pop-up windows or hyperlinks. Then highlight new hot buttons within the expanded pop-ups or hyperlinks for additional hypertext documents. Create four or five such layers.

    If you have the computer capability to do so, use your computer system. If you don't have the computer capabilities, use 3×5 index cards for the layered pop-ups or hyperlinks.

4. Using our suggestions in this chapter, rewrite Figure 13.12 as an effective online help screen. To do so, reformat the information for access on a smaller screen and create hot-button words or phrases. Determine which of the hot buttons can be expanded with a short pop-up window and which require a longer hyperlink. Research any data with which you are unfamiliar to provide additional information for your hypertext surfer.

Because your company is located in the East Side Commercial Park, the outside power is subject to surges and brownouts. These are caused when industrial motors and environmental equipment frequently start and stop. The large copiers in your office also cause power surges that can damage electronic equipment.

As a result of these power surges and outages, your three network servers have become damaged. Power supplies, hard drives, memory and monitors have all been damaged. The repair costs for 2005 damages totaled $55,000.

Power outages at night also have caused lost data and failed backups and file transfers. Not being able to shut down the servers during power failures is leading to system crashes. Data loss requires time-consuming tape restorations. If the failures happen before a scheduled backup, the data are lost, which requires new data entry time. Ten percent of your data loss can not be restored. This will lead to the loss of customer confidence in the reliability of your services.

Finally, staff time losses of $40,000 are directly related to network downtime. Data-entry staff are not able to work while the servers are being repaired. These staff also must spend extra time reentering lost data. Fifteen percent of the data management staff time is now spent recovering from power-related problems.

To solve your problems, you need to install a BACK-UPS 9000 system. It has the configuration and cost-effectiveness appropriate for your current needs.

Figure 13.12   Flawed Text That Needs Revision

Web Sites

1. Create a corporate Web site. To do so, make up your own company and its product or service. Your company's service could focus on dog training, computer repair, basement refinishing, vent cleaning, Web site construction, child care, auto repair, personalized aerobic training, or online haute cuisine. Your company's product could be paint removers, diet pills, interactive computer games, graphics software packages, custom-built engines, flooring tiles, or duck decoys. The choice is yours. To create this Web site, follow our writing process. Prewrite by listing answers to reporter's questions. Then sketch your site through storyboarding. Next, draft your Web site, using HTML coding. Finally, rewrite by adding, deleting, simplifying, moving, reformatting, enhancing, correcting, and making sure your links work.

2. Create a Web site for your technical writing, business writing, or professional writing class. To create this Web site, follow our writing process. Prewrite by listing answers to reporter's questions. Then sketch your site through storyboarding. Next, draft your Web site, using HTML coding. Finally, rewrite by adding, deleting, simplifying, moving, reformatting, enhancing, correcting, and making sure your links work.

3. Create a Web site for your high school, your college, your fraternity or sorority, your church or synagogue, or your professional organization. To create this Web site, follow our writing process. Prewrite by listing answers to reporter's questions. Then sketch your site through storyboarding. Next, draft your Web site using HTML coding. Finally, rewrite by adding, deleting, simplifying, moving, reformatting, enhancing, correcting, and making sure your links work.

4. Create a personal Web site for yourself or for your family. To create this Web site, follow our writing process. Prewrite by listing answers to reporter's questions. Then sketch your site through storyboarding. Next, draft your Web site using HTML coding. Finally, rewrite by adding, deleting, simplifying, moving, reformatting, enhancing, correcting, and making sure your links work.

5. Research several Web sites, either corporate or personal. Use our Web Site Usability Checklist to determine which sites excel and which sites need improvement. Then write a report (see Chapter 16 for report criteria) justifying your assessment. In this report, clarify exactly what makes the sites

successful. To do so, you could use a table, listing effective traits and giving examples from the Web sites to prove your point. Next, explain why the unsuccessful sites fail. To do so, again use a table to list effective traits and give examples from the inferior sites to show why they are unsuccessful. Finally, suggest ways in which the unsuccessful sites could be improved.

WEB WORKSHOP

1. Web site design is challenging. Some Web sites are outstanding; others do not fare as well. Check out *http://www.webpagesthatsuck.com/* and *http://www. webpractices.com/samplesites.htm*, two URLs that assess flawed Web sites. Do you agree with their assessments? Write a memo to your instructor or give an oral presentation explaining your decisions.

2. Access any company's Web site and study the site's content, layout (color, graphics, headings, use of varying font sizes and types, etc.), links (internal and external), ease of navigation, tone, and any other considerations you think are important. Then, determine how the Web site could be improved if you were the site's webmaster. Once you have made this determination, write a memo recommending the changes that you believe will improve the site. In this memo,

   • Analyze the Web site's current content and design, focusing on what is successful and what could be improved

   • Provide feasible alternatives to improve the site

   • Recommend changes

3. The Internet offers you many resources for finding out more about how to design successful Web sites. Look at Web Page Design for Designers (*http://www.wpdfd.com/*), which provides a Web design FAQ feature. Top Ten Mistakes in Web Design (*http://www.useit.com/alertbox/9605.html*) offers you exactly what the name implies. The Web Design Group *http://www. htmlhelp.com/*) provides an FAQ archive, information on Web design elements, and feature articles. Finally, Web Design Guide (*http://dreamink. com/*) provides effective Web design principles, tutorials, a step-by-step beginner's guide, and Web resources.

   Review any of these sites for more information on Web design principles. What new and interesting information have you discovered? For example, what questions or concerns do most Web designers seem to have? What mistakes are common to Web design? Write a memo to your instructor or give an oral presentation to share your findings.

QUIZ QUESTIONS

1. What is the Internet?

2. What is an intranet?

3. What is an extranet?

4. Why is the use of online help growing?

5. What types of information are given in a pop-up or hyperlink?

6. How can you achieve effective online help screens?

7. How can you achieve a pleasant tone in your help screen?

8. In what three ways can you design an effective screen?

9. Why should you be concerned with length in your help screen?

10. What are characteristics of successful Web sites?

11. Why should you avoid "noise" on your Web site?

12. What is a home page?

13. What is included on a home page?

14. What is a hypertext link or button?

15. What information should be included in a linked page?

16. How can you achieve effective document design on a Web page?

17. Which highlighting techniques should you avoid on a Web site?

18. What is HTML coding?

19. Do you need to know HTML coding to create a Web site? Why or why not?

20. Why is it helpful to know HTML coding when you create a Web site?

Research

Criteria for Writing Research Reports
When writing from research, recognize audience, use an effective style, and format the research report for ease of access.

Process
Follow a step-by-step process to create successful research documents:

- Use prewriting techniques.
- Use libraries, online sources, and the Internet.
- Take notes.
- Outline.
- Draft your report.
- Cite your sources.

Research skills are important in your school or work environment. You may want to perform research to better understand a technical term or concept; locate a magazine, journal, or newspaper article for your supervisor; or find data on a subject to prepare an oral or written report. Technology is changing so rapidly that you must know how to do research to stay up-to-date.

You can research information using online catalogs; online indexes and databases; CD-ROM indexes and databases; reference books in print, online, and CD-ROM format; and by using Internet search engines and directories. Reference sources vary and are numerous, so this chapter discusses only research techniques.

When you complete this chapter, you will be able to

- Locate information in the library and online
- Read and analyze sources of information for your report
- Form an idea for your research report
- Write a research report
- Document your sources of information

CRITERIA FOR WRITING RESEARCH REPORTS

As with all types of technical writing, writing from research requires that you

- Recognize your audience
- Use an effective style, appropriate to your reader and purpose
- Use effective formatting techniques for reader-friendly ease of access

447

Audience

When writing a research report, you first must recognize the level of your audience. Are your readers high-tech, low-tech, or lay? If you are writing to your boss, consider his or her level of technical expertise. Your boss probably is at least a low-tech reader, but you must determine this based on your own situation. Doing so will help you determine the amount of technical definition necessary for effective communication.

You must decide next whether your reader will understand the purpose of your research report. This will help you determine the amount of detail needed and the tone to take (persuasive or informative). For instance, if your reader has requested the information, you will not need to provide massive amounts of background data explaining the purpose of your research. Your reader has probably helped you determine the scope and purpose. On the other hand, if your research report is unsolicited, your first several paragraphs must clarify your rationale. You will need to explain why you are writing and what you hope to achieve.

If your report is solicited, you probably know what your reader plans to do with your research. He or she will use it for a briefing, an article to be written for publication, a technical update, and so forth. Thus, your presentation will be informative. If your research report is unsolicited, however, your goal is to persuade your reader to accept your hypotheses substantiated through your research.

Effective Style

Research reports should be more formal than many other types of technical writing. In a research report, you are compiling information, organizing it, and presenting your findings to your audience using documentation. Because the rules of documentation are structured rigidly (to avoid plagiarism and create uniformity), a research report is also rigidly structured.

The tone need not be stuffy, but you should maintain an objective distance and let the results of your research support your contentions. Again, considering your audience and purpose will help you decide what style is appropriate. For example, if your boss has requested your opinion, then you are correct in providing it subjectively. However, if your audience has asked for the facts—and nothing but the facts—then your writing style should be more objective.

Formatting

Reading a research report is not always an easy task. As the writer, you must ensure that your readers encounter no difficulties. One way to achieve this is by using effective formatting. Reader-friendly ease of access is accomplished when you use highlighting techniques such as bullets, numbers, headings, subheadings, and graphics (tables and figures).

In addition, effective formatting includes the following:

- Overall organization (including an introduction, discussion, and conclusion)
- Internal organization (various organizational patterns, such as problem/solution, comparison/contrast, analysis, and cause/effect)
- Parenthetical source citations
- Works cited—documentation of sources

Each of these areas is discussed in detail throughout this chapter.

The Writing Process

The Writing Process		
Prewriting	Writing	Rewriting

• Examine your purposes
• Determine your goals
• Consider your audience
• Gather your data
• Determine how the content will be provided
	• Organize the draft according to some logical sequence that your readers can follow easily
	• Format the content to allow for ease of access
		• Revise
		  ° Add missing details
		  ° Delete wordiness
		  ° Simplify word usage
		  ° Enhance the tone of your communication
		  ° Reformat your text for ease of access
		  ° Practice the speech or review the text
		• Proofread
		  ° Correct errors

Writing a research report, as with other types of technical correspondence, is easiest when you follow a process. Rather than just wandering into a library and hoping that the correct book or periodical will leap off a shelf and into your hands or expecting your online search to reveal useful information immediately, approach your research systematically. Prewrite (to gather your data and determine your objectives), write a draft, and rewrite to ensure that you meet your goals successfully. The writing process is dynamic, with the steps frequently overlapping.

Prewriting Research Techniques

1. *Select a general topic (or the topic you have been asked to study)*—Your topic may be a technical term, phrase, innovation, or dilemma. If you're in emergency medical technology, for example, you might want to focus on the current problems with emergency medical service. If you're in telecommunications, you might write about satellite communication. If you work in electronics, select a topic such as robotics. If your field is computer science, you could focus on artificial intelligence.

2. *Spot-check sources of information*—Check a library or online sources to find material that relates to your subject. A quick review of your library's online periodical databases, such as ProQuest, Infotrac, or other equivalent sources, will help you locate periodical articles. Your library

may also have a print edition of the *Reader's Guide to Periodical Literature* and other similar specialized periodical indexes. Most libraries now have online catalogs to help you easily search for books on your topic. A keyword search of Internet metasearch engines will give you an idea of how much information might be readily available through the Internet.

3. *Establish a focus*—After you have chosen a topic for which you can find available source material, decide what you want to learn about your topic. A focus statement can guide you. In other words, if you are interested in emergency medical technology, you might write a focus statement such as the following:

> **example**
>
> I want to research current problems with emergency medical services, including variances in training required, delays in timely response to emergency calls, and limited number of vehicles available.

For telecommunications, you could write:

> **example**
>
> I want to research satellite communication maintenance, reliability, and technical innovations.

If you are an electronics technician, you might write:

> **example**
>
> I want to discover the uses, impact on employment, and expenses of robotics applications.

Finally, if you are a computer technician, you might write:

> **example**
>
> I want to discover the pros and cons regarding artificial intelligence.

With focus statements such as these, you can begin researching your topic, concentrating on articles pertinent to your topic.

4. *Research your topic*—You may feel overwhelmed by the prospect of research. But there are many sources that, once you know how to use them, will make the act of research less overwhelming.

Books

All books owned by a library are listed in catalogs, usually online. Books can be searched in online catalogs in a variety of ways: by title, by author, by subject, by keyword, or by using some combination of these. No matter how you search for a book, the resulting record will look something like the following example.

Artificial intelligence : robotics and machine evolution / David Jefferis.

Database:	DeVry Institute of Technology
Main Author:	Jefferis, David.
Title:	Artificial intelligence : *robotics* and machine evolution / David Jefferis.
Primary Material:	Book
Subject(s):	*Robotics*—Juvenile literature.
	Artificial intelligence—Juvenile literature.
	Robotics.
	Robots.
	Artificial intelligence.
Publisher:	New York : Crabtree Pub. Co., 1999.
Description:	32 p. : col. ill. ; 29 cm.
Series:	Megatech
Notes:	Includes index.
	An introduction to the past, present, and future of artificial intelligence and *robotics,* discussing early science fiction predictions, the dawn of AI, and today's use of robots in factories and space exploration.
Database:	DeVry Institute of Technology
Location:	Dallas Main Stacks
Call Number:	TJ211.2.J44 1999
Number of Items:	1
Status:	Not Changed

Periodicals

Use online, CD-ROM, or print periodical indexes to find articles on your topic. Online indexes can be searched in a variety of ways: by title, by author, by subject, by keyword, or by using a combination of these. No matter how you search, the resulting record will look something like the following example.

DH Pro Eric Carter

| Help | ? |

Bicycling; Emmaus; Mar 2001; Andrew Juskaitis;

Volume:	42
Issue:	2
Start Page:	14
ISSN:	00062073
Subject Terms:	Athletes
	Bicycling
	Organizations
	Bicycle racing

Personal Names: <u>Carter, Eric</u>
Abstract:
*Juskaitis discusses bicycling with downhill national champion Eric Carter. Only after riding with Carter did he realize he's also at the forefront of changing the face of gravity-fed mountain **bike racing**. Carter has joined with promoter Rick Sutton to push cycling's race organizations to implement four-to-six-rider racing on highly specialized courses.*

Full Text: . . . (omitted here for copyright issues)

The preceding example from an online periodical database tells us that an article on the subjects **Athletes, Bicycling, Organizations,** and **Bicycle racing** can be found in the March 2001 issue of *Bicycling* magazine, volume 42, issue 2, beginning on page 14. The article is titled "DH Pro Eric Carter" and was written by Andrew Juskaitis. The article contains information about a person named Eric Carter. An abstract, or summary of the article, is given, followed by the full text of the article itself.

Indexes to General, Popular Periodicals   Most libraries provide access to at least one of a number of indexes covering popular, nontechnical literature and newspaper articles from a variety of subject fields. There are online, CD-ROM, and print counterparts for most of these. The online and CD-ROM indexes provide the full text of many of the articles.

> **Examples**
> - *Reader's Guide to Periodical Literature*
> - *Periodicals Research I or II*
> - *Ebscohost*
> - *SIRS Researcher* (emphasizes social issues)
> - *Newsbank* (emphasizes newspaper articles)

Indexes to Specialized, Scholarly, or Technical Periodicals   Many libraries provide access to one or more specialized indexes covering literature from a variety of disciplines. There are online, CD-ROM, and print counterparts for most of these. The online and CD-ROM indexes provide the full text of many of the articles.

> **Examples**
> - **Applied Science & Technology Index.** Covers engineering, aeronautics and space sciences, atmospheric sciences, chemistry, computer technology and applications, construction industry, energy resources and research, fire prevention, food and the food industry, geology, machinery, mathematics, metallurgy, mineralogy, oceanography, petroleum and gas, physics, plastics, the textile industry and fabrics, transportation, and other industrial and mechanical arts.
> - **Business Periodicals Index.** Covers major U.S. publications in marketing, banking and finance, personnel, communications, computer technology, and so on.
> - **ABI/Inform.** Covers business and management.
> - **General Science Index.** Covers the pure sciences, such as biology and chemistry.

- **Social Sciences Index.** Covers psychology, sociology, political science, economics, and other social sciences topics.
- **ERIC (*Education Resources Information Center*).** Provides bibliography and abstracts about educational research and resources. Available for free through the Internet.
- **MEDLINE.** Covers medical journals and allied health publications.
- **PsycINFO.** Covers psychology and behavioral sciences.
- **NEXIS/LEXIS.** Includes the full text of newspaper articles, reports, transcripts, law journals, and legal reporters and other reference sources in addition to general periodical articles.
- **SIRS (*Social Issues Resources*).** Provides bibliography and full-text articles about social issues.

The Internet

Perhaps one of the largest sources of research available today is the World Wide Web. Millions of documents from countless sources are found on the Internet. You can find material on the Internet published by government agencies, organizations, schools, businesses, or individuals (see Table 14.1). The list of options grows daily. For example, nearly all newspapers and news organizations have online Web sites.

How do you find information online? Use directories, search engines, or metasearch engines.

spotlight

Web Resources

TABLE 14.1	**A SAMPLE OF INTERNET RESEARCH SOURCES**				
Search Engines, Directories, Metasearch Engines	Online References	Online Libraries	Online Newspapers	Online Magazines	Online Government Sites
Yahoo	*Webster's Dictionary*	Library of Congress	*New York Times*	*National Geographic*	United Nations
Excite	*Roget's Thesaurus*	New York Public Library	CNN	*HotWired*	The White House
Lycos	*Britannica Online*	Cleveland Public Library	*USA Today*	*Atlantic*	The IRS
Go.com	*Encyclopedia*	Gutenberg Project	*Washington Post*	*The New Republic*	U.S. Postal Service
AltaVista	*Encyclopedia Smithsonian*	Most city and university libraries	*Kansas City Star*	*U.S. News Online*	FirstGov
MetaSearch	*The Internet Almanac*		Most city newspapers	*Time Magazine*	Most states' supreme courts, legislatures, executive offices, and local governments
MetaCrawler	*The Old Farmer's Almanac*			*Ebony Online*	
Google				*Slate*	
Britannica.com					
Northern Light					
Ask Jeeves					
Dogpile					

Directories   Directories, like Yahoo, AltaVista, HotBot, and Excite, let you search for information from a long list of predetermined categories, including the following:

Arts	Government	Politics and Law
Business	Health and Medicine	Recreation
Computers	Hobbies	Science
Education	Money and Investing	Sports
Entertainment	News	Society and Culture

To access any of these areas, click on the appropriate category and then "drill down," clicking on each subcategory until you get to a useful site.

Search Engines   Search engines, like Google, Northern Light, or Ask Jeeves, let you search millions of Web pages by keywords. Type a word, phrase, or name in the appropriate blank space and press the Enter key. The search engine will search through documents on the World Wide Web for "hits," documents that match your criteria. One of two things will happen: Either the search engine will report "no findings," or it will report that it has found thousands of sites that might contain information on your topic.

In the first instance, "no findings," you'll need to rethink your search strategy. You may need to check your spelling of the keywords or find synonyms. For example, if you want to research information about online writing, you could try typing "writing online," "online writing," "electronic writing," "writing electronically," and other similar terms. In the second instance, finding too many hits, you'll need to narrow your search. For example, if you are researching illegalities in baseball, you cannot type in "baseball." That's too broad. Instead, try "Shoeless Joe Jackson," "Pete Rose," "crime in baseball," "baseball scandals," and so on.

Metasearch Engines   A metasearch engine, like Dogpile, lets you search for a keyword or phrase in a group of search engines at once, saving you the time of searching separately through each search engine.

Researching the Internet presents at least two problems other than finding information. First, is the information you have found trustworthy? Paperbound newspapers, journals, magazines, and books go through a lengthy publication process involving editing and review by authorities. Not all that's published on the Internet is so professional. Be wary. What you read online needs to be filtered through common sense. Second, remember that although a book, magazine, newspaper, or journal can exist unchanged in print form for years, Web sites change constantly. A site you find today online will not necessarily be the same tomorrow. That's the nature of electronic communication.

In addition to these sources, you can consult the following for help: U.S. government publications, databases, and your reference librarian.

5. *Read your researched material and take notes*—Once you have researched and located a source (whether it is a book, magazine, journal, or newspaper article), study the material. For a book, use the index and the table of contents to locate your topic. When you have found it, refer to the pages indicated and skim, reading selectively. For shorter documents, such as magazine or journal articles or online materials, you can read closely. Reading and rereading the source material is an essential step in understanding your researched information. After you have studied the document thoroughly, go through it page by page and briefly summarize the content.

For a short magazine article, you can make marginal notes on a photocopy. For instance, read a paragraph and then briefly summarize its main point(s) in several words. Such notations are valuable because they are easy to make and provide a clear and concise overview of the article's focus.

You can also take notes on 3 × 5-inch cards. If you do this, be sure to write only one fact, quotation, or paraphrase per card, along with the author's name. This will let you organize the information later according to whatever organizational sequence (chronological narrative, analysis by importance, comparison/contrast, problem/solution, cause/effect, etc.) you prefer. Prepare a bibliography of your sources on 3 × 5-inch cards as well, one source per card.

The following are examples of a bibliography card and a note card with quotation, respectively.

example

Stephens, Guy M. "To Market, to Market." *Satellite Communications.* November 1987: 15–16.

example

Future need for C-band

"'We believe there is going to be an important need for C-band capacity into the '90s, into the next century, because it's an ideal way to serve many users, particularly video users,' Koehler said."
Stephens, "To Market," p. 15.

On a *summary* note card, you condense original material by presenting the basic idea in your own words. You can include quotations if you place them in quotation marks, but do not alter the organizational pattern of the original. A summary is shorter than the original.

example

Future need for C-band

C-band is going to be increasingly important in the '90s because of the increasing number of video owners.
Stephens, "To Market," p. 15.

On a *paraphrase* note card, you restate the original material in your own words without condensing. The paraphrase is essentially the complete version rewritten. A paraphrase is the same length as the original.

example

Future need for C-band

The author asserts the continued relevance of C-band even in the 1990s and beyond. C-band is the top performer, especially for people who use videos.
Stephens, "To Market," p. 15.

6.   *Isolate the main points*—After you complete the analysis of the document (whether you use note cards and/or marginal comments), isolate the main points discussed in the books or periodicals. You will find that, of

the major points in an article or book, sometimes only three or four ideas will be relevant for your topic. Choose the ones discussed repeatedly or those that most effectively develop the ideas you want to pursue.

7. *Write a statement of purpose*—Once you have chosen two to four main ideas from your research, write a purpose statement that expresses the direction of your research. For example, one student wrote the following statement of purpose after performing research on superconductivity:

> **example**
>
> The purpose of this report is to reveal the future for this exciting product, which depends on further progress of technology, the development of easily accessible and cost-efficient materials, and industry's need for the final product.

8. *Create an outline*—After you have written a purpose statement, formulate an outline. An outline will help you organize your paragraphs and ensure that you stay on track as you develop your ideas through quotes and paraphrases. Figures 14.1 and 14.2 are examples of topic and sentence outlines.

Writing

You now are ready to write your research report.

1. *Review your research*—Prior to writing your report, look back over your research sources to make sure that you're satisfied with what you've found. Do you have enough information to develop your points thoroughly? Is the information you've found what you want? If not, it's time to do more research. If you're content with what you've discovered in your research, you can start drafting your text.

2. *Organize your report effectively*—When you are ready to write, provide an introductory paragraph, discussion (body) paragraphs, a conclusion or recommendation, and your works cited page.

Introduction

Begin with something to arouse your reader's interest. This could include a series of questions, an anecdote, a quote, or data pertinent to your topic. Then use this to lead into your statement of purpose.

Discussion

The number of discussion paragraphs will depend on the number of divisions and the amount of detail necessary to develop your ideas. Use quotes and paraphrases to develop your content. Students often ask how much of a research report should be *their* writing, as opposed to researched information. A general rule is to lead into and out of every quotation or paraphrase with your own writing. In other words,

- Make a statement (your sentence).
- Support this generalization with a quotation or paraphrase (referenced material from another source).
- Provide a follow-up explanation of the referenced material's significance (your sentence).

        I.   Sensors used to help robots move
             A. Light
             B. Sound
             C. Touch
       II.   Touch sensor technology (microswitches)
             A. For gripping
             B. For maintaining contact with the floor
      III.   Optical sensors (LED/phototransistors)—Like bowling alley
             foul-line sensors
             A. Less bulky/connected to computer interface
             B. Not just for gripping, but for locating objects by
                 following this sequence:
                 1. Scan gripper to locate object
                 2. Move gripper arm left and right to center object
                 3. Move gripper forward to grasp
                 4. Close gripper
             C. Problem—What force to use for gripping?
      IV.   Force sensors
             A. Spring and microswitch
             B. Optical encoder discs—Microprocessors determine
                 speed of discs to determine force necessary
             C. Integrated circuits with strain gauge and pressure-
                 sensitive paint
             D. Pressure sensors built with conductive foam
       V.   Conclusion

Figure 14.1   Topic Outline

Conclusion/Recommendations

In a final paragraph, summarize your findings, draw a conclusion about the significance of these discoveries, and recommend future action.

Citing Sources

On a final page, provide an alphabetized list of your research sources. (We discuss this documentation later in this chapter.)

   3.   *Document your sources correctly*—Your readers need to know where you found your information and from which sources you are quoting or paraphrasing. Therefore, you must document this information. Correct documentation is essential for several reasons:

- You must direct your readers to the books, periodical articles, and online reference sources that you have used in your research report. If your readers want to find these same sources, they depend only on your documentation. If your documentation is incorrect, readers will be confused. Instead, you want your readers to be able to rely on the correctness and validity of your research.

- Do *not* plagiarize. *Plagiarism* is the appropriation (theft!) of some other person's words and ideas without giving proper credit. Writers are often guilty of unintentional plagiarism. This occurs when you incorrectly alter part of a quotation but still give credit to the writer. Your

I. Because robots must move, they need sensors. These sensors could include light sensors, sound sensors, and touch sensors.
II. Touch sensors can have the following technology:
  A. Microswitches can be used for gripping.
  B. Microswitches can also be used for maintaining contact with the floor. This would keep the robot from falling down stairs, for example.
III. Optical sensors might be better than microswitches.
  A. LED/phototransistors are less bulky than switches.
  B. When connected to a computer interface, optical sensors also can help a robot locate an object as well as grip it.
  C. Here is the sequence followed when using optical sensors:
    1. The robot's grippers scan the object to locate it.
    2. The robot moves its gripper arms left and right to center the object.
    3. The robot moves forward to grip the object.
    4. The robot closes the gripper.
  D. The only problem faced is what force should be used when gripping.
IV. Force sensors can solve this problem.
  A. A combination spring and microswitch can be used to determine the amount of force required.
  B. Optical encoder discs can be used also. A microprocessor determines the speed of the disc to determine the required force.
  C. Integrated circuits with strain gauges and pressure-sensitive paint can be used to determine force.
  D. Another pressure sensor can be built from conductive foam.
V. To conclude, all these methods of tactile sensing comprise a field of inquiry important to robotics.

Figure 14.2   Sentence Outline

quotation must be *exactly the same* as the original word, sentence, or paragraph. You cannot haphazardly change a word, a punctuation mark, or the ideas conveyed. Even if you have cited your source, an incorrectly altered quotation constitutes plagiarism.

• On the other hand, if you intentionally use another person's words and claim them as your own, omitting quotation marks and source citations, you have committed theft. This is dishonest and could raise questions about your credibility and the credibility of your research. Teachers, bosses, and colleagues will have little, if any, respect for a person who purposely takes another person's words or ideas. It is essential, therefore, for you to cite your sources correctly.

To document your research correctly, you must (a) provide parenthetical source citations and (b) supply a Works Cited page (Modern Language Association) or a References page (American Psychological Association).

Sample MLA Works Cited Page

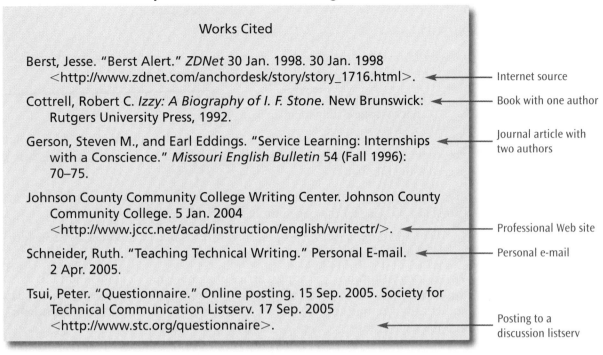

Works Cited

Berst, Jesse. "Berst Alert." *ZDNet* 30 Jan. 1998. 30 Jan. 1998   ←——— Internet source
    <http://www.zdnet.com/anchordesk/story/story_1716.html>.

Cottrell, Robert C. *Izzy: A Biography of I. F. Stone.* New Brunswick: ←——— Book with one author
    Rutgers University Press, 1992.

Gerson, Steven M., and Earl Eddings. "Service Learning: Internships   ←——— Journal article with two authors
    with a Conscience." *Missouri English Bulletin* 54 (Fall 1996):
    70–75.

Johnson County Community College Writing Center. Johnson County
    Community College. 5 Jan. 2004
    <http://www.jccc.net/acad/instruction/english/writectr/>.   ←——— Professional Web site

Schneider, Ruth. "Teaching Technical Writing." Personal E-mail.   ←——— Personal e-mail
    2 Apr. 2005.

Tsui, Peter. "Questionnaire." Online posting. 15 Sep. 2005. Society for
    Technical Communication Listserv. 17 Sep. 2005
    <http://www.stc.org/questionnaire>.   ←——— Posting to a discussion listserv

Sample APA References Page

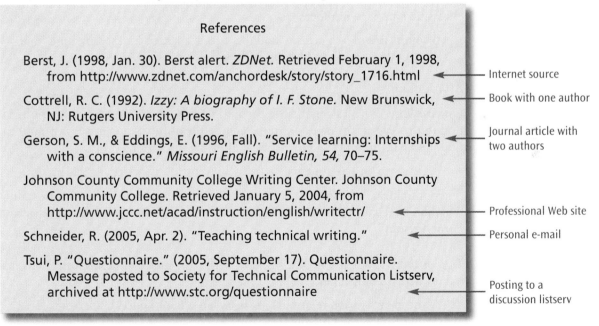

References

Berst, J. (1998, Jan. 30). Berst alert. *ZDNet.* Retrieved February 1, 1998,
    from http://www.zdnet.com/anchordesk/story/story_1716.html   ←——— Internet source

Cottrell, R. C. (1992). *Izzy: A biography of I. F. Stone.* New Brunswick,   ←——— Book with one author
    NJ: Rutgers University Press.

Gerson, S. M., & Eddings, E. (1996, Fall). "Service learning: Internships   ←——— Journal article with two authors
    with a conscience." *Missouri English Bulletin, 54,* 70–75.

Johnson County Community College Writing Center. Johnson County
    Community College. Retrieved January 5, 2004, from
    http://www.jccc.net/acad/instruction/english/writectr/   ←——— Professional Web site

Schneider, R. (2005, Apr. 2). "Teaching technical writing."   ←——— Personal e-mail

Tsui, P. "Questionnaire." (2005, September 17). Questionnaire.
    Message posted to Society for Technical Communication Listserv,   ←——— Posting to a discussion listserv
    archived at http://www.stc.org/questionnaire

Parenthetical Source Citations   The Modern Language Association (MLA) and the American Psychological Association (APA) use a simplified form for source citations. Before 1984, footnotes and endnotes were used in research reports. In certain instances, this form of documentation is still correct. If your boss or instructor requests footnotes or endnotes, you should still use these forms. However, the most modern approach to source citations requires only

tech link

For online help with MLA style, go to *http://owl. english.purdue.edu/ handouts/research/ r_mla.html.*

tech link

For online help with APA style, go to *http://owl. english.purdue.edu/ handouts/research/ r_apa.html*.

that you cite the source of your information parenthetically after the quotation or paraphrase.

MLA Format

One author   After the quotation or paraphrase, parenthetically cite the author's last name and the page number of the information.

> "Viewing the molecular activity required state-of-the-art electron microscopes" (Heinlein 193).

Note that the period follows the parenthesis, not the quotation. And, note that no comma separates the name from the page number and that no lowercase *p* precedes the number.

Two authors   After the quotation or paraphrase, parenthetically cite the authors' last names and the page number of the information.

> "Though *Gulliver's Travels* preceded *Moll Flanders*, few scholars consider Swift's work to be the first novel" (Crider and Berry 292).

Three or more authors   Writing a series of names can be cumbersome. To avoid this, if you have a source of information written by three or more authors, parenthetically cite one author's name, followed by *et al.* (Latin for "and others") and the page number.

> "Baseball isn't just a sport; it represents man's ability to meld action with objective—the fusion of physicality and spirituality" (Norwood et al. 93).

Anonymous works   If your source has no author, parenthetically cite the shortened title and page number.

> "Robots are more accurate and less prone to errors caused by long hours of operation than humans" ("Useful Robots" 81).

APA Format

One author   If you do not state the author's name or the year of the publication in the lead-in to the quotation, include the author's name, year of publication, and page number in parenthesis, after the quotation.

> "Izzy's stay in Palestine was hardly uneventful" (Cottrell, 1992, p. 118).

(Page numbers are included for quoted material. The writer determines whether page numbers are included for source citations of summaries and paraphrases.)

Two authors   When you cite a source with two authors, always use both last names with an ampersand (&).

> "Line charts reveal relationships between sets of figures" (Gerson & Gerson, 1992, p. 158).

Three or more authors   When your citation has more than two authors but fewer than six, use all the last names in the first parenthetical source citation. For subsequent citations, list the first author's name last followed by *et al.*, the year of publication, and for a quotation, the page number.

> "Two-party politics might no longer be the country's norm next century" (Conners et al., 1993, p. 2).

Anonymous works   When no author's name is listed, include in the source citation the title or part of a long title and the year. Book titles are underlined or italicized, and periodical titles are placed in quotation marks.

> Two-party politics might be a thing of the past (*Winning Future Elections*, 1992).

> Many memos and letters can be organized in three paragraphs ("Using Templates," 1994).

Works Cited   Parenthetical source citations are an abbreviated form of documentation. In parentheses, you tell your readers only the names of your authors and the page numbers on which the information can be found. Such documentation alone would be insufficient. Your readers would not know the names of the books, the names of the periodicals, or the dates, volumes, or publishing companies. This more thorough information is found on the Works Cited page or References page, a listing of research sources alphabetized either by author's name or title (if anonymous). This is the last page of your research report.

Your entries should follow MLA or APA standards.

MLA Works Cited

A book with one author

Cottrell, Robert C. *Izzy: A Biography of I. F. Stone.* New Brunswick: Rutgers University Press, 1992.

A book with two or three authors

Tibbets, Charlene, and A. M. Tibbets. *Strategies: A Rhetoric and Reader.* Glenview: Scott, Foresman and Company, 1988.

A book with four or more authors

Nadell, Judith, et al. *The Macmillan Writer.* Boston: Allyn and Bacon, 1997.

A book with a corporate authorship

Corporate Credit Union Network. *A Review of the Credit Union Financial System: History, Structure, and Status and Financial Trends.* Kansas City: U.S. Central, 1986.

A translated book

Phelps, Robert, ed. *The Collected Stories of Colette.* Trans. Matthew Ward. New York: Farrar, Straus Giroux, 1983.

An entry in a collection or anthology

Irving, Washington. "Rip Van Winkle." *Once Upon a Time: The Fairy Tale World of Arthur Rackham.* Ed. Margery Darrell. New York: Viking, 1972. 13–36.

A signed article in a journal

Gerson, Steven M., and Earl Eddings. "Service Learning: Internships . . . with a Conscience." *Missouri English Bulletin* 54 (Fall 1996): 70–75.

A signed article in a magazine

Kroll, Jack. "T. Rex Redux." *Newsweek* 26 May 1997: 74–75.

A signed article in a newspaper

Hoffman, Donald. "Bank Consigned to Vault of Gloom." *The Kansas City Star* 24 Oct. 1988: C1.

An unsigned article

"Diogenes Index." *Time* 23 Sep. 1996: 22.

Encyclopedias and almanacs

"Rocket." *The World Book Encyclopedia.* 1979 ed. Chicago: World Book.

Computer software

PFS: Write Sampler. Computer software. Software Pub. Corp., 1984.

Internet source

Berst, Jesse. "Berst Alert." *ZDNet* 30 Jan. 1998. 1 Feb. 1998 *<http://www.zdnet.com/anchordesk/story/story_1716.html>*.

E-mail

Schneider, Ruth. "Teaching Technical Writing." Personal E-mail. 2 Apr. 1998.

CD-ROM

McWard, James. "Graphics On-line." TW/Inform. CD-ROM. New York: EduQuest, 1998.

Personal web site

Mohr, Ellen. Home page. 29 Dec. 2003 *<http://www.jccc.net/home/depts/1504>*.

Professional web site

Johnson County Community College Writing Center. Johnson County Community College. 5 Jan. 2004 <http://www.jccc.net/acad/instruction/english/writectr/>.

Posting to a discussion listserv

Tsui, Peter. "Questionnaire." Online posting. 15 Sep. 2005. Society for Technical Communication Listserv. 17 Sep. 2005 *<http://www.stc.org/questionnaire>*.

APA References

A book with one author

Cottrell, R. C. (1992). *Izzy: A biography of I. F. Stone.* New Brunswick, NJ: Rutgers University Press.

A book with two authors

Tibbets, C., & Tibbets, A. M. (1988). *Strategies: A rhetoric and reader.* Glenview, IL: Scott, Foresman.

A book with three or more authors

Nadell, J., McNeniman, L., & Langan, J. (1997). *The Macmillan writer.* Boston: Allyn & Bacon.

A book with a corporate authorship

Corporate Credit Union Network. (1986). *A review of the credit union financial system: History, structure, and status and financial trends.* Kansas City, MO: U.S. Central.

A translated book

Phelps, R. (Ed.). (1983). *The collected stories of Colette* (M. Ward, Trans.). New York: Farrar, Straus & Giroux.

An entry in a collection or anthology

Irving, W. (1972). Rip Van Winkle. In M. Darrell (Ed.), *Once upon a time: The fairy tale world of Arthur Rackham* (pp. 13–36). New York: Viking Press.

A signed article in a journal

Gerson, S. M., & Eddings, E. (1996, Fall). Service learning: Internships . . . with a conscience. *Missouri English Bulletin*, 54, 70–75.

A signed article in a magazine

Kroll, J. (1997, May 26). T. rex redux. *Newsweek*, 74–75.

A signed article in a newspaper

Hoffman, D. (1988, October 24). Bank consigned to vault of gloom. *The Kansas City Star*, p. C1.

Schneider, R. (2005, April). "Teaching technical writing.

An unsigned article

Diogenes index. (1996, September 23). *Time*, 22.

Encyclopedias and almanacs

Rocket. (1979). *The world book encyclopedia*. Chicago: World Book.

Computer software

PFS: Write Sampler [Computer software]. (1984). Software Pub. Corp.

Internet source

Berst, J. (1998, January 30). Berst alert. *ZDNet*. Retrieved February 1, 1998, from *http://www.zdnet.com/anchordesk/story/story_1716.html*

CD-ROM

McWard, J. (1998). Graphics on-line [CD-ROM]. TW/Inform. New York: EduQuest.

Personal web site

Mohr, E. Home page. Retrieved December 29, 2003, from *http://www.jccc. net/home/depts/1504*

Professional web site

Johnson County Community College Writing Center. Johnson County Community College. Retrieved January 5, 2004, from *http://www.jccc. net/acad/instruction/english/writectr/*

Posting to a discussion listserv

Tsui, P. "Questionnaire." (2005, September 17). Questionnaire. Message posted to Society for Technical Communication Listserv, archived at *http://www.stc.org/questionnaire*

Alternative Style Manuals   Although MLA and APA are popular style manuals, others are favored in certain disciplines. Refer to these if you are interested or required to do so.

- *U.S. Government Printing Office Style Manual*. Washington, DC: Government Printing Office, 1973.
- *The Chicago Manual of Style*, 15th edition. Chicago: University of Chicago Press, 1994.

- Turabian, Kate L. *A Manual for Writers of Term Papers, Theses, and Dissertations.* Chicago: University of Chicago Press, 1973.

4. *Develop your ideas*—You have learned how to organize your report (through an introduction, discussion, and conclusion/recommendation) and how to document your sources of research (through parenthetical source citations and a works cited or references page). Writing your research report also requires that you use your research effectively to develop your ideas. Successful use of research demands that you correctly quote, paraphrase, or summarize. Summaries are discussed in Chapter 15.

Rewriting

As with all types of writing, drafting the text of your research report is only the second stage of the writing process. To ensure that your report is effective, revise your draft as follows:

1. *Add new detail for clarity and persuasiveness*—Too often, students and employees assume that they have developed their content thoroughly when, in fact, their assertions are general and vague. This is especially evident in research reports. You might provide a quotation to prove a point, but is this documentation sufficient? Have you truly developed your assertions? If an idea within your report seems thinly presented, either add another quotation, paraphrase, or summary for additional support or explain the significance of the researched information.

2. *Delete dead words and phrases and researched information that does not support your ideas effectively*—Good writing in a work environment is economical writing. Thus, as always, your goal is to communicate clearly and concisely. Delete words that serve no purpose, maintaining a low fog index. In addition, review your draft for clarity of focus. The goal of a research report is not to use whatever researched information you've found wherever it seems valid. Instead, you want to use quotations, paraphrases, and summaries only when they help develop your statement of purpose. If your research does not support your thesis, it is counterproductive and should be eliminated. In the rewriting stage, delete any documented research that is tangential or irrelevant.

3. *Simplify your words for easy understanding*—The goal of technical writing is to communicate, not to confuse. Write to be understood. Don't say *grain-consuming animal units* if you mean *chickens*. Don't call the July 2000 stock market crash a *fourth-quarter equity retreat*.

4. *Move information within your report to ensure effective organization*—How have you organized your report? Did you use a problem/solution format? Did you use comparison/contrast or cause/effect? Is your report organized as a chronological narrative or by importance? Whichever method you've used, you want to be consistent. To ensure consistency, rewrite by moving any information that is misplaced.

5. *Reformat your text for reader-friendly ease of access*—Look at any technical journal. You will notice that the writers have guided their readers through the text by using headings and subheadings. You will also notice that many journals use graphics (pie charts, bar graphs, line drawings, flowcharts, etc.) to clarify the writer's assertions. You should do the same.

To help your readers follow your train of thought, reformat any blocks of wall-to-wall words. Add headings, subheadings, itemized lists, white space, and graphics.

6. *Correct any errors*—This represents your greatest challenge in writing a research report. You not only must be concerned with grammar and mechanics, as you are when writing a memo, letter, or report, but also with accurate quoting, paraphrasing, summarizing, parenthetical source citations, and works cited.

When revising, pay special attention to these concerns. If you quote, paraphrase, or summarize incorrectly, you run the risk of plagiarizing. If you fail to provide correct parenthetical source citations or works cited, you will make it impossible for your readers to find these same sources of information in their research or to check the accuracy of your data. Research demands accuracy and reliability.

CHAPTER HIGHLIGHTS

1. You can research a topic either in a library or online at your computer.
2. You need to consider the audience's level of technical knowledge when you write a research paper.
3. Reader-friendly highlighting techniques help your audience access information in your research report.
4. Narrowing a topic can help you find sources of information.
5. A focus statement lets you determine the direction of your document.
6. Use discrimination and consider the source when you research on the Internet.
7. Careful source citations help you avoid plagiarism.
8. The Modern Language Association (MLA) and the American Psychological Association (APA) are two widely used style manuals for citing sources.
9. On a summary note card, condense the original material by using your own words.
10. On a paraphrase note/card, restate the original material in your own words without condensing.

EXERCISES

1. Correctly format and alphabetize a Works Cited page that contains the following entries:

   - An anonymously written magazine article
   - A magazine article signed by two authors
   - A journal article signed by one author
   - A book with three or more authors
   - A book with an editor
   - A signed newspaper editorial
   - An online document
   - A CD-ROM document
   - An e-mail message
   - A professional Web site

2. Summarize in one sentence any paragraph from this textbook. Provide a parenthetical source citation and works cited information.

3. Read a one- or two-page article from a magazine, journal, or online source. Then practice note taking. Writing in the margins or between paragraphs, briefly note the key point(s) made in each paragraph. (These notes can be limited to one or two words.)

4. Using a one- or two-page article from a magazine or journal, practice note taking on 3 × 5-inch cards. To do so, first write the correct works cited information on one card. Then, on separate cards, take notes about approximately four key ideas discussed in the article. Write only one note per card; give the card a title for future reference; provide either quotations or paraphrases; and then write the author's name, the article's title, and the page number on the bottom of the card.

5. Select a technical topic from your major field or your job and write a research report. You might want to consider a controversy in your area of interest (such as the greenhouse effect, hazardous waste management, or computer viruses) or the impact of a technical innovation (such as micromachines or the Internet).

6. Many communities have recycling projects that allow residents to recycle paper products, cans, and plastic. Not all businesses recycle, however. Research the benefits of recycling, determine how a business or businesses could implement a corporate recycling plan, and write a report recommending action based on your research.

7. In today's global economy, understanding and accommodating multiculturalism and cross-culturalism in business is important. Research the following:

   - The unique challenges that cultural diversity presents to businesses.
   - How companies have responded to these challenges.

Write a report recommending why and how a business or businesses can help employees develop cultural awareness.

8. Many companies track the time their employees spend either surfing the Web or sending and receiving personal e-mail while at work. Research the following:

- Software that companies use to track employee electronic communication usage.

- Corporate guidelines for employee use of company-owned electronic communication hardware and software.

- The legal and ethical ramifications of an employee's private use of corporate-owned e-mail and Internet access.

- The legal and ethical ramifications of an employer eavesdropping on an employee's Web usage.

Write a research report on your findings and provide a corporate guideline for both employee and employer electronic communication responsibilities.

9. Corporate training is big business. Many companies hire outside consultants or staff company training departments to teach employees new skills. These could include training workshops on diet and exercise to improve work efficiency, techniques for avoiding e-mail viruses or screening e-mail spam, resume writing for transitional employees, time management, leadership skills, improved oral presentations, improved customer service skills, basic word processing, or business accounting for non-accountants.

What training class does your company need? Research possible topics and training approaches. How have other companies offered this training? What benefits do employees derive from this training? How does the company benefit? What are the costs for this training (personnel, time, equipment, etc.)? Then, write a proposal or an instructional training module based on your research.

10. Entrepreneurialism is one of the fastest growing sectors in business. Many people are opening their own businesses. What does it take to open your own business? Before you can write an effective business plan and seek financing from a bank, you must research the project.

Choose a new business venture, selling a product or service of your choice. What would it cost to open this business? What would be your best location, or should your business be online? What certifications or licensing is needed? How many personnel would you need? What equipment is necessary? Who would be your clientele?

Based on research, write a proposal, appropriate for presentation to a bank. In this proposal, present your business plan for a new entrepreneurial opportunity.

WEB WORKSHOP

1. FirstGov.com allows you to research a wide variety of topics, such as education and jobs, benefits and grants, consumer protection, environment and energy, science and technology, and public safety and health. This Web site

also provides information on breaking news. Access FirstGov.com and research a topic relevant to your career goals. Write a memo or report to your instructor summarizing your findings.

2. Go to an online news magazine, such as *Slate*, *Time*, or *U.S. News Online*. Type a topic of interest in the magazine's search engine. Research this topic and make a brief oral presentation to your class about the information you have gathered.

3. Access a search engine such as Google, Ask Jeeves, or Dogpile. Type in a topic relating to your major field. For example, for computer information technology, medical records and health information, or accounting, look for job openings in your field to learn about salary ranges, benefits, application requirements, etc. Report your findings either in an oral presentation to the class or in a memo to your instructor.

4. Using an online newspaper, such as *The New York Times*, *CNN*, or *USA Today*, research business and technology news. Find out the major news stories of the day which relate to your career path. Report your findings in an oral presentation to the class or in a memo to your instructor.

Quiz Questions

1. In what places can you locate researched information?

2. Explain why you need to consider audience when you write a researched report.

3. What is the difference between solicited and unsolicited research?

4. What is the appropriate writing style of a research report?

5. How can you create a reader-friendly researched report?

6. How does a focus statement direct you when you write a research report?

7. What types of topics can you research on search engines?

8. Why must you document your sources?

9. What is the difference between a summary note card and a paraphrase note card?

10. How does APA differ from MLA when you cite a book with one author?

The Summary

Chapter Preview

Criteria for Writing Summaries
A good summary covers all the highlights of a much longer presentation or written document. Following organizational hints allows you to prepare an effective summary.

Process
You can write your summary more easily by following a process.

Process Log
A student-written process log serves as a useful guide.

You might be required to write a summary if, for example, your boss is planning to give a presentation at a civic meeting (Better Business Bureau, Rotary Club, or Businesswomen's Association) and needs some up-to-date information for his or her presentation. Bosses often are too busy to perform research, wade through massive amounts of researched data, or attend a conference at which new information might be presented. Therefore, your boss asks you to research the topic and provide the information in a shortened version. He or she wants you to provide a summary of your findings.

Similarly, your boss might need to give a briefing to upper-level management. Again, time is a problem. How can your boss get the appropriate information and digest it rapidly? The answer is for *you* to read the articles, attend the conference or meeting, and then summarize your research. At other times, you will be asked to summarize your own writing. For example, if you are writing a long report, you might include an **"executive summary,"** an overview of the report's key points (we discuss this in Chapter 17).

As a class assignment, your teacher might want you to write a summary of a formal research report. The summary, though shorter than a research report, would still require that you practice research skills, such as reading, note taking, and writing paraphrases.

To write a summary, you'll need to study the research material. Then, in condensed form (a summary is no more than 5 to 15 percent of the length of the original source), you'll report on the author's main points. Preparing a summary puts to good use your research, analyzing, and writing skills.

The following will help you write a summary:

- Criteria for writing summaries
- Process to follow
- Process log with examples

CRITERIA FOR WRITING SUMMARIES

A well-constructed summary, though much shorter than the original material being summarized, highlights the author's important points. A summary is like a *Reader's Digest* approach to writing. Although the summary will not cover every fact in the original, after reading the summary you should have a clear overview of the original's main ideas. There are several criteria for accomplishing this.

Overall Organization

As with any good writing, a summary contains an introduction, a discussion, and a conclusion.

Introduction

Begin with a topic sentence. This sentence will present the primary focus of the original source and list the two or three major points to be discussed. You must also tell your reader what source you are summarizing. You can accomplish this in one of three ways. You can either list the author's name and article title in the topic sentence; preface your summary with a works cited notation providing the author's name, title, publication date, and page numbers; or follow your topic sentence with a footnote and give the works cited information at the bottom of the page.

> ### example
>
> **Name and Title in Topic Sentence**
>
> In "Robotics: Tactile Sensing" (*Radio Electronics*, August 1986: 71–72), Mark J. Robillard states that because robots move to do their jobs, they need to be equipped with an assortment of tactile sensors.

> ### example
>
> **Works Cited Prefacing Topic Sentence**
>
> Robillard, Mark J. "Robotics: Tactile Sensing." *Radio Electronics*. August 1986: 71–72.
>
> Because robots must move around to do their jobs, they need to be equipped with an assortment of tactile sensors.

> ### example
>
> **Topic Sentence Followed by Explanatory Footnote**
>
> Because robots move around to do their jobs, they need to be equipped with an assortment of tactile sensors.*
>
> _____
> _____
> _____
> _____
> _____
> _____
> _____
> _____
>
> *This paragraph summarizes Mark J. Robillard's "Robotics: Tactile Sensing," *Radio Electronics*, August 1986: 71–72.

Discussion

In this section, briefly summarize the main points covered in the original material. To convey the author's ideas, you can paraphrase, using your own words to restate the author's point of view.

Conclusion

To conclude your summary, you can either reiterate the focus statement, reminding the reader of the author's key ideas; highlight the author's conclusions regarding his or her topic; or state the author's recommendations for future activity.

Internal Organization

How will you organize the body of your discussion? Because a summary is meant to be objective, you should present not only what the author says but also how he or she organizes the information. For example, if the author has developed his or her ideas according to a problem/solution format, your summary's discussion should also be organized as a problem/solution. This would give your audience the author's content and method of presentation. Similarly, if the author's article is organized according to cause/effect, comparison/contrast, or analysis (classification/division), this would determine how you would organize your summary.

Development

To develop your summary, you'll need to focus on the following:

- **Most important points**—Because a summary is a shortened version of the original, you can't include all that the author says. Thus, you should include only the two or three key ideas within the article. Omit irrelevant details, examples, explanations, or descriptions.
- **Major conclusions reached**—Once you've summarized the author's key ideas, then state how these points are significant. Show their value or impact.
- **Recommendations**—Finally, after summarizing the author's major points and conclusions, you'll want to tell your audience if the author recommends a future course of action to solve a problem or to avoid potential problems.

Style

The summary, like all technical writing, must be clear, concise, accurate, and accessible. Therefore, you'll want to abide by the fog index. Watch out for long words and sentences. Avoid technical jargon. Most important, be sure that your summary truly reflects the author's content. Your summary must be an unbiased presentation of what the author states and include none of your opinions.

Length

As mentioned earlier, the summary will be approximately 5 to 15 percent the length of the original material. To achieve this desired length, omit references to the author (after the initial reference in the works cited or topic sentence). You'll also probably need to omit some types of material:

- Past histories
- Definitions
- Complex technical concepts

- Statistics
- Tables and figures
- Tangential information (such as anecdotes and minor refutations)
- Lengthy examples
- Biographical information

Audience Recognition

You must consider your audience in deciding whether to omit or include information. Although you usually would omit definitions from a summary, this depends on your audience. Technical writing is useless if your reader does not understand your content. Therefore, determine whether your audience is high tech, low tech, or lay.

Grammar and Mechanics

As always, flawed grammar and mechanics will destroy your credibility. Not only will your reader think less of your writing and research skills, but also errors in grammar and mechanics might threaten the integrity of your summary. Your summary will be inaccurate and, therefore, invalid.

PROCESS

The Writing Process		
Prewriting	Writing	Rewriting
• Examine your purposes • Determine your goals • Consider your audience • Gather your data • Determine how the content will be provided	• Organize the draft according to some logical sequence that your readers can follow easily • Format the content to allow for ease of access	• Revise ° Add missing details ° Delete wordiness ° Simplify word usage ° Enhance the tone of your communication ° Reformat your text for ease of access ° Practice the speech or review the text • Proofread ° Correct errors

Writing your summary will be easiest if you follow the same step-by-step process detailed throughout this text. Remember that the writing process is dynamic, with the steps frequently overlapping.

Prewriting

To gather your data and determine your objectives, do the following:

Locate Your Periodical Article, Book Chapter, or Report

If you are summarizing a meeting or seminar, take notes. If your boss or instructor gives you your data, one major obstacle has been overcome. If, however, you need to visit a library or go online to research your topic, refer to Chapter 14 for helpful hints on research skills.

Write Down Your Works Cited or References Information

Once you've found your research material, *immediately* write down the author's name, the article's title, the name of the periodical or book in which this source was found, the date of publication, and the page numbers. Doing so will save you frustration later. For example, if you lose your copy of the article and have not documented the source of your information, you will have to begin your research all over again. Instead, spend a few minutes at the beginning of your assignment documenting your sources. This will ensure that you do not spend several harried hours the night before the summary is due frantically searching for the missing article. (Chapter 14 provides the correct works cited or references format.)

Read the Article to Acquire a General Understanding of Its Content

Determine exactly what the author's thesis is and what main points are discussed. You can take notes in one of two ways. Either you can underline important points, or you can make marginal or interlinear notes on a photocopy. For a summary of an article or chapter in a book, we suggest that you write notes directly on the photocopy. To do so, read and reread a paragraph. After you've read and understood the paragraph, write a one- or two-word notation in the margin that sums up that paragraph's main idea.

Writing

Once you've gathered your data and determined your author's main ideas, you're ready to write a rough draft of your summary.

Review Your Prewriting

Look back over your marginal notes and your underlining. Have you omitted any significant points? If you have, now is the time to include them. Have you included any ideas that are insignificant or tangential? If so, delete these to ensure that your summary is the appropriate length.

Write a Rough Draft, Using the Sufficing Technique

Once you've reviewed the data you have gathered, quickly draft the text of your summary. Follow the criteria for writing an effective summary discussed earlier in this chapter, including

- A topic sentence clearly stating the author's main idea. Also provide the works cited information here, in a title, or in an explanatory footnote.
- Organization paralleling the author's method of organization (comparison/contrast, argument/persuasion, cause/effect, analysis, chronology, spatial, importance, etc.).

- Clear transitional words and phrases.
- Development through paraphrases. Restate the author's ideas in your own words.
- A conclusion in which you reiterate the main points discussed, state the significance of the author's findings, or recommend a future course of action.

Rewriting

Once you've written a rough draft of your summary, it's time to revise. Follow the six-step revision techniques to ensure that your finished product will be acceptable to your readers.

Add Detail for Accuracy

When looking over your draft, be sure that you've covered all of the author's major points. If you haven't, now is the time to add any omissions. Be sure that your summary accurately covers the author's primary assertions.

Delete Unnecessary Information, Biased Comments, and Dead Words and Phrases

Deleting is especially important in a summary. Because a summary must be no more than 5 to 15 percent of the length of the original source, your major challenge is brevity. Therefore, review your draft to see if you have included any of the author's points that are *not* essential. Such nonessential information may include side issues that are interesting but not mandatory for your reader's understanding.

Another type of information you'll want to delete will be complex technical theories. Although such data might be valuable, a summary is not the vehicle for conveying this kind of information. If you have included such theories, delete them. These deletions serve two purposes. First, your summary will be stronger since it will focus only on key ideas and not on tangential arguments. Second, the summary will be briefer due to your deletions.

Delete biased comments. If you inadvertently have included any of your attitudes toward the topic, remove these biases. A summary should not present your ideas; it should reflect only what the author says.

Finally, as with all good technical writing, conciseness is a virtue. Long words and sentences are not appreciated. Reread your draft and delete any unnecessary words and phrases. Strive for an average of 15-word sentences and one- or two-syllable words. We're not suggesting that every multisyllabic word is incorrect, but keep them to a minimum. If an author has written about telecommunications or microbiology, for example, you shouldn't simplify these words. You should, however, avoid long words that aren't needed and long sentences caused by wordy phrases.

Simplify "Impressive" Words and Complex Technical Terms

Good technical writing doesn't force the reader to look up words in a dictionary. Your goal in a summary is to present complex data in a brief and easily understandable package. This requires that you avoid difficult words. In a summary, *difficult* means two things. First, as with all good technical writing, you want to use words that readers understand immediately. Don't write *supersede* when you mean *replace*. Don't write *remit* when you mean *pay*. Second, you'll also need to simplify technical terms. To do so, you can define them parenthetically or merely replace the technical term with its definition.

Move Information Within the Summary to Parallel the Author's Organization

Does your summary reflect the order in which the author has presented his or her ideas? Your summary must not only tell the reader what the author has said but also how the author has presented these ideas. Make sure that your summary does this. If your summary fails to adhere to the author's organization, move information around. Cut and paste. Maintain the appropriate comparison/contrast, argument/persuasion, or cause/effect sequence.

Correct Errors

Reread your draft and look for grammatical and mechanical errors. Don't undermine your credibility by misspelling words or incorrectly punctuating sentences. In addition, accurate information is mandatory. You might be writing this summary for your boss, who is giving a speech at a civic meeting or briefing upper-level management. Your boss is trusting you to provide accurate information. Therefore, to ensure that your boss is not embarrassed by errors, review your text against the original source. Make sure that all the information is correct.

Avoid Biased Language

Biased language is always inappropriate and potentially offensive to readers.

As one of our regular rewriting tips, we ask you to *enhance* the text by adding pronouns and positive words. For summaries, however, such enhancements would be incorrect. Because the summary must only restate what the author has written and be devoid of your attitude, total objectivity is necessary.

Use the following Summary Checklist to help you write an effective summary.

Summary Checklist

___ **1.** Does your summary provide the works cited information for the article that you're summarizing?

___ **2.** Is the works cited information correct?

___ **3.** Does your summary begin with an introduction clearly stating the author's primary focus?

___ **4.** Does your summary's discussion section explain the author's primary contentions and omit secondary side issues?

___ **5.** In the discussion section, do you explain the author's contentions through pertinent facts and figures while avoiding lengthy technicalities?

___ **6.** Is your content accurate? That is, are the facts that you've provided in the summary exactly the same as those the author provided to substantiate his or her point of view?

___ **7.** Have you organized your discussion section according to the author's method of organization?

___ **8.** Did you use transitional words and phrases?

___ **9.** Have you omitted direct quotations in the summary, depending instead on paraphrases?

___ **10.** Does your conclusion either reiterate the author's primary contentions, reveal the author's value judgment, or state the author's recommendations for future action?

___ **11.** Is your summary completely objective, avoiding any of your own attitudes?

___ **12.** Have you used an effective technical writing style, avoiding long sentences and long words?

___ **13.** Are your grammar and mechanics correct?

___ **14.** Have you avoided biased language?

Process Log

Following is one student's successful summary of an article, including the student's rough draft, revisions, and finished copy.

Prewriting

First, the student found an article, wrote down the required works cited information, read the article, and then took marginal notes and underlined key points.

Writing

The student wrote a rough draft from the notes and underlining (Figure 15.1).

Rewriting

Student peer evaluators suggested revisions (Figure 15.2 on page 478).

The student made these changes and prepared a finished version of the summary (Figure 15.3 on page 479).

In "Robotics: Tactile Sensing" (Radio Electronics, August 1986), Mark J. Robillard states, "Most robots must move around to accomplish their tasks" (71). The information that is needed to accomplish these tasks is gathered through an assortment of sensors—touch sensors, sound sensors, and light sensors. Depending on what type of tactile information is obtained through these sensors, objectives of the robot can be met. A robot's gripper could crush an object or not exert enough force to hold on to an object if it doesn't have sensors to determine the amount of force needed. Microswitches are used in the form of touch sensors. They are the simplest form of sensors used for this purpose. To eliminate weight and space used for switches, LED/phototransistor pairs can be used. "If you've ever been bowling, that setup should look familiar" (71). The phototransistor is then interfaced to a computer or other type of controller. The pairs of sensors provide more than just a there/not there signal. "The amount of light that is reflected provides an indication of how close the object is" (72). This approach is patented by "Heath's Hero 2000" (72) and uses optical encoder disks. Integrated circuits that use "strain gauges and pressure sensitive pain" (72) are yet another way to detect the amount of force applied to an object.

All of these different methods of allowing a robot to interface with the real world "comprise a field of inquiry that is as large as robotics itself" (72).

Figure 15.1   Rough Draft

Could you separate Bibliography from your

topic In "Robotics: Tactile Sensing" (<u>Radio Electronics</u>, August 1986), Mark J.

sentence? *— underline periodical titles*

Robillard states, "Most robots must move around to accomplish their tasks" (71). *Avoid quotes— paraphrase instead*

The information ~~that is~~ needed to accomplish these tasks is gathered through

Repetitious

an assortment of <u>sensors</u>—touch <u>sensors</u>, sound <u>sensors</u>, and light <u>sensors</u>.

Depending on what type of tactile information is obtained through these sen-

sors, objectives of the robot can be met. A robot's gripper could crush an object

Do you need this? or not exert enough force to hold on to an object if it doesn't have sensors to

determine the amount of force needed. Microswitches are used in the form

— Combine these sentences of touch sensors. They are the simplest form of sensors used for this purpose.

To eliminate weight and space used for switches, LED/phototransistor pairs can

be used. "If you've ever been bowling, that setup should look familiar" (71). *Don't quote*

The phototransistor is then interfaced to a computer or other type of con-

— Combine these sentences troller. The pairs of sensors provide more than just a there/not there signal.

Don't quote "The amount of light that is reflected provides an indication of how close the

— Do you need this? object is" (72). ~~This approach is patented by "Heath's Hero 2000"~~ (72) and uses

optical encoder disks. Integrated circuits that use "strain gauges and pressure

Don't quote sensitive pain" (72) are yet another way to detect the amount of force applied

to an object.

— Write just one ¶ All of these different methods of allowing a robot to interface with the real

Don't world "comprise a field of inquiry that is as large as robotics itself" (72). *quote*

Figure 15.2   Rough Draft with Editing

Robillard, Mark J. "Robotics: Tactile Sensing." *Radio Electronics* August 1986: 71–72.

Robots, which must move to do their jobs, require an assortment of tactile sensors. These sensors help the robots locate items and use the appropriate amount of force for gripping an object. Microswitches, in the form of touch sensors, are the simplest sensing devices. These microswitches can be used to grip an object. In addition, microswitches can help the robot maintain contact with the ground to avoid falling down stairs. LED/phototransistors, similar to those used on bowling alley foul lines, are less bulky than microswitches. The phototransistor, when interfaced with a computer, provides more than a there/not there signal. The light reflected indicates the exact placement of the object. The above sensors, however, have difficulty gauging the appropriate force required for gripping. This problem could be solved by using optical encoder disks and integrated circuits to determine the appropriate force. All of the above tactile sensing devices constitute an important part of successful robotics.

Figure 15.3   Revised Summary

CHAPTER HIGHLIGHTS

1. A summary is a compressed version of a much longer document or speech.
2. You can summarize chapters, books, speeches, reports, and material from the Internet.
3. A well-written summary is about 5–15 percent the length of the original.
4. Include an introduction, a discussion, and a conclusion to give your summary a coherent design.
5. Include source citations when appropriate.
6. The writing process helps you create an effective summary.
7. Delete biased comments from a summary because a summary should not reflect your ideas.
8. Avoid using long words and sentences in a summary.
9. Organize your summary in the same way the author organized the original source.
10. In a summary focus only on key ideas and not on tangential arguments.

EXERCISES

1. Locate an article that interests you (one within your field of expertise, your degree program, or an area that you would like to pursue). Study this article and summarize it according to the criteria provided in this chapter.

2. Locate an article that interests you. After reading it, take marginal notes (one- to three-word notations per paragraph) highlighting the article's key points.

3. Once you've read an article and made marginal notes, write either a topic or a sentence outline (discussed in Chapter 14).

4. Read three to five articles. Then determine what method of organization the authors have used. Have the authors used analysis? Others might use division, focusing on parts of a whole, comparison/contrast, argument/persuasion, cause/effect, and so on.

5. Many textbooks begin or end chapters with summaries. Find such a summary in one of your textbooks. Then read the accompanying chapter. Is the summary effective? If so, why? If not, why not? If the summary is ineffective, how would you rewrite it?

6. After attending a lecture, meeting, or conference, summarize its content. Provide the speaker's name, the location of the presentation, and the date of presentation for the source citation.

WEB WORKSHOP

1. Using the Internet, research 10 companies in your major field. Write a summary of each home page and connecting links.
2. Many companies provide annual stockholder's reports online. Access a Web browser and find a company's online stockholder's report. Using the techniques discussed in this chapter, summarize the report.
3. Many companies issue press releases or news releases online to update their employees or stakeholders of important corporate information. Access a Web browser and find a company's press release. Using the techniques discussed in this chapter, summarize your findings.
4. Access a Web browser to research articles on current issues, such as *outsourcing, telecommuting, globalization, multiculturalism, flextime, microrobotics, job sharing, virtual teams, best practices,* and *retooling.* Summarize your findings by following the techniques presented in this chapter.

QUIZ QUESTIONS

1. What constitutes a well-constructed summary?

2. Explain the function of a topic sentence in a summary.

3. In what ways can you organize a summary?

4. What types of material are omitted from a summary?

5. What key aspects of writing style are evident in a summary?

6. Approximately how long is a summary?

7. What is an executive summary?

8. Why should you avoid biased comments in a summary?

9. What two primary goals do you achieve when you delete unnecessary information or biased comments from a summary?

10. Why should you avoid complex technical terms in a summary?

ACTIVITIES

CHAPTER

16 Reports

Chapter Preview

Criteria for Writing Reports
Consider format, development, and style when you write reports.

Types of Reports
- Trip reports
- Progress reports
- Lab reports
- Feasibility/recommendation reports
- Incident reports
- Investigative reports
- Meeting minutes

Process
The writing process—prewriting, writing, and rewriting—helps you prepare effective reports.

Writing at Work

EFA Incorporated

Education For All . . . You're Always at Home at Our College

Cindy Kaye is director of administrative technology at **EFA (Education For All) Incorporated,** a proprietary university system with branches in 35 cities and a home office in Philadelphia, Pennsylvania.

EFA has 49,000 students nationwide and offers traditional, on-campus bachelor's degrees, as well as nontraditional degree programs online. Students can earn degrees in accounting, telecommunications management, biomedical technology, electronic engineering technology, computer information systems, and computer engineering technology.

Cindy has been traveling to Philadelphia from her university site in Miami, Florida, four days a week for a year. She is part of a team being trained on EFA's new administrative software. Her team consists of faculty, staff, and administrators. The software they are learning will be used to manage systems nationwide. It will allow for

- Electronic registration
- Online grading and submission of end-of-term grades
- E-mail for faculty, staff, and students
- Discussion groups
- Course information
- Online counseling
- Employee benefits
- Online coursework
- Community news

Cindy must document her activities weekly, monthly, and biannually. First, as the project team's recording secretary, Cindy keeps her team's weekly **meeting minutes.**

Next, she must record her travel expenses and the team's achievements, necessitating weekly **trip reports** submitted to her dean at her home school site.

As a member of her team's technology impact task force subcommittee, Cindy also has been asked to study technology options and the extent to which changes will influence EFA's academic procedures. This means that she will write a **feasibility/recommendation report** following her study to justify the implementation of the new technology systems the team decides on.

Finally, when her project is completed, Cindy will collaborate with her team members to write a **progress report** for EFA's board of directors.

Though Cindy's area of expertise is computer information systems, her job requires much more than programming or overseeing the networking of her corporation's computer systems. Cindy's primary job has become communication with colleagues and administrators. Writing reports is a major component of this job requirement.

At one time or another, you'll be asked to write a report. Reports can vary in length. Generally, a shorter report (approximately one to five pages) will be formatted differently than a longer report (more than five pages long). This chapter focuses on the design of shorter reports. Chapter 17 discusses the design of longer reports, using a proposal as an example. Your reports will satisfy one or all of the following needs:

- Supply a record of work accomplished
- Record and clarify complex information for future reference
- Present information to a large number of people
- Record problems encountered
- Document schedules, timetables, and milestones
- Recommend future action
- Document current status
- Record procedures

CRITERIA FOR WRITING REPORTS

Although there are many different types of reports and individual companies have unique demands and requirements, certain traits, including format, development, and style, are basic to all report writing.

Organization

Every report should contain four basic units: heading, introduction, discussion, and conclusion/recommendations.

Heading

The heading includes the date on which the report is written, the names of the people to whom the report is written, the names of the people from whom the report is sent, and the subject of the report (as discussed in Chapter 5, the subject line should contain a *topic* and a *focus*).

```
DATE:      August 13, 2005
TO:        Shelley Stine
FROM:      Julie Jones
SUBJECT:   REPORT ON TRIP TO SOUTHWEST REGIONAL
           CONFERENCE ON ENGLISH (FORT WORTH, TEXAS)
```

Introduction

The introduction supplies an overview of the report. It can include three optional subdivisions:

- **Purpose**—a topic sentence(s) explaining why you are submitting the report (rationale, justification, objectives) and exactly what the report's subject matter is.
- **Personnel**—names of others involved in the reporting activity.
- **Dates**—what period of time the report covers.

---

example

> Objectives—I attended the National Electronic Packaging Conference in Anaheim, California, to review innovations in vapor phase soldering.
> Dates—September 26–30, 2005
> Personnel—Susan Lisk and Larry Rochelle

---

tech link

Go to *http://www.prenhall. com*/gerson for Web links, samples, and interactive activities.

Some businesspeople omit the introductory comments in writing reports and begin with the discussion. They believe that introductions are unnecessary because the readers know why the reports are written and who is involved.

These assumptions are false for several reasons. First, it is false to assume that readers will know why you're writing the report, when the activities occurred, and who was involved. Perhaps if you are writing only to your immediate supervisor, there's no reason for introductory overviews. However, even in this situation you might have an unanticipated reader because

- **Immediate supervisors change**—they are promoted, fired, retire, or go to work for another company.
- **Immediate supervisors aren't always available**—they're sick for the day, on vacation, or off-site for some reason.

Second, avoiding introductory overviews assumes that your readers will remember the report's subject matter. This is false because reports are written not just for the present, when the topic is current, but for the future, when the topic is past history. Reports go on file—and return at a later date. At that later date,

- You won't remember the particulars of the reported subject matter.
- Your colleagues, many of whom weren't present when the report was originally written, won't be familiar with the subject.
- You might have outside, lay readers who need additional detail to understand the report.

An introduction—which seemingly states the obvious—is needed to satisfy multiple readers, readers other than those initially familiar with the subject matter, and future readers who are unaware of the original report.

Discussion

The discussion of the report summarizes your activities and the problems you encountered. This is the largest section of the report and involves development, organization, and style (more on these later).

Conclusion/Recommendations

The conclusion allows you to sum up, to relate what you've learned, or to state what decisions you have made regarding the activities reported. The recommendations allow you to suggest future action, to state what you believe you and your company should do next.

> The conference was beneficial. Not only did it teach me how the computer can save us time and money, but also I received hands-on training. Because the computer can assist our billing and inventory control, let's buy and install three terminals in book-keeping before our next quarter.

Development

Now that you know what subdivisions are traditional in reports, the next questions are, "What do I say in each section? How do I develop my ideas?"

First, answer the reporter's questions.

1. *Whom* did you meet or contact, who was your liaison, who was involved in the accident, who was on your technical team, and so on?
2. *When* did the documented activities occur (dates of travel, milestones, incidents, etc.)?
3. *Why* are you writing the report and why were you involved in the activity (rationale, justification, objectives)? Or, for a lab report, for example, why did the electrode, compound, equipment, or material act as it did?
4. *Where* did the activity take place?
5. *What* were the steps in the procedure, what conclusions have you reached, or what are your recommendations?

Second, when providing the foregoing information, *quantify!* Do not hedge or be vague or imprecise. Specify to the best of your abilities with photographic detail.

The following justification is an example of vague, imprecise writing.

> **Installation of the machinery is needed to replace a piece of equipment deemed unsatisfactory by an Equipment Engineering review.**
>
> *Before*

Which machine are we purchasing? Which piece of equipment will it replace? Why is the equipment unsatisfactory (too old, too expensive, too slow)? When does it need to be replaced? Where does it need to be installed? Why is the installation important? A department supervisor will not be happy with the preceding report. Instead, supervisors need information *quantified*, as follows:

> The *exposure table* needs to be installed by *9/05* so that we can *manufacture printed wiring products with fine line paths and spacing (down to 0.0005 inch).* The table will replace the *outdated printer* in *Dept. 76.* Failure to install the table *will slow the production schedule by 27%.*
>
> *After*

Note that the italicized words and phrases provide detail by quantifying.

Style

Style includes conciseness, simplicity, and highlighting techniques. As already discussed, you achieve conciseness by eliminating wordy phrases. Say *consider* rather than *take into consideration;* say *now* rather than *at this present time.* You achieve simplicity by avoiding old-fashioned words: *utilize* becomes *use, initiate* becomes *begin, supersedes* becomes *replaces.*

The value of highlighting has already been shown in this chapter. The parts of reports reviewed earlier use headings (Introduction, Discussion, Conclusion/Recommendation). Graphics can also be used to help communicate content, as evident in the following example. A recent demographic study of Kansas City predicted growth patterns for Johnson County (a large county south of Kansas City):

Before

Johnson County is expected to add 157,605 persons to its 1980 population of 270,269 by the year 2010. That population jump would be accompanied by a near doubling of the 96,925 households the county had in 1980. The addition of 131,026 jobs also is forecast for Johnson County by 2010, more than doubling its employment opportunity.

This report is difficult to access readily. We are overloaded with too much data. Luckily, the report provided a table (Table 16.1) for easier access to the data. Through highlighting techniques (tables, white space, headings), the demographic forecast is made accessible at a glance.

After

TABLE 16.1 JOHNSON COUNTY PREDICTED GROWTH BY 2010			
	Population	Households	Employment
1980	270,269	96,925	127,836
2010	427,874	192,123	258,862
% change	+58.3%	+98.2%	+102%

TYPES OF REPORTS

All reports include a heading (date, to, from, subject), an introduction, a discussion, and conclusion/recommendations. However, different types of short reports customize these generic components to meet specific needs. Let's look at the criteria for seven common types of reports: trip reports, progress reports, lab reports, feasibility/recommendation reports, incident reports, investigative reports, and meeting minutes.

Trip Reports

Purpose and Examples

A *trip report* allows you to report on job-related travel. When you leave your work site and travel for job-related purposes, your supervisors not only require that you document your expenses and time while off-site, but they also want to be kept up-to-date on your work activities. For example, you might be engaged in work-related travel as follows:

- **Information Technology**—You go to a conference to learn about the latest hardware and software technologies for the workplace. There, you meet with vendors, participate in hands-on technology workshops, and learn what other companies are doing to manage their technology needs. When you return, you write a trip report documenting your activities.

- **Heating, Ventilating, and Air Conditioning**—One of your clients is building an office site. Your company has been hired to install their heating, ventilating, and air conditioning (HVAC) system. You travel to your client's home office to meet with other contractors (engineering and architectural) so all team members can agree on construction plans. At the conclusion of your job-related travel, you will write a trip report about your meeting.

- **Engineering**—To be a responsible member of your community, your company has partnered with the local school district. One of your jobs is to visit the city's high schools and discuss engineering job preparedness. To do so, you give an oral presentation about the engineering job market, desired job skills, and the importance of technical communication in the workplace. Upon returning to your office, your supervisor wants a trip report not only to document how the company is working with the community but also to learn how else the company can meet community needs.

- **Biomedical Equipment Sales**—Four days a week, every week, you are on the road making sales calls. Each month, you must document your job-related travel to show that you are making your quota and to receive recompense for travel expenditures.

Criteria

Following is an overview of what you will include in an effective trip report.

1. *Heading*

   Date

   To

   From

   Subject (topic + focus)

2. *Introduction (overview, background)*

   Purpose. In the purpose section, document the date(s) and destination of your travel. Then comment on your objectives or rationale. What motivated the trip, what did you plan to achieve, what were your goals, why were you involved in job-related travel?

   You might also want to include these following optional subheadings:

   Personnel. With whom did you travel?

   Authorization. Who recommended or suggested that you leave your work site for job-related travel?

3. *Discussion (body, findings, agenda)*

   Using subheadings, document your activities. This can include a review of your observations, contacts, seminars attended, or difficulties encountered.

4. *Conclusion/recommendations*

Conclusion. What did you accomplish—what did you learn, whom did you meet, what sales did you make, what of benefit to yourself, colleagues, or your company occurred?

Recommendations. What do you suggest next? Should the company continue on the present course (status quo) or should changes be made in personnel or in the approach to a particular situation? Would you suggest that other colleagues attend this conference in the future, or was the job-related travel not effective? In your opinion, what action should the company take?

Figure 16.1 presents an example of a trip report.

Progress Reports

Purpose and Examples

A *progress report* lets you document the status of an activity, explaining what work has been accomplished and what work is remaining. Supervisors and customers want to know what progress you are making on a project, whether you are on schedule, what difficulties you might have encountered, and what your plans are for the next reporting period. Because of this, your audience might ask you to write progress (or activity or status) reports—daily, weekly, monthly, quarterly, or annually.

- **Biomedical Technology**—You and your team are developing a new heart monitor. This entails researching, patenting, building, testing, and marketing. You have been working on this project for months. What is your status? A progress report will tell your investors and supervisors where you stand, if you are on schedule, and when the project will conclude.

- **Hospitality Management**—The city's convention center is considering new catering options. Your job has been to compare and contrast catering companies to see which one or ones would best be suited for the convention center's needs. The deadline is arriving for a decision. What is the status? Whom have you considered, what are their prices and food choices, what additional services do they offer, and so forth? You need to submit a progress report so management can determine what the next steps should be.

- **Project Management**—Your company is renovating its home office. Many changes have occurred. These include new carpeting, walls moved to create larger cubicles, the construction of larger conference rooms, a new cafeteria and fitness center, and improved lighting. Other changes are still in progress, such as increased parking spaces, exterior landscaping, a child care center, and handicapped accessibility. The supervisor wants to know when these renovations will be concluded. You need to write a progress report to quantify what has occurred, what work is remaining, and when work will be finished.

- **Automotive Technology**—Your company recently suffered negative publicity due to product failures. As manufacturing supervisor, you have initiated new procedures for automotive manufacturing to improve your product quality. How are these procedural changes going? Your company CEO needs an update. To provide this information, you must write a progress report.

DATE: February 26, 2005
TO: Pat Berry
FROM: Debbie Rulo
SUBJECT: TRIP REPORT—RENTON WEST SEMINAR ON
           ELECTRONIC PACKAGING

The focus

The topic

INTRODUCTION

On Tuesday, February 23, 2005, I attended the Renton West National Electronic Packaging Seminar, held in Ruidoso, New Mexico. My goal was to acquire hands-on training and to learn new techniques for electronic packaging for our Telemetry Department. Also in attendance from our department were Richard DiBono and Bill Cole.

The introduction answers the reporter's questions—who, what, when, where, and why.

DISCUSSION

Different heading levels and highlighting techniques are used to make the information more accessible.

Richard, Bill, and I attended the following seminars:

- *Production Automation*
  This two-hour workshop was presented by Dr. Wang Ng, a noted scientist from Southwest Texas State University. During Dr. Ng's presentation, we reviewed four foam-encapsulation automated techniques for electronic packaging and received hands-on training. Dr. Ng worked individually with each seminar participant.

- *Vapor Phase Soldering*
  The hour-long presentation was facilitated by Garth Nelson, a manufacturing supervisor from Spark Welding and Soldering, Inc. (Colorado Springs, Colorado). Mr. Nelson showed a slide presentation on techniques for processing double-sided chip components.

- *Electronics for Extreme Temperatures*
  This hour-long presentation was led by Randy Towner and Leanna Wilson, professors at the University of Nevada, Las Vegas. Their scientific data on packaging under temperature extremes was accompanied by a workbook and a multimedia presentation.

- *Robotics and You*
  Denise Pakula, Canyon Electronics, Tempe, Arizona, spoke about her company's robotics packaging

Figure 16.1   Trip Report

A new-page notation helps your readers avoid losing or misplacing pages.

equipment. This half-hour presentation focused on the various machines Canyon manufactures and its diverse capabilities.

CONCLUSION/RECOMMENDATION

Every presentation we attended was beneficial. However, the following information will clarify which workshop(s) would profit our company the most:

The *conclusion* sums up the report by focusing on primary findings.

1. Dr. Ng's program was the most useful and informative. His interactive presentation skills were superb, including hands-on activities, small-group discussions, and individual instruction. Richard, Bill, and I suggest that you invite Dr. Ng to our site for further consultation and training.

The *recommendation* suggests action to be taken.

2. Vapor phase soldering is costly. We could not pay back our investment within this quarter. In addition, Garth Nelson's training techniques are outdated—training videos are not as interactive as our personnel request in their annual training evaluations.

3. The Towner/Wilson scientific data on packaging under temperature extremes would benefit our supervisors, who are interested in up-to-date information. However, our new hires would be overwhelmed by the data. If we considered inviting these professors in for consultation, we would have to assess audience carefully.

Recommendation

4. Canyon Electronics' half-hour presentation was geared more to sales than to instruction. Ms. Pakula's focus was on new models of robotics equipment. Our current equipment is satisfactory. If the Purchasing Department is looking for new vendors, we could suggest that they contact Canyon.

Recommendation

Figure 16.1 *Continued*

Criteria

Following is an overview of what you will include in an effective progress report.

1. *Heading*

   Date

   To

   From

   Subject: Include the topic about which you are reporting and the reporting interval (date). (Please see the example on page 491.)

2. *Introduction (overview, background)*

   Objectives: These can include the following:

   - Why are you working on this project (what's the rationale)?
   - What problems motivated the project?
   - What do you hope to achieve?
   - Who initiated the activity?

   Personnel: With whom are you working on this project (i.e., work team, liaison, contacts)?

   Previous activity: If this is the second, third, or fourth report in a series, remind your readers what work has already been accomplished. Bring them up-to-date with background data or a reference to previous reports.

3. *Discussion (findings, body, agenda)*

   Work accomplished: Using subheadings, itemize your work accomplished either through a chronological list or a discussion organized by importance.

   Work remaining: Tell your reader what work you plan to accomplish next. List these activities, if possible, for easy access.

   A visual aid, such as a Gantt chart or a pie chart, fits well after these two sections. The chart will graphically depict both work accomplished and work remaining. (See Chapter 9 for discussions of Gantt and pie charts.)

   Problems encountered: Inform your reader(s) of any difficulties encountered (late shipments, delays, poor weather, labor shortages) not only to justify your possibly being behind schedule but also to show the readers where you'll need help to complete the project.

4. *Conclusion/Recommendations*

   Conclusion: Sum up what you've achieved during this reporting period and provide your target completion date.

   Recommendations: If problems were presented in the discussion, you can recommend changes in scheduling, personnel, budget, or materials that will help you meet your deadlines.

Figure 16.2 presents an example of a progress report.

Lab Reports

Purpose and Examples

A *lab report* lets you document the status of and findings from a laboratory experiment, procedure, or study. Professionals in electronics, engineering, medical fields, the computer industry, and other technologies often rank the ability to

DATE: April 2, 2005
TO: Buddy Ramos
FROM: Pat Smith
SUBJECT: FIRST QUARTERLY REPORT—PROJECT 80
           CONSTRUCTION

INTRODUCTION

The *introduction* explains *why* the report was written and *what* topic will be discussed.

In response to your December 10, 2004, request, following is our first quarterly report on Project 80 Construction (Downtown Airport). Department 93 is in the start-up phase of our company's 2003 build plans for the downtown airport and surrounding site enhancements. These construction plans include the following:

1. Airport construction—terminals, runways, feeder roads, observation tower, parking lots, maintenance facilities.

2. Site enhancements—northwest and southeast collecting ponds, landscaping, berms, and signage.

DISCUSSION

<u>Work Accomplished</u>
In this first quarter, we've completed the following:

1. *Subcontractors*—Toby Summers and Karen Kuykendahl worked with our primary subcontractors (Apex Engineering and Knoblauch and Sons Architects). Toby and Karen arranged site visitations and confirmed construction schedules. This work was completed January 12, 2005.

2. *Permits*—Once site visitations were held and work schedules agreed upon, Karen and Toby acquired building permits from the city. They accomplished this task on January 20, 2005.

The *discussion* provides quantified data and dates for clarity.

3. *Core Samples*—Core sample screening has been completed by Department 86 with a pass/fail ratio of 76.4 percent pass to 23.6 percent fail. This meets our goal of 75 percent. Sample screening was completed January 30, 2005.

Figure 16.2   Progress Report

4. *Shipments*—Timely concrete, asphalt, and steel beam shipments this quarter have provided us a 30-day lead on scheduled parts provisions. Materials arrived February 8, 2005.

5. *EPA Approval*—EPA agents have approved our construction plans. We are within guidelines for emission controls, pollution, and habitat endangerment concerns. Sand cranes and pelicans nest near our building site. We have agreed to leave the north plat (40 acres) untouched as a wildlife sanctuary. This will cut into our parking plans. However, because the community will profit, we are pleased to make this concession. EPA approval occurred on February 15, 2005.

<u>Work Remaining</u>
To complete our project, we need to accomplish the following:

1. *Advertising*—Our advertising department is working on brochures, radio and television spots, and highway signs. Advertising's goal is to make the construction of a downtown airport a community point of pride and civic celebration.

2. *Signage*—With new roads being constructed for entrance and exit, our transportation department is working on street signage to help the public navigate our new roads.

   In addition, transportation is working with advertising on signage designs for the downtown airport's two entrances. These signs will juxtapose the city's symbol (a flying pelican) with an airplane taking off. The goal is to create a logo that simultaneously promotes the preservation of wildlife and suggests progress and community growth.

Figure 16.2   *Continued*

communicate as highly as they do their technical skills. Conclusions drawn from a technical procedure are worthless if they reside in a vacuum. The knowledge acquired from a laboratory activity *must* be communicated to colleagues and supervisors so they can benefit from your discoveries. You write a lab report after you have performed the lab to share your findings.

- **Biomedical Technology**—You have performed a pathology study on tissue, reviewed a radiological scan, or drawn blood. What have you found? To help nurses and doctors provide the best patient care, you must write a lab report documenting your findings.

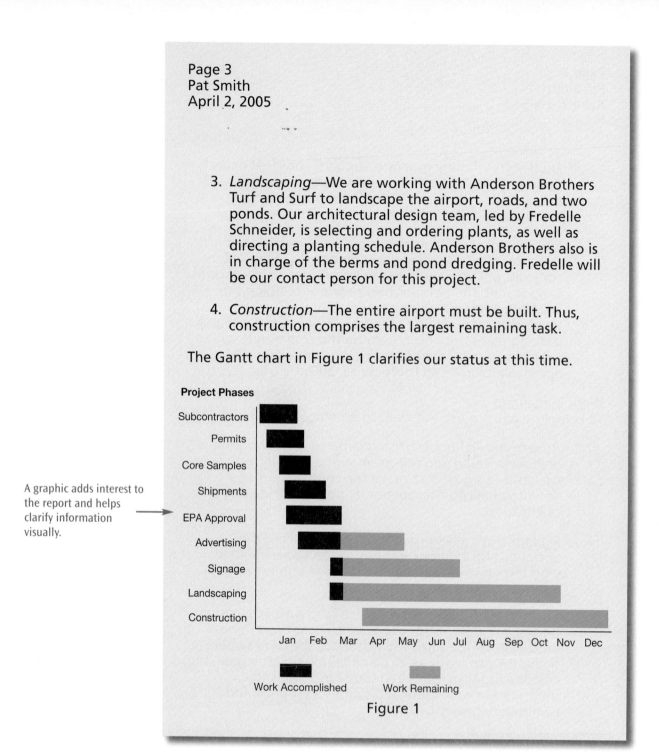

Page 3
Pat Smith
April 2, 2005

3. *Landscaping*—We are working with Anderson Brothers Turf and Surf to landscape the airport, roads, and two ponds. Our architectural design team, led by Fredelle Schneider, is selecting and ordering plants, as well as directing a planting schedule. Anderson Brothers also is in charge of the berms and pond dredging. Fredelle will be our contact person for this project.

4. *Construction*—The entire airport must be built. Thus, construction comprises the largest remaining task.

The Gantt chart in Figure 1 clarifies our status at this time.

Project Phases

A graphic adds interest to the report and helps clarify information visually.

Figure 1

Figure 16.2   *Continued*

- **Electronics**—Your company manufactures Global Positioning Systems (GPSs) to correctly inform a user of his or her exact location. The GPS receiver must compare the time a signal is transmitted by a satellite with the time it is received. Your company's receptors are malfunctioning as are the units' electronic maps. Why? Your job is to study the electronic systems on randomly selected GPS units and write a lab report documenting your findings.

Problems Encountered

Core samples are acceptable throughout most of our construction site. However, the area set aside for the northwest pond had a heavy rock concentration. We believed this would cause no problem. Unfortunately, when Anderson Brothers began dredging, it hit rock, which had to be removed with explosives. Because this northwest pond is near the sand crane and pelican nesting sites, EPA requested that we wait until the birds were resettled. The extensive rock removal and wait for wildlife resettlement has slowed our progress. We are behind schedule on this phase. This schedule delay and increased rock removal will affect our budget.

CONCLUSION/RECOMMENDATION

Though we have just begun this project, we still are approximately 15 percent of the way toward our goal. We anticipate a successful completion, especially since deliveries have been timely.

Only the delays at the northwest pond site present a problem. We are behind schedule and over cost. With additional personnel to speed the rock removal and increased funds, we can meet our target dates. Darlene Laughlin, our city council liaison, is the person to see about corporate investors, city funds, and big-ticket endowments. With your help, and Darlene's cooperation, we should meet our build schedules.

Figure 16.2 *Continued*

- **Information Technology**—Customers are calling your company's 1–800 hotline almost daily, complaining about hard drive error readings. This is bad for business and profitability. To solve these hard drive malfunctions, you must study units to find the problem. Then, you will write a lab report to document your discoveries.

You write a lab report after you've performed a laboratory test to share with your readers

- Why the test was performed
- How the test was performed
- What the test results were
- What follow-up action (if any) is required

Criteria

The following are components of a successful lab report.

1. *Heading*

   Date

   To

   From

   Subject (topic + focus)

2. *Introduction (overview, background)*

   Purpose: Why is this report being written? To answer this question, provide any or all of the following:

   - The rationale (What problem motivated this report?)
   - The objectives (What does this report hope to prove?)
   - Authorization (Under whose authority is this report being written?)

3. *Discussion (body, methodology)*

   How was the test performed? To answer this question, provide the following:

   - Apparatus (What equipment, approach, or theory have you used to perform your test?)
   - Procedure (What steps—chronologically organized—did you follow in performing the test?)

4. *Conclusion/recommendations*

   Conclusion. The conclusion of a lab report presents your findings. Now that you've performed the laboratory experiment, what have you learned or discovered or uncovered? How do you interpret your findings? What are the implications?

   Recommendations. What follow-up action (if any) should be taken?

You might want to use graphics to supplement your lab report. Schematics and wiring diagrams are important in a lab report to clarify your activities, as shown in Figure 16.3.

Feasibility/Recommendation Reports

Purpose and Examples

A *feasibility/recommendation report* accomplishes two goals. First, it studies the practicality of a proposed plan. Then, it recommends action. Occasionally, your company plans a project but is uncertain whether the project is feasible. Will the plan work, does the company have the correct technology, will the idea solve the problem, or is there enough money? One way a company determines the viability of a project is to perform a feasibility study, to document the findings, and then to recommend the next course of action.

DATE: July 18, 2005
TO: Dr. Jones
FROM: Sam Ascendio, Lab Technician
SUBJECT: LAB REPORT ON THE ACCURACY OF DECIBEL
            VOLTAGE GAIN (A) MEASUREMENTS

INTRODUCTION

Purpose
Technical Services has noted inaccuracies in recent measurements. In response to its request, this report will present results of tested A (gain in decibels) of our ABC voltage divider circuit. Measured A will be compared to calculated A. This will determine the accuracy of the measuring device.

DISCUSSION

Apparatus
- Audio generator
- Decade resistance box
- 1/2 W resistor: four 470 ohm, two 1 kilohm, 100 kilohm
- AC millivoltmeter

Procedure
1. Figure A shows a voltage divider. For each value of R (resistances) in Table 1, voltage gain was calculated (table attached).
2. An audio generator was adjusted to give a reading of 0 dB for input voltage.
3. Output voltage was measured on the dB scale. This reading is the measured A and is recorded in Table 1.
4. Step 3 was repeated for each value of R listed in Table 1.
5. Figure B shows three cascaded voltage dividers. A1 = v2/v1, A2 = v3/v2, and A3 = v4/v3. These voltage gains added together give the total voltage, as recorded in Table 2.
6. The circuit in Figure B was connected.
7. Input voltage was set at 0 dB on the 1-V range of the AC millivoltmeter.
8. Values V2, V3, and V4 were read and recorded in Table 3.

The *discussion* includes the equipment used and a step-by-step procedure.

Figure 16.3   Lab Report

- **Manufacturing**—Your company is considering the purchase of new equipment but is concerned that the machinery will be too expensive, the wrong size for your facilities, or incapable of performing the desired tasks. You need to research and analyze the options, determining which equipment best suits your company's needs. Then, you will recommend purchase.

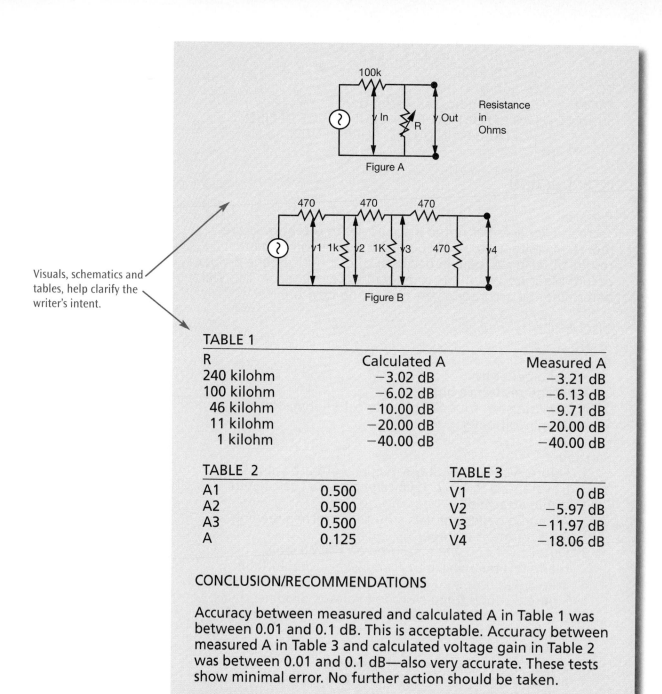

Visuals, schematics and tables, help clarify the writer's intent.

Figure A

Resistance in Ohms

Figure B

TABLE 1

R	Calculated A	Measured A
240 kilohm	−3.02 dB	−3.21 dB
100 kilohm	−6.02 dB	−6.13 dB
46 kilohm	−10.00 dB	−9.71 dB
11 kilohm	−20.00 dB	−20.00 dB
1 kilohm	−40.00 dB	−40.00 dB

TABLE 2

A1	0.500
A2	0.500
A3	0.500
A	0.125

TABLE 3

V1	0 dB
V2	−5.97 dB
V3	−11.97 dB
V4	−18.06 dB

CONCLUSION/RECOMMENDATIONS

Accuracy between measured and calculated A in Table 1 was between 0.01 and 0.1 dB. This is acceptable. Accuracy between measured A in Table 3 and calculated voltage gain in Table 2 was between 0.01 and 0.1 dB—also very accurate. These tests show minimal error. No further action should be taken.

Figure 16.3   *Continued*

- **Accounting**—Your company wants to expand and is considering new locations. The decision makers, however, are uncertain whether the market is right for expansion. Are interest rates good? Are local property taxes and sales taxes too high? Will the city provide tax rebate incentives for your company's growth? You need to study the feasibility of expansion at this time and report your recommendations.
- **Web Design**—Your company wants to create a Web site to market your products and services globally. The company CEO wants this Web site to be unique—different from the competitors' sites. The CEO wants to

be sure that online checkout is easy, that pricing is cost effective, that products are depicted in a visually appealing way, and that the site loads quickly. How will you make your Web site stand out from the competition? You must write a feasibility report to present the options as well as to offer your recommendations.

- **Health Management**—It is time to update your health information system. With increasingly complex insurance and regulatory challenges, your current system is outdated. What are your options? You could install software to help code and classify patient records. You could hire consultants to help comply with in-patient and outpatient regulations. You could outsource your patient load to home health care agencies. You could upgrade your intranet system to provide decision makers with more accurate information. A feasibility report is needed to study the options before you recommend changes.

Criteria

One way a company determines the viability of a project is to perform a feasibility study and then write a feasibility report documenting the findings. The following are components of an effective feasibility report.

1. *Heading*

   Date

   To

   From

   Subject (topic + focus)

SUBJECT: FEASIBILITY REPORT ON XYZ PROJECT
            (focus)   **(topic)**

2. *Introduction (overview, background)*

   Objectives: Under this subheading, you can answer any of the following questions:

   - What is the purpose of this feasibility report? Until you answer this question, your reader doesn't know. As mentioned earlier in this chapter, it's false to assume prior knowledge on the part of your audience. One of your responsibilities is to provide background data. To answer the question regarding the report's purpose, you should provide a clear and concise statement of intent.

   - What problems motivated this study? To clarify for your readers the purposes behind the study, *briefly* explain either

     —what problems cause doubt about the feasibility of the project (i.e., is there a market, is there a piece of equipment available that would meet the company's needs, is land available for expansion?).

     —what problems led to the proposed project (i.e., current equipment is too costly or time consuming, current facilities are too limited for expansion, current net income is limited by an insufficient market).

   - Who initiated the feasibility study? List the name(s) of the manager(s) or supervisor(s) who requested this report.

Personnel: Document the names of your project team members, your liaison between your company and other companies involved, and your contacts at these other companies.

3. *Discussion (body, findings)*

Under this subheading, provide accessible and objective documentation.

Criteria: State the criteria upon which your recommendation will be based. Criteria are established so you have a logical foundation for comparison of personnel, products, vendors, costs, options, schedules, and so on.

Analysis: In this section, compare your findings against the criteria. In objectively written paragraphs, develop the points being considered. You might want to use a visual such as a table to organize the criteria and to provide easy access.

4. *Conclusion/recommendations*

Conclusion: In this section, you go beyond the mere facts as evident in the discussion section: You state the significance of your findings. Draw a conclusion from what you have found in your study. For example, state that "Tim is the best candidate for director of personnel" or "Site 3 is the superior choice for our new location."

Recommendations: Once you have drawn your conclusions, the next step is to recommend a course of action. What do you suggest that your company do next? Which piece of equipment should be purchased, where should the company locate its expansion, or is there a sufficient market for the product?

Figure 16.4 presents an example of a feasibility/recommendation report.

Incident Reports

Purpose and Examples

An *incident report* documents an unexpected problem that has occurred. This could be an automobile accident, equipment malfunction, fire, robbery, injury, or even problems with employee behavior. In this report, you will document what happened. If a problem occurs within your work environment that requires analysis (fact finding, review, study, etc.) and suggested solutions, you might be asked to prepare an incident report (also called a trouble report or accident report), as follows:

- **Sales**—One of your sales representatives has been involved in a car accident while on job-related travel. The sales representative or his or her supervisor must document this problem.

- **Biomedical Technology**—A CAT scan in the radiology department is not functioning correctly. This has led to the department's inability to read x-rays. To avoid similar problems, you need to report this incident.

- **Hospitality Management**—An oven in your restaurant caught fire. This not only injured one of your cooks but also damaged the oven, requiring that it be replaced with more fire-resistant equipment.

- **Retail**—A customer was hurt while in your showroom. Incidents also could include employees who are not abiding by company policy. Maybe one of your retail locations has experienced a burglary. The police have been contacted, but as site manager, you believe the problem could have been avoided with better in-store security. Your incident report will document the event and show how to avoid future problems.

DATE: August 13, 2005
TO: Pat Hobby, Supervisor—Project Management
FROM: Nick Adams, Purchasing Agent
SUBJECT: FEASIBILITY STUDY FOR TECHNOLOGY PURCHASES

INTRODUCTION

Purpose
The purpose of this report is to study which technology will best meet your department's communication needs and budget. After analyzing the feasibility of various technologies, Purchasing will recommend the most cost-effective technology options.

Problem
According to your memo dated August 1, 2005, your department needs new communication technologies for the following reasons:

- Your department has hired three new employees, increasing your headcount to 10 project managers.
- Currently, your department has only five pagers. This allows too few of your employees to communicate with home office while they are on job-related travel.
- Pagers limit your employees' communication options.
- With our company's expanded projects, your department's job-related travel is increasing.
- To accommodate your personnel growth and the company's expansion, Project Management requires more and better portable means of communication technology. These could include laptop computers, handheld computers, and cell phones. ◄— The *introduction* explains what issue has led to this report.

Personnel
Our vendor contacts are as follows:

Electek	Tech On the Go	Mobile Communications
Steve Ross	Jay Rochlin	Karen Allen
stever1@electek.com	jrochlin@tog.com	karen.allen@mobcom.net

Figure 16.4   Feasibility/Recommendation Report

DISCUSSION

In the *discussion*, explain the criteria used to determine which options are feasible solutions to the problem.

Criteria

The following criteria were considered to determine which communication technology would best meet Project Management needs:

1. *Use*—While on job-related travel, Project Management personnel need to be accessible to home office. Pagers can fill this need. However, your employees also need to be able to access corporate intranet forms and data, send and receive e-mail, and compose reports. To accomplish these tasks, pagers are insufficient. Thus, your employees need either (or combinations of)
   - Cell phones with wireless access and e-mail functions
   - Handheld computers with wireless Internet access and Microsoft Office Suite software (complete with Excel, Word, and Outlook)
   - Laptop computers with wireless Internet access and Microsoft Office Suite software

2. *Maintenance*—Due to the expanded use of the communication tools, we need to purchase equipment and software complete with either quarterly or biannual service agreements (at no extra charge).

3. *Service Personnel*—The service people should be certified to repair and maintain whatever hardware we purchase. In addition, the vendors must also be able to train our personnel in software usage.

4. *Warranties*—The warranties should be for at least one year with options for renewal.

5. *Cost*—The total allowable to your department is $5,000.

Analysis

Needs Assessment

Purchasing agrees that the Project Management Department's communication needs exceed its current technology. Not only are the department's pagers insufficient in number, but also they do not allow the Project Management personnel to access corporate e-mail, the Internet, or word processing packages. Either cell phones, handhelds, or laptops are necessary.

Figure 16.4   *Continued*

Vendor Evaluation

- **Electek**—Having been in business for 10 years, this company is staffed by highly trained technicians and sales staff. All Electek employees are certified for software training. The company promises a biannual maintenance package and subcontracted personnel if its employees cannot repair hardware problems. It offers manufacturer's guarantees with extended service warranties costing only $100 a year for up to 5 years. Electek offers 20 percent customer incentives for purchases of over $2,000.
- **Tech On the Go**—This company has been in business for two years. TOG provides only subcontracted service technicians for hardware repair. TOG's employees are certified in software training. The owners do not offer extended warranty options beyond manufacturer's guarantees. No special customer pricing incentives are offered, though TOG sells retail at a wholesale price.
- **Mobile Communications**—Having been in business for 5 years, Mobile has certified technicians and sales representatives. All repairs are provided in-house. The company offers quarterly maintenance at a fee of $50 ($200 per year). Mobile offers a customer incentive of 10 percent discounts on purchases over $5,000.

In the *discussion*, provide specific details to prove the feasibility of the plan or project.

Cost Analysis

- **Laptops**—Project Management requests one laptop per departmental employee. Our analysis has determined that the most affordable laptops we can purchase (with the requested software and wireless Internet connections) would cost $1,500 per unit. Thus, 10 laptops would cost $15,000. Even with discounts, this exceeds your department's budget.
- **Cell phones**—Outstanding values are possible. For only $100, your department could purchase state-of-the-art equipment. Thus, your entire staff could have access to portable communication for only $1,000. However, what you gain in expense, you lose in functionality. Even though cell phones allow for e-mail and Internet access, monitor size limits readability. More important, submitting reports is severely limited.

Figure 16.4   *Continued*

Page 4
Nick Adams
August 13, 2005

- **Handhelds**—This communication option offers the best compromise. Handhelds, at approximately $350 per unit, are more expensive than cell phones but cheaper than laptops. More important, with "thumb-typing capabilities" and larger monitor size and resolution, handhelds will provide your employees more portable communication capabilities. Not only will handhelds allow for e-mail and Internet access, but also your Project Management staff can create reports using handheld computers. Finally, at $350 times 10 employees, you stay within your departmental budget.

The following table compares the three vendors we researched on a scale of 1:3, 3 representing the highest score.

Table 1 Criteria Comparison

Criteria	Electek	TOG	Mobile
Maintenance	3	2	3
Personnel	3	3	3
Warranties	3	2	2
Cost	3	2	2
Total	**12**	**9**	**10**

Graphics depict your findings more clearly.

CONCLUSION

Laptops and cell phones are not feasible options to achieve your communication technology needs. Laptops are too expensive, and cell phones do not provide the communication tools you require. Handheld computers, in contrast, are cost effective and provide you word processing, e-mail, and Internet access. Furthermore, the handhelds are portable yet provide a monitor size and resolution that allows for readability.

All three vendors have the technology you need. However, TOG and Mobile do not meet the Purchasing Department's criteria. In particular, these companies do not provide either the maintenance packages, warranties, or pricing we require.

The *conclusion* sums up your findings, explaining the feasibility of a course of action—why a plan should or should not be pursued.

Figure 16.4 *Continued*

Page 5
Nick Adams
August 13, 2005

RECOMMENDATION

Given the combination of cost, maintenance packages, warranties, and service personnel, Electek is our best choice for purchase. In addition, Purchasing suggests that handheld computers will best meet the Project Management Department's needs for portable, effective communication technology. You can purchase wireless handheld computers for all 10 employees within your department's budget. They then will be able to access the corporate e-mail, Internet, and intranet systems and be able to create quality documentation using improved word processing software.

In the *recommendation*, explain what should happen next.

Figure 16.4   *Continued*

Criteria

Engineering environments requiring maintenance reports rarely provide employees with easy-to-fill-in forms. To write an incident report when you have not been given a printed form, include the following components:

1. *Heading*

    Date

    To

    From

    Subject (topic + focus)

example

SUBJECT: REPORT ON CHILLED WATER LEAKS IN D/823
              (focus)   (topic)

2. *Introduction*

    Purpose: In this section, document when, where, and why you were called to perform maintenance. What motivated your visit to the scene of the problem?

    Personnel: Who was involved, and *what* role do you play in the report? That could entail listing all of the people involved in the accident or event. These might be people injured, as well as police or medical personnel answering an emergency call.

    In addition, *why* are you involved in the activity? Are you a supervisor in charge of the department or employee? Are you a police officer or medical personnel writing the report? Are you a maintenance employee responsible for repairing the malfunctioning equipment?

3. *Discussion (body, findings, agenda, work accomplished)*

   Using subheadings or itemization, quantify what you saw (the problems motivating the report). You should list your findings in chronological order. Be specific. Include the

   - Make or model of the equipment involved
   - Police departments or hospitals contacted
   - Names of witnesses
   - Witness testimonies (if applicable)
   - Extent of damage—financial and physical
   - Graphics (sketches, schematics, diagrams, layouts, etc.) depicting the incident visually
   - Follow-up action taken to solve the problem

4. *Conclusion/recommendations*

   Conclusion: Explain what caused the problem.

5. *Recommendations.* Relate what could be done in the future to avoid similar problems.

   Figure 16.5 presents an example of an incident report.

Investigative Reports

Purpose and Examples

As the word "investigate" implies, an *investigative report* asks you to examine the causes behind an incident. Something has happened. The report does not just document the incident. It focuses more on why the event occurred. You might be asked to investigate causes leading up to a problem in the following instances:

- **Security**—You work in a bank's security department. You are responsible for investigating theft, burglary, fraud, vandalism, check kiting, and other banking illegalities. One of your clients, a college student at the local university, reports losing her purse at a campus party. Within hours of the theft, checks bearing her name are showing up across the city. Your job now is to investigate the incident and report your findings.

- **Engineering**—A historic, 100-year-old bridge crossing your city's river is buckling. The left lane is now two inches higher than the right lane, and expansion joints are separating beyond acceptable specifications. You must visit the bridge site, inspect the damage, and report on the causes for this construction flaw.

- **Medicine**—As radiographic technologist, you have administered a bone scan to a patient. This scan detected a shadowy area in the right medial humerus. This medical problem could be arthritis or inflammation. However, the shadow also could indicate a metastatic condition. With the help of a pathologist and radiologist, you must submit an investigative report explaining the causes for this aberration and suggesting a follow-up procedure.

- **Computer Technology**—You work in a college's technology department. Primarily, your job is to help faculty members with their technology applications. The college requires that all student grades be kept on a newly installed, campuswide database and then be submitted

DATE: October 16, 2005
TO: Chris Sutliff, Bank Security Supervisor
FROM: Tom Warren, Senior Security Agent
SUBJECT: INCIDENT REPORT ON SEPTEMBER ATM ROBBERY

INTRODUCTION

Purpose. On Sunday, October 15, 2005, a drive-up ATM at our Leawood, Kansas, bank branch (14562 Pheasant Ridge) was broken into. A total of $5,000 in twenty-dollar-bill denominations was stolen.

The *introduction* focuses on who, what, when, and where.

Personnel. As Security Agent III, responsible for bank branches in southern Johnson County, I am reporting this incident.

Upon receiving the electronic break-in alarm, the Leawood office contacted the Leawood Police Department and me. Police officer, Sergeant John Armstrong, is in charge of the case. The police case file number is 24516.

Following is a report on my findings.

DISCUSSION

Agenda

A chronological organization for the report's *discussion* helps the reader follow your ideas.

8:15 P.M.	The ATM burglary was electronically reported to the Leawood bank branch office and to the Leawood Police Department.
8:18 P.M.	The Leawood Police Department received a 911 call from a bank customer who witnessed a car speeding away from the drive-up ATM, and then saw the damage.
8:37 P.M.	Leawood police officers Shirley Chandley and Rob Hotchiss arrived at the scene. Officer Chandley interviewed the witness (Mrs. Jill Stinson, 912-555-7879, 15011 Stillbrook, Leawood, KS). Mrs. Stinson reported seeing a dark two-door sedan speed away as she was driving up to the ATM. She was unaware of the car's make or model. When she arrived at the ATM, she said "it looked like someone had driven into it."
8:48 P.M.	I arrived at the site of the incident. After showing the officers my bank identification, I was allowed to open the ATM. I collected the receipts register and discovered that the cash vault had been depleted. When full, the vault maintains a $5,000 balance.

Witness testimony and contact information provide details.

Figure 16.5   Incident Report

Page 2
Warren
October 16, 2005

| 8:50 p.m. | I turned off the ATM operating system and activated the on-screen message "Closed—Please use an ATM from one of our other 36 locations, or call 1-800-ATM-BANK for help." |
| 9:02 p.m. | The police officers secured the ATM site with crime-scene yellow tape. I thanked Mrs. Stinson for her help, offering her our standard Good Citizen reward. The officers and I left the scene. |

CONCLUSION

The *conclusion* focuses on cause.

The police officers informed me that they had received notices of three other similar break-ins that day at ATMs located throughout the metro area. Each break-in used the same method of operation (MO). The break-ins were caused by a car ramming into the units, shorting out the electronic lock circuitries. This allowed the perpetrator to pry open the ATMs' front panels with a sharp object (knife, screwdriver, crossbar, etc.). Our ATM revealed the same MO.

RECOMMENDATIONS

The *recommendation* focuses on the next course of preventative action.

We should not experience a similar break-in at other ATMs. Twenty-five of our other ATMs have already been retrofitted to withstand car rammings. Aware of this potential danger, our security department has installed two cement pillars 5 feet in front of every ATM and 6 feet from either side of the units. These cement protective pillars (CPPs) protect the ATMs from rammings while still allowing for customer access. The Leawood unit and 10 others had not yet been retrofitted. This retrofit was scheduled for Wednesday, October 18, 2005. We will now replace the ATM in question with a new model and ensure its security with the cement protection devices today. We also will fast-track the retrofitting of all other models.

The diagram below shows a before-and-after overhead view of our new ATM protection devices.

Simple sketches clarify the problem and recommend a solution.

Before

Front Panel

After

Front Panel

CPPs

Figure 16.5 *Continued*

electronically when the semester ends. For some reason, faculty cannot access their students' records for grade inputting. You must investigate the causes behind this technology glitch and solve the problems—*now!* The semester grades are due within 24 hours.

Criteria

Following is an overview of what you might include in an effective investigative report.

1. *Heading*

   Date

   To

   From

   Subject (topic + focus)

2. *Introduction (overview, background)*

   Purpose. In the purpose section, document the date(s) of the incident. Then comment on your objectives or rationale. What incident are you reporting on and what do you hope to achieve in this investigation?

   You might also want to include these following optional subheadings:

   Location. Where did the incident occur?

   Personnel. Who was involved in the incident? This could include those with whom you worked on the project or those involved in the situation.

   Authorization. Who recommended or suggested that you investigate the problem?

3. *Discussion (body, findings, agenda)*

   This is the major part of the investigation. Using subheadings, document your findings. This can include the following:

   - A review of your observations. This includes physical evidence, descriptions, lab reports, testimony, and interview responses. Answer the reporter's questions: who, what, when, where, why, and how.
   - Contacts—people interviewed
   - Difficulties encountered
   - Techniques, equipment, or tools used in the course of the investigation
   - Test procedures followed, organized chronologically

4. *Conclusion/recommendations*

   Conclusion. What did you accomplish? What did you learn? What discoveries have you made regarding the causes behind the incident? Who or what is at fault?

   Recommendations. What do you suggest next? Should changes be made in personnel or in the approach to a particular situation? What training is required for use with the current technology, or should technology be changed? What is the preferred follow-up for the patient or client? How can the problem be fixed?

Figure 16.6 illustrates an investigative report.

The *subject line* provides a topic (Frog Creek Wastewater Pollution) and a focus (Investigative Report).

DATE: September 15, 2005
TO: Bowstring City Council; Arrowhead School District 234
FROM: Mike Moore, Frog Creek Wastewater Treatment Plant Director of Public Relations
SUBJECT: INVESTIGATIVE REPORT ON FROG CREEK WASTEWATER POLLUTION

Introduction

Background. On September 7, 2005, teachers at Arrowhead Elementary School reported that over a five-day period (September 2–6), approximately 20 students complained of nausea, lightheadedness, and skin rashes. On the fifth day, the Arrowhead administration called 911 and the Arrowhead School District (ASD 234) in response to this incident. Bowstring City paramedics treated the children's illnesses, suggesting that the problems might be due to airborne pollutants. The Bowstring City Council contacted the Frog Creek Wastewater Treatment Plant (FCWTP) to investigate the causes of this problem. This report is submitted in response to Bowstring City's request.

Personnel. Mike Moore (director of public relations), Sue Cottrell (Wastewater engineer), and Fred Mittleman (Wastewater engineer)

The *findings* use chronological organization to document the incidents.

Findings

Impact on Schoolchildren. Arrowhead Elementary School administrators reported the following:

- Monday, September 2—two children reported experiencing nausea.
- Tuesday, September 3—two children reported experiencing nausea, and one child experienced lightheadedness.
- Wednesday, September 4—three children experienced skin rashes.
- Thursday, September 5—two children complained of nausea, one child was lightheaded, and two children showed evidence of skin rashes.
- Friday, September 6—two children reported nausea, three reported skin rashes, and two lightheadedness.

After Friday's occurrences, Arrowhead Elementary School administrators called 911. Bowstring paramedics reported that (with parental approval) the children were treated with antacids for nausea, antihistamines for skin rashes, and oxygen for their lightheadedness. No other incidents were reported in the neighborhood surrounding the school.

Figure 16.6 Investigative Report

Pollutants: Frog Creek is usually characterized by low alkalinity (generally less than 30 milligrams per liter [<30 mg/l]). Inorganic fertilizer nutrients (phosphorus and nitrogen) are also generally low (<20 mg/l), with limited algae growth.

Despite normally low readings, in late summer, with heat and rain, these readings can escalate. Higher algae-related odors above the 3–6 picometer thresholds, along with increased alkalinity (<50 mg/l) can create health problems for youth, elderly, or anyone with respiratory illnesses.

Wastewater engineers Sue Cottrell and Fred Mittleman took samples of Frog Creek on September 7–10. These studies showed that algae, alkalinity, and fertilizer were higher than usual.

- Algae readings: 4 picometers
- Alkalinity readings: <45 mg/l
- Phosphorus and nitrogen readings: <25 mg/l

On September 11–14, our engineers rechecked Frog Creek, finding that the chemical levels had returned to a normal, acceptable range.

Atmospheric Factors. The above elevated readings were caused by three factors (heat, rain, and northeasterly winds).

- **Heat**—On the days of the Arrowhead Elementary School incident, the temperature ranges were 92–95 degrees Fahrenheit (F), unusually high for early September. Algae and chemical growth increases in temperatures above 84°F.
- **Rain**—In addition, on September 4–6, Bowstring City received 2 inches of rain, swelling Frog Creek to 3 inches above its normal levels. Studies show that rain-swollen creeks and rivers lead to increased pollutants, as creek bottom silt rises.
- **Wind**—On September 4–5, a prevailing northeasterly wind blew from Frog Creek toward Arrowhead Elementary School's playground.

The *findings* not only investigate the causes of the incident but also document with specific details.

To achieve a readable format, the text is made accessible through highlighting techniques—boldface and italicized subheadings, and bulleted details.

Figure 16.6   *Continued*

Page 3
Mike Moore
September 15, 2005

Conclusions

Frog Creek normally has acceptable levels of algae, alkalinity, and fertilizer nutrient levels. The heat and higher water levels temporarily led to elevated pollutant readings. These levels subsequently returned to normal. Wind directions during the school incidents also had an impact on the children's illnesses. On follow-up questionnaires, FWCTP employees found that the school children's ailments had subsided.

The Arrowhead Elementary School situation appears to have been an isolated incident due to atmospheric changes.

Recommendations

Frog Creek is constantly monitored for safety. However, rain, wind, and heat will continue to affect its chemical levels. In unusual situations (children playing outside, the wind blowing from Frog Creek toward the elementary school's playground, and a combination of heat and rain), similar results could occur. Children especially susceptible to airborne pollutants could experience nausea, rashes, and lightheadedness.

FCWTP's, HAZMAT employees would be happy to work with parents and teachers.

In a one-hour workshop, presented during the school day or at a Parent-Teacher Organization meeting, FCWTP will provide the following information:

- Scientific data about stream and creek pollutants
- The effects of rain, wind, and heat on creek chemicals
- Useful preventive medical emergency techniques

This information would explain real-world applications for science classes, as well as provide valuable health tips for parents and teachers.

The *conclusion* provides options for the readers and a positive tone appropriate for the intended audience. ──▶ Please let us know if you would like to benefit from this free-to-the-public workshop. We would be happy to schedule one at your convenience.

Figure 16.6   *Continued*

Meeting Minutes

Purpose and Examples

Meeting minutes document the results of a meeting—what was discussed, proposed, voted on, and planned for future meetings. As the recording secretary for your project team, a community organization, your city council, or a departmental meeting, your job is to record the meeting's minutes. In meeting minutes, you record who attended the meeting, when it began, when it ended, and where it took place. You report on the topics discussed, decisions arrived at, and plans for the future.

Criteria

Meeting minutes should include the following key components:

1. *Heading*

   In most reports, your heading will include the subheadings *Date, To* (the recipients of the report), and *From* (the name of the person submitting the report). You *do not need these components* in minutes, however. The date is provided in the *introductory* part of the minutes (discussed later). *To* is a given because the minutes are for the committee being reported on. *From* is taken care of at the end of the minutes, when the reporting secretary signs the minutes.

   Subject lines are still valid in meeting minutes. As noted throughout this chapter, subject lines include a *topic* and a *focus*.

   SUBJECT: SAFETY COMMITTEE APRIL MEETING MINUTES
             (topic)   (focus)

2. *Introduction*

   Include the following:

   - *Date, Time, and Place.* At the beginning of the minutes, list the date on which the meeting is held, what time the meeting began, and where the meeting was located.

     September 7, 2005

     7:00 P.M.

     Conference Room C

   - **Attendees.** List the names of those who attended the meeting.

   - **Approval of Last Meeting's Minutes.** After asking participants to read the last meeting's minutes, vote to accept them.

3. *Discussion (findings, agenda)*

   This is the most important part of your minutes. In this long section, you report on the agenda items. Recording secretaries have shared with us a major concern. They believe that their job requires them to report *every* word that has been spoken. This is not true. In fact, *Robert's Rules of Order* says exactly the opposite:

   - a reporting secretary should keep a record of the proceedings, stating what was done and *not* what was said. Your job is not to report every committee members' comments. Instead, focus on decisions made, conclusions arrived at, issues confronted, opposing points of view, and votes taken.

   - If resolutions are agreed upon, the reporting secretary must document the wording of this resolution exactly.

- Finally, all content within this discussion section must be reported objectively, without "criticism, favorable or otherwise." (*Robert's Rules of Order Revised*)

4. *Conclusion*
   - **Old/New Business.**   The last topic of a meeting's agenda should be a review of any "old" topics still unresolved and needing further discussion. Similarly, toward the end of a meeting, you will need to report on new topics, perhaps ones that will need to be covered in future meetings.
   - **Next Meeting.**   Report when the committee will meet next, providing the date, time, and location.
   - **Time of Adjournment.**   Report when the meeting ended.
   - **Signature.**   A typical statement reads "Respectfully submitted by _____." You can sign your name beneath the typed signature (unless the minutes will be submitted electronically).

Figure 16.7 shows an example of meeting minutes.

Organization

Typically, meeting minutes are organized chronologically. When you write your minutes, you can organize the content according to the sequential order of the topics discussed. This sequential organization allows readers to follow a meeting's agenda. However, minutes also can be organized according to importance. Though chronology is an easy approach to documenting a meeting, your audience might better understand the meeting's focus if you document which topics received the greatest discussion or which topics will have the largest impact.

Layout

Reporting secretaries often contend that they must write essaylike reports, following a paragraph format. This is neither true nor advisable. In fact, an essaylike report, including wall-to-wall words, will turn off your readers. Instead, use headings, subheadings, and bulleted lists to make the minutes more accessible.

**Employee Benefits Project Team
Meeting Minutes**

Date: April 12, 2005

Time: 7:00 P.M.–9:30 P.M.

Objectives. In this second project team meeting, the goal was to discuss employee benefits options that could be added to the company's cafeteria plan.

Attendees. Stacy Helgoe, Christine Pieburn, Darren Rus, Andrew McWard, Bill Lamb, Jessica Studin

Team Leader. Phyllis Goldberg, director of Employee Benefits

Agenda

1. Flextime—Stacy and Christie reported on flextime options. They stated that in a survey dated March 15, 76 percent of employees favored flextime benefits. These included beginning our workdays at 7:00 A.M. and extending work hours until 6:30 P.M. Doing so would allow employees to either arrive for work earlier than standard or work later than usual. This work-schedule flexibility would allow employees to manage child care needs, appointments, and other concerns that might fall within the traditional workday.

   The downside of flextime, according to management, includes increased security coverage, increased utilities (heating, lighting, etc.), and inconsistent coverage of office hours. The benefits of flextime include increased employee morale, less congestion in the parking lots at traditionally prime times, and employee empowerment.

   Our project team voted 6–1 to promote flextime to management at next month's meeting.

2. Stock Options—At last month's meeting, Darren was asked to study the possibility of adding stock options to our cafeteria plan. He met with Accounting, our external auditor, and three proprietary stock brokerage companies (Bull and Bear, Market Trend, and BuyItNow). Please see the attached brokerage firm proposals. The proposals offered exciting information.

   However, both Accounting and our auditor convinced Darren that this year would not be the best time to offer our employees stock options. Our profit margin ratio is down, we cannot meet our quarterly forecasts, and we cannot distribute dividends to stockholders.

   Future European expansion plans should increase our revenues. Therefore, our project team voted 5–2 to table this issue until the end of next quarter.

Figure 16.7   Meeting Minutes

Page 2
Meeting Minutes
April 12, 2005

3. Personal Leave Days—Currently, our employees are allotted five sick leave days a year. Our proposal asks for an additional two personal leave days. These days would have to be used within the calendar year. The personal leave days would not roll over, nor could they be banked. Andrew McWard surveyed the employees regarding this topic. Ninety-eight percent of those surveyed favored the personal leave proposal.

   Our project team voted 6–1 to promote personal leave days as a cafeteria plan option.

4. Sick Bank—Last year, a survey suggested that employees might be willing to share their unused sick leave days with other employees facing catastrophic illnesses or family emergencies. This is a common practice at other corporations of our size. This not only gives employees a feeling of ownership but also is a method of team building.

   Bill revisited this topic with employees. He attended departmental meetings across the company, asking for input. Bill reported that employees do not seem to favor this proposal at the moment. His informal survey shows only a 30 percent interest.

   Unless these numbers change, our committee sees no reason to pursue a sick bank. We voted 4–3 against proposing a sick bank option.

5. Short- and Long-Term Disability—Jessica visited three insurance carriers (In-need Insurance, Evergreen Insurance Inc., and Employee Associated General). Her goal was to get quotes on short- and long-term disability insurance. Our employee survey ranked this as a number one priority.

   Our committee agreed with the survey and highly recommended this as an add-on to our employee cafeteria plan (7–0 vote). Without this benefit, employees with illnesses longer than five days or any banked hours have no salary or job insurance. This fact will negatively impact both employee retention and recruitment. Most major firms offer this benefit. To stay competitive, we must do so also.

Old Business. Andrew reminded us that the company holiday party and family picnic needed to be planned. We will contact the hospitality committee to determine their status.

New Business. In next month's meeting, we will focus on ways to implement our recommendations.

   Respectfully submitted by Jessica Studin, Recording Secretary

Attachment:
  Brokerage Firm Proposals
  Insurance Proposals

Figure 16.7 *Continued*

PROCESS

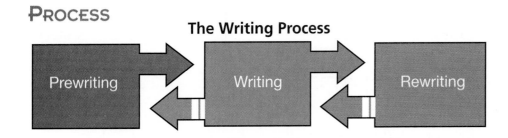

The Writing Process

The Writing Process		
Prewriting	Writing	Rewriting

• Examine your purposes
• Determine your goals
• Consider your audience
• Gather your data
• Determine how the content will be provided
	• Organize the draft according to some logical sequence that your readers can follow easily
	• Format the content to allow for ease of access
		• Revise
		  ° Add missing details
		  ° Delete wordiness
		  ° Simplify word usage
		  ° Enhance the tone of your communication
		  ° Reformat your text for ease of access
		  ° Practice the speech or review the text
		• Proofread
		  ° Correct errors

Now that you know the criteria for reports in general and for specific types of reports (trip reports, progress reports, lab reports, feasibility/recommendation reports, incident reports, investigative reports, and meeting minutes), the next step is to construct these reports. How do you begin? As always, *prewrite, write,* and *rewrite.* Remember that the process is dynamic and the steps frequently overlap.

Prewriting

We have presented several techniques for prewriting—reporter's questions, clustering/mind mapping, flowcharting, and brainstorming/listing. An additional technique is called *branching.*

As with flowcharting and mind mapping, branching allows you to depict information graphically so you can not only gather data but also visualize it. This type of prewriting benefits both left-brain and right-brain people—those who are linear (outline oriented) as well as those who are more graphically attuned. Figure 16.8 shows an example of branching.

Branching is ideally suited for short reports. You can focus on whether the primary subject is a trip report, progress report, lab report, feasibility/recommendation report, incident report, investigative report, or meeting minutes in the *main idea.* In addition, the *subordinate points* easily correspond to the introduction, discussion, and conclusion/recommendations. Finally, you can develop your ideas more specifically in the *subheadings.* Figure 16.9 shows an example of branching for a trip report.

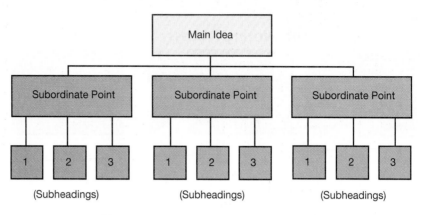

Figure 16.8   Branching

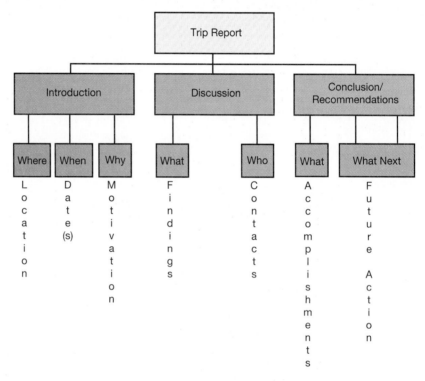

Figure 16.9   Branching (Trip Report)

You can see in Figure 16.9 how branching meshes with reporter's questions. Branching allows you to sketch out your organization and visualize your content. The reporter's questions help you make that content factual, precise, specific, and quantified.

Writing

Once you've sketched an outline for your report using branching, the next step is to write the text. To write your report, do the following:

Reread Your Prewriting

Review what you've sketched in branching. Determine whether you've covered all important information. If you believe you've omitted any significant points, add them for clarity. If you've included any irrelevant ideas, omit them for conciseness.

Assess Your Audience

Your decisions regarding points to omit or include will depend on your audience. If your audience is familiar with your subject, you might be able to omit background data. However, if you have multiple audiences or an audience new to the situation, you'll have to include more data than you might have assumed necessary in your prewriting.

Draft the Text

Focusing on your major headings (introduction, discussion, conclusion/ recommendations), use the sufficing technique mentioned in earlier chapters. Just get your ideas down on paper in a rough draft without worrying about grammar.

Organize Your Content

You can organize the discussion portion of your report using any one of these four modes: *chronology, importance, comparison/contrast*, and *problem/ solution*. Which of these modes you use depends on your subject matter.

For example, if the subject of your trip report is a price check of sales items at two stores, comparison/contrast would be appropriate, as in Figure 16.10.

If the subject of your incident report is a site evaluation, chronology might work in the report's discussion, as in the following example:

Example
Chronology

	DISCUSSION
8:00 A.M.	I arrived at the site and met with the supervisor to discuss procedures.
9:00 A.M.	We checked the water tower for possible storm damage. Only 10 shingles were missing.
10:00 A.M.	We checked the irrigation channel. It was severely damaged, the wall cracked in six places and water seeping through its barriers. Surrounding orchards will be flooded.
11:00 A.M.	We checked fruit bins. No water had entered.
1:00 P.M.	We checked the freezer units. The storm had disrupted electricity for four hours. All contents were destroyed.
2:00 P.M.	The supervisor and I returned to his office to evaluate our findings.

As discussed in earlier chapters, chronology is an easy method of organization to use and to follow. However, it is not always your most successful choice. Chronology inadvertently buries key data. For instance, in the preceding report findings, the most important discoveries occur at 10:00 A.M. and at 1:00 P.M., hidden in the middle of the list. To avoid making your readers guess where the

TO: Meagan Clem
FROM: Mary Jane Post
DATE: January 12, 2005
SUBJECT: PRICE CHECK REPORT—HANDY SANDY HARDWARE

INTRODUCTION

On Thursday, January 8, 2005, I compared our sale prices on plumbing items with those of Handy Sandy Hardware, 1000 W. 29th St., Newtown, Wisconsin.

DISCUSSION

Item	Hughes's Sale Price	Handy Sandy's Sale Price
1/2" copper tubing	$ 4.89 (10')	$ 4.99 (10')
3/4" copper tubing	10.50 (10')	9.99 (10')
1/2" CPVC pipe	2.39 (10')	2.99 (10')
3/4" CPVC pipe	3.49 (10')	5.99 (10')
1 1/2" PVC pipe	3.99 (10')	4.99 (10')
Acme 800 faucet	40.95	39.95
Acme 700 faucet	22.95	25.95

CONCLUSION

On five of the seven items (71.4 percent), Hughes's had the lower price.

RECOMMENDATIONS

We should continue to compare our prices to Handy Sandy's. We should also try to lower any prices that exceed theirs.

The table uses comparison contrast for organization.

Figure 16.10 Trip Report Organized by Comparison/Contrast

important information is, use the third method of organization—importance—in which you list the most important point first, lesser points later, as in the following example:

Example
Importance

DISCUSSION

1. Freezer units: Electricity was out for four hours. All contents were destroyed.
2. Irrigation channel: The wall was cracked in six places. Seepage was significant and sure to affect the orchards.
3. Water tower: Minor damage to shingles.
4. Fruit bins: No damage.

If you're writing a progress report to document a problem and suggest a solution, use problem/solution to organize your data, as follows:

Example
Problem/Solution

DISCUSSION

Problem—Production schedules on our M23 and B19 are behind three weeks due to machinery failures. Our numerical control device no longer maintains tolerance. This is forcing us to rework equipment, which is costing us $200 per reworked piece.

Solution—We must purchase a new numerical control device. The best option is an Xrox 1234. This machine is guaranteed for five years. Any problems with tolerance during this period are covered. Xrox either will correct the errors on site or provide us a loaner until the equipment is fixed. To avoid further production delays, we must purchase this machine by 1/18/05.

Rewriting

After a gestation period in which you let the report sit so you can become more objective about your writing, retrieve the report and rewrite. Perfect your text by using the following rewriting techniques:

Add Detail for Clarity

Have you answered the reporter's questions as thoroughly as needed? Don't assume your readers will know the why's and wherefore's. Spell it all out, exactly. You often have multiple readers, many of whom do not know your motivations or objectives. They need clarity.

Delete Dead Words and Phrases for Conciseness

For example, in your recommendations, don't say, "Acme's opinion is that apparently the proposed ideas will not successfully supersede those already implemented." Instead, simply write, "Recommendations: No further action is required."

Simplify Old-Fashioned Words and Phrases

What does *pursuant* mean? How about *issuance of this report?* The ultimate goal of technical writing is to communicate, not to confuse.

Move Information Within Your Discussion for Emphasis

Make sure that you've either maintained chronology (if you're documenting an agenda) or used importance to focus on your main point. If you've confused the two, cut and paste—move around information to achieve your desired goals.

Reformat Your Text for Accessibility

Highlight your key points with underlining or boldface. Use graphics to assist your readers. Don't overload your readers with massive blocks of impenetrable text.

Enhance Your Text for Style

Be sure to quantify when needed, personalize with pronouns for audience involvement, and be positive. Stress words like *benefit, successful, achieved,* and *value* rather than bowing to negative words like *cannot, failed, confused,* or *mistaken.*

Proofread and Correct the Report for Grammatical and Contextual Accuracy

Refer to ch. 3 for proofreading tips. See ch. 19 for proofreader's marks.

Avoid Biased Language

Foreman should be *supervisor, chairman* should be *chairperson* or *chair,* and *Mr. Swarth, Mr. James,* and *Sue* must be *Mr. Swarth, Mr. James,* and *Ms. Aarons.*
  Use the following checklist to help you in writing your short report.

Report Checklist

___ **1.** Does your subject line contain a topic and a focus? If you write only "SUBJECT: TRIP REPORT" or "SUBJECT: FEASIBILITY/ RECOMMENDATION REPORT," you have not communicated thoroughly to your reader. Such a subject line merely presents the focus of your correspondence. But what's the topic? To provide both topic and focus, you need to write "SUBJECT: TRIP REPORT ON SOLVENT TRAINING COURSE, ARCO CORPORATION—3/15/05" or "SUBJECT: FEASIBILITY REPORT ON COMPANY EXPANSION TO BOLKER BLVD."

___ **2.** Does the introduction explain the purpose of the report, document the personnel involved, or state when and where the activities occurred?

___ **3.** When you write the discussion section of the report, do you quantify what occurred? In this section, you must clarify precisely. Supply accurate dates, times, calculations, and problems encountered.

___ **4.** Is the discussion accessible? To create reader-friendly ease of access, use highlighting techniques, such as headings, boldface,

continued

continued

underlining, and itemization. You also might want to use graphics, such as pie charts, bar charts, or tables.

___ 5. Have you selected an appropriate method of organization in your discussion? You can use chronology, importance, comparison/contrast, or problem/solution to document your findings.

___ 6. Does your conclusion present a value judgment regarding the findings presented in the discussion? The discussion states the facts; the conclusion decides what these facts mean.

___ 7. In your recommendations, do you tell your reader what to do next or what you consider to be the appropriate course of action?

___ 8. Have you maintained a low fog index for readability?

___ 9. Have you effectively recognized your audience's level of understanding (high tech, low tech, lay, management, subordinate, colleague) and written accordingly?

___ 10. Is your report accurate? Correct grammar and calculations make a difference. If you've made errors in spelling, punctuation, grammar, or mathematics, you will look unprofessional.

PROCESS LOG

Let's look at how one student used the writing process (prewriting, writing, and rewriting) to construct her progress report.

Prewriting

First, the student used branching so she could visually gather data and determine objectives (Figure 16.11).

Writing

Next, the student wrote a quick draft without worrying about correctness. She focused on the main units of the information she discovered in prewriting (Figure 16.12).

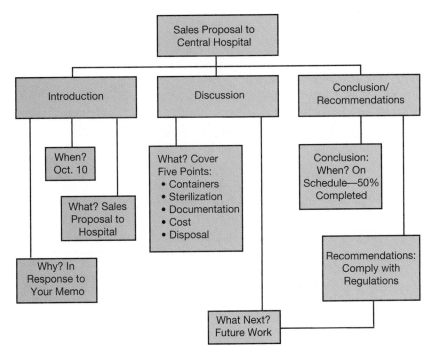

Figure 16.11 Student's Branching

November 6, 2005

TO: Carolyn Jensen
FROM: Shuan Wang.
SUBJECT: PROGRESS REPORT.

Purpose: This is a progress report on the status of a sales proposal you requested in your memo of October 10, 2005. The objective of this proposal is the sale of a total program for infectious waste control and disposal to a Central Hospital. The proposal will cover the following five areas:

1. Containers
2. Steam sterilization
3. Cost savings
4. Landfill operating
5. Computerized documentation

Work Completed: The following is a list of items that are finished on the project.

Containers
I have finished a description of the specially designed containers with disposable biohazard bag.

Steam sterilization
I have set a instructions for using the Biological Detector, model BD 12130 as a reliable indicator of sterilization of infectious wastes.

Cost savings
Central Hospital sent me some information on installing a pathological incinerator as well as the constant personnel and maintenance costs to operate them. I ran this information through our computer program "Save-Save." The results show that a substantial savings will be realized by the use of our services.

Future Work: I have an appointment to visit Central Hospital on November 7, 2005. At that time I will make on-site evaluations of present waste handling practices at Central Hospital. After carefully studing this evaluations, I will report to you on needed changes to comply with governmental guidelines.

Conclusion: The project is proceeding on schedule. Approximately 50 percent of work is done (see graph). I don't see any problem at this time and should be able to meet the target date.

Figure 16.12   Student's Rough Draft

Rewriting

After drafting this report, the student submitted her copy to a peer review group, which helped her revise her text (Figure 16.13 on pages (525–526).

After discussing the suggested changes with her peer review group, the student revised her draft and submitted her finished copy (Figure 16.14 on pages 527–528).

November 6, 2005

TO: Carolyn Jensen   *Needs topic*
FROM: Shuan Wang₍ₓ₎   *& focus*
SUBJECT: PROGRESS REPORT₍ₓ₎

— Redundant

(Purpose:) This is a progress report on the status of a sales proposal you requested in your memo of October 10, 2005. The objective of this proposal is the sale of a total program for infectious waste control and disposal to a Central Hospital. The proposal will cover the following five (areas:) *weak word*

Reformat- set in italics on a separate line and add 1st level headings like "Introduction"

wordy?

1. Containers
2. Steam sterilization
3. Cost savings
4. Landfill operating   *— Needs new title*
5. Computerized documentation

Reformat   *wordy*
(Work Completed:) The following is a list of items that are finished on the project.

Containers
I have finished a description of the specially designed containers with disposable biohazard bag. *when?*

Steam sterilization *Huh?*
I have (set a instructions) for using the Biological Detector, model BD 12130 as a reliable indicator of sterilization of *when* infectious wastes. *wordy*

Cost savings   *vague*
Central Hospital sent me some information on installing a pathological incinerator as well as the constant personnel and maintenance costs to operate them. I ran this information through our computer program "Save-Save." The results show that a substantial savings will be realized by the use of our services. *vague*

Figure 16.13   Peer Group Revisions

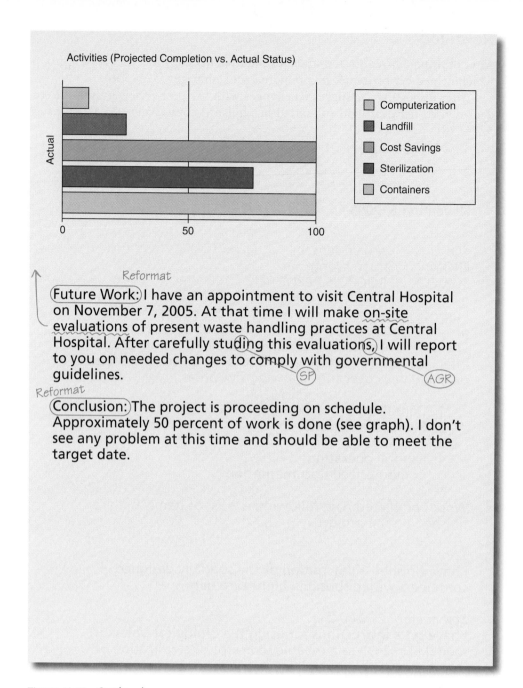

(Future Work:) I have an appointment to visit Central Hospital on November 7, 2005. At that time I will make on-site evaluations of present waste handling practices at Central Hospital. After carefully studing this evaluations, I will report to you on needed changes to comply with governmental guidelines. (SP) (AGR)

Reformat

(Conclusion:) The project is proceeding on schedule. Approximately 50 percent of work is done (see graph). I don't see any problem at this time and should be able to meet the target date.

Figure 16.13 *Continued*

INTEROFFICE CORRESPONDENCE

DATE: November 6, 2005
TO: Carolyn Jensen
FROM: Shuan Wang
SUBJECT: PROGRESS REPORT ON CENTRAL HOSPITAL
          SALES PROPOSAL

INTRODUCTION

Purpose
In response to your October 10, 2005, request, following is a report on the status of our Central Hospital sales proposal. This proposal will present Central Hospital our total program for infectious waste control and disposal. The proposal will cover these five topics:

- Containers
- Steam sterilization
- Cost savings
- On-site evaluations
- Landfill disposal

DISCUSSION

Work Completed
1. Containers—On October 5, I finished a description of the specially designed containers with biohazard bags.
2. Steam sterilization—On November 4, I wrote instructions for using our Biological Detector, model BG 12130. This detector measures infectious waste sterilization levels.
3. Cost savings—Central Hospital sent me its cost charts for pathological incinerator expenses and maintenance costs. I used our computer program "Save-Save" to evaluate these costs. The program results show that our services can save Central Hospital $15,000 per year on annual incinerator costs, after a two-year break-even period.

Figure 16.14   Final Copy

Page 2
Shuan Wang
November 6, 2005

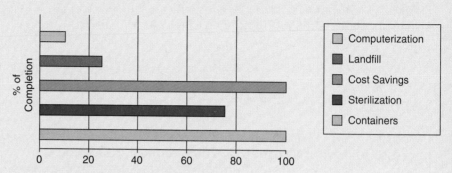

Figure 1   Completion Status of Planned Activities

Work Remaining
1. On-site evaluations—When I visit Central Hospital on November 11, 2005, I will evaluate their present waste handling system. After studying these evaluations, I will report changes necessary for government compliance.
2. Landfill disposal—I will contact the Environmental Protection Agency on November 13, 2005, to receive authorization for disposing sterilized waste at our sanitary landfill.

CONCLUSION

The project is proceeding on schedule. Approximately 55 percent of our work is completed. I see no problems at this time. We should meet our December 31, 2005, target date.

Figure 16.14   *Continued*

CHAPTER HIGHLIGHTS

1. Reports are used to document many different occurrences on the job.
2. Use headings, such as "Introduction," "Discussion," and "Conclusion/recommendations," when designing your report.
3. Trip reports document work-related travel.
4. Progress reports recount work accomplished and work remaining on a project.
5. Lab reports document the findings from a lab analysis.
6. Feasibility/recommendation reports are used to determine the viability of a proposed project.
7. Branching is a visual prewriting technique that will help you write effective reports.
8. An incident report documents an unexpected problem that has occurred.
9. An investigative report asks you to examine the causes behind an incident.
10. Meeting minutes document the results of a meeting.

CASE STUDIES

In small groups, read the following case studies and write the appropriate report.

Acme Aerospace

You manage an engineering department at Acme Aerospace. Your current department supervisor is retiring. Thus, you must recommend the promotion of a new supervisor to the company's executive officer, Kelly Adams. You know that Acme seeks to promote individuals who have the following traits:

- Familiarity with modern management techniques and concerns, such as TQM (total quality management), teamwork, global economics, and the management of hazardous materials.
- An ability to work well with colleagues (subordinates, lateral peers, and management).
- Thorough knowledge of one's areas of expertise.

You have the following candidates for promotion. Using the information provided about each and the criteria for feasibility/recommendation reports discussed in this chapter, write your report recommending your choice for a new supervisor.

A. *Pat Jefferson.* Pat has worked for Acme for 12 years. In fact, Pat, has worked up to a position as a lead engineer by having started as an assembler, then working in test equipment, quality control, and environmental safety and health (ESH). As an engineer in ESH, Pat was primarily in charge of hazardous waste disposal. Pat's experience is lengthy, although Pat has only taken two years of college coursework and one class in management techniques. Pat is well liked by all colleagues and is considered to be a team player.

B. *Kim Kennedy.* Kim is a relatively new employee at Acme, having worked for the company for two years. Kim was hired directly out of college after earning an MBA degree from the Mountaintop College School of Management. As such, Kim is extremely familiar with today's management climate and modern management techniques. Kim's undergraduate degree was a BS in business with a minor in engineering. Currently, Kim works in the engineering department as a departmental liaison, communicating the engineering department's concerns to Acme's other departments. Kim has developed a reputation as an excellent coworker who is well liked by all levels of employees.

C. *Chris Clinton.* Chris has a BS degree in engineering from Poloma College and an MBA degree from Weatherford University. Prior to working for Acme, Chris served on the IEEE (Institute of Electrical and Electronic Engineering) Commission for Management Innovation, specializing in global concerns and total quality management. In 1991, Chris was hired by Acme and since then has worked in various capacities. Chris is now lead engineer in the engineering department. Chris has earned high scores on every yearly evaluation, especially regarding knowledge of engineering. Whenever you have needed assistance with new management techniques, Chris has been a valued resource. Chris's only negative points on evaluations have resulted from difficulties with colleagues, some of whom regard Chris as haughty.

Whom will you recommend for supervisor? Write your feasibility/recommendation report stating your decision.

AAA Computing

You are the accountant at AAA Computing, a retail store specializing in computer hardware. Your boss states that all new computers sold should be accompanied by an optional service contract. This service will be held by an outside vendor. Your job is to research several vendors and write a feasibility/recommendation report recommending the best choice for your company. To do so, you know your boss will emphasize the following criteria:

- Years in business/expertise—to be sure that customers receive quality service, the vendor should have a good track record and be familiar with computer hardware innovations.

- Quick response/turnaround—because many of your customers depend on their computers for daily business operations, the vendor should provide on-site service for minor repairs and 24-hour turnaround service on major repairs.

- Cost-effective pricing—the less the vendor charges, the more you can mark up the service cost. Thus, AAA can receive more profit.

The following vendors have proposed their services. Using the information provided about each company and the criteria for feasibility reports provided in this chapter, write your report recommending a vendor for AAA's service contracts.

A. *QuickBit.* This company has 16 months' experience in computer hardware service. QuickBit, although without a lengthy track record, is co-owned by three graduates of the Silicon Valley Institute of Technology's renowned BA program in computer information services. The three individuals are very knowledgeable about computers, having received superior instruction and hands-on training using today's most up-to-date technology. Because

QuickBit is small, it can provide immediate, 24-hour on-call service. Furthermore, QuickBit owns a 12-wheeler truck fully stocked with parts. Therefore, all service can be handled on-site. QuickBit charges $50 per hour plus parts.

B. *ROM on the Run.* This company has been in business for 10 years. It employs 50 servicepeople, has a lengthy track record, and has provided service for ARC Telecommunications, Capital Bank, Helping Hand Hospital, and the State Penitentiary at Round Rock. Because of its years in business and the fact that all employees have at least two-year certificates from vocational colleges, it charges $85.50 per hour plus parts. ROM on the Run promises on-site service within 24 hours on all service requests.

C. *You Bet Your Bytes (YBYB).* YBYB has been in business for four years, employs 10 servicepeople (all of whom have at least a BS degree in computer systems), and specializes in retail outlets. YBYB charges $70 per hour plus parts. It advertises that it responds within 30 minutes for on-site service calls. Furthermore, it advertises that most parts are carried on its service trucks. Any repairs requiring unavailable parts can be made within 24 hours if the service call is received by 3:30 P.M. Calls after that time require two working days.

Which company do you recommend? Write the feasibility/recommendation report stating your choice.

Telecommunications R Us

Your company, Telecommunications R Us (TRU), has experienced a 45 percent increase in business, a 37 percent increase in warehoused stock, and a 23 percent increase in employees. You need more room. Your executive officer, Polina Gertsberg, has asked you to research existing options. To do so, you know you must consider these criteria:

- **Ample space for further expansion.** Gertsberg suggests that TRU could experience further growth upward of 150 percent. You need to consider room for parking, warehouse space, additional offices, and a cafeteria—approximately 20,000 sq ft total.
- **Cost.** Twenty million dollars should be the top figure, with a preferred payback of five years at 10 percent.
- **Location.** Most of your employees and customers live within 15 miles of your current location. This has worked well for deliveries and employee satisfaction. A new location within this 15-mile radius is preferred.
- **Aesthetics.** Ergonomics suggest that a beautiful site improves employee morale and increases productivity.

After research, you've found three possible sites. Based on the following information and on the criteria for feasibility/recommendation reports discussed in this chapter, write your report recommending a new office site.

A. *Site 1 (11717 Grandview).* This four-story site, located 12 miles from your current site, offers three floors of finished space equaling 18,000 sq ft. The fourth floor is an unfinished shell equaling an additional 3,000 sq ft. As is, the building will sell for $19 million. If the current owner finishes the fourth floor, the addition would cost $4 million more. For the building as is, the owner asks for payment in five years at 12 percent interest. If the fourth floor is finished by the owner, payment is requested in seven years at

10 percent. The building has ample parking space but no cafeteria, although a building next door has available food services. Site 1 is nestled in a beautifully wooded area with hiking trails and picnic facilities.

B. *Site 2 (808 W. Blue Valley).* This one-story building offers 21,000 sq ft that includes 100 existing offices, a warehouse capable of holding 80 storage bins that measure 20 ft tall × 60 yd long × 8 ft wide, and a full-service cafeteria. Because the complex is one story, it takes up 90 percent of the lot, leaving only 10 percent for parking. Additional parking is located across an eight-lane highway that can be crossed via a footbridge. The building, located 18 miles from your current site, has an asking price of $22 million at 8.75 percent interest for five years. Site 2 has a cornfield to its east, the highway to its west, a small lake to its north where flocks of geese nest, and a strip mall to its south.

C. *Site 3 (1202 Red Bridge Avenue).* This site is 27 miles from your current location. It has three stories offering 23,000 sq ft, a large warehouse with four-bay loading dock, and a cafeteria with ample seating and vending machines for food and drink. Because this site is located near a heavily industrialized area, the asking price is $15 million at 7.5 percent interest for five years.

Which site do you recommend? Write your feasibility/recommendation report stating your choice.

EFA

Incorporated

Education For All . . . You're Always
at Home at Our College

This week's EFA (Education For All) project team meeting focused on electronic registration. Cindy Kaye, recording secretary, needs to write the team's meeting minutes. Read the following details from the meeting and write Cindy's minutes for her.

To accomplish this task, review the criteria for minutes provided in this chapter, use highlighting techniques to make the minutes more readable, and omit any information that does not need to be reported.

> **example**
>
> The meeting was held on Tuesday, March 18, 2005, in Conference Room C at EFA's home office in Philadelphia. The meeting began at 8:15 A.M. and concluded at 11:20 A.M. The following six people attended: Cindy Kaye (recording secretary), Martha Collins (vice president, Academic Affairs), John Tomkins (vice president, Finances), Larry Rochelle (English professor), Sandra Warner (accounting professor), and Rhonda Berg (counselor).
>
> Rhonda, representing the counseling department, led the discussion. Her primary concerns for today's meeting were as follows:
>
> - Firewall protection for student enrollment
> - 24/7 access for electronic registration and payment for enrollment
> - 24/7 technical help for students and faculty with technology-related questions
> - Readily available counseling assistance for distance-learning students
> - Improved online information regarding course content
>
> First, to protect students' rights of confidentiality, Rhonda raised the issue of firewall protection. Her suggestions included providing students with personal identifi-

cation numbers (PIN) and college-unique e-mail addresses. John questioned whether it was financially feasible to create college-unique e-mail addresses, or if students should be allowed to use any e-mail addresses they might already have. Rhonda countered with two facts: not all students have their own e-mail, and sending messages to a student's e-mail outside the college's system could lead to anyone reading the contents. That would be a Buckley Amendment infraction, denying confidentiality. Instead, a PIN-accessible college e-mail address would achieve confidentiality. To maintain legal responsibilities, Rhonda assured the team, EFA had to create college e-mail addresses for all students, regardless of expense. The team voted on this issue, 5–1 with John stating that the team was acting somewhat irresponsibly in terms of the college's finances.

Next, the team focused on 24/7 coverage for technical help, electronic registration, and payment. Cindy, as director of Administrative Technology, stated that it would be cost prohibitive to hire 24/7 technology support staff. However, she told the team that her tech staff could provide personal assistance during their usual work hours (8:00 A.M.–5:00 P.M., Monday–Friday; 9:00 A.M.–2:00 P.M., Saturday).

They also could create an e-mail hotline for 24/7 questions, which would be answered within 24 hours. Though this would not allow for immediate, person-to-person attention, it would provide better turnaround than the college currently offered. The entire team agreed with this conclusion. Cindy said she would meet with her technology staff to set this practice in motion. They would create an e-mail hotline system within two weeks.

Rhonda said that 24/7 registration and payment was an easier goal to achieve. Her counseling staff would work with Cindy's staff to move all paper registration information (enrollment booklets) online. Because the text was already kept on disks, Cindy assured the group that moving the hard-copy text online would require no more than creating a shell in the college's Web site and then copying and pasting the information.

John contended that the task might be more daunting than Cindy and Rhonda assumed. Not only did he believe that moving the files would require more than copying and pasting, but more important, he asserted that online payment was a huge challenge. The college would have to make arrangements with credit card vendors to allow for online payment, credit card fraud was a concern, and bookkeeping would be stressed by new procedures. Rhonda stated that none of these issues would be a problem. She had just returned from a conference where these issues were discussed. She gave the team handouts, showing that online payment could be managed seamlessly. Sandra, our accounting professor, agreed with Rhonda. After a vote of 5–1, with only John voting "No," the team agreed to proceed with Rhonda and Cindy's suggestions.

The next topic of discussion was counseling help for online registration. The counseling staff was most comfortable helping students face-to-face, of course, but, as Rhonda stated, "The times, they are a changin'." Her staff, she said, wanted to help students regardless of their location. According to Rhonda, all they needed was up-to-date information on any distance-learning courses. This would include answers to common questions, such as "are tests administered online or on campus," "how many on-campus visits are required for distance learning courses," "is teacher–student communication achieved via e-mail, telephone, 'snail mail,' or a combination of the three," "how many assignments are required," and "is there a textbook"? Furthermore, because 24/7 counseling help was an impossibility, students would need to know that counseling would be available for help either during usual office hours (Monday–Friday, 8:00 A.M.–8:00 P.M., and Saturdays (9:00 A.M.–2:00 P.M.) or through e-mail (answered within 24 hours of the e-mail request). No vote was needed on this topic.

The final topic involved improving online information regarding courses. This proved to be a contentious topic. The counseling staff contended that they could help students only if the counselors had thorough information about quizzes, grading techniques, mandatory on-campus meetings, due dates, required textbooks, and so on. The faculty representatives, however, were not in favor of mandating

any information from faculty. They said that mandatory information would not be achievable for the following reasons: Mandating anything of faculty would require a vote and acceptance by 75 percent of the faculty members, according to Faculty Association guidelines. Next, faculty could not tell counselors which books they would be requiring in course. This would be an impossibility because different sections taught by different faculty required different textbooks. Finally, the faculty representatives were opposed to mandated home pages, templates, grading techniques, or due dates because these would need to be decided on months in advance of the semesters. Faculty, they stated, rarely if ever organized their course requirements this far in advance.

Cindy, John, and Rhonda disagreed with the faculty stance. Martha, Larry, and Sandra were resolute in their point of view. Our team was at a stalemate regarding this issue. It was tabled for further discussion.

Though a number of key issues were decided upon, the meeting ended somewhat unhappily.

EXERCISES

1. Write a progress report. The subject of this report can involve a project or activity at work. Or, if you haven't been involved in job-related projects, write about the progress you're making in this class or another course you're taking. Write about the progress you're making on a home improvement project (refinishing a basement, constructing a deck, painting and papering a room). Write about the progress you're making on a hobby (rebuilding an antique car, constructing a computer, or making model trains, etc.). Whatever your topic, first prewrite (using branching), then write a draft, and, finally, rewrite, revising the text. Abide by all the criteria presented in this chapter regarding progress reports.

2. Write a lab report. The subject of this report can involve a test you're running at work or in one of your classes. Whatever topic you select, follow the three stages in the writing process to construct your report: prewrite (using branching), write a draft, and then revise the draft in rewriting. Use the criteria regarding lab reports presented in this chapter to help you write the report.

3. Write a feasibility/recommendation report. You can draw your topic either from your work environment or home. For example, if you and your colleagues were considering the purchase of new equipment, the implementation of a new procedure, expansion to a new location, or the marketing of a new product, you could study this idea and then write a report on your findings. If nothing at work lends itself to this topic, then consider plans at home. For example, are you and your family planning a vacation, the purchase of a new home or car, the renovation of your basement, or a new business venture? If so, study this situation. Research car and home options, study the market for a new business, and get bids for the renovation. Then write a feasibility/recommendation report to your family documenting your findings. Whether your topic comes from business or home, gather your data in prewriting (using branching), draft your text in writing, and then revise in rewriting. Follow the criteria for feasibility/recommendation reports provided in this chapter to help you write the report.

4. Write an incident report. You can select a topic either from work or home. If you have encountered a problem at work, write an incident report documenting the problem and providing your solutions to the incident. If nothing has happened at work lending itself to this topic, then look at home. Has your car broken down, did the water heater break, did you or any members of your family have an accident of any sort, did your dog or cat knock over the vase your mother-in-law gave you for Christmas? Consider such possibilities, and then write an incident report documenting the incident. Follow the criteria for incident reports provided in this chapter, and use writing process techniques (prewriting, writing, and rewriting).

5. Write meeting minutes. To do so, attend a meeting. You could go to a meeting at work, on your campus, at a community organization, or at your city hall. Take notes and transfer these notes into minutes. To do so, follow the guidelines provided in this chapter. Then, bring the minutes to your class to share with other students. Exchange the minutes and assess everyone's achievements. Which minutes are most readable? Which minutes achieve objectivity by focusing on what is done versus what is said? Which minutes are most complete, including date, time, attendees, location, and meeting content?

6. Find examples of lab reports, trip reports, progress reports, feasibility/recommendation reports, incident reports, investigate reports, or meeting minutes. You can find these at your company or visit another job site. Bring these reports to class. Then, using the criteria presented in this chapter, form small groups and discuss whether the reports are successful or unsuccessful. If they are good, specify how and why. If they are flawed, discuss what's wrong, prescribe solutions, and rewrite the reports to improve them.

7. Look around your company or visit another job site and find examples of short reports not discussed in this chapter. Bring these reports to class. Then, using the criteria for short reports presented in this chapter, form small groups and discuss how these unique reports differ from or abide by the criteria we have presented. Also discuss whether the reports you have found are successful or unsuccessful. If they are good, discuss why and how. If they are flawed, talk about how they could be improved and rewrite them.

8. Interview an employee and a supervisor in your profession or the field of your choice. Ask these people why they write reports, what their reports are supposed to accomplish, who their audiences are, what length reports are preferred in their company, how many reports they write in a week or month, and so on. Use your imagination regarding the questions you ask. Then, once you've received your answers, form small groups and convey your findings to your classmates. To do so, either present an oral presentation or—better yet—write a group report!

9. As a class, take a field trip. Visit a publishing firm, see a play, go to a museum, hear a guest speaker on campus, or interview a professional technical writer, for example. Then, in small groups, write a trip report about your observations. Once the reports are written, compare and contrast the writing. Which group of students (or which student) has written the best report? To make this judgment, specify why the report is successful. Doing so will provide for the class a model of good

writing, as agreed on by class consensus. Class members can replicate this report for future success. In addition, discuss how less successful reports fell short of the desired objectives. This will help show students what to avoid in future report-writing activities.

10. Revision is the key to good writing. An example of a seriously flawed progress report that needs revising is shown in Figure 16.15. To improve this report, form small groups and first decide what's missing according to this chapter's criteria for good progress reports. Then use the rewriting techniques presented in this chapter to rewrite and revise the report.

WEB WORKSHOP

1. More and more, companies and organizations are putting report forms online. The reason for doing this is simple—ease of use. For example, look at the following screen capture:

Google™   Advanced Search   Preferences   Language Tools   Search Tips

online trip report form   Google Search

Web | Images | Groups | Directory | News |
Searched the web for **online trip report form**.   Results **1 - 10** of about **493,000**. Search took

Hiking **Trip Report Form**
Online Trip Report Form. Destination: Date of Hike: Leader: Number of Hikers: Hike Category: Easy, ...
www.mountaineerhikes.org/leaders/trip_report.html - 7k - Cached - Similar

Robertson Group Fuel Tax System
... tax **form** management **online** fuel taxes **online** trucking **online** support trucking trips trucking news resource record **trip** quarter quarterly fuel **report** legal mid ...
www.truckingfueltax.com/index.cfm - 15k - Cached - Similar pages

NOAA Fisheries - GPEA Forms
... Some are web-based applications in which the **form** information is entered directly into a ... wpd) and Microsoft Word files (.doc) can be filled in **online** and faxed ...
www.nmfs.noaa.gov/gpea_forms/southeast.htm - 30k - Cached - Similar pages

Go online and access online report forms. All you need to do is use any Internet search mechanism, and type in "online_____report form." (In the space provided, type in "trip," "progress," "incident," "investigative," or "feasibility/recommendation.") Once you find examples, evaluate how they are similar to and different from the written reports shown in this chapter. Share your findings with others in your class, either through oral presentations or written reports.

2. Every company writes reports—and you can find examples of them online. Use the Internet to access Report Gallery, found at http://www.reportgallery.com/. Report Gallery bills itself as "the largest Internet publisher of annual reports." From this Web site, you can find the annual reports from 2,200 companies, including many Fortune 500 companies.

Study the annual reports from 5 to10 different companies. What do these reports have in common? How do they differ from each other? Discuss the reports' page layout, readability, audience involvement, content, tone, and development. How are the reports similar to and different from those discussed in this chapter?

Share your findings with others in your class, either through oral presentations or written reports.

Site Visit—Alamo Manufacturing
November 1

Sam, I visited our Alamo site and checked on the following:

1. Our plant facilities suffered some severe problems due to the recent wind and hail storms. The west roof lost dozens of shingles, leading to water damage in the manufacturing room below. The HVAC unit was submerged by several feet of water, shorting out systems elsewhere in the plant. In addition, our north entryway awning was torn off its foundation due to heavy winds. This not only caused broken glass in our front entrance door and a few windows bordering the entrance, but also the entryway driveway now is blocked for customer access.

2. Maintenance and security failed to handle the problems effectively. Security did not contact local police to secure the facilities. Maintenance responded to the problems far too many hours late. This led to additional water and wind damage.

3. Luckily, the storm hit early in the day, before many of our employees had arrived at work. Still, a few cars were in the parking lot, and they suffered hail damage. Are we responsible?

4 The storm, though hurting manufacturing, will not affect sales. Our 800-lines and e-mail system were unaffected. But, we might have problems with delivery if we can't fix the entryway impediment. I think I have some solutions. We could reroute delivery to one of our new plants, or maybe we should consider direct delivery to our sales staff (short term at least). Any thoughts?

5. Meanwhile, I have gotten on Maintenance's case, and it is working on repairs. Brownfield HVAC service has given us a quote on a new sump pump system that has failsafe programs (not too bad an extra cost, given what we can save in the long run). Plainview Windows & Doors is already on the front entrance problem. It promises replacement soon.

6. As for future plans, I think we need to reevaluate both Maintenance and Security. Training is an option. We could also add new personnel, reconsider our current management in those departments, or maybe just a few, good, hard, strongly stated comments from you would do the trick.

Anyway, we're up and running again. Upfront repair costs are covered by insurance, and long-term costs are minimal because sales weren't hurt. Our only remaining challenges are preventive, and that depends on what we do with our Maintenance and Security staff. Let me know what you think.

Figure 16.15   Flawed Report

Quiz Questions

1. What are some purposes of a report?

2. What are some common types of reports?

3. List the four basic units of a report.

4. What types of material are included in the Discussion section?

5. Explain how the reporter's questions can help you prepare your report.

6. How will a Gantt chart assist your reader in a progress report?

7. How does branching help you to prewrite a report?

8. When do you write a trip report?

9. When do you write a progress report?

10. When do you write a lab report?

11. When do you write a feasibility/recommendation report?

12. When do you write an incident report?

13. When do you write an investigative report?

14. What content do you include in meeting minutes?

15. How do visual aids assist the reader in a report?

16. What do you include in the heading?

17. What do you include in the conclusion?

18. What do you include in the recommendation?

19. What is the writing style in a report?

20. How do you satisfy multiple readers in the introduction?

Proposals

Writing at Work

Bellaire Educational Supplies/Technologies
Your **_BEST_** buy in the market!

BEST (Bellaire Educational Supplies/ Technologies) manufactures and markets school supplies for all levels of education (K-12, private or parochial, and college).

These supplies include the following:

- Desks and chairs
- Modular computer workstations
- Blackboards and whiteboards
- Markers, chalk, and erasers
- Overhead projectors (both mobile and fixed computer projection systems)
- Pull-down screens
- Pens and pencils
- Grade books

BEST produces catalogs and mails fliers and brochures, but the majority of its large-ticket sales (to entire school systems) comes from external proposals. These proposals, which usually are written in response to letters of inquiry or as follow-ups to person-to-person sales meetings, focus on the following components:

- Cover letters prefacing the proposal to personalize the enclosed text as well as direct the audience to key parts of the proposal
- Table of contents itemizing the major headings and subheadings within the text
- A directory of tables and figures
- An executive summary, geared toward low-tech readers with limited time to read the entire proposal

- An introduction highlighting the reader's unique educational supplies problems and explaining how BEST will meet the client's needs
- Text that details costs, services, delivery, maintenance, guarantees, and product quality
- A conclusion summing up the proposal's key points
- A glossary to define unfamiliar words, acronyms, and abbreviations

Many of BEST's proposals are actually hard-copy printouts of PowerPoint slides. BEST's sales force and technical writers have concluded that most readers want less text and easy-to-read documentation, complete with ample graphics. PowerPoint is the solution to this need. Not only can BEST's sales staff use PowerPoint to make oral presentations about its products, but they also can print out the PowerPoint slides as 11 × 8 1/2-inch landscape pages and bind them as handouts. The hard-copy pages are easy to create, easy to read, and meet the audience's need for conciseness.

BEST's products, once used, sell themselves. However, the proposals open the door. Through proposals, BEST has found a perfect way to present clear, concise, and thorough documentation of the company's assets.

In the preceding chapter, we discussed various types of reports, including trip reports, progress reports, lab reports, incident reports, feasibility/recommendation reports, investigative reports, and meeting minutes. The majority of these reports will be limited to no more than five pages. A report on your job-related travel would rarely require multipage documentation. Similarly, most progress reports address only daily, weekly, or monthly activities. The same applies to most lab reports. Because the subject matter will be limited, the report will be short.

However, in some instances, your subject matter might be so complex that a shorter report will not suffice. For example, your company asks you to write a report proposing the purchase of a new facility. You will have to write a longer report—an *internal proposal* for your company's management. Or, perhaps your company is considering offering a new service or manufacturing a new product. Your responsibility is to write an *external proposal* selling the benefits of this new corporate offering to a prospective client.

External proposals are also written in response to RFPs (*requests for proposals*). Often, companies, city councils, and state or federal agencies need to procure services from other corporations. A city, for example, might need extensive road repairs. To receive bids and analyses of services, the city will write an RFP, specifying the scope of its needs. Competing companies will respond to this RFP with an external proposal.

In each of these instances, you ask your readers to make significant commitments regarding employees, schedules, equipment, training, facilities, and finances. Only a long report, possibly complete with research, will convey your content sufficiently and successfully. (We discuss research writing in Chapter 14.)

To help you write your proposals effectively, this chapter provides the following: criteria for proposals, sequential process for writing proposals, and sample proposals.

CRITERIA FOR PROPOSALS

To guide your readers through a proposal, you will need to provide the following:
- Title page
- Cover letter
- Table of contents

The Winning Elements of Business Proposals

As owner and principal of Pied Piper Internet Solutions, Darrell Zahorsky says that there may be times in your small business life when your company will have to submit a business proposal to gain business from a larger corporation or government contact.

What is A Business Proposal?

Unlike a business plan, which is written to run your company and raise capital, a business proposal is an unsolicited or solicited bid for business. There are two types of business proposals that can help you gain more business to grow your company.

Solicited Business Proposal: A corporation or government body is seeking a business to fulfill a project or complete a task and thereby, allows companies to bid for the project. An open bid is placed on the market with other companies competing for an interview spot.

Unsolicited Business Proposal: At some point, your small business may want to do business with a larger company or forge a joint venture. A well-written business proposal can win the hearts and minds of your target audience.

If you need to write a business proposal to win a bid, you will need to know the key winning elements of a successful proposal. Make sure your proposal stands out in the stack of competitor proposals by including the following 5 elements:

5 Key Elements of Winning Business Proposals

- Solutions: After you have written a lead paragraph on the company's needs and problems, follow up with a solid presentation of how your business can provide solutions. The key here is to promise solutions you can deliver.

- Benefits: All winning business proposals clearly outline for the company the benefits to be gained from doing business with you. If your small business can offer complete confidentiality and meet tight deadlines, state it in your benefits section.

- Credibility: This is often the overlooked portion of a business proposal, but all winning proposals glow with credibility. If you have worked with clients in the same field or have an award-winning business, then third-party endorsements will build credibility.

- Samples: A business proposal with samples and evidence of your ability to deliver is vital to gaining the winning bid. A small sample of your work can show your ability to do the job.

- Targeted: A winning business proposal is all about communication. Speak in a language spoken by your intended audience. If the proposal evaluators are from an engineering background or financial department, use the appropriate jargon.

Source: Zohorsky 2004

- List of illustrations
- Abstract (or executive summary)
- Introduction
- Discussion (the body of the proposal)
- Conclusion/recommendation

- Glossary
- Works cited (or references) page (if you're documenting research; this is discussed in Chapter 14)
- Appendix

Title Page

The title page serves several purposes. On the simplest level, a title page acts as a dust cover or jacket keeping the report clean and neat. More important, the title page tells your reader the

- Title of the proposal (thereby providing clarity of intent)
- Name of the company, writer, or writers submitting the proposal
- Date on which the proposal was completed

If the external proposal is being mailed outside your company to a client, you also might include on the title page the audience to whom the report is addressed. If the internal proposal is being submitted within your company to peers, subordinates, or supervisors, you might want to include a routing list of individuals who must sign off or approve the proposal.

Following are two sample title pages. Figure 17.1 is for an internal proposal; Figure 17.2 is for an external proposal.

PROPOSED CABLE TRANSMISSION NETWORK
FROM CHEYENNE, WY
TO
HARTFORD, CT

Prepared by: _____ Date: _____
Pete Niosi
Network Planner

Reviewed by: _____ Date: _____
Leah Workman
Manager, Capital Planning

Recommended by: _____ Date: _____
Greg Foss
Manager, Facilities

Recommended by: _____ Date: _____
Shirley Chandley
Director, Implementation Planning

Approved by: _____ Date: _____
Ralph Houston
Vice President, Network Planning

Figure 17.1   Title Page for an Internal Proposal

PROPOSAL TO MAINTAIN COMPUTER EQUIPMENT

For
Acme Products, Inc.
2121 New Tech Avenue
Bangor, ME

Submitted by
Thomas Brasher
Engineering Technician

August 13, 2005

Figure 17.2 Title Page for an External Proposal

Cover Letter

Your cover letter prefaces the proposal and provides the reader an overview of what is to follow. It tells the reader

- Why you are writing
- What you are writing about (the subject of this proposal)
- What exactly of importance is within the proposal
- What you plan to do next as a follow-up
- When the action should occur
- Why that date is important

Each of these points is discussed in greater detail in Chapter 6.

Table of Contents

Proposals are read by many different readers, each of whom will have a special area of interest. For example, the managers who read your proposals will be interested in cost concerns, timeframes, and personnel requirements. Technicians, in contrast, will be interested in technical descriptions and instructions. Not every reader will read each section of your proposal.

Your responsibility is to help these different readers find the sections of the proposal that interest them. One way to accomplish this is through a table of contents. The table of contents should be a complete and accurate listing of the main *and* minor topics covered in the proposal. In other words, you don't want just a brief and sketchy outline of major headings. This could lead to page gaps; your readers would be unable to find key ideas of interest. In the table of contents on page 544, we can see that the proposal section contains approximately eight pages of data. What is covered in those eight pages? Anything of value? We don't know. The same applies to the appendix, which covers four pages. What is in this section?

```
              FLAWED TABLE OF CONTENTS
                  Table of Contents
            List of Illustrations . . . . . . . . . . . . . . . . . . . . . . .iv
            Abstract . . . . . . . . . . . . . . . . . . . . . . . . . . . . . .v
    1.0     Introduction . . . . . . . . . . . . . . . . . . . . . . . . . . .1
    2.0     Proposal . . . . . . . . . . . . . . . . . . . . . . . . . . . . .3
    3.0     Conclusion . . . . . . . . . . . . . . . . . . . . . . . . . . .10
    4.0     Appendix . . . . . . . . . . . . . . . . . . . . . . . . . . . .11
    5.0     Glossary . . . . . . . . . . . . . . . . . . . . . . . . . . . .15
```

In contrast, an effective table of contents fleshes out this detail so your readers know exactly what is covered in each section. By providing a thorough table of contents, you will save your readers time and help them find the information they want and need. Figure 17.3 is an example of a successful table of contents.

In the example, note that the actual pagination (page 1) begins with the introductory section. Page 1 begins with your main text, not the front matter. Instead, information prior to the introduction is numbered with lowercase Roman numerals (i, ii, iii, etc.). Thus, the title page is page i, and the cover letter is page ii. However, you never print the numbers on these two pages. Therefore, the first page with a printed number is the table of contents. This is page iii, with the lowercase Roman numeral printed at the foot of the page and centered.

List of Illustrations

If your proposal contains several tables or figures, you will need to provide a list of illustrations. This list can be included below your table of contents, if there is room on the page, or on a separate page. As with the table of contents, your list of illustrations must be clear and informative. Don't waste your time and your reader's time by providing a poor list of illustrations like the one below.

```
              FLAWED LIST OF ILLUSTRATIONS
                  List of Illustrations
    Fig. 1                                                    2
    Fig. 2                                                    4
    Fig. 3                                                    5
    Fig. 4                                                    5
    Fig. 5                                                    9
    Table 1                                                  3
    Table 2                                                  6
```

The example provides your reader with very little information. All the reader can ascertain from this list is that you've used some figures and tables. However, the reader will have no idea what purpose each illustration serves. Instead of supplying such a vague list, you should accompany the table and figure numbers with descriptive titles.

Table of Contents

Figure 17.3   Successful Table of Contents

SUCCESSFUL LIST OF ILLUSTRATIONS
List of Illustrations

Abstract (or Executive Summary)

As mentioned earlier, a number of different readers will be interested in your proposal. One group of those readers will be management—supervisors, managers, and highly placed executives. How do these readers' needs differ from others? Because these readers are busy with management concerns and might have little technical knowledge, they need your help in two ways: they need information quickly, and they need it presented in low-tech terminology. You can achieve both these objectives through an abstract or executive summary.

The abstract is a brief overview of the proposal's key points geared toward a low-tech reader. If the intended audience is composed of upper-level management, this unit might be called an executive summary. To accomplish the required brevity, you should limit your abstract to approximately 3 to 10 sentences. These sentences can be presented as one paragraph or as smaller units of information separated by headings. Each proposal you write will focus on unique ideas. Therefore, the content of your abstracts will differ. Nonetheless, abstracts should focus on the following: (a) the *problem* necessitating your proposal, (b) your suggested *solution*, and (c) the *benefits* derived when your proposed suggestions are implemented. These three points work for external as well as internal proposals.

For example, let's say you are asked to write an internal proposal suggesting a course of action (limiting excessive personnel, increasing your company's work force, improving your corporation's physical facilities, etc). First, your abstract should specify the problem requiring your planned action. Next, you should mention the action you are planning to implement. This leads to a brief overview of how your plan would solve the problem, thus benefiting your company.

If you were writing an external proposal to sell a client a new product or service, you would still focus on problem, solution, and benefit. The abstract would remind the readers of their company's problem, state that your company's new product or service could alleviate this problem, and then emphasize the benefits derived.

In each case, you not only want to be brief, focusing on the most important issues, but also you should avoid high-tech terminology and concepts. The purpose of the abstract is to provide your readers with an easy-to-understand summary of the entire proposal's focus. Your executives want the bottom line, and they want it quickly. They don't want to waste time deciphering your high-tech hieroglyphics. Therefore, either avoid all high-tech terminology completely or define your terms parenthetically.

The following is an example of a brief, low-tech abstract from an internal proposal.

ABSTRACT

Due to deregulation and the recent economic recession, we must reduce our workforce by 12 percent.

Our plan for doing so involves

- Freezing new hires
- Promoting early retirement
- Reassigning second-shift supervisors to our Desoto plant
- Temporarily laying off third-shift line technicians

Achieving the above will allow us to maintain production during the current economic difficulties.

Introduction

Your introduction should include two primary sections: (1) purpose, and (2) problem.

Purpose

In one to three sentences, tell your readers the purpose of your proposal. This purpose statement informs your readers *why* you are writing or *what* you hope to achieve. This statement repeats your abstract to a certain extent. However, it's not redundant; it's a reiteration. Although numerous people read your report, not all of them read each line or section of it. They skip and skim.

The purpose statement, in addition to the abstract, is another way to ensure that your readers understand your intent. It either reminds them of what they have just read in the abstract or informs them for the first time if they skipped over the abstract. Your purpose statement is synonymous with a paragraph's topic sentence, an essay's thesis, the first sentence in a letter, or the introductory paragraph in a shorter report.

The following is an effective purpose statement.

> Purpose: The purpose of this report is to propose the immediate installation of the 102473 Numerical Control Optical Scanner. This installation will ensure continued quality checks and allow us to meet agency specifications.

Problem

Whereas the purpose statement should be limited to one to three sentences for clarity and conciseness, your discussion of the problem must be much more detailed.

For example, if you are writing an internal proposal to add a new facility, your company's current work space must be too limited. You have got a problem that must be solved. If you are writing an external proposal to sell a new piece of equipment, your prospective client must need better equipment. Your proposal will solve the client's problem.

Your introduction's focus on the problem, which could average one to two pages, is important for two reasons. First, it highlights the importance of your proposal. It emphasizes for your readers the proposal's priority. In this problem section, you persuade your readers that a problem truly exists and needs immediate attention.

Second, by clearly stating the problem, you also reveal your knowledge of the situation. The problem section reveals your expertise. Thus, after reading this section of the introduction, your audience should recognize the severity of the problem and trust you to solve it.

One way to help your readers understand the problem is through the use of highlighting techniques, especially headings and subheadings.

Figure 17.4 provides a sample introduction stating purpose and problem.

Discussion

The discussion section of your proposal constitutes its body. In this section, you sell your product, service, or suggested solution. As such, the discussion section represents the major portion of the proposal, perhaps 85 percent of the text.

1.0 INTRODUCTION

1.1 Purpose

This is a proposal for a storm sewer survey for Yakima, Washington. First, the survey will identify storm sewers needing repair and renovation. Then it will recommend public works projects that would control residential basement flooding in Yakima.

1.2 Problem

1.2.1. *Increased Flooding*

Residential basement flooding in Yakima has been increasing. Fourteen basements were reported flooded in 2004, whereas 83 residents reported flooded basements in 2005.

1.2.2. *Property Damage*

Basement flooding in Yakima results in thousands of dollars in property damage. The following are commonly reported as damaged property:
  a. Washers
  b. Dryers
  c. Freezers
  d. Furniture
  e. Furnaces
Major appliances cannot be repaired after water damage. Flooding can also result in expensive foundation repairs.

1.2.3. *Indirect Costs*

Flooding in Yakima is receiving increased publicity. Flood areas, including Yakima, have been identified in newspapers and on local newscasts. Until flooding problems have been corrected, potential residents and businesses may be reluctant to locate in Yakima.

1.2.4. *Special-Interest Groups*

Citizens over 55 years old represent 40 percent of the Yakima, Washington, population. In city council meetings, senior citizens with limited incomes expressed their distress over property damage. Residents are unable to obtain federal flood insurance and must bear the financial burden of replacing flood-damaged personal and real property. Senior citizens (and other Yakima residents) look to city officials to resolve this financial dilemma.

Figure 17.4   Proposal Introduction

What will you focus on in this section? Because every proposal will differ, we can't tell you exactly what to include. However, your discussion can contain any or all of the following:

- Analyses
  —Existing situation
  —Solutions
  —Benefits
- Technical descriptions of mechanisms, tools, facilities, or products
- Technical instructions
- Options
  —Approaches or methodologies
  —Purchase options
- Managerial chains of command (organizational charts)
- Biographical sketches of personnel
- Corporate and employee credentials
  —Years in business
  —Satisfied clients
  —Certifications
  —Previous accomplishments
- Schedules
  —Implementation schedules
  —Reporting intervals
  —Maintenance schedules
  —Delivery schedules
  —Completion dates
  —Payment schedules
  —Projected milestones (forecasts)
- Cost charts

You will have to decide which of these sections will be geared toward high-tech readers, low-tech readers, or a lay audience. Once this decision is made, you will write accordingly, defining terms as needed. However, one way to handle multiple audience levels is through a glossary (see page 551).

In addition to audience recognition, you should also enhance your discussion with figures and tables for clarity, conciseness, and cosmetic appeal.

Conclusion/Recommendation

As with shorter reports, you must sum up your proposal, providing your readers with a sense of closure. The conclusion can restate the problem, your solutions, and the benefits to be derived. In doing so, remember to quantify. Be specific—state percentages and amounts.

Your recommendation will suggest the next course of action. Specify when this action will or should occur and why that date is important.

The conclusion/recommendation section can be made accessible through highlighting techniques, including headings, subheadings, underlining, boldface, itemization, and white space.

3.0 CONCLUSION

Our line capability between San Marcos and LaGrange is insufficient. Presently, we are 23 percent under our desired goal. Using the vacated fiber cables will not solve this problem because the current configuration does not meet our standards. Upgrading the current configuration will improve our capacity by only 9 percent and still present us the risk of service outages.

4.0 RECOMMENDATION

We suggest laying new fiber cables for the following reasons. They will

- Provide 63 percent more capacity than the current system
- Reduce the risk of service outages
- Allow for forecasted demands when current capacity is exceeded
- Meet standard configurations

If these new cables are laid by September 1, 2005, we will predate state tariff plans to be implemented by the new fiscal year.

Figure 17.5 Proposal Conclusion/Recommendation

Your conclusion/recommendation, like your abstract, will be read primarily by executives. Thus, write to a low-tech reader.

Figure 17.5 is an example of a successfully written conclusion/recommendation from an internal proposal.

Glossary

Because you will have numerous readers with multiple levels of expertise, you must be concerned about your use of high-tech language (abbreviations, acronyms, and terms). Although some of your readers will understand your terminology, others won't. However, if you define your terms each time you use them, two problems will occur: you will insult high-tech readers, and you will delay your audience as they read your text. To avoid these pitfalls, use a glossary.

A glossary is an alphabetized list of high-tech terminology placed after your conclusion/recommendation. When your first high-tech, unfamiliar abbreviation, acronym, or term is used, follow it with an asterisk (*). Then, at the bottom of the page, in a footnote, write

example

*This and subsequent terms followed by an asterisk are defined in the glossary beginning on page _____.

GLOSSARY

BSCE	Bachelor of Science, Civil Engineering
Drainage studies	The study of moving surface or surplus water
Gb	Gigabyte
Interface	Communication with other agencies, entities, and systems to discuss subjects of common interest
POP	Place of purchase
Smoke test	Test using underground smoke bombs to give visual, above-ground signs of sewer pipe leaks
Video scanner	A portable video camera that examines the inside surface of sewer pipes
Water parting	A boundary line separating the drainage districts of two streams
Watershed area	An area bounded by a water parting and draining into a particular watercourse

Figure 17.6   Glossary

Subsequent high-tech terms should be followed by an asterisk, but you don't have to place the footnote at the bottom of each page.

A glossary is invaluable. Readers who are unfamiliar with your terminology can turn to the glossary and read your definitions. Those readers who understand your word usage can continue to read without stopping for unneeded information.

Figure 17.6 is a sample glossary.

Works Cited (or References)

If you have used research to write your proposal, you will need to include a works cited page. This page documents the sources (books, periodicals, interviews, computer software, etc.) you have researched and quoted or paraphrased. For more about conducting research, see Chapter 14.

Appendix

A final, optional component is an appendix. Appendices allow you to include any additional information (survey results, tables, figures, previous report findings, relevant letters or memos, etc.) that you have not built into your proposal's main text.

The contents of your appendix should not be of primary importance. Any truly important information should be incorporated within the proposal's main text. Valuable data (proof, substantiation, or information that clarifies a point) should appear in the text where it is easily accessible. Information provided within an appendix is buried, simply because of its placement at the end of the report. You don't want to bury key ideas. An appendix is a perfect place to file nonessential data that provides documentation for future reference.

Before you submit your proposal, be sure you have included all the necessary information by consulting the Proposal Checklist that appears on page 557.

PROCESS

The Writing Process

The Writing Process		
Prewriting	Writing	Rewriting
• Examine your purposes • Determine your goals • Consider your audience • Gather your data • Determine how the content will be provided	• Organize the draft according to some logical sequence that your readers can follow easily • Format the content to allow for ease of access	• Revise ° Add missing details ° Delete wordiness ° Simplify word usage ° Enhance the tone of your communication ° Reformat your text for ease of access ° Practice the speech or review the text • Proofread ° Correct errors

Proposals, which include descriptions, instructions, cost analyses, scheduling assessments, and personnel considerations, are more demanding than other kinds of technical correspondence. Therefore, writing according to a process approach is even more important than for other kinds of writing discussed in this textbook. For your proposal, you will have more data to gather, more information to organize, and more text to revise. To help you tackle these tasks,

- Prewrite
- Write
- Rewrite

Remember that the writing process is dynamic and these steps often overlap.

Prewriting

Throughout this textbook, we have provided numerous prewriting techniques geared toward helping you gather data and determine your objectives. Any or all of these can be used while prewriting your proposal. For example, you might want to use

- *Brainstorming/listing* to outline the key components of your proposal or for your technical description, if you plan to include one
- *Reporter's questions* (*who, what, when, where, why*, and *how*) to help you gather data for any of the sections in your proposal
- *Flowcharting* to organize procedures and schedules
- *Branching* and *mind mapping* to organize your managerial sections (organizational charts for chains of command, personnel responsibilities, etc.)

In addition to these prewriting techniques, you might need to perform research prior to writing. For instance, let's say your engineering firm is submitting a proposal to an out-of-state corporation. Prior to bidding on the job, you will need to find out what kinds of certifications or licenses this state requires. To do so, you will have to research the state's requirements and read state laws regarding construction certifications and licenses. (We discuss techniques for doing research in Chapter 14.)

Perhaps you will need to survey residents, clients, or personnel to get their ideas regarding proposed changes. If so, here is what you should do.

Surveys

Before you can take a survey, you must decide what questions to ask. Brainstorming/listing would be an excellent way to gather such data. Follow this procedure:

1. Rapidly jot down whatever questions come to mind regarding your topic.
2. Don't editorialize at this point. Don't try to organize the list; don't delete any questions that emerge.
3. Once you have a list of possible topics, review the list.
   a. Add any omissions (keep in mind your reporter's questions during this stage).
   b. Delete any redundancies or irrelevant ideas.
   c. Organize the list according to some rational order.

Your next step is to format the list into a survey questionnaire, as in Figure 17.7.

Writing

After gathering your data and organizing your thoughts through prewriting, your next step is to draft your proposal.

Review Your Prewriting

Double-check your brainstorming/listing, mind mapping, flowcharting, reporter's questions, interviews, research, and surveys. Is all the necessary information there? Do you have what you need? If not, now is the time to add new detail or research your topic further. In contrast, perhaps you've gathered more information than necessary. Maybe some of your data are irrelevant, contradictory, or misleading. Delete this information. Focus your attention on the most important details.

STUDENT AND STAFF QUESTIONNAIRE
FOR
PROPOSED COLLEGE DAY-CARE CENTER

1. Are you male or female? ____________
2. Are you married? ____________
3. Are you in a single- or dual-income family? _________
4. Are you a student or a staff member? ____________
5. How many children do you have? ____________
6. What are their ages? _____ _____ _____ _____
7. Would you be interested in having a day-care center at this college? ____________
8. How much would you be willing to pay per hour for child care at this college? ____________
9. Do you think a day-care center would improve attendance at this college? ____________

If student:

10. Are you enrolled full time or part time at this college? _____
11. Do you attend mornings, afternoons, evenings, or a combination of the above? (Please specify.) ________
12. Would you be willing to work in this day-care center part time? ____________

If staff:

13. What hours do you work at this college (primarily mornings, afternoons, evenings, or a combination of the above)? (Please specify.) ____________
14. What hours would you need to use the day-care center? ________________________

Optional:

Name: _______________________________
Address: _____________________________
Phone: ______________________________
Thank you for your assistance.

Figure 17.7 Questionnaire

Organize the Data

Each of your proposal's sections will require a different organizational pattern. Following are several possible approaches:

- Abstract: problem/solution/benefit
- Introduction: cause/effect (The problem unit is the cause of your writing; the purpose statement represents the effect.)

- Main text: This unit will demand many different methods of organization, including
  —Analysis (cost charts, approaches, managerial chains of command, personnel biographies, etc.)
  —Chronology (procedures, scheduling)
  —Spatial (descriptions)
  —Comparison/contrast (options—approaches, personnel, products)
- Conclusion/recommendation: analysis and importance (Organize your recommendations by importance to highlight priority or justify need.)

Write Using Sufficing Techniques

When you draft your text, don't worry about correct grammar, highlighting techniques, or graphics. Just get down the information as rapidly as you can. You can revise the draft during rewriting, the third stage of the writing process.

Format Your Writing According to the Criteria for Effective Proposals

To format your proposal, include the following components:

- Title page
- Cover letter
- Table of contents
- List of illustrations
- Abstract (or executive summary)
- Introduction
- Discussion (the body of the proposal)
- Conclusion/recommendation
- Glossary
- Works cited (or references) page
- Appendix

Rewriting

After you have written a rough draft of your proposal, the next step is to revise it—fine-tune, hone, sculpt, and polish your draft.

Add Detail for Clarity

In addition to rereading your rough draft and adding a missing *who, what, when, where, why,* and *how* where necessary, add your graphics. Go back to each section of your proposal and determine where you could use any of the following:

- Tables. Your cost section lends itself to tables.
- Figures. Your introduction's problem analysis and any of the main text sections could profit from the following figures:
  —Line charts (excellent for showing upward and downward movement over a period of time. A line chart could be used to show how a company's profits have decreased, for example.)
  —Bar charts (effective for comparisons. Through a bar or grouped bar chart, you could reveal visually how one product, service, or approach is superior to another.)

—Pie charts (excellent for showing percentages. A pie chart could help you show either the amount of time spent or amount of money allocated for an activity.)

—Line drawings (effective for technical descriptions)

—Photographs (effective for technical descriptions)

—Flowcharts (a successful way to help readers understand procedures)

—Organizational charts (excellent for giving an overview of managerial chains of command)

We discuss each of these types of graphics in detail in Chapter 9.

Another important addition to your proposal in this rewriting stage is a glossary. Now that you've written your rough draft, go back over each page to decide which abbreviations, acronyms, and high-tech terms must be placed in your glossary. After each of these high-tech words or phrases, add an asterisk. After the first abbreviation, acronym, or high-tech word or phrase, add a footnote stating that subsequent words or phrases followed by an asterisk will be defined in the glossary. Then write your glossary and add it to your proposal.

Delete Dead Words and Phrases for Conciseness

Because proposals are long, you've already made your reader uncomfortable. People don't like wading through massive pages of text. Anything you can do to help your reader through this task will be appreciated. Avoid making the proposal longer than necessary. (Chapter 3 gives detailed explanations for achieving conciseness.)

Simplify Old-Fashioned Words

A common adage in technical writing is, "Write the way you speak." In other words, try to be more conversational in your writing. We aren't suggesting that you use slang or colloquialisms. But a word like *supersede* should be simplified: instead, write *replace*. Check your text to see whether you've used words that are unnecessarily confusing.

Deleting and simplifying, when used together, will help you lower your fog index. This is especially important in your abstract and conclusion/recommendation sections, which are geared toward low-tech management. Review these sections of your proposal to determine if you have written at the appropriate level. If you have used terminology that is too high tech, simplify.

Move Information

Each section of your proposal will use a different organizational method. Your abstract, for example, should be organized according to a problem/solution/benefit approach. Your introduction will be organized according to cause (the problem) and effect (the purpose). The organization of your main text will vary from section to section. You might analyze according to importance, set up schedules and procedures chronologically, describe spatially, and so forth.

In rewriting, revise your proposal to ensure that each section maintains the appropriate organizational pattern. To do so, move information around (cut and paste).

Reformat for Reader-Friendly Ease of Access

If you give your readers wall-to-wall words, they will doze off while attempting to wade through your proposal. To avoid this, revise your proposal by reformatting. Indent to create white space. Add headings and subheadings. Itemize ideas, boldface key points, and underline important words or phrases. Using graphics (tables and figures) will also help you avoid long, overwhelming blocks of text. By reformatting your proposal, you will make the text inviting and ease reader access.

Enhance the Tone of Your Proposal

Although you want to keep your proposal professional, remember that people write to people. Your reader is a human being, not a machine. Therefore, to achieve a sense of humanity, enhance the tone of your text by using pronouns and positive, motivational words. These are important in your statements of benefit and recommended courses of action. *Sell* your ideas.

Correct Errors

If you're proposing a sale, you must provide accurate figures and information. Proposals, for example, are legally binding. If you state a fee in the proposal or write about schedules, your prospective client will hold you responsible; you must live up to those fees or schedules. Make sure that your figures and data are accurate.

Reread each of your numbers, recalculate your figures, and double-check your sources of information. Correct errors prior to submitting your report. Failure to do so could be catastrophic for your company or your client.

Of course, you must also check for typographical, mechanical, and grammatical errors. If you submit a proposal containing such errors, you will look unprofessional and undermine your company's credibility.

Avoid Biased Language

As noted throughout this textbook, both men and women read your writing. Don't talk about *foremen, manpower,* or *men and girls.* Change *foremen* to *supervisors, manpower* to *workforce* or *personnel,* and *men and girls* to *men and women.* Don't address your cover letter to *gentlemen.* Either find out the name of your readers or omit the salutation according to the simplified letter style discussed in Chapter 6. In addition, remember that your audience will be diverse. Avoid cultural or age-related biases.

Proposal Checklist

Have you included the following in your proposal?

___ **1.** Title page (listing title, audience, author or authors, and date)

___ **2.** Cover letter (stating why you're writing and what you're writing about; what exactly you're providing the readers; what's next—follow-up action)

___ **3.** Table of contents (listing all major headings, subheadings, and page numbers)

___ **4.** List of illustrations (listing all figures and tables, including their numbers and titles, and page numbers)

___ **5.** Abstract (stating in low-tech terms the problem, solution, and benefits)

___ **6.** Introduction (providing a statement of purpose and a lengthy analysis of the problem)

___ **7.** Discussion (solving the readers' problem by discussing topics such as procedures, specifications, timetables, materials/equipment, personnel, credentials, facilities, options, and costs)

___ **8.** Conclusion (restating the benefits and recommendation for action)

___ **9.** Glossary (defining terminology)

___ **10.** Appendix (optional additional information)

Figure 17.8 (pages 558–570) illustrates a sample external proposal that you can use as a model for your writing. Figure 17.9 (pages 571–586) shows a sample internal proposal.

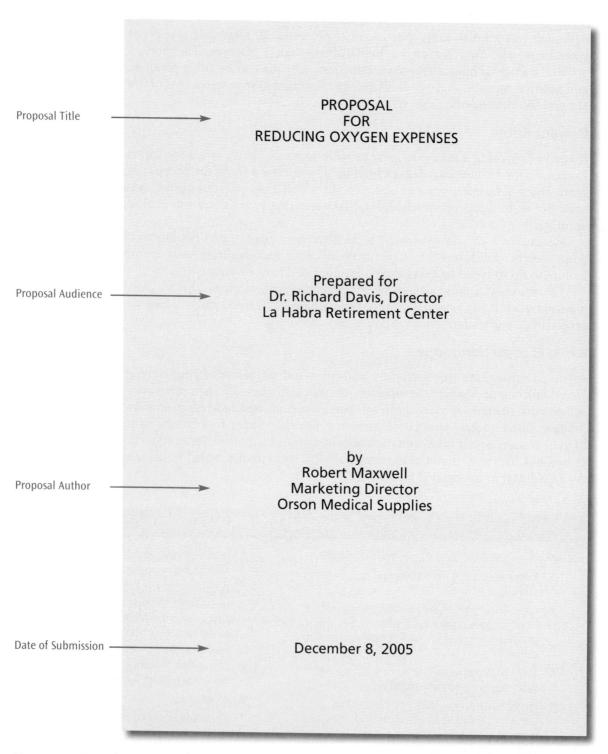

Proposal Title

PROPOSAL
FOR
REDUCING OXYGEN EXPENSES

Proposal Audience

Prepared for
Dr. Richard Davis, Director
La Habra Retirement Center

Proposal Author

by
Robert Maxwell
Marketing Director
Orson Medical Supplies

Date of Submission

December 8, 2005

Figure 17.8   External Proposal

Orson Medical Supplies
"Improving the Quality of Life"
12345 College Blvd.
Overland Park, OR 90091
976-988-2000

December 8, 2005

Dr. Richard Davis, Director
La Habra Retirement Center
220 Cypress
La Habra, CA 90631

Dear Dr. Davis:

Submitted for your review is our proposal regarding the
Electronic Demand Cannula oxygen-saving system. This
document is in response to your September 25, 2005, letter
and our subsequent discussions.

Within our report, you will find the following supporting
materials geared toward your requests:

- Product specifications . . . . .pages 4–5
- Operating instructions . . . . .page 6
- Qualifications and experience . . . . .page 7
- Cost . . . . .page 8
- Warranty . . . . .page 10

Thank you for your interest in our product. We look forward to
serving you and will call within two weeks to finalize
arrangements.

Sincerely,

Robert Maxwell

Robert Maxwell
Marketing Director

Enclosure: Proposal

Itemized cover letter body
focusing on key points
within the proposal and
the appropriate page
numbers

Figure 17.8   *Continued*

TABLE OF CONTENTS

LIST OF ILLUSTRATIONS

Headings, subheadings, and page numbers to help the audience find information

Figure and table numbers plus titles

iii

Figure 17.8 *Continued*

ABSTRACT

Expenses for medical oxygen have increased steadily for several years. Now the federal government is reducing the amount of coverage that Medicare allows for prescription oxygen.

These cost increases can be reduced through the use of our new Electronic Demand Cannula (EDC). The EDC delivers oxygen to the patient only when the patient inhales. Oxygen does not flow during the exhalation phase. Therefore, oxygen is conserved.

This oxygen-saving feature can reduce your oxygen expenses by as much as 50 percent. Patients who use portable oxygen supplies can enjoy prolonged intervals between refilling, thus providing more freedom and mobility.

Emphasizes problems generating the proposal

Shows how the proposal can solve the problems

Highlights the benefits derived

iv

Figure 17.8   *Continued*

1.0 INTRODUCTION

1.1 PURPOSE

This is a proposal to sell the new Electronic Demand Cannula (EDC)* to the La Habra Retirement Center, La Habra, California. This bid to sell offers you a special discount when you purchase our EDCs in the quantities suggested in this proposal.

1.2 PROBLEMS

The IEEE numbering system (1.2, 1.2.1)

1.2.1 *High Costs*
Since 2000, the price of medical-grade oxygen has sky-rocketed. It cost $10 per 1,000 cubic feet (cu ft) in 2000. Today, medical-grade oxygen costs $26 per 1,000 cu ft. In fact, you can expect next year's oxygen expenses to double the amount you spent this year.

Headings and subheadings to aid reader access

1.2.2 *Governmental/Insurance Involvement*
Many factors have contributed to this soaring cost, including demand, product liability, and inflation. However, two factors contributed the most. First, legislation reduced the amount that Medicare pays for prescription oxygen. Second, few insurance companies offer programs covering long-term prescription oxygen. Therefore, you, or your patients, must pay the additional expenses.

1.2.3 *Decreased Quality of Service*
Because prescription oxygen has risen in cost so dramatically, few medical service companies can produce affordable EDCs and stay competitive. Since 2001, according to *Medical Digest Bulletin,* 80 percent of medical service vendors have gone out of business. Your ability to receive quality service at an affordable price has diminished.

*This and subsequent terms marked by an asterisk (*) are defined in the glossary.

1

Figure 17.8   *Continued*

2.0 DISCUSSION

2.1 IMPLEMENTATION OF ELECTRONIC DEMAND CANNULA

Because the price of oxygen will not go down, you must try to use less while obtaining the same clinical benefits.

Orson Medical Supplies, a leader in oxygen-administering technology, proposes the implementation of our new EDC*. Using state-of-the-art electronics, the EDC senses the patient's inspiratory effort.* When a breath is detected, the EDC dispenses oxygen through the patient's cannula.* The patient receives oxygen only when he or she needs it.

Continuously flowing cannulas waste gas during exhalation and rest. Clinical studies have proved that 50 percent of the oxygen used by cannula patients is wasted during that phase. These same tests also revealed that blood oxygen saturation* does not significantly vary between continuous and intermittent flow cannulas. The patient receives the same benefit from less oxygen. Table 1 explains this in greater detail.

Table 1
BLOOD OXYGEN SATURATION
USING THE ELECTRONIC DEMAND CANNULA
VERSUS
CONTINUOUS FLOW CANNULAS

Prescribed Flowrate (L/min)*	Breaths per Minute (bpm)*	Blood Oxygen Saturation %	
		Intermittent	Continuous
0.5	12	96%	98%
1	12	98%	99%
2	12	99%	100%
3	12	100%	100%
4	12	100%	100%

A table to add visual appeal and to make complex information easier to understand

We have included a technical description of the EDC to help explain how this system will benefit oxygen cannula users.

2

Figure 17.8 *Continued*

2.2 TECHNICAL DESCRIPTION

The EDC* is an oxygen-administering device that is designed to conserve oxygen. The EDC is composed of six main parts: oxygen inlet connector, visual display indicators (LEDs),* power switch, patient connector, AC adapter connector, and high-impact plastic case (see Figures 1 and 2).

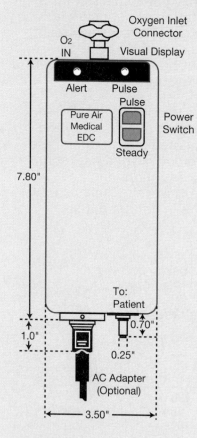

OXYGEN INLET CONNECTOR: The oxygen inlet connector is a DISS No. 1240 (Diameter index safety system)* and is made of chrome-plated brass.

VISUAL DISPLAY: Two LEDs* provide visual indications of important functions. Alarm functions are monitored by a red LED, Motorola No. R32454. An indication of each delivered breath is given by the pulse display, which is a yellow LED, Motorola No. Y32454.

POWER SWITCH: The power switch is an ALCO No. A72-3 slide switch. The dimensions are 0.5" × 0.30": button height is 0.20". Electrical Specifications: Dry contact rating is 1 amp, contract resistance is 20 milliohms, and the life expectancy is 100,000 actuations.

Figure 1 EDC Front View

3

Figure 17.8 *Continued*

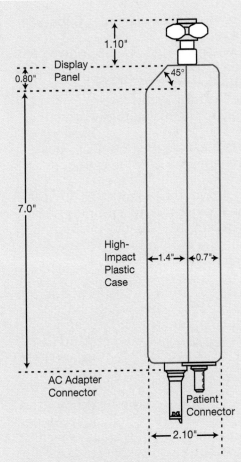

PATIENT CONNECTOR: Attachment of the patient cannula system is made at the patient connector, which is located at the bottom of the case. The white nylon connector, Air Logic No. F-3120-85, is a 10-32, UNF male threaded, straight barbed connector for 1/8" ID flexible tubing.

AC ADAPTER CONNECTOR: An optional AC adapter* and battery charger assembly, part number PA-32, plugs into the AC adapter connector, which is located at the bottom left-hand side of the case. The connector is a male, D-subminiature, 12-pin flush insert supplied by Dupont Connector Systems. The part number is DCS: 68237009.

HIGH-IMPACT PLASTIC CASE: The case housing is made from an impact-resistant, flame-retardant, oxygen-compatible ABS plastic*.

Figure 2 EDC Side View

2.3 OPERATING INSTRUCTIONS

The EDC is an oxygen-saving and administering device (see Figures 1 and 2). By following these five easy steps, you will be able to enjoy the benefits of intermittent demand oxygen.

WARNING: Federal law prohibits the sale or use of this device without the order of a physician.

4

Figure 17.8   *Continued*

1. Attach your oxygen supply to the Oxygen Inlet Connector located at the top of the case.
2. Move the Pulse-Steady Switch to the Pulse position to begin intermittent demand flow.
3. Connect your nasal cannula to the Patient Outlet Connector located at the bottom of the case.
4. Adjust your oxygen supply to the oxygen flow prescribed by your physician.
5. Put on your nasal cannula and breathe normally. The pulse light will turn on when a breath is delivered.

You are now ready to conserve oxygen by as much as 50 percent. Should you have the need to go back to continuous flow, just push the Pulse-Steady Switch to the "steady" position.

2.4 QUALIFICATIONS AND EXPERIENCE

Qualifications and experience to help persuade the reader of your expertise and ability to do the job

Orson Medical Supplies has been an international leader in the field of respiratory therapy since 1975. Pure Air introduced the first IPPB* respirator on the market. In 1985, responding to the needs of doctors and therapists, we produced the first life support volume ventilator, the VV-1. The VV-1 became the industry standard by which all other ventilators were measured.

In 1990, Orson Medical Supplies introduced the first computer-controlled life support system, the VV-2. Technology developed for this product has found application in other areas as well. Recently, we introduced one such product, the Electronic Demand Cannula.

Orson Medical Supplies is located in Overland Park, Oregon. The main manufacturing and engineering facility employs 450 people. Regional sales and service branch offices are located throughout the United States.

2.5 PERSONNEL

Each of our engineering facilities is staffed by trained technicians ready to answer your questions. The following individuals have been assigned to La Habra Retirement Center:

5

Figure 17.8 *Continued*

Randy Draper
Randy (BS, Electrical Engineering, South Central Texas University, 1992) has worked at Orson Medical Supplies since 1992. In 11 years at Orson, Randy has been promoted from service technician to supervisor. Randy specializes in developing new medical equipment. He has supervised the development teams that worked on the X29 respirator, the Z284-00 ventilator, and the Omega R-449 sphygmomanometer. Randy was lead development specialist for the Pure Air EDC.

Randy will be in charge of your account. Please contact him directly regarding any questions you might have about the EDC.

Ruth Bressette
Ruth (BS, Mechanical Engineering, Pittsburgh State University, 1997) has worked at Orson Medical Supplies since 1998. She has risen in our company from service technician to manager of troubleshooting/maintenance. Ruth has received the highest-level certification (Master Technician) offered by the IEEE for service on every piece of equipment developed, manufactured, and sold by Orson.

Ruth will be the manager of your Orson equipment maintenance and troubleshooting crew. Her responsibility is to ensure that your equipment is kept in outstanding working condition. She will schedule maintenance checks and promptly assign technicians to troubleshoot potential malfunctions.

Douglas Loeb
Doug (AA, Electrical Engineering Technology, Plainview Community College, 1992) is one of our most accomplished troubleshooters. Having worked at Orson for 11 years, Doug is commended annually for his speed, accuracy, and skill. Your equipment is in good hands with Doug. He will be your primary troubleshooter and maintenance person.

2.6 COST

Orson Medical Supplies is pleased to offer our Pure Air EDC at cost-effective pricing. Table 2 explains the benefits you'll derive when purchasing in quantity.

6

Figure 17.8 *Continued*

Table 2 LIST PRICE VS. DISCOUNT PRICE FOR EDC			
List Price	Quantity Warranty*	Extended *per EDC*	Total Cost
$350	1–9 units	$75	$425
Discount Price			
$310	10–24 units	$60	$370
$300	25+ units	$50	$350

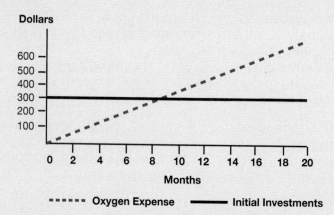

Visuals, such as tables and figures, to break up monotonous text and help persuade the audience

As you can see, Orson is happy to offer you substantial savings when you purchase our Pure Air EDC in volume. At these prices, and assuming normal use, the oxygen cost savings will exceed your initial investment in less than one year, as shown in Figure 3.

Figure 3 Return on Investment

2.7 WARRANTY

Orson Medical Supplies warrants this product to be free of manufacturing defects for a one-year period after the original date of consumer purchase. This warranty does not include damage done to the product due to accident, misuse, improper installation or operation, or unauthorized repair. This warranty also does not include replacement of

*Orson's extended warranty is discussed in Section 2.8.

7

Figure 17.8   *Continued*

parts due to normal wear. If the product becomes defective within the warranty period, we will replace or repair it free of charge.

This warranty gives you special legal rights. You may also have other rights that vary from state to state. Some states do not allow exclusions or limitations of incidental damage, so the above limitations may not apply to you.

2.8 EXTENDED WARRANTY

In addition to the coverage provided in our unconditional warranty, you might want to take advantage of our extended warranty package. For the prices provided in Table 2, Orson Medical Supplies will extend the warranty to cover a three-year period after the original date of consumer purchase. With this three-year extended warranty, Orson not only covers manufacturing defects but also replaces worn parts free of charge.

The extended warranty does not cover damage to the product resulting from accident, misuse, improper installation or operation, or unauthorized repair.

For answers to any of your questions regarding repair, replacement, warranty, or extended warranty, please call 1-800-555-ORSN, or write to Manager, Customer Relations, Orson Medical Supplies, 12345 College Blvd., Overland Park, OR 90091.

3.0 CONCLUSION

3.1 MAJOR CONCERN

Prescription oxygen expenses are escalating whereas government support has been reduced. The cost increase to the patient and the health care facility will be enormous.

3.2 RECOMMENDATION

To offset the inevitable rise of oxygen expenses, we recommend the use of the Electronic Demand Cannula.

8

Figure 17.8 *Continued*

4.0 GLOSSARY

Alphabetized glossary defining jargon →

ABS plastic	Acrylonitrile butadiene styrene—a durable and long-lasting plastic
AC adapter	A remote power supply used to convert alternating current to direct current
Blood oxygen saturation	The partial pressure of oxygen in alveolar blood recorded in percent
Cannula	A small tube inserted into the nose, specifically for administering oxygen
Electronic Demand Cannula (EDC)	An electronically controlled device that dispenses oxygen only when triggered by an inspiratory effort
Inspiratory effort	The act of inhaling
Light-emitting diode (LED)	A solid-state semiconductor device that produces light when current flows in the forward direction

LIST OF ACRONYMS/ABBREVIATIONS

bpm	Breaths per minute
DISS	Diameter index safety system
EDC	Electronic Demand Cannula
IPPB	Intermittent positive pressure breathing
L/min	Liters per minute

9

Figure 17.8   *Continued*

$\mathcal{B}$io
$\mathcal{S}$taffing

Your One-Stop Shop for Biomedical Needs

Technology Support Proposal

Submitted to
Leann Towner
Chief Financial Officer

By
Jonathan Bacon
Manager, Information Technology Department

June 12, 2005

$\mathcal{B}$
$\mathcal{S}$

Figure 17.9   Internal Proposal

Bio Staffing

Your One-Stop Shop for Biomedical Needs

DATE: June 12, 2005
TO: Leann Towner, Chief Financial Officer
FROM: Jonathan Bacon, Information Technology Manager
SUBJECT: PROPOSAL FOR NEW CORPORATE TECHNOLOGY
           SUPPORT

Leann, in response to your request, enclosed is a proposal to update BioStaffing's technology. The IT Department has researched this topic thoroughly and is happy to recommend a revised technology policy, improved hardware and software, and a new vendor.

Among detailed analyses of our current system and proposed options, this proposal presents the following:

1. BioStaffing's current technology challenges 1–4
2. Technology replacement policy options 5–6
3. Hardware and software purchasing suggestions 6–7
4. Technology vendor recommendations 8–10
5. Benefits to BioStaffing 11

Thank you, Leann, for giving the IT Department a chance to present this proposal. We are confident that our suggestions will maintain BioStaffing's competitive edge and increase employee satisfaction with better hardware and software. If I can answer any questions, either call me (ext. 3625) or e-mail me at jbacon@biostaff.com.

Figure 17.9 *Continued*

Table of Contents

List of Illustrations

iii

Figure 17.9 *Continued*

1.0 Executive Summary

1.1 Problem

BioStaffing's current technology policies, hardware and software, and vendor agreements are inefficient. Our technology policies are not responsive to our evolving needs, our hardware and software are outdated, and our current vendor is changing ownership.

1.2 Solution

To solve these problems, the Information Technology (IT) Department suggests the following:

- Upgrading our technology needs at least biannually instead of the current five-year technology changeover policy
- Purchasing new computers, printers, scanners, and digital cameras
- Hiring a new vendor to supply and repair our hardware and software

1.3 Benefits

Reviewing our technology needs frequently will help BioStaffing avoid costly repairs due to overused equipment. A biannual replacement schedule also will allow us to stay current with hardware and software advancements.

In addition, purchasing new computers, printers, scanners, and digital cameras will help our staff more effectively meet client needs.

Hiring a new technology vendor is key to these goals. Our current vendor agreement ends June 30, 2005. By hiring a new vendor with better turnaround, pricing, maintenance, and merchandise, BioStaffing will increase customer and employee satisfaction.

iv

Use bulleted points to help readers access information (note that the bullets are grammatically parallel)

Emphasize reader benefit to sell the proposal

Figure 17.9 *Continued*

2.0 Introduction

2.1 Purpose

The purpose of this proposal is to improve BioStaffing's current technology. By revising our policies, purchasing new hardware and software, and hiring a new technology vendor, we can increase both customer and worker satisfaction.

A clear statement of purpose acts like a thesis

2.2 Problem

2.2.1 *Technology Policies*
Since 1987, we have replaced hardware and software on a five-year, rotating basis. This was an effective policy initially. In the late 1980s through the early 1990s, costs were high for new computers and software upgrades, and our technology needs were limited. Therefore, it made sense financially to change our hardware and software on an ongoing, regular basis.

Background information helps clarify the problem

However, times have changed. The current policy is not responsive to our needs, as follows:

- *Prices Have Gone Down.* Last decade's technology prices were cost prohibitive. BioStaffing correctly chose to upgrade hardware and software only every five years. When every desktop computer cost over $2,000, for example, replacing them often did not make good business sense.

  Today, however, costs are more affordable and flexible. First, desktop computers can be purchased for approximately $1,500. Laptops can be bought for around $1,000. More important, handheld computers (Personal Digital Assistants [PDAs], and pocket PCs), with wireless Internet access, can be bought for under $500 (Cooper 2005).

  In addition, vendors making bulk purchases will offer BioStaffing incentives. We can buy computers on an as-needed basis cost effectively by taking advantage of special sales pricing.

  In contrast, our current five-year hardware and software replacement policy is not responsive. It disregards today's changes in technology pricing, and it disallows us from taking advantage of dealer incentives.

Comparison/contrast helps readers understand the problem

- *Repair Costs Have Gone Up.* By replacing technology only every five years, BioStaffing has resorted to repairing and retrofitting outdated equipment. Costs for these repairs have increased more than fivefold over the last 10 years, as noted in the following line graph (Figure 1).

1

Figure 17.9   *Continued*

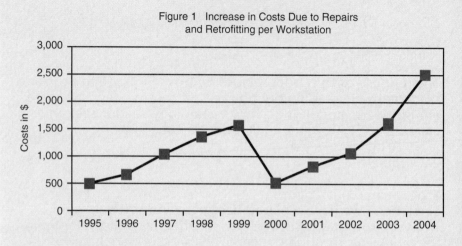

Figure 1   Increase in Costs Due to Repairs and Retrofitting per Workstation

(Source: Norton 2005)

NOTE:
The dip between 1999 and 2000 occurred at the five- year replacement schedule.

- *BioStaffing's Technology Needs Have Expanded.* Our current five-year policy was based on our limited technology needs. In 1987, all BioStaffing needed for corporate communication purposes was one computer per department and one dot matrix printer located on every floor. No one in 1987 envisioned how technology would change communication needs.

  Today, every BioStaffing employee needs a desktop computer, Internet access, and his or her own laser printer. This is necessary due to our employees' increased correspondence responsibilities and the emergence of e-mail as a primary communication vehicle. In addition, to be productive, our employees need a fax machine and scanner in every department, at least one laptop computer per department (to be checked out for travel purposes), and handheld computers for work off site. Finally, to improve our user manuals, we need five digital cameras (using up-to-date JPEG* images of our equipment to replace outdated line drawings).

  We cannot wait for the completion of a five-year replacement cycle to make these purchases. Instead, technology purchases must be made as needed.

2.2.2 *Outdated Hardware and Software*
We need to purchase new technology items because our current hardware and software are outdated.

*This and other terms marked with an asterisk (*) are defined in the glossary.

Analysis clarifies the reader's need

2

Figure 17.9   *Continued*

- *Hardware Limitations.* Our current hardware is limited in several key ways:
  1. We have *insufficient memory* in our computers. Most of our computers have only 2.0 GHz* processors with approximately 256 MB* of RAM* versus the current minimal standard 3.0 GHz and 512 MB. This negatively impacts speed of document retrieval.
  2. Our computers' *monitor sizes are limited.* The majority of our 15-inch monitors provide only 13.8 inches of viewable screen. This is sufficient for word processing but not for the creation of our newsletters, annual reports, and Web site. Correspondence with more graphical content requires larger viewable screens. Thus, we need to upgrade to 17-inch flat panel monitors (15.9-inch viewable screens), if not 19-inch (18-inch viewable screens) plasma panels. Doing so would allow our corporate communication personnel to create more visually appealing documents.
  3. Our *inkjet printers are slow and produce poor quality documents.* Our current black and white printers produce only 7 ppm* with a dpi* resolution of 1440 × 720. In contrast, new black and white laser printers will produce up to 20 ppm at a dpi resolution of 5760 × 1440. Thus, we would gain speed for quicker turnaround time as well as improved readability.

Specificity of detail shows the writer's knowledge of the situation and highlights the pressing need

To clarify the importance of ppm, look at Figure 2. This bar chart shows the difference 20 ppm versus 7 ppm makes in terms of time. The figure is based on the 4,000 pages of documentation BioStaffing produces every day.

At 7 ppm (times 60 minutes per hour), BioStaffing personnel can print 420 pages per hour. When we divide the 4,000 pages per day by our current 420-page hourly capacity, you can see that printing requires **9.5 hours.**

In contrast, at 20 ppm (times 60 minutes per hour), improved printers can produce 1,200 pages per hour. Thus, at 4,000 pages divided by 1,200 pages per hour, we can accomplish the same printing task in only **3.3 hours.** *Laser printers can save our company over 6 hours of lost time each day.*

Figure 2   Time Spent Printing
Based on PPM

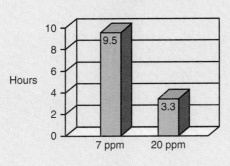

3

Figure 17.9   *Continued*

4. We have *no laptop computers, no handheld computers, no scanners, and no digital cameras.* Without portable computers, our employees engaged in work-related travel cannot communicate effectively with customers or BioStaffing personnel. The lack of scanners and digital cameras is hurting our graphic designers' abilities to create quality documentation.

- *Software Limitations.* Desktop publishing options have expanded rapidly. We could only word process in-house in the late 1980s. We outsourced our marketing collateral (brochures, newsletters, annual reports, etc.). Now our needs have changed. Today, to create in-house proposals for clients, annual reports to our stockholders, and a Web site, we need improved software. Updated software will allow BioStaffing to import data via spreadsheets; enhance text with tables, figures, and downloadable clip art; and automatically generate tables of contents, glossaries, and indexes. To send newsletters to our employees and stakeholders, we need improved graphical and layout capabilities.

  Our current software does not allow us to create the quality documentation BioStaffing requires. We are losing our competitive edge due to poor, outdated software.

2.2.3 *Vendor Changes*

Our current technology vendor is changing ownership. After having been in business for 20 years, Business Sourcing is being sold to an offshore company, International Technologies. Our technology agreement will end, as of June 30, 2005. We must renegotiate a technology contract.

This offers us a window of opportunity. We have maintained a long-standing relationship with Business Source because it also was one of our primary clients. Though it did not necessarily provide us the best pricing, we continued the arrangement for business purposes. Now, through a new vendor, we can improve quality of service as well as pricing.

4

Figure 17.9   *Continued*

3.0 Discussion

3.1 Revised Technology Replacement Policy

We must revise BioStaffing's technology replacement policy. It is not responsive to our growing technology needs. Following are two proposed options for your review:

3.1.1 *Option 1—Quarterly Assessments/Replacements*

The IT Department believes technology is changing so rapidly that quarterly assessments are necessary—for every department. Each quarter, we suggest that every department submit a Technology Needs Assessment (TNA). In this TNA, the departments would

- Prioritize their technology needs
- Explain how the technology will be used (emphasizing value added to BioStaffing)
- List the make and model of the desired hardware and software
- Provide costs

Then, a panel (composed of representatives from all departments) will review the reports. Based on a 10-point scale (10 = essential; 1 = non essential), the panel will determine which proposals will be supported that quarter.

> **NOTE:**
> To ensure that all departments needs are treated equally, any department that was not given technology upgrades in the prior review would receive two additional "carryover points."

Option 1 requires competition on the part of each department. In addition, it will entail added expense for BioStaffing. However, the importance of technology in our business mandates this level of responsiveness.

3.1.2 *Option 2—Biannual Rotating Replacements*

Although Option 2 is less responsive than Option 1 for technology needs, it would be more cost effective for the company and more fair for all departments.

We propose that BioStaffing divide its 12 departments into two categories. The Red team would receive technology upgrades one year (even-numbered years), whereas the Blue team would receive technology upgrades the next year (odd-numbered years).

IT has divided the departments into teams alphabetically for fairness, as indicated in Table 1.

Presenting options gives the reader standards for comparison

5

Figure 17.9   *Continued*

Table 1 BIANNUAL ROTATING TECHNOLOGY REPLACEMENTS

Red Team (*even*-numbered years)	Blue Team (*odd*-numbered years)
Accounting	Legal
Administration	Manufacturing
Administrative Services	Maintenance
Benefits	Personnel
Corporate Communication	Sales
Information Technology	Shipping and Receiving

3.2 Hardware Purchases

After researching department requirements, IT determined that BioStaffing needs to upgrade business application hardware as shown in Table 2.

Table 2 BUSINESS APPLICATION HARDWARE REQUIREMENTS

Hardware	Minimum Specifications (prices approximate)
15 desktop computers	Intel Pentium 4 processor, 2.6 GHz 512 MB RAM 80 GB HD* CD* drives 17-inch color monitor (15.9-inch viewable) $1,000 each
12 laptop computers	Intel Celeron processor, 2.2 GHz 128 MB RAM 20 GB* HD ROM* drive $700 each
50 handheld computers	32 MB RAM Hotsyncs to Outlook Wireless e-mail and Internet access Thumb-typable keyboard Compatible with Word, PowerPoint, and Excel $150 each
5 laser printers	16 MB RAM 20 ppm Letter, legal, #10 envelopes 133 MHz* processor 250-sheet paper tray $400 each
3 scanners	2400 × 4800 dpi resolution Adapter for slides, negatives Auto retouch, enhancement Copy, scan, e-mail, and file $170 each
5 digital cameras	JPEG file format Auto focus 3X optical zoom 32 MB flash card $200 each

Source: "Hardware/Software" 2005.

6

Figure 17.9   *Continued*

3.2.1 *Hardware Allocation Analysis*

Based on survey results from each department, IT suggests that the hardware listed in Table 2 be allocated as follows ("Allocation" 2005):

- **Accounting**—2 desktops, 1 laptop, 5 handhelds, and 1 laser printer
- **Administration**—1 laptop and 1 desktop, and 4 handhelds
- **Administrative Services**—3 desktops, 1 laptop, 1 handheld, and 1 laser printer
- **Benefits**—1 desktop, 1 laptop, 4 handhelds, and 1 scanner
- **Corporate Communication**—1 laptop, 7 handhelds, 1 scanner, 1 laser printer, and 2 digital cameras
- **Information Technology**—1 laptop, 7 handhelds, 1 scanner, 1 laser printer, and 2 digital cameras
- **Legal**—1 desktop, 1 laptop, 2 handhelds, and 1 scanner
- **Manufacturing**—2 desktops, 1 laptop, 2 handhelds, and 1 laser printer
- **Maintenance**—1 desktop, 1 laptop, and 2 handhelds
- **Personnel**—2 desktops, 1 laptop, and 4 handhelds
- **Sales**—2 desktops, 1 laptop, 10 handhelds, and 1 digital camera
- **Shipping and Receiving**—1 desktop, 1 laptop, and 2 handhelds

The above allocations are based on department size, need, location, and recent upgrades. For example, departments located on the same floors can share equipment. Manufacturing, Shipping and Receiving, and Maintenance would share a laser printer. Accounting would share a laser printer with Benefits and Legal. Administration and Administrative Services would share a laser printer.

Neither Corporate Communication nor Information Technology are requesting new desktop computers because they received upgrades last year.

Only Corporate Communication, Information Technology, and Sales need digital cameras to complete their jobs.

Similarly, the scanners are necessary only in Legal, Information Technology, Corporate Communication, and Benefits.

3.3 Software Purchases

To improve our business communication and create professional-looking marketing collateral, BioStaffing needs to upgrade software as illustrated in Table 3.

7

Figure 17.9 *Continued*

Table 3
SOFTWARE REQUIREMENTS FOR BUSINESS
COMMUNICATION AND MARKETING COLLATERAL

Purposes	Software Required	Approximate Cost
Letters, Memos, Reports	Microsoft Office Suite 2003 (Word, Excel, PowerPoint)	$300
Brochures, Newsletters, Annual Reports	Microsoft Office Suite Quark Express Adobe Photoshop	Listed above $650 $250
Oral Presentations, Proposals	PowerPoint (included in Microsoft Office Suite)	Listed above
Web Design	Macromedia Suite (Dreamweaver, Flash, ColdFusion)	$950
Online Help	RoboHelp	$50

3.3.1 *Cost-Effective Purchasing*
The software shown in Table 3 is cost effective for these reasons:

- These software purchases benefit every department in BioStaffing.
- With site licenses, we can purchase only one software package and install it on all computers throughout the company.
- We are proposing the purchase of the most up-to-date, state-of-the-art software. Therefore, Information Technology anticipates that we will not need to upgrade these purchases for approximately five years.

3.4 Technology Vendor Options

Because BioStaffing's vendor agreement with Business Sourcing ends June 30, 2005, Information Technology has researched new vendor options. Our primary criteria included *pricing, quality, delivery time, maintenance, warranties,* and *customer service.*

Based on these criteria, Table 4 presents an overview of our findings:

8

Figure 17.9 *Continued*

Table 4 TECHNOLOGY VENDOR OPTIONS

Companies	Pricing	Quality	Delivery Time	Maintenance	Warranties	Customer Service
BizTech Warehouse	20% bulk discount Hardware: $18,730 Software: $1,760	Remanufactured name-brand hardware	Same-day delivery at no extra cost	24-hour maintenance options at additional hourly cost—negotiable; 8:00 A.M.–5:00 P.M. maintenance for products under warranty	90 days; renewable at negotiable price	24/7 hotline; fleet of maintenance trucks
Technologies Today	No discount Hardware: $23,400 Software: $2,200	New name-brand hardware	Same-day delivery for home office; overnight FedEx for other sites	Manufacturer's maintenance (1-800 number for nearest dealership)	1-year manufacturer's warranty	8:00 A.M.–5:00 P.M. M–F office help
SOS (Super-Saver Office Supplies)	10% bulk discount Hardware: $21,000 Software: $1,980	In-house store-brand hardware	Overnight FedEx	1-800 number for nearest dealership	3-year extended warranties, parts and labor	8:00 A.M.–5:00 P.M. M–F office help; answering service on weekends

3.4.1 *Assessment of Vendor Options*

- **Pricing**—BizTech Warehouse provides the best pricing. This is an especially important factor when we consider the rapidly changing nature of technology and the constant need to upgrade.
- **Quality**—Technologies Today provides new, name-brand hardware. This is appealing. However, remanufactured hardware can be as good as new. We do not have enough data on SOS's store-brand hardware to judge its value.
- **Delivery Time**—With 25 locations nationwide, including Topeka, our home office's city, BizTech provides same-day delivery. This is an important factor considering our plans to open new office locations.
- **Maintenance**—Again, BizTech provides the most prompt maintenance service. It comes at a cost, but its vendors have assured us that we will receive a substantial discount as well as product rebates if we choose their company.
- **Warranties**—SOS's three-year warranty is outstanding, as is Technologies Today's 1-year guarantee. However, if we plan to upgrade our hardware biannually, a three-year warranty loses its value.
- **Customer Service**—BizTech's 24/7 accessibility is perfect for our workplace. We run our services around the clock. The other vendors do not provide the kind of customer service our business requires.

9

Figure 17.9   *Continued*

Information Technology evaluated the vendor options, giving each company points for their services, as follows:

5 = Excellent	4 = Good	3 = Average
2 = Poor	1 = Unacceptable	

Based on this scale, Figure 3 shows our evaluation findings:

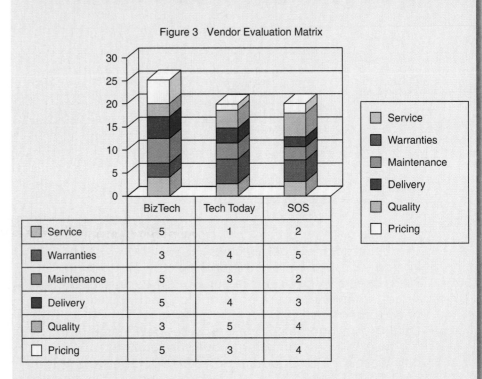

Figure 3   Vendor Evaluation Matrix

	BizTech	Tech Today	SOS
☐ Service	5	1	2
☐ Warranties	3	4	5
☐ Maintenance	5	3	2
☐ Delivery	5	4	3
☐ Quality	3	5	4
☐ Pricing	5	3	4

Despite its limited warranties and remanufactured hardware, BizTech Warehouse would be our choice for a new technology vendor. Its pricing, maintenance, delivery, and customer service are superior to the other vendors we researched.

10

Figure 17.9   *Continued*

4.0 Conclusion

4.1 Conclusion

BioStaffing's technology needs are critical.

- *Repair costs* due to old hardware have risen fivefold per workstation, from as low as $500 to as high as $2,500.
- *Outdated printers,* producing only 7 pages per minute, require up to 9.5 hours a day to print BioStaffing's 4,000 pages of documentation.
- *Insufficient software* for business communication applications and marketing collateral are hurting our competitive edge.
- *Expired technology vendor contracts* need to be renegotiated.

Key to the above challenges is our current five-year technology replacement policy. Technology advances demand a more responsive policy.

Remind the readers of the problems necessitating the proposal

4.2 Recommendation

The Information Technology Department, based on research and surveys, proposes the following:

- *A quarterly or biannual technology replacement policy*—this would allow BioStaffing to meet employee technology needs more responsively as well as benefit from vendor pricing incentives.
- *New communication hardware*—this would include desktop, laptop, and handheld computers; laser printers; scanners; and digital cameras.
- *Upgraded software*—to improve business communication needs, all computers must have access to Microsoft Office Suite. For our marketing collateral (brochures, newsletters, annual reports, and Web site), we need new software including Quark Express, Adobe Photoshop, RoboHelp, Flash, and Dreamweaver.
- *A new technology vendor*—based on our research, we suggest that BizTech Warehouse will best meet our technology needs.

Summarize your recommendation for the audience

4.3 Benefits

These changes must occur before June 30, 2005, when our current vendor contract ends. By acting now, BioStaffing will benefit in several ways. We can save up to 20 percent on hardware, software, and maintenance costs. We will maintain our competitive edge in the marketplace. Most important, our employee satisfaction and productivity will increase as they work with the latest technology and software upgrades.

Highlight the benefits

11

Figure 17.9   *Continued*

5.0 Glossary

Acronym/Abbreviation	Definition
CD	compact disk
dpi	dots per inch
GHz	gigahertz
GB	gigabyte
HD	hard drive
JPEG	Joint Photographic Experts Group
MB	megabyte
MHz	megahertz
ppm	pages per minute
ROM	Read Only Memory
RAM	Random Access Memory

6.0 Works Cited

"Allocation of Technology Needs." Survey. May 1, 2005.

Cooper, Jack. AAA Computers. Interview. May 15, 2005.

"Hardware/Software Specifications and Costs." *PCWorld.*
May 12, 2005. http://www.pcworld.com/.

Norton, Susan. *BioStaffing Annual Report.* Jan. 15, 2005.
May 31, 2005. http://www.BioStaff.com/Annualreport.htm.

12

Figure 17.9   *Continued*

CHAPTER HIGHLIGHTS

1. You might have multiple readers for a proposal. Consider your audiences' needs. To communicate with different levels of readers, include abstracts or executive summaries, glossaries, and parenthetical definitions.

2. A proposal could include the following:
   - title page
   - cover letter
   - list of illustrations
   - abstract (or executive summary)
   - introduction
   - discussion
   - conclusion
   - recommendation
   - glossary
   - works cited (or references)
   - appendix
3. Subheadings will make your proposal more accessible.
4. You could use a questionnaire to generate content for your proposal.
5. Using visual aids will make content more accessible in a proposal.

CASE STUDIES

1. The technical writing department at Bellaire Educational Supplies/Technologies (BEST) needs new computer equipment. Currently, the department has outdated hardware, outdated word processing software, an outdated printer, and limited graphics capabilities. Specifically, the department is using computers with 12-inch black-and-white monitors, hard drives with only 256 KB of memory, and one, 10 MB hard disk drive. The word processing package used is WordPro 3.0, a version created in 1996. Since then, WordPro has been updated four times; the latest version is 6.5. The department printer is a black-and-white Amniprint dot matrix machine. The current word processing package has no clip art. To create art, the department must go off-site to a part-time graphic artist who charges $35 an hour, so the department uses very few graphics.

   Because of these problems, the company's user manuals, reports, and sales brochures are being poorly reviewed by customers. Further, BEST has no Web site for product advertisement or company recognition. The bottom line: BEST is falling behind the curve, and profits are off 27 percent from last year.

   As technical writing department manager, you have consulted with your five staff members (Jim Nguyen, Mario Lozano, Mike Thurmand, Amber Badger, and Maya Liu) to correct these problems. As a team, you have decided the company needs to purchase new equipment:

   - *Six new personal computers.* Each computer must have a 17-inch color monitor, 32 MB of memory, a 2 GB hard drive, a 12-speed CD-ROM drive, and a VGA graphics card.
   - *Two laser printers.* These must have a print speed of 24 ppm, resolution of 600×600 dpi, and 4 MB of memory, expandable to 132 MB.
   - *Word processing software.* WordPro 6.5 with these capabilities: voice-activated annotations, typing, and correcting; automatic

footnoting and endnoting; envelope labeling; grammar and spell checking; thesaurus; help options; automatic index generating; 50 or more scalable fonts.

- *Graphics software.* For professional-quality newsletters and brochures, BEST needs the capability for quick demonstrations and design tips; a layout checker with at least 10 online views; 20 true type fonts, each scalable; a table creation toolbar; 2,000 clip art images; and a logo creator with 50 border design options.
- *Scanner.* To increase your graphics potential, BEST also needs a flatbed scanner with these specifications: 300 dpi image resolution; 155 ppm gray-scale scanning capability; 8-bit, 256 gray-scale color support; and approximately 8 1/2 × 12-inch bed size.

Using the criteria provided in this chapter, write an internal proposal to BEST's CEO, Jim McWard. In this proposal, explain the problem, discuss the solution to this problem, and then highlight the benefits derived once the solution has been implemented. These benefits will include increased productivity, better public relations, increased profits, and less employee stress. Develop these points thoroughly, and provide Mr. McWard the names of vendors for the required hardware and software. To find these vendors, you could search the Internet.

2. Employees at **Gulfview Architectural and Engineering Services** are complaining about the company's poor meetings. Co-workers are missing meetings, arriving at meetings late, and leaving early. Some employees dominate the meetings by speaking long and loudly. Other workers, many of whom have valuable ideas to share, choose not to speak during the meetings. Instead, they e-mail their comments later or discuss their feelings only to a few co-workers.

For example, one team member, Caroline Jensen, has missed at least one meeting a month. Occasionally, she misses two or three meetings in a row complaining of childcare issues. These have forced her to use the company's flextime option, allowing her to come to work later than usual, at 9:00 A.M., though the meetings tend to begin at 8:00 A.M.

Another employee, Guy Stapleton, tends to talk a lot during the meetings. He has good things to say, but he speaks his mind very loudly and interrupts others as they are speaking. He also elaborates on his points in great detail, even when the point has been made.

A third team member, Rosa Martinez, almost never provides her input during the meetings. She will e-mail comments later or talk to people during breaks. Her comments are valid and on-topic, but not everyone gets to hear what she says.

A fourth team member, Ling Tsung, is very impatient during the meetings. This is evident from his verbal and nonverbal communication. He grunts, slouches, drums on the table, and gets up to walk around while others are speaking.

A fifth employee, Sharon Mitchell, is overly aggressive. She is confrontational, both verbally and physically. Sharon points her finger at people when she speaks, raises her voice to drown out others as they speak, and uses sarcasm as a weapon. Sharon also crowds people, standing very close to them when speaking.

When meetings do not go as desired, work is not accomplished, and people's feelings are hurt. As a human resources associate, you have been asked to write an internal proposal suggesting a solution to this problem.

Your boss, Tom Rodriguez, Director of Human Resources, has suggested the following criteria:

- Cost—no more than $5,000 including supplies.
- Time—a maximum of 16 work hours.
- Instructional delivery—e-learning, lecture format, hands-on activities, or multimedia presentations.
- References required.
- Workbooks—participants should be able to practice meeting techniques during the session.
- Post-workshop assessment. Gulfview needs a matrix or method to determine if the workshop has succeeded.

You have decided that Gulfview should hire a consultant to facilitate training sessions on improved meeting decorum and procedures. Three companies look promising.

- SMG Training can provide an 8-hour workshop focusing on meeting skills. This company will charge $1,500 for the session and provide a workbook for each participant. The workshop is provided online as e-learning. The participants can view the workshop asynchronously. The workshop is "canned": it will not be customized to Gulfview's needs, based on SMG's assumption that "most problems are common." In addition, the workshop does not offer post-training checkups to correct any lingering problems. SMG comes highly recommended for its expertise and cost effectiveness.
- ProForm, a second training company, charges $3,000 for their meeting workshop. Unlike SMG, ProForm's training facilitator would interview Gulfview personnel to survey their needs, provide an 8-hour workshop customized to the survey findings, and then offer a post-workshop "meeting wellness checkup." This company is an industry leader and is well-known for its attention to details.
- MeetRite, the final option, charges the most ($5,500), but they offer a guarantee: "Improved meetings, or your money back." They visit a meeting to assess meeting procedures, provide a two-day workshop on meeting decorum, offer one-on-one tutorials/consultations to meeting participants, and visit a second meeting where they make suggestions for improvement. You have not spoken to anyone who has ever used their service or attended their sessions.

Assignment

Write an internal proposal based on the information provided and any additional information you find through research (online, in libraries, through surveys, or in interviews). Recommend one of the companies and explain your rationale. Incorporate visual aids to improve access and document design. To write the proposal, follow the criteria provided in this chapter.

3. **Quick and Sure Delivery** has not been either quick or sure lately. Customer complaints are up 23 percent this quarter (QSD only receives an average of three complaints a month). Clients are telling customer service representatives that deliveries are arriving up to 10 hours later

than promised. In addition, delivered goods are being left unattended outside homes and businesses. This has led to damages due to rain, and on at least five instances, delivered packages have been stolen.

Since QSD guarantees that packages will be handed directly to a home or business owner and never left unattended, these occurrences are actionable under the law. More importantly, QSD's reputation is being harmed. Already, word is getting around, and customers are taking their business elsewhere. Business is down 12 percent this month, when compared to last year at the same time (QSD made 1,578 deliveries during the month last year, charging an average of $27 per delivery). Something must be done.

As Delivery Supervisor, you have been asked by the company's CEO, Kelly Cordes, to write an internal proposal suggesting solutions to these problems. You have researched the problems and concluded that a combination of the following could improve QSD's business:

- Improved training for delivery drivers. This should include techniques for handling packages, providing customer service, maintaining trucks, and logging procedures.
- Supervisor "ride alongs." Requiring supervisors to ride with delivery employees would ensure that training techniques were being followed. In addition, these "ride alongs" would allow the supervisors to get first-hand knowledge about potential problems.
- Customer surveys. At the moment, all QSD has is anecdotal information about complaints. Customer surveys would quantify the problems. You have found the following from a first round of surveys: 15 percent of those surveyed complained about late package arrivals; 18 percent complained about damage; 28 percent complained about rude drivers; 5 percent complained about packages left unattended.
- Tracking technology. Currently, drivers log their deliveries by hand in a paper notebook. Competitors, like SameDay Delivery and Quick-to-You, use handheld computers to log their deliveries. These computerized logs allow the companies precise tracking.
- Delivery teams. One possible solution is to organize all the drivers into teams headed by supervisors. For 200 drivers, QSD would need 20 teams. This promises more rapid delivery with better supervision.
- Changed delivery routes. Late deliveries are caused by routes with extreme distances. If routes could be limited, this would ensure more timely deliveries. Currently, delivery routes range up to 50 square miles. Industry standards are about 25 square miles.
- Truck packing procedures. Many packages are damaged in the warehouse and during truck loading procedures. Employees need proper training on improved package handling and loading procedures.

Assignment

Write an internal proposal based on the information provided and any additional information you find through research (online, in libraries, through surveys, or in interviews). Incorporate visual aids to improve access and document design. To write the proposal, follow the criteria provided in this chapter.

4. You own **AppleKorp**, a party and entertainment company catering to kindergarten through middle school-aged youth. You plan and present events for groups of 10 to 150 people. AppleKorp can work small events in a home or backyard, as well as larger events in parks, auditoriums, and gymnasiums.

For children between ages 6 to 10, AppleKorp sets up games and game equipment, including the following:

- Magic shows
- Balloon sculptures
- Moon jumps
- Face painting
- Soft gymnastic equipment
- And more . . .

For children between ages 11 to 14, AppleKorp sets up games and game equipment and provides additional services, including the following:

- Batting cages (completely enclosed, with helmets provided)
- Climbing walls (with safety harnesses and rope attendants)
- Outdoor basketball hoops
- Outdoor volleyball nets
- DJ and party music
- Sumo Wrestling (with inflatable body padding)

Parties for groups of 10 to 50 cost $10 per person for two hours of entertainment. Each additional hour costs a flat fee of $50 per hour.

Parties for groups of 51–100 cost $5 per person for two hours of entertainment. Each additional hour costs a flat fee of $60 per hour.

Parties for groups of 101–150 cost $5 per person for two hours of entertainment. Each additional hour costs a flat fee of $75 per hour. All costs include setup and takedown of equipment.

Your audience for this party service includes elementary and middle/junior high schools, churches, synagogues, mosques, and associations such as scouting.

AppleKorp is owned and run by Ron and Susan Apple. Both Ron and Susan are ex-elementary school teachers, with certifications in CPR. Ron is a former youth soccer, basketball, and softball coach, and Susan also teaches aerobics. The Apples have owned and operated AppleKorp for five years. They have worked with the Blue River School District, Oakpark School District, Elm Street Daycare Centers, and various churches.

Three additional employees work with the Apples. Mike Smith is a retired fire fighter, skilled in first aid. Jean Cottrell is a retired nurse. Henry Ortega is a singer, dancer, juggler, and all-around entertainer.

AppleKorp's motto is "Safe Fun at Affordable Prices."

Assignment

Write an external proposal based on the information provided and any additional information you find through research (online, in libraries, or in interviews). Incorporate visual aids to improve access and document design. Your audience—schools, churches, and not-for-profit youth organizations—is looking for cost-effective, safe entertainment. To write the proposal, follow the criteria provided in this chapter.

ACTIVITIES

5. You own **Buzz Electronics Co.**, 4256 Crawfish Blvd., Rouex, LA 65221. Mr. and Mrs. Allan Thibodeux, 3876 Spanish Moss Drive, Bayside, LA 65223, have asked you to give them a bid on electrical work for a new family room they are adding to their home.

You and Mr. and Mrs. Thibodeux have gone over the couple's electrical needs, including the following. The room, which will measure 18' (east to west) by 15' (north to south), should have four 110 V outlets for 3 lamps, a clock, a radio, a CD and DVD player, and a TV. The family wants the four 110 V outlets to be placed equidistant throughout the room.

The client wants two 220 V outlets. One 220 V outlet will go by the southwest window on the west wall where the family plans to put a window air conditioning unit. The window will be located 3' in from the south wall. There will be another window on the west wall, located 3' in from the north wall. A third picture window, measuring 6' wide by 4' high, will be centered on the south wall. The other 220 V outlet must be placed on the east wall, where the family plans to put home office equipment (computer, printer, scanner, and fax machine). Their office desk will sit 5' from the door leading into the room. The door will be built on the east wall where it comes to a corner meeting the south wall.

Centered in the ceiling, the family wants electrical wiring for a fan with a light package. In addition to this light, the family also wants a light mounted on the east wall above the desk area, so wiring is needed there, approximately 5–6' up from the floor.

The family wants two light switches in the room: one by the door and one on the north wall, approximately 6' in from where the north and west walls meet. The Thibodeuxs plan to have a couch and lamp in that area for reading. They want both to be double switches, one to control the fan and ceiling lights; the other to control additional floor and ceiling lights in the room. All light switches need rheostats for dimming.

Finally, the Thibodeuxs plan to have a whole-house vacuum system installed in the walls, and they have asked you if you can provide this service.

Buzz Electronics has been in business since 1995. The company has worked with thousands of satisfied customers, including both residential and business owners. Buzz has long-standing contracts for service with Acme Construction, J&L Builders, Food-to-Go Groceries, the City of Bayou Bend, LA, and Ross and Reed Auto Showroom.

As owner of Buzz Electronics, you have an A.S. in Electronics from Sandy Shoal Community College, Sandy Shoal, LA. You are ETA-I (Electronics Technicians Association International) Certified; NASTec (National Appliance Service Technician) Certified; and a Certified Industrial Journeyman. You have eight employees, all of whom also are Certified Industrial Journeymen.

Assignment

Write a short (3–5 pages) external proposal—bid for contract. To do so, study the Thibodeux's electrical needs, list the parts you will need to complete the job, estimate the time for your labor—including setup, work performed, and cleanup. Then, provide a price quote. You might need to research the wiring and equipment needed for this job, either online, in technical journals, or in parts catalogues. Include *everything* you will need to complete this job. Remember, what you write in your proposal is legally binding. You cannot add on parts and labor after the fact and expect payment or a happy customer. Follow the guidelines provided in the textbook for short reports and proposals.

EXERCISES

1. Write an external proposal. To do so, create a product or a service and sell it through a long report. Your product can be an improved radon detection unit, a new fiber optic cable, safety glasses for construction work, bar codes for pricing or inventory control, an improved four-wheel steering system, computer graphics for an advertising agency, and so on. Your service may involve dog grooming, automobile servicing, computer maintenance, home construction (refinishing basements, building decks, room additions, and so on), freelance technical writing, at-home occupational therapy, or telemarketing. The topic is your choice. Draw from your job experience, college coursework, or hobbies. To write this proposal, follow the process provided in this chapter—prewrite, write a draft, and then rewrite to revise.

2. Write an internal proposal. You can select a topic from either work or school. For example, your company or department is considering a new venture. Research the prospect by reading relevant information. Interview involved participants or survey a large group of people. Once you have gathered your data, document your findings and propose to management the next course of action. If you choose a topic from school, you could propose a day-care center, on-campus bus service, improved computer facilities, tutoring services, coed dormitories, pass/fail options, and so on. Write an internal proposal to improve your company's Web site; expand or improve the security of your company's parking lot; create or improve your company's physical site security in light of post 9-11 issues; improve policies for overtime work; improve policies for hiring diversity; improve your company's policies for promotion. Research your topic by reading relevant information or by interviewing or surveying students, faculty, staff, and administration. Once you have gathered your data, document your findings and recommend a course of action. In each instance, be sure to prewrite, write, and rewrite.

3. Find a previously written proposal (at work or one already submitted by a prior student) and improve it according to the criteria presented in this chapter. Add or change any or all of the report's components, including the cover letter, table of contents, list of illustrations, abstract, introduction, main text, conclusion/recommendation, glossary, and so forth.

4. Find a previously written proposal (at work or one already submitted by a prior student) and improve it by adding graphics where needed. You can use tables or figures to clarify a point and to make the report more cosmetically appealing.

WEB WORKSHOP

By typing "RFP," "proposal," "online proposal," or "online RFP" in an Internet search engine, you can find tips for writing proposals and requests for proposals (RFPs), software products offered to automatically generate e-proposals

and winning RFPs, articles on how to write proposals, samples of RFPs and proposals, and online RFP and proposal forms.

To perform a more limited search, type in phrases like "automotive service RFP," "computer maintenance RFP," "desktop publishing RFP," "web design RFP," and many more topics. You will find examples of both proposals and RFPs from businesses, school systems, city governments, and various industries.

To enhance your understanding of business and industry's focus on proposal writing, search the Web for information on RFPs and proposals. Using the criteria in this chapter and your knowledge of effective technical writing techniques, analyze your findings.

- How do the online proposals or RFPs compare to those discussed in this textbook, in terms of content, tone, layout, and so on?
- What information provided in the textbook is missing in the online discussions?
- What are some of the industries that are requesting proposals, and what types of products or services are they interested in?

  a. Report your findings, either in an oral presentation or in writing (e-mail message, memo, letter, or report).

  b. Imagine that you are requesting a proposal for a product or service. Create your own online form to meet this need.

  c. Rewrite any of the proposals that you think can be improved, using the criteria in this chapter as your guide.

  d. Respond to an online RFP by writing a proposal. To complete this assignment, go online to research any information you need for your content.

QUIZ QUESTIONS

1. Explain the purpose of an internal proposal.

2. Explain the purpose of an external proposal.

3. How do you number information prior to the introduction?

4. What is an abstract?

5. What do you include in a purpose statement?

6. What are the typical components of the discussion section?

7. What do you include in the recommendation?

8. Explain why you include a glossary.

9. What can you include in an appendix?

10. What type visuals can you include in a proposal?

11. Explain the writing style in a proposal.

12. How can you format a proposal for reader-friendly ease of access?

13. In what ways can you enhance the tone of your proposal?

14. In what ways can you organize the different sections of your proposal?

15. Why should you avoid biased writing in a proposal?

Oral Communication

Writing at Work

TechStop, with 12 locations throughout the state, sells DVDs, VCRs, TVs, audio components, computers, and computer peripherals. Shuan Wang is the Vice President of Customer Service. Lately, Shuan has been receiving complaints from customers about poor service.

For example, one customer, Carolyn Jensen, owns a large electronics company. She has done business with TechStop for six years. During that time, Carolyn's company has purchased over 1,000 pieces of equipment, including computers, printers, paper items, cell phones, pagers, and radios for a fleet of trucks.

Carolyn called a local TechStop to complain about problems her company was having with over a dozen printers. Her company's text looks fine on screen but hard copies print differently than they look. The company cannot print three different sized envelopes (as had been promised by the sales person when Carolyn purchased the printers). Graphics are not printing in color, though the color ink cartridges are full. Finally, printers are stopping their print jobs before all pages of a document have been printed. She purchased the printers 12 months and two weeks ago. The printer's in-store warranty was for 12 months, guaranteeing full replacement of parts and labor coverage.

Carolyn realizes that the warranty expired two weeks ago. However, she contends that several facts should negate this deadline. The deadline expired over a Christmas weekend, a severe snowstorm left many homes and businesses without power for several days, and other, more pressing business activities required her attention.

595

When she informed the sales help of her situation and her company's long-standing patronage of TechStop, the salesperson said, "Sorry, lady. The warranty's just no good anymore. Hey, we've all got problems. Anyway, there's no way my boss will even listen to you with this complaint. He told me to leave him alone when he's busy."

Unfortunately, Shuan has heard of other such complaints regarding rude and unresponsive sales personnel. Shuan fears that TechStop's staff is developing an overall corporate disregard for customer satisfaction.

To address this issue for the entire sales staff at TechStop's numerous locations throughout the state, Shuan plans to give a formal oral presentation on the following:

- sales etiquette and customer interaction
- store policy regarding customer satisfaction
- the impact on poor customer relations
- the consequences of failing to handle customers correctly

Shuan only hopes that his actions aren't too late.

Many people, even the seemingly most confident, are afraid to speak in front of others. This chapter offers techniques to make your oral communication experience rewarding rather than frightening.

You may have to communicate orally with your peers, your subordinates, your supervisors, and the public. Oral communication is an important component of your business success because you will be required to speak

- On an *everyday basis* to colleagues, customers, and vendors
- *Informally* to co-workers and clients
- *Formally* to large and small groups

EVERYDAY ORAL COMMUNICATION

"Hi. My name is Bill. How may I help you?" Think about how often you have spoken to someone today or this week at your job. You constantly speak to customers, vendors, and coworkers face-to-face, on the telephone, or by leaving messages on voice mail.

- If you work an 800 hotline, your primary job responsibility is oral communication.
- When you return the dozens of calls you receive or leave voice-mail messages, each instance reveals your communication abilities.
- As an employee, you must achieve rapport with your coworkers. Much of your communication to them will be verbal. What you say impacts your working relationships.

Every time you communicate orally, you reflect something about yourself and your company. The goal of effective oral communication is to ensure that your verbal skills make a good impression and communicate your messages effectively.

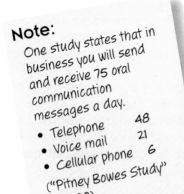

Note:
One study states that in business you will send and receive 75 oral communication messages a day.
- Telephone   48
- Voice mail   21
- Cellular phone   6

("Pitney Bowes Study" 2000)

TELEPHONE AND VOICE MAIL

You speak on the telephone dozens of times each week. When speaking on a telephone, make sure that you do not waste either your time or your listener's time. See Figure 18.1 for Ten Tips for Telephone and Voice Mail Etiquette.

Ten Tips for Telephone and Voice Mail Etiquette

1. Know what you are going to say before you call.

2. Speak clearly. Enunciate each syllable.

3. Avoid rambling conversationally.

4. Avoid lengthy pauses.

5. Leave brief messages.

6. Avoid communicating bad news.

7. Repeat your phone number twice, including the area code.

8. Offer your e-mail address as an option.

9. Sound pleasant, friendly, and polite.

10. If a return call is unnecessary, say so.

Figure 18.1 Ten Tips for Telephone and Voice Mail Etiquette

tech link
Go to *http://www.inserve. com/r_VmailTips.html* to see "Top Ten Tips for Using Voicemail Professionally & Effectively."

INFORMAL ORAL PRESENTATIONS

As a team member, manager, supervisor, employee, or job applicant, you often will speak to a coworker, a group of colleagues, or a hiring committee. You will need to communicate orally in an informal setting for several reasons:

- Your boss needs your help preparing a presentation. You conduct research, interview appropriate sources, and prepare reports. When you have concluded your research, you might be asked to share your findings with your boss in a brief, informal oral presentation.

- Your company is planning corporate changes (staff layoffs, mergers, relocations, or increases in personnel). As a supervisor, you want to provide your input in an oral briefing to a corporate decision maker.

- At a departmental meeting, you are asked to report orally on the work you and subordinates have completed and to explain future activities.

- In a team meeting, you participate in oral discussions regarding agenda items.

- Your company is involved in a project with coworkers, contractors, and customers from distant sites. To communicate with these individuals, you participate in a teleconference, orally sharing your ideas.

- You are applying for a job. Your interview, though not a formal, rehearsed presentation, requires that you speak effectively before a hiring committee.

See Figure 18.2 for Ten Tips for Video and Teleconferences.

tech link
Go to *http://www.business. att.com/default/?pageid= tele_tips_exts&branchid= vmeetings* for teleconferencing tips.

tech link
Go to *http://tns.its.psu.edu/ userGuides/videoConferencing /guide/guide-tips.html* for a videoconferencing orientation guide.

FORMAL PRESENTATIONS

You might need to make a formal presentation for the following reasons:

- Your company asks you to visit a civic club meeting and to provide an oral presentation to maintain good corporate or community relations.

> ## Ten Tips for Video and Teleconferences
>
> 1. Make sure participants know the conference date, time, time zone, and expected duration.
> 2. Make sure participants have printed materials before the conference.
> 3. Ensure that equipment has good audio quality.
> 4. Choose a private, quiet location.
> 5. Consider arrangements for participants with hearing impairments. You might need a TTY (text telephone) system or simultaneous transcription in a chat room.
> 6. Introduce all participants.
> 7. Direct questions and comments to specific individuals.
> 8. Do not talk too loudly, too softly, or too rapidly.
> 9. Turn off cell phones and pagers.
> 10. Limit side conversations.

Figure 18.2   Ten Tips for Video and Teleconferences

- Your company asks you to represent it at a city council meeting. You will give an oral presentation explaining your company's desired course of action or justifying activities already performed.

- Your company asks you to represent it at a local, regional, national, or international conference by giving a speech.

- A customer has requested a proposal. In addition to writing this long report, you and several coworkers also need to make an oral presentation promoting your service or product to the potential customer.

Types of Formal Oral Presentations

Three types of formal oral presentations include the following:

1. *The Memorized Speech*—The least effective type of oral presentation is the memorized speech. This is a well-prepared speech which has been committed to memory. Although such preparation might make you feel less anxious, too often these speeches sound mechanical and impersonal. They are often stiff and formal, and allow no speaker-audience interaction.

2. *The Manuscript Speech*—In a manuscript speech, you read from a carefully prepared manuscript. The entire speech is written on paper. This may lessen speaker anxiety and help you to present information accurately, but such a speech can seem monotonous, wooden, and boring to the listeners.

3. *The Extemporaneous Speech*—Extemporaneous speeches are probably the best and most widely used method of oral communication. You carefully prepare your oral presentation by conducting necessary research, and then you create a detailed outline. However, you avoid writing out the complete presentation. When you make your presentation, you rely on

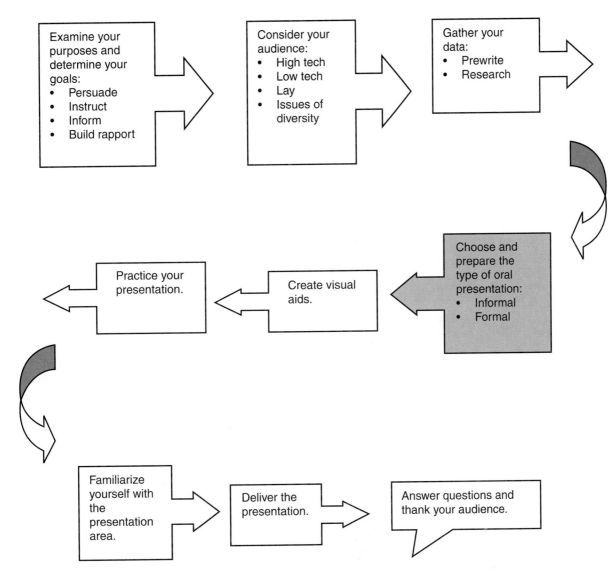

Figure 18.3   The Oral Presentation Process
Adapted from "Communications: Oral Presentations" 2004 *http://www.surrey.ac.uk/Skills/pack/pres.html*.

notes or PowerPoint slides with the major and minor headings for
reference. This type of presentation helps you avoid seeming dull and
mechanical, allows you to interact with the audience, and still ensures that
you correctly present complex information.

See Figure 18.3 for the Oral Presentation Process

PARTS OF A FORMAL ORAL PRESENTATION

A formal oral presentation consists of an introduction, a discussion (or body for
development), and a conclusion.

Introduction

The introduction should arouse and capture your audience's attention and interest. This is the point in the presentation where you are drawing in your listener,
hoping to create enthusiasm and a positive impression.

To create a positive impression, you could address your audience politely by saying "Good morning" or "Good afternoon." You can welcome participants to the conference or seminar; you can thank them for inviting you to speak.

Lead-ins to Arouse Reader Interest

You can use a variety of openings to capture your audience's attention, such as the following.

An Anecdote   Anecdotes are short, interesting, and relevant stories. Your audience needs to be drawn in quickly.

For example, at a technical writing workshop for engineers, the facilitator began as follows:

An Anecdote—Short, Interesting, and Relevant

"Recently I met an engineer who told me this story. Bob had been hired for his engineering expertise. After all, that was what he had trained to do; that was his educational background. On his first month on the job, he needed to write a progress report. He wrote it and turned it in to his boss. A few days later, the boss called him in to his office and said, 'That was an excellent report! It was clear, concise, and easy to read. Thanks for a great job.' Little by little, things started changing for Bob. Other engineers came to him for help, his boss asked him to write important reports, and Bob was promoted rapidly. Bob is now a manager, having moved up the ladder faster than some of his colleagues for whom written communication was more of a challenge. When I asked Bob to explain his success, he simply replied, "I had added value. In addition to my engineering talents, I could write well."

This short anecdote is relevant. It shows the audience how they can benefit from what will be discussed in the oral presentation.

This short anecdote is relevant. It shows the audience how they can benefit from what will be discussed in the oral presentation.

A Question or a Series of Questions   Asking questions involves the audience immediately. The training facilitator in the preceding example could have begun the workshop as follows:

Questions to Involve Your Audience

"How many reports do you write each week or month? How often do you receive and send e-mail messages to customers and colleagues? How much time do you spend on the telephone? Face it; technical writing is a larger part of your engineering job than you ever imagined."

These three questions are both personalized and pertinent. Through the use of the pronoun *you*, the facilitator speaks directly to each individual in the audience. By focusing on the listeners' job-related activities, the questions directly lead into the topic of conversation—technical writing.

A Quotation from a Famous Person   The training facilitator in the previous example could have begun his speech with a quote from Warren Buffet, a famous businessperson.

A Quote to Arouse Audience Interest

"How important is effective technical writing? Just listen to what Warren Buffet has to say on the topic:

> For more than forty years, I've studied the documents that public companies file. Too often, I've been unable to decipher just what is being said or, worse yet, had to conclude that nothing was being said. . . . Perhaps the most common problem . . . is that a well-intentioned and informed writer simply fails to get the message across to an intelligent, interested reader. In that case, stilted jargon and complex constructions are usually the villain. . . . When writing Berkshire Hathaway annual reports, I pretend that I'm talking to my sisters. I have no trouble picturing them: Though highly intelligent, they are not experts in accounting or finance. They will understand plain English, but jargon may puzzle them. My goal is simply to give them the information I would wish them to supply me if our positions were reversed. To succeed, I don't need to be Shakespeare; I must, though have a sincere desire to inform.

"That's what I want to impart to you today: good writing is communication that is easy to understand. If simple language is good enough for Mr. Buffet, then that should be your goal."

(Buffet 1998, 1–2)

(Margin note: This introduction combines a question, a quote, and an anecdote simultaneously.)

Facts and Figures   Facts are another good way to involve your audience objectively and with quantifiable impact. Notice how the training facilitator begins the speech this time.

Facts

"How important is communication in your work? If you are like most people, you are spending lots of time writing and speaking on the job. A Pitney Bowes study tell us the following:

You talk on the phone and listen to voice-mail messages **75** times each day.

Between pagers, faxes, and e-mail, you are involved in electronic communication **68** times a day.

You write and read **33** hard-copy memos and letters daily.

You even read and write **30** sticky notes each day.

That's 206 pieces of communication a day. And you thought you went to school to learn to become an engineer!"

This introduction includes a question and a quote, but the real impact is made by the facts. This will persuade your audience of the importance of your presentation.

Thesis (Overview of Key Points)   After you have captured your audiences' attention, you need to clarify the topic of your presentation. To do so, provide a *thesis statement* or a clear *overview of your key points*. With this statement (one or two sentences), you let your audience know exactly what you plan to talk about.

For example, look at how the workshop facilitator could have begun the formal presentation, after arousing the audiences' attention:

Thesis Statement for Formal Oral Presentation

"What we are going to talk about, and in this order, is the importance of *clarity, conciseness, ease of access, audience, organization,* and *accuracy.* These are key parts of effective technical writing."

Discussion (or Body)

After you have aroused your listeners' attention and clarified your goals, you have to prove your assertions. In the *discussion* section of your formal oral presentation, provide details to support your thesis statement. These details can be presented in a variety of ways including the following.

Comparison/Contrast   In your presentation, you could compare different makes of office equipment, employees you are considering for promotion, different locations for a new office site, vendors to supply and maintain your computers, different employee benefit providers, and so forth. Comparison/contrast is a great way to make value judgments and provide your audience options.

Problem/Solution   You might develop your formal oral presentation by using a problem-to-solution analysis. For example, you might need to explain to your audience why your division needs to downsize. Your division has faced problems with unhappy customers, increased insurance premiums, decreased revenues, and several early retirements of top producers. In your speech, you can then suggest ways to solve these problems ("We need to downsize to lower outgo and ultimately increase morale"; "Let's create a 24-hour, 1–800 hotline to answer customer concerns"; "We should compare and contrast new employee benefits packages to find creative ways to lower our insurance costs").

Argument/Persuasion   Almost every oral presentation has an element of argument/persuasion to it—as does all good written communication. You will usually be persuading your audience to do something based on the information you share with them in the presentation.

Importance   Prioritizing the information you present from least to most important (or most important to least) will help your listeners follow your reasoning more easily. To ensure the audience understands that you are prioritizing, provide verbal cues. These include simple words like *first, next, more important,* and *most important.* Do not assume that these cues are remedial or obvious. Remember, sometimes it is hard to follow a speaker's train of thought. Good speakers realize this and give the audience verbal signposts, reminding the listeners exactly where they are in the oral presentation and where the speaker is leading them.

Chronology   A chronological oral presentation can outline for your audience the order of the actions they need to follow. For example, you might need to prepare a yearly evaluation of all sales activities. Provide your audience with target deadlines and with the specific steps they must follow in their reports each quarter.

Maintaining Coherence   To maintain coherence, guide your audience through your speech as follows:

- **Use clear topic sentences.** Let your listeners know when you are beginning a new, key point: "Next, let's talk about the importance of *conciseness* in your technical writing."

- **Restate your topic often.** Constant restating of the topic is required because listeners have difficulty retaining spoken ideas. A reader can refer to a previously discussed point by turning back a page or two. Listeners do not have this option.

   Furthermore, a listener is easily distracted from a speech by noises, room temperature, uncomfortable chairs, or movement inside and outside the meeting site. Restating your topic helps your reader maintain focus. "Repeat major points. Reshow visuals, repeat points and ideas several times during your presentation. Put them in your summary, too" (O'Brien 2003).

- **Use transitional words and phrases.** This helps your listeners follow your speech. Some good transitional words to consider using frequently in your presentation include *first, second, third, therefore, moreover, furthermore, for example, another idea is,* and *in conclusion.*

Conclusion

Conclude your speech by restating the main points, by recommending a future course of action, or by asking for questions or comments.

Restating Main Points

"Therefore, as I have mentioned throughout this presentation, the key to successful technical writing is *clarity, conciseness, ease of access, audience, organization,* and *accuracy.* If you can accomplish these goals, your communication will succeed."

Recommending Future Action

"Now that we have discussed important aspects of successful technical writing, what should you do next? Here are some helpful tips:

1. Let others read your text or listen to your speech before a formal presentation.

2. Use your word processing spell checkers and grammar checkers (but don't trust them!).

3. Put yourself in your audiences' shoes. Will they understand the acronyms you have used, for example?

4. Memorize the phone number to your local college's grammar hotline.

5. Above all, do not fear communication. The techniques I have shared with you today will help you succeed."

Though the previous example might seem self-evident, a polite speaker should leave time for a few follow-up questions from the audience. Gauge your time well, however. You do not want to bore people with a lengthy discussion after a lengthy speech. You also do not want to cut short an important question-and-answer session. If you have given a controversial speech that you know will trouble some members of the audience, you owe them a chance to express their concerns.

VISUAL AIDS

Most speakers find that visual aids enhance their oral communication. "Visuals have the greatest, longest lasting impact—show as much or more visually as what you say. Use pictures; use color. Use diagrams and models" (O'Brien 2003).

Although PowerPoint slide shows, graphs, tables, flip charts, and overhead transparencies are powerful means of communication, you must be the judge of whether visual aids will enhance your presentation. Avoid using them if you think they will distract from your presentation or if you lack confidence in your ability to create them and integrate them effectively. However, with practice, you probably will find that visuals add immeasurably to the success of most presentations.

Table 18.1 lists the advantages, disadvantages, and helpful hints for using visual aids. For all types of visual aids, practice using them before you actually make your presentation. When you practice your speech, incorporate the visual aids you plan to share with the audience.

TABLE 18.1 VISUAL AIDS—ADVANTAGES AND DISADVANTAGES			
Type	Advantages	Disadvantages	Helpful Hints
Chalkboards	Are inexpensive.	Make a mess.	Clean the board well.
	Help audiences take notes.	Can be noisy.	Have extra chalk.
	Allow you to emphasize a point.	Make you turn your back to the audience.	Stand to the side as you write.
	Allow audiences to focus on a statement.	Can be hard to see from a distance.	Print in large letters.
	Help you be spontaneous.	Can be hard to read if your handwriting is poor.	Write slowly.
	Break up monotonous speeches.		Avoid talking with your back to the audience.
			Don't erase too soon.
Chalkless Whiteboards	Same as above.	Are expensive.	Use blue, black, or red ink.
		Require unique, erasable pens.	Cap pens to avoid drying out.
		Can stain clothing.	Use pens made especially for these boards.
		Some pens can be hard to erase if left on the board too long.	Erase soon after use.
		Pens that run low on ink create light, unreadable impressions.	

continued

| Table 18.1 | Continued | | |

Type	Advantages	Disadvantages	Helpful Hints
Flip Charts	Can be prepared in advance. Are neat and clean. Can be reused. Are inexpensive. Are portable. Help you avoid a nonstop presentation. Allow for spontaneity. Help audiences take notes. Allow you to emphasize key points. Encourage audience participation. Allow easy reference by turning back to prior pages. Allow highlighting with different colors.	Are limited by small size. Require an easel. Require neat handwriting. Won't work well with large groups. You can run out of paper. Markers can run out of ink.	Have two pads. Have numerous markers. Use different colors for effect. Print in large letters. Turn pages when through with an idea so audience will not be distracted. Don't write on the back of pages where print bleeds through.
Overhead Transparencies	Are inexpensive. Can be used with lights on. Can be prepared in advance. Can be reused. Can be used for large audiences. Allow you to return to a prior point. Allow you to face audiences.	Require an overhead projector. Can be hard to focus. Require an electrical outlet and cords. Can become scratched and smudged. Can be too small for viewing. Bulbs burn out.	Use larger print. Protect the transparencies with separating sheets of paper. Frame transparencies for better handling. Turn off the overhead to avoid distractions. Focus the overhead before beginning your speech. Keep spare bulbs. Don't write on transparencies. Face the audience.
Slides (Slide Shows)	Are portable. Slides are easy to protect. Can be used for large groups. Can be prepared in advance. Are entertaining and colorful. Allow for later reference.	Require dark rooms. Hurt speaker–audience interaction (eye contact). Can be expensive. Can be challenging to create. Require screen, machinery, and electrical outlets. Slides can get out of order. Dark room makes taking notes challenging. Machinery can malfunction.	Use a pointer. Use a remote control for freedom of movement. Check slides to make sure none are out of order. Check working condition of the equipment.

continued

Type	Advantages	Disadvantages	Helpful Hints

TABLE 18.1 CONTINUED

Type	Advantages	Disadvantages	Helpful Hints
Videotapes	Allow instant replay. Can be freeze-framed for emphasis. Can be economically duplicated. Can be rented or leased inexpensively. Are entertaining.	Require costly equipment (monitor and recorder). Are bulky and difficult to move. Can malfunction. Require dark rooms. Require compatible equipment. Deny easy note taking. Deny speaker–audience interaction.	Practice operating the equipment. Avoid long tapes.
Films	Are easy to use. Are entertaining. Have many to choose from. Can be used for large groups.	Require dark rooms. Require equipment and outlets. Make note taking difficult. Deny speaker–audience interaction. Can malfunction and become dated.	Use up-to-date films. Avoid long films. Provide discussion time. Use to supplement the speech, not replace it. Practice with the equipment.
PowerPoint Presentations	Are entertaining. Offer flexibility, allowing you to move from topics with a mouse click. Can be customized and updated. Can be used for large groups. Allow for speaker–audience interaction. Can be supplemented with handouts easily generated by PPt. Can be prepared in advance. Can be reused. Allow you to return to a prior slide. Can include animation and hyperlinks.	Require computers, screens, and outlets. Work better with dark rooms. Computers can malfunction. Can be too small for viewing. Can distance the speaker from the audience.	Practice with the equipment. Bring spare computer cables. Be prepared with a backup plan if the system crashes. Have the correct computer equipment (cables, monitors, screens, etc.). Make backup transparencies. Practice your presentation.

POWERPOINT PRESENTATIONS

One of the most powerful oral communication tools is visual—Microsoft PowerPoint (PPt). Whether you are giving an informal or formal oral presentation, your communication will be enhanced by PowerPoint slides.

Today, you will attend very few meetings where the speakers do not use PowerPoint slides. PowerPoint slides are used frequently because they are simple to use, economical, and transportable. Even if you have never created a slide show before, you can use the templates in the software and the autolayouts to develop your own slide show easily. An added benefit of PowerPoint slides is that you can print them and create handouts for audience members.

Benefits of PowerPoint

When you become familiar with Microsoft PowerPoint, you will be able to achieve the following benefits:

1. Choose from many different presentation autolayouts and designs.

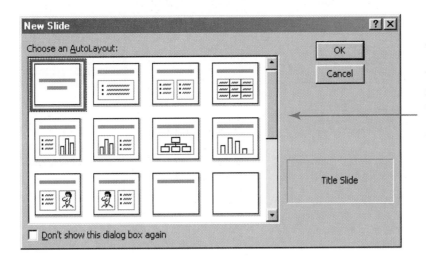

PPt autolayouts let you vary your presentation slides. You can include graphics, tables, columns, and bulleted points to enhance your text.

2. Create your own designs and layouts, changing colors and color schemes from preselected designs.

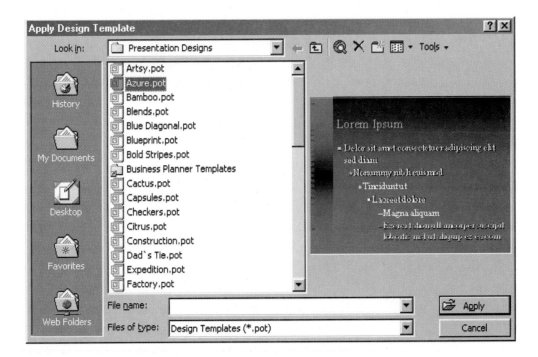

3. Add, delete, or rearrange slides as needed. By left-clicking on any slide, you can copy, paste, or delete it. By left-clicking between any of the slides, you can add a new blank slide for additional information.

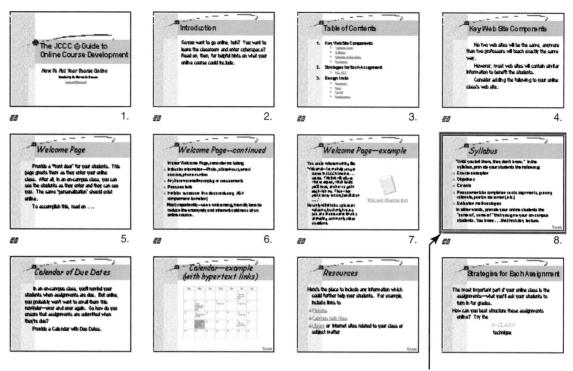

By left-clicking on any slide, you can copy, paste, or delete it. Also, by left clicking between any of the slides, you can add a new black slide for additional information.

4. Insert art from the Web, add images, or create your own drawings.

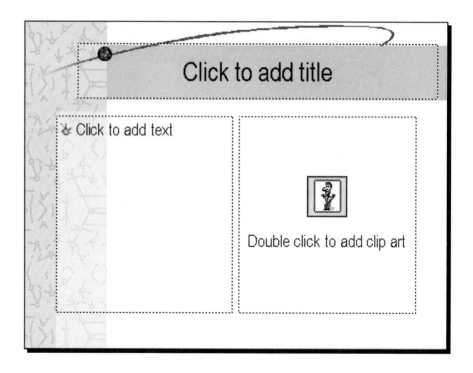

5. Use Microsoft PowerPoint's Help menu to learn how to include sound and video in the presentation.

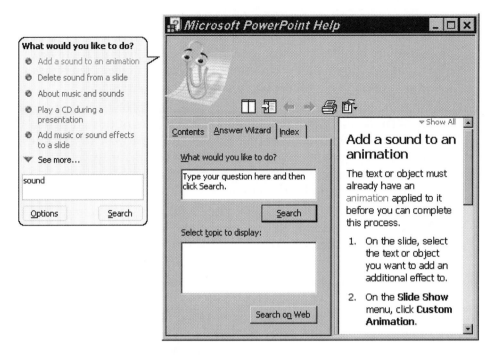

6. Add hyperlinks either to slides within your PowerPoint presentation or to external Web links.

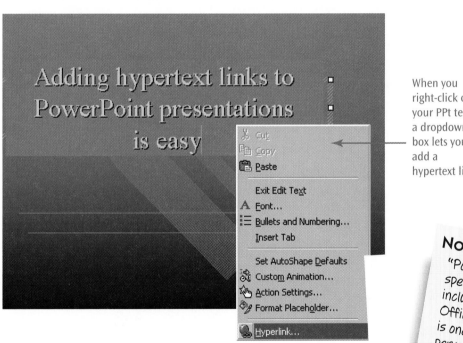

When you right-click on your PPt text, a dropdown box lets you add a hypertext link.

Note

"PowerPoint, the public-speaking application included in the Microsoft Office software package, is one of the most pervasive and ubiquitous technological tools ever concocted. In less than a decade, it has revolutionized the worlds of business, education, science and communications."
(Keller 2004)

Tips for Using PowerPoint

To make it easy for you to add PowerPoint slides to your presentations, consider the following hints.

1. *Create optimal contrast*—Use dark backgrounds for light text or light backgrounds for dark text. Avoid using red or green text (individuals who are color blind cannot see these colors). You should use color for emphasis only.

2. *Choose an easy-to-read font size and style*—Use common fonts, such as Times New Roman, Courier, or Arial. Arial is considered to be the best to use because sans serif fonts (those without feet) show up best in PowerPoint. Use no more than three font sizes per slide. Use at least a 24-point font size for text and 36-point font for headings.

3. *Limit the text to six or seven lines per slide and six or seven words per line*—Think 6 × 6. Two or more short, simple lines of text are better than one slide with many words. Also, use no more than 40 characters per line (a character is any letter, punctuation mark, or space). You can accomplish these goals by creating a screen for each major point discussed in your oral presentation.

4. *Use headings for readability*—To create a hierarchy of headings, use larger fonts for a first-level heading and smaller fonts for second-level headings. Each slide should have at least one heading to help the audience follow your thoughts.

5. *Use emphasis techniques*—To call attention to a word, phrase, or idea, use color (sparingly), boldface, all caps, or arrows. Use a layout that includes white space. Include figures, graphs, pictures from the Web, or other line drawings.

6. *End with an obvious concluding screen*—Often, if speakers do not have a final screen that *obviously* ends the presentation, the speakers will click to a blank screen and say, "Oh, I didn't realize I was through," or "Oh, I guess that's it." In contrast, an obvious ending screen will let you as the speaker end graciously—and without surprise.

7. *Prepare handouts*—Give every audience member a handout, and leave room on the handouts for note taking.

8. *Avoid reading your screens to your audience*—Remember they can read and will become quickly bored if you read slides to them. Speakers lose their dynamism when they resort to reading slides rather than speaking to the audience.

9. *Elaborate on each screen*—PowerPoint should not replace you as the speaker. In contrast, PowerPoint should add visual appeal, while you elaborate on the details. Give examples to explain fully the points in your presentation.

10. *Leave enough time for questions and comments*—Instead of rushing through each slide, leave sufficient time for the audience to consider what they have seen and heard. Both during and after the PowerPoint presentation, give the audience an opportunity for input.

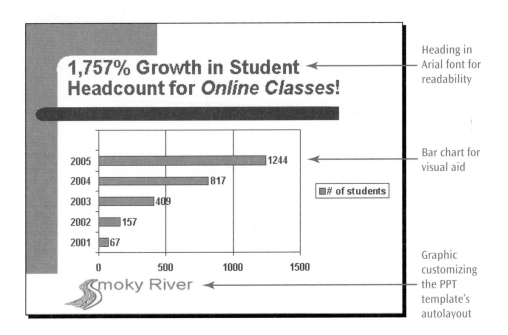

1,757% Growth in Student Headcount for *Online Classes*!

Heading in Arial font for readability

Bar chart for visual aid

Graphic customizing the PPT template's autolayout

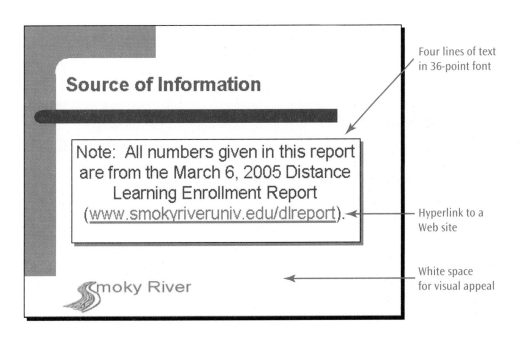

Source of Information

Four lines of text in 36-point font

Note: All numbers given in this report are from the March 6, 2005 Distance Learning Enrollment Report (www.smokyriveruniv.edu/dlreport).

Hyperlink to a Web site

White space for visual appeal

PowerPoint design → template with autolayout and customized color

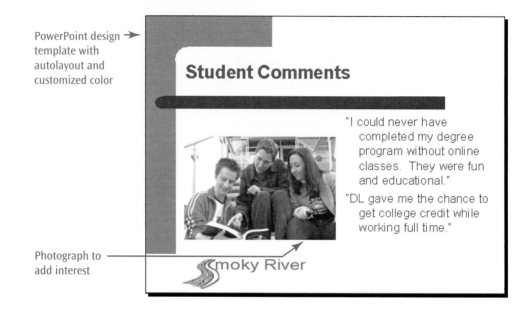

Photograph to add interest

Student Comments

"I could never have completed my degree program without online classes. They were fun and educational."

"DL gave me the chance to get college credit while working full time."

Smoky River

PowerPoint Slides Checklist	
__ **1.** Does the presentation include headings for each slide?	__ **6.** Did you use special effects effectively versus overusing them?
__ **2.** Have you used an appropriate font size for readability?	__ **7.** Have you limited text on each screen (remembering the 6×6 rule)?
__ **3.** Did you choose an appropriate font type for readability?	__ **8.** Did you size your graphics correctly for readability, avoiding ones that are too small or too complex?
__ **4.** Did you limit yourself to no more than three different font sizes per screen?	__ **9.** Have you used highlighting techniques (arrows, color, white space) to emphasize key points?
__ **5.** Has color been used effectively for readability and emphasis, including font color and slide background?	__**10.** Have you edited for spelling and grammatical errors?

THE WRITING PROCESS

The Writing Process		
Prewriting	Writing	Rewriting
• Examine your purposes • Determine your goals • Consider your audience • Gather your data • Determine how the content will be provided	• Organize the draft according to some logical sequence that your readers can follow easily • Format the content to allow for ease of access	• Revise ° Add missing details ° Delete wordiness ° Simplify word usage ° Enhance the tone of your communication ° Reformat your text for ease of access ° Practice the speech or review the text • Proofread ° Correct errors

As with writing memos, letters, e-mail messages, and reports, approach your oral communication project as a step-by-step process. Doing so will allow you to express yourself with confidence. Follow the writing process to organize your presentation effectively.

- Prewriting
- Writing
- Rewriting (presenting)

Prewriting

Prewriting gets you started with your presentation. Similar to prewriting for written communication, when you prewrite for a speech, you should accomplish the following.

Consider the Purpose

Determine why you are making an oral presentation. Ask yourself questions like

- Are you selling a product or service to a client?
- Do you want to inform your audience of the features in your newly created software?
- Are you persuading your audience to increase corporate spending to enhance a benefits package?
- Are you giving a speech for one of your college classes?

- Has your boss asked for your help in preparing a presentation? After you research the content, will you have to present your information orally?
- Are you a supervisor justifying workforce cuts to your division?
- Did a customer request information on solutions to a problem?
- Are you representing your company at a conference by giving a speech?
- Are you running for an office on campus and giving a speech about your candidacy?
- At the division meeting, are you reporting orally on the work you and your team have completed and the future activities you plan for the project?

Determining the purpose of a presentation will ensure that you choose the appropriate content.

Inform   For example, when your speech is to *inform*, you want to update your listeners. Such a speech could be about new tax laws affecting listeners' pay, new management hirings or promotions, or budget constraints or cutbacks. Speeches that inform do not necessarily require any action on the part of your audience. Your listeners cannot alter tax laws, change hiring or promotion practices, or prevent cutbacks. The informative speech keeps your audience up-to-date.

Persuade   On the other hand, some speeches *persuade*. You will speak to motivate listeners. For instance, you might give an oral presentation about the need to hold more regular and constructive meetings. You might tell your audience that teamwork will enhance productivity. Maybe you are giving an oral presentation about the value of quality controls to enhance product development. In each instance, you want your audience to leave the speech ready and inspired to act on your suggestions.

Instruct   You might speak to *instruct*. In an instruction, you will teach an audience how to follow procedures. For instance, you could speak about new sales techniques, ways to handle customer complaints, implementation of software, manufacturing procedures, or how to prepare for on-campus interviews. When you instruct, your goal is both to inform and persuade. You will inform your audience how to follow steps in the procedure. In addition, you will motivate them, explaining why the procedure is important.

Build Trust   Finally, you might give an oral presentation to *build trust*. Let's say you are speaking at an annual meeting. Your goal not only might be to inform the audience of your company's status, but also to instill the audience with a sense of confidence about the company's practices. You could explain that the company is acting with the audience's best interests in mind. Similarly, in a departmental meeting, you might speak to build rapport. As a supervisor, you will want all employees to feel empowered and valued. Speaking to build trust will accomplish this goal.

Consider Your Audience

When you plan your oral presentation, consider your audience. Ask yourself questions such as the following:

- Are you speaking to a high-tech, low-tech, or lay audience?
- Are you speaking up to supervisors?
- Are you speaking down to subordinates?
- Are you speaking laterally to peers?

- Are you speaking to the public?
- Are you addressing multiple audience types (supervisors, subordinates, and peers)?
- Is your audience friendly and receptive or hostile?
- Are you speaking to a captive audience (one required to attend your presentation)?
- Is your audience diverse in terms of culture, gender, or age?
- Will you need translators for those with hearing impairments?

Considering your audience's level of knowledge and interest will help you prepare your presentation. You should consider whether or not your audience needs terms defined and what tone you should take. You cannot communicate effectively if your audience fails to understand you or if your tone offends or patronizes them. Plan how you will design your content and style to communicate most effectively with your audience.

Presentation Plan   A presentation plan, like the example shown in Figure 18.4, can help you accomplish these goals.

Presentation Plan

Topic: _____

Objectives:
- What do you want your audience to believe or do as a result of your presentation?
- Are you trying to persuade, instruct, inform, build trust, or combinations of the above?

Development: What main points are you going to develop in your presentations?

1.

2.

3.

Organization: Will you organize your presentation using *analysis, comparison/contrast, chronology, importance,* or *problem/solution?*

Visuals: Which visual aids will you use?

Figure 18.4   Oral Presentation Plan

Gather Information

The best delivery by the most polished professional speaker will lack credibility if the speaker has little of value to communicate. As you plan your presentation, you must study and research your topic thoroughly before you package it.

You can rely on numerous sources when you research a topic for an oral presentation. For example, you could use any of the following sources:

- Interviews
- Questionnaires and surveys
- Visits to job sites
- Conversations in meetings or on the telephone
- Company reports
- Internet research
- Library research including periodicals and books
- Market research

Using information from company reports or other sources such as the Internet, books, or periodicals requires that you read and document your research (discussed in Chapter 14). Gathering information through interviews, questionnaires, surveys, or conversations, on the other hand, requires help from other people. To ensure that you get this assistance, consider doing the following:

- Ask politely for their assistance.
- Explain why you need the interview and information.
- Explain how you will use the information.
- Make a convenient appointment for the interview or to fill in the survey.
- Come prepared. Research the subject matter so you will be prepared to ask appropriate questions. Write your interview questions or the survey before you meet with the person. Take the necessary paper, forms, pencils, pens, laptop, electronic Notepads, handheld PDAs, or recording devices you will need for the meeting.
- When the interview ends or the individuals complete the survey, thank them for their assistance.

Figure 18.5 shows a sample questionnaire used by a team. They were researching the feasibility of adding a day-care center to their university campus for use by students and staff.

Writing the Presentation

After you obtain your information, your next step is to write a draft and consider visual aids for the presentation. The writing step in the communication process lets you use the research you gathered in the planning stage. When you organize your information, you will determine whether or not additional material is needed or if you can delete some of the material you have gathered.

Don't write out the complete text of the presentation. Too often when people have a complete text in front of them, they rely too heavily on the written words. They end up reading most of the paper to the audience rather than speaking more conversationally. Instead of writing out a complete copy of the presentation, use an outline or note cards to present your speech.

**Student and Staff Questionnaire
for Proposed Day-Care Center**

1. Are you male or female?
2. Are you in a single- or double-income family?
3. Are you a student or staff?
4. How many children do you have?
5. What are the ages?
6. Would you be interested in having a day-care center on campus?
7. How much would you be willing to pay per hour for child care at this center?
8. Do you think a day-care center would increase enrollment at this university?
9. How many hours per week would you enroll your child/children?
10. What hours of operation should the day-care center cover?
11. What credentials should the day-care providers have?
12. What should be the number of children per classroom?

If student:
13. Are you enrolled full time or part time at the university?
14. Do you attend mornings, afternoons, evenings, weekends, or a combination of the above (please specify)?
15. Would you be willing to work in the day-care center part time?

If staff:
16. What hours do you work at the university?
17. What hours would you need to use for day care at the center?

Additional comments (if any):

Thank you for your assistance.

Figure 18.5   Sample Research Questionnaire

Outline

You may want to write a more detailed outline focusing on your speech's major units of discussion and supporting information. A skeleton speech outline (Figure 18.6) provides you with a template for your presentation.

Skeleton Outline

Title:

Purpose:

  I. Introduction

    A. Attention getter:

    B. Focus statement:

  II. Body

    A. First main point:

      1. Documentation/subpoint:

        a. Documentation/subpoint:

        b. Documentation/subpoint:

      2. Documentation/subpoint:

      3. Documentation/subpoint:

    B. Second main point:

      1. Documentation/subpoint:

      2. Documentation/subpoint:

        a. Documentation/subpoint:

        b. Documentation/subpoint:

      3. Documentation/subpoint:

    C. Third main point:

      1. Documentation/subpoint:

      2. Documentation/subpoint:

      3. Documentation/subpoint:

        a. Documentation/subpoint:

        b. Documentation/subpoint:

  III. Conclusion

    A. Summary of main points:

    B. Recommended future course of action:

Figure 18.6   Skeleton Speech Outline

Note Cards

If you decide that presenting your speech from the outline will not work for you, consider writing highlights on 3 × 5-inch note cards. Avoid writing complete sentences or filling in the cards from side to side. If you write complete sentences, you will be tempted to read the notes rather than speak to the audience. If you fill in the cards from side to side, you will have trouble finding key ideas. Write short notes (keywords or short phrases) that will aid your memory when you make your presentation.

Sample 3 × 5-inch Note Cards

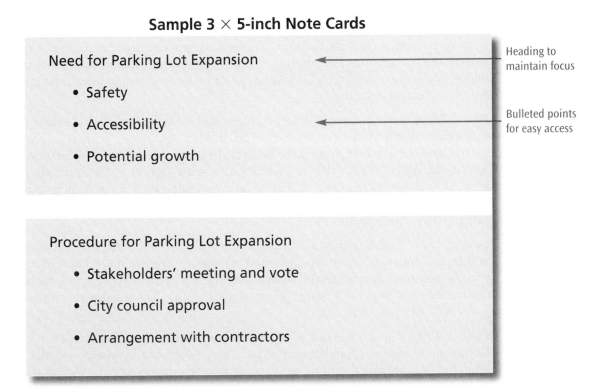

Need for Parking Lot Expansion — Heading to maintain focus

- Safety
- Accessibility — Bulleted points for easy access
- Potential growth

Procedure for Parking Lot Expansion

- Stakeholders' meeting and vote
- City council approval
- Arrangement with contractors

Rewriting the Presentation

In the rewriting step of the writing process, consider all aspects of style, delivery, appearance, and body language and gestures. Then, most important, practice. Even if you have excellent visual aids and well-organized content, if you fail to deliver effectively, your audience could miss your intended message.

Style

As with good writing, effective oral communication demands clarity and conciseness. To achieve clarity, stick to the point. Your audience does not want to hear about your personal life or other irrelevant bits of information. You need to maintain focus on the topic. Concise oral presentations depend on the same skills evident in concise writing—word and sentence length. Trim your sentences of excess words (12–15 words per sentence is still the preferred length).

Remember to speak so that your audience can understand you and your level of vocabulary. You should speak to communicate rather than to impress your listeners.

Delivery

Effective oral communicators interact with and establish a dynamic relationship with their audiences. The most thorough research will be wasted if you are unable to create rapport and sustain your audience's interest. Although smaller audiences are usually easier to connect with, you can also establish a connection with much larger audiences through a variety of delivery techniques.

Eye Contact   Avoid keeping your eyes glued to your notes. You will find it easy to speak to one individual because you will naturally look him or her in the eye. The person will respond by looking back at you.

With a larger audience, whether the audience has 20 or 200-plus people, keeping eye contact is more difficult. Try looking into different people's eyes as you move through your presentation (or look slightly above their heads if that makes you more comfortable). Most of the audience has been in your position before and can sympathize.

Rate   Because your audience wants to listen and learn, you need to speak at a rate slow enough to achieve those two goals. Determine your normal rate of speech, and cut it in half. *Slow* is the best rate to follow in any oral presentation. You could speed up your delivery when you reach a section of less interesting facts. Slow down for the most important and most interesting parts. Match your rate of speaking to the content of your speech, just as actors vary their speech rate to reflect emotion and changes in content.

Enunciation   Speak each syllable of every word clearly and distinctly. Rarely will an audience ask you to repeat something even if they could not understand you the first time. It is up to you to avoid mumbling. Remember to speak more clearly than you might in a more conversational setting. Slowing down your delivery will help you enunciate clearly.

Pitch   When you speak, your voice creates high and low sounds. That's *pitch*. In your presentation, capitalize on this fact. Vary your pitch by using even more high and low sounds than you do in your normal, day-to-day conversations. Modulate to stress certain keywords or major points in your oral presentations.

Pauses   One way to achieve a successful pace is to pause within the oral presentation. Pause to ask for and to answer questions, to allow ideas to sink in, and to use visual aids or give the audience handouts. These pauses will not lengthen your speech; they will only improve it.

A well-prepared speech will allow for pauses and will have budgeted time effectively. Know in advance if your speech is to be 5 minutes, 10 minutes, or an hour long. Then plan your speech according to time constraints, building pauses into your presentation. Practice the speech beforehand so you can determine when to pause and how often.

Emphasis   You will not be able to underline or boldface comments you make in oral presentations. However, just as in written communication, you will want to emphasize key ideas. Your body language, pitch, gestures, and enunciation will enable you to highlight words, phrases, or even entire sentences.

Interaction with Listeners   You might need your audience to be active participants at some point in your oral presentation, so you will want to encourage this response. Your attitude and the tone of your delivery are key elements contributing to an encouraging atmosphere.

Conflict Resolution   You might be confronted with a hostile listener who either disagrees with you or does not want to be in attendance. You need to be prepared to deal with such a person. If someone disagrees with you or takes issue with a comment you make, try these responses:

- "That is an interesting perspective."
- "Thanks for your input."
- "Let me think about that some more and get back to you."
- "I have got several more ideas to share. We could talk about that point later, during a break."

If you are confronted with a challenge,

- Put it off until later so it does not distract from your presentation.
- Let the situation diffuse.
- Give yourself some more time to think about it.
- Give the person time to cool off.

The important point to remember is to not allow a challenging person to take charge of your presentation. Be pleasant but firm and maintain control of the situation. You will be unable to please all of your listeners all of the time. However, you should not let one unhappy listener destroy the effect of your presentation for the rest of the audience.

Appearance

When you speak to an audience, they see you as well as hear you. Therefore, avoid physical distractions. For example, avoid wearing clothes or jewelry that might distract the audience. You might be representing your company or trying to make a good impression for yourself when you speak, so dress appropriately.

Body Language and Gestures

During an oral presentation, nonverbal communication can be as important as verbal communication. Your appearance and attitude are important. In addition, the way you present your speech through your movements and tone of voice will affect your listeners. If you are enthusiastic about your topic, your listeners will respond enthusiastically. If you are bored or ambivalent, your tone and mannerisms will reflect your attitude. If you are negative, your tone will communicate negativity to your audience.

To communicate effectively, be aware of your body language and your gestures:

- Avoid standing woodenly. Move around somewhat, scanning the room with your eyes, stopping occasionally to look at one person. Remember to look at all parts of the room as you make the oral presentation.
- If you are nervous and your hands shake, try holding onto a chair back, lectern, the top of the table, or a paper clip.
- Use hand motions to emphasize ideas and provide transitions. For instance, you could put one finger up for a first point, two fingers up for a second point, and so forth.
- Avoid folding your arms stiffly across your chest. This projects a negative, defensive attitude.

Postspeech Question and Answers

After your presentation, be prepared for a question-and-answer session. Politely invite your audience to participate by saying, "If you have any questions, I am happy to answer them."

When an audience member asks a question, make sure everyone hears it. If not, you can repeat the question. If you fail to understand the question, ask the audience member to repeat it and clarify it.

If you have no answer, tell the audience. You could say that you will research the matter and get back with them. Faking an answer will only harm your credibility and detract from the overall effect of your presentation. Another valid option is to ask the audience what they think regarding the question or if they have any possible solutions or answers to the question. This is not only a good way to answer the question but also to encourage audience interaction.

Practice

Practice your speech including manipulation of your visual aids so you use them at appropriate times and places during the presentation. As you practice, you will grow more comfortable and less dependent on your note cards or outline. Use the Effective Oral Presentation Checklist to determine if you are sufficiently prepared for your oral presentation.

You will find that the more you practice, the more comfortable you feel. Practicing will help you achieve the following:

- Decrease your fear
- Process your thoughts
- Become more comfortable with the topic
- Pronounce troublesome words
- Decide what to emphasize and how to emphasize it
- Enhance verbal and nonverbal cues
- Rearrange your content
- Add further details
- Know when to use your visual aids

Effective Oral Presentation Checklist

___ **1.** Does your speech have an introduction,
- arousing the audience's attention?
- clearly stating the topic of the presentation?

___ **2.** Does your speech have a body,
- explaining what exactly you want to say?
- developing your points thoroughly?

___ **3.** Does your speech have a conclusion, suggesting
- what is next?
- explaining when (due date) a follow-up should occur?
- stating why that date is important?

___ **4.** Does your presentation provide visual aids to help you make and explain your points?

___ **5.** Do you modulate your pace and pitch?

___ **6.** Do you enunciate clearly so the audience will understand you?

___ **7.** Have you used body language effectively,
- maintaining eye contact?
- using hand gestures?
- moving appropriately?
- avoiding fidgeting with your hair or clothing?

___ **8.** Have you prepared for possible conflicts?

___ **9.** Do you speak slowly and remember to pause so the audience can think?

___**10.** Have you practiced with any equipment you might use?

Chapter Highlights

1. Effective oral communication ensures that your message is conveyed successfully and that your verbal skills make a good impression.

2. You will communicate verbally on an everyday basis, informally and formally.

3. Some points to consider when you use the telephone are to script your telephone conversation, speak clearly and slowly, and avoid rambling.

4. To prepare for a teleconference, plan ahead, check equipment for audio quality, and choose a quiet and private location.

5. A formal oral presentation consists of an introduction, discussion, and conclusion.

6. An effective introduction might include an anecdote, question, quote, facts, or a startling comment.

7. In the discussion section of a formal oral presentation, organize your content according to comparison/contrast, problem/solution, argument/persuasion, importance, and chronology.

8. Conclude your formal speech by restating your main points, recommending a future course of action, or by asking for questions and comments.

9. One of the most powerful oral communication tools is Microsoft PowerPoint.

10. Some benefits of PowerPoint slides include different autolayouts and design templates, the ability to add and delete slides easily, hypertext links, and the incorporation of sound and graphics.

11. When you use PowerPoint slides, ensure contrast between the font color and background, choose an easy-to-read font size and type, use few words and lines per slide, use headings for navigation, and prepare print handouts for your audience.

12. During a PowerPoint presentation, avoid merely reading the slides, choosing instead to elaborate on each slide with additional details.

13. The writing process of prewriting, writing, and rewriting allows you to create effective oral communication.

14. When you prewrite to plan your presentation, consider purpose, recognize audience, and gather information.

15. To write your presentation, use a presentation plan, outline, note cards, and visual aids.

16. Many visual aids can enhance your oral presentation. Choose from chalkboards, chalkless whiteboards, flip charts, overhead projections, PowerPoint, videos, and film.

17. To perfect your oral presentation (the third part of the communication process), work on your delivery, including eye contact, rate of speech, enunciation, pitch, pauses and emphasis, interaction with the audience, and conflict resolution.

18. Appearance, body language, and gestures will impact your oral presentation.

19. Restate main points frequently in an oral presentation.

20. Be prepared for questions and discussion after you conclude your oral presentation.

CASE STUDIES

Read the **TechStop** scenario that begins this chapter. In this scenario, a customer is treated poorly by a salesperson. The Vice President of Customer Service, Shuan Wang, has received reports of other problems with customer-sales staff interaction.

Shuan needs to make a formal oral presentation to all sales personnel to improve customer service.

Assignment

Based on the scenario at the beginning of the chapter, outline the content for Shuan's oral presentation as follows:

- Determine what kind of introduction should be used to arouse the audience's interest.

- Provide a thesis statement.

- Develop information to teach employees sales etiquette and customer interaction, store policy regarding customer satisfaction, the impact on poor customer relations, and the consequences of failing to handle customers correctly. Research these topics (either online, in a library, or through interviews) to find your content.

- Organize the speech's content with appropriate transitions to aid coherence.

- Determine which types of visual aids would work best for this presentation.

Halfmoon outdoor equipment is an online wholesaler of hiking, biking, and boating gear. Their staff is located in four northeastern states. Halfmoon's CEO, Montana Wildhack, wants his employees to excel in communication skills, oral and written. He has hired a consultant to train the employees. To accomplish this training, the consultant will provide a teleconference. Fifty employees will meet with the consultant in a training room, while another 150 Halfmoon staff members view the training session from their different work locations.

During the presentation, the consultant highlights his comments by writing on a chalkless whiteboard, using green and yellow markers. He attempts humor by suggesting that Halfmoon's thermal stocking caps would be a perfect gift for one of the seminar's balding participants, Joe. In addition, the consultant has participants break into small groups to role play. They take mock customer orders, field complaints, and interact with vendors, using a prepared checklist of do's and don'ts.

At one point, when a seminar participant calls in a question, the consultant says, "No, that would be wrong, wrong, wrong! Why would you ever respond to a customer like that? Common sense would dictate a different response." During a break, the consultant does not mute his clip-on microphone. To conclude the session, the consultant asks for questions and comments. When he cannot provide a good answer, he asks the participants to suggest options, and he says he will research the question and place an answer on his corporate Web site.

Assignment

Using the guidelines in the chapter, discuss the speaker's oral communication skills. Explain what went wrong and why. Explain what was effective.

Gullwing, Texas

Jessica Studin is the information technology manager for the City of Gullwing, Texas. Every month she updates the city's decision makers and constituents about Gullwing's technology needs. To do so, she makes an oral presentation at the city's board meeting. Her immediate audience includes the mayor, other city officials, invited guests, and any city residents in attendance. In addition, Jessica's oral presentation is televised locally to all of Gullwing's citizens.

To make her oral presentations, Jessica works from a presentation plan, speaks from an outline, and uses PowerPoint slides to enhance her speech.

This month, Jessica must focus on financial and technology needs confronting the city. Specifically, Gullwing's city hall has old and unreliable computers, limited printers, software that is outdated, and a Web site that needs revision. Her challenge is to solve these problems in a fiscally responsible manner.

Some possible options include the following:

- Purchasing new laser printers that can be shared by several offices
- Finding new hardware and software providers who will offer government discounts
- Hiring a Web designer

Assignment

Using the guidelines presented in this chapter, help Jessica make her oral presentation. To do so,

- Organize the above information according to a logical order. This could include problem/solution, cause/effect, or importance.
- Create an outline for Jessica's speech. You might need to research the above topics further to find more detailed information.
- Create a PowerPoint presentation, using the above information and any additional information you consider necessary.

Applying What You've Learned

After reading the following scenarios, answer these questions:

- Would your oral presentation be everyday, informal, or formal?
- Would you use a videoconference, teleconference, or face-to-face meeting?
- Which visual aids would work best for your presentation?
- Is your oral presentation goal to persuade, instruct, inform, or build trust?

In addition, complete any or all of these assignments:

- Write a presentation plan
- Write an outline for the presentation
- Write a brief, introductory lead-in to arouse your listeners' interest
- Create a questionnaire

A. You work at FlashCom Electric. Your company has created a new interface for modems. Your boss has asked you to make a presentation to sales representatives from 20 potential vendors in the city. The oral presentation will explain to the vendors the benefits of your product, the sales breaks you will offer, and how the vendors can increase their sales.

B. After working for Friendly's, a major discount computer hardware and software retailer, for over a year, you have created a new organizational plan for their vast inventory. In an oral presentation, you plan to show the CEO and board of directors why your plan is cost effective and efficient.

C. You are the manager of an automotive parts supply company, Plugs, Lugs, & More. Your staff of 10 in-store employees lacks knowledge of the store's new merchandise, has not been meeting sales goals, and does not always treat customers with respect and care. It's time to address these concerns.

D. As CEO of your Engineering/Architectural firm (Levin, Lisk, and Lamb), you must downsize. Business is decreasing and costs are rising. To ensure third-quarter profitability, 10 percent of the staff must be laid off. That will amount to a layoff of over 500 employees. You now must make an oral presentation to your stockholders and the entire workforce (located in three states) at your company's annual meeting to report the situation.

E. Your Quality Circle Team is composed of account representatives, management, engineers, and technicians. You have been asked to research new ways to improve product delivery. You have concluded your research. Now, you need to share your suggestions with your 15 team members.

F. Your company needs to hire a new technical writer. The search committee has narrowed the options to six potential candidates. You want to invite only three of the candidates for interviews.

G. A mid-level manager e-mailed an entire department of 18 employees, complaining about their work ethic. Though the manager had a right to complain, the e-mail was sarcastic, sexist, and offensive. As upper-level management, your job is to deal with the problem created by the writer's poor communication skills.

H. Your technical communication company, WriteNeeds, has written a proposal to bid on a local convention center's communication business (fliers, newsletters, brochures, or Web site). Now, the company needs to follow up with an oral presentation to the convention center's management team. In this presentation, you will present examples of your prior work, costs, credentials of your writers and desktop publishers, and recommendations from satisfied customers.

I. Your hospital staff of 200 employees, including nurses, doctors, biomedical records personnel, and technicians, needs to be told about new federal regulations regarding patient privacy.

EXERCISES

Evaluating an Oral Presentation

Listen to a speech. You could do so on television; at a student union; at a church, synagogue, or mosque; at a civic event, city hall meeting, or community organization (Lions Club, Rotary Club, or Boy Scouts); at a company activity; or in your classroom. Answer the following questions:

1. What was the speaker's goal? Did the speaker try to persuade, instruct, inform, or build trust? Explain your answer by giving examples.

2. What type of introduction did the speaker use to arouse listener interest (anecdote, question, quote, facts, or startling comments)? Give examples to support your decision.

3. Were the introductions successful? If so, why and how? If not, what type of introductory method would have been better?

4. What visual aids were used in the presentation? Were these visual aids effective? Explain your answers.

5. How did the speaker develop the assertions? Did the speaker use analysis, comparison/contrast, argument/persuasion, problem/solution, or chronology? Give examples to prove your point.

6. Was the speaker's delivery effective? Use the following Presentation Delivery Rating Sheet in Table 18.2 to assess the speaker's performance by placing check marks in the appropriate columns.

TABLE 18.2 PRESENTATION DELIVERY RATING SHEET			
Delivery Techniques	Good	Bad	Explanation
Eye Contact			
Rate of Speech			
Enunciation			
Pitch of Voice			
Use of Pauses			
Emphasis			
Interaction with Listeners			
Conflict Resolution			

Giving an Oral Presentation

1. Find an advertised job opening in your field of expertise or degree program. You could look either online, in the newspaper, at your school's career placement center, or at a company's human resource office. Perform a mock interview for this position. To do so, follow this procedure:
   - Designate one person as the applicant.
   - He or she should prepare for the interview by making a list of potential questions (ones that the applicant believes he or she will be asked, as well as questions to ask the search committee).
   - Designate others in the class to represent the search committee.
   - The mock search committee should prepare a list of questions to ask the applicant.

2. Research a topic in your field of expertise or degree program. The topic could include a legal issue, a governmental regulation, a news item, an innovation in the field, or a published article in a professional journal or public magazine. Make an oral presentation about your findings.

3. Research job opportunities in your degree program. Are jobs decreasing or increasing? What is the pay scale? Which companies are hiring in your field? What skills are required for the job? Make an oral presentation about your findings.

4. Interview employees at your job site or at any company of your choice to find out the following:
   - The estimated amount of time they spend in oral communication on the job
   - Whether their oral communication is everyday, informal, or formal
   - If their oral communication goals are to persuade, instruct, inform, or build trust
   - Which types of visual aids they use
   - Whether they have participated in teleconferences or videoconferences

   After concluding your interviews, share your findings with your class in an oral presentation.

5. For assignments 2–4 above ("Giving an Oral Presentation"), create an outline or note cards for your speech.

Creating a PowerPoint Slide Presentation

For assignments 2–4 above ("Giving an Oral Presentation"), create PowerPoint slides to enhance your oral communication. To do so, follow the guidelines provided in this chapter.

Role Playing

TechStop sells DVDs, VCRs, TVs, audio components, computers, and computer peripherals. Such stores often receive customer complaints when equipment malfunctions. Practice face-to-face encounters, telephone conversations, or leaving voice-mail messages.

To do so, divide the class into three groups: complaining customers, customer service representatives, and supervisors.

The customers should script a complaint, drawn from personal experience, and share it with those students acting as customer service representatives. The customer service reps will respond extemporaneously. The supervisors will assess the oral communication, based on this chapter's guidelines.

Assessing PowerPoint Slides

After reviewing the following Microsoft PowerPoint slides, determine which are successful, which are unsuccessful, and explain your answers, based on the guidelines provided in this chapter.

A Library Perspective on Distance Learning

1. Library online research sites are rapidly evolving.
2. Instructors engaged in online course development should visit with library staff for updates on resources.
3. Many traditional library services are available to serve distance learning students, including course reserves, interlibrary loan, reference, reciprocal borrowing policies, document delivery, access to online databases and collections.
4. Many new services, electronic books and journals, are available from libraries.
5. The Internet is not the online equivalent of an academic library.

A Technology Perspective on Online Education

Students—24-hour Call Centers answer student hardware/software needs

Faculty—The Tech Center helps faculty with course creations and tech resources

ACTIVITIES

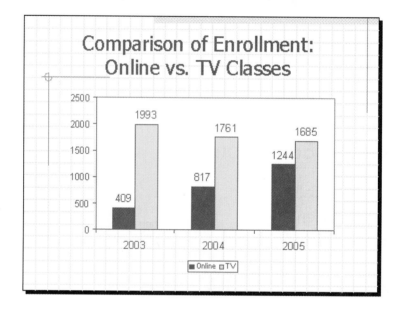

As **DISTANCE LEARNING** classes have risen in popularity, many *diverse areas* of interest across campuses have converged.

These include counseling, library services, technology, faculty, administration, continuing education, etc.

To ensure student success, *Distance Learning Coordinating Committees* need to draw from these multiple disciplines and collaborate on solutions to arising technology challenges.

Comparison of Enrollment: Online vs. TV Classes

2003: Online 409, TV 1993
2004: Online 817, TV 1761
2005: Online 1244, TV 1685

■ Online □ TV

QUIZ QUESTIONS

1. What are three ways you can achieve successful telephone conversations?

2. Why would you participate in a teleconference?

3. How can you create an effective teleconference or videoconference?

4. How many parts are in a formal oral presentation?

5. In what ways can you arouse audience attention in an oral presentation?

6. What does a thesis statement in your formal oral presentation achieve?

7. How can you organize the details in the discussion section of a formal oral presentation?

8. How can you achieve coherence in a formal oral presentation?

9. How can you conclude a formal oral presentation?

10. What are benefits of using Microsoft PowerPoint slide presentations?

11. How can you create an effective PowerPoint slide presentation?

12. What can you do during a PowerPoint presentation to make it more effective?

13. When prewriting a formal oral presentation, you should focus on which three considerations?

14. Instead of writing out an entire formal oral presentation, how can you prepare your speech?

15. In addition to PowerPoint, what are four other types of visual aids?

Grammar, Punctuation, Mechanics, and Spelling

Chapter Preview

Grammar Rules
A review of grammar rules will help you write correctly.

Punctuation
Learning punctuation rules will help you avoid errors.

Mechanics
Abbreviations, capitalization, and number usage are an important part of technical writing.

Spelling
Included is a list of commonly misspelled words.

It is essential to correctly organize and develop your memo, letter, or report. However, no one will be impressed with the quality of your work, or with you, if your writing is riddled with errors in sentence construction or punctuation. Your written correspondence is often your first contact with business associates. Many people mistakenly believe that only English teachers notice grammatical errors and wield red pens, but businesspeople as well take note of such errors and may see the writer as less competent.

We were working recently with a young executive who is employed by a branch of the federal government. This executive told us that whenever his supervisor found a spelling error in a subordinate's report, this report was paraded around the office. Everyone was shown the mistake and had a good laugh over it, and the report was then returned to the writer for correction. Our acquaintance assured us that all of this was in good-natured fun. However, he also said that employees quickly learned to edit and proof their work to avoid such public displays of their errors. He went on to say that his dictionary was well thumbed and always on his desk.

Your writing at work may not be exposed to such scrutiny by coworkers. Instead, your writing may go directly to another firm, and those readers will see your mistakes. To avoid this problem, you must evaluate your writing for grammar, punctuation, and spelling errors. If you don't, your customers and colleagues will.

This chapter focuses on

- Grammar rules
- Punctuation
- Rules for effective mechanics
- Spelling
- Proofreader's marks

In addition, the chapter provides hands-on experience with grammar, punctuation, and spelling in the form of memos and letters needing correction.

To understand the fundamentals of grammar, you must first understand the basic components of a sentence.

A correctly constructed sentence consists of a subject and a predicate (some sentences also include a phrase or phrases).

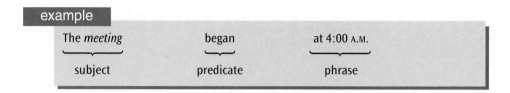

The *meeting*	began	at 4:00 A.M.
subject	predicate	phrase

Subject: The *doer* of the action; the subject usually precedes the predicate.

Predicate: The *action* in the sentence

He	ran	to the office to avoid being late.
doer	action	phrases

If the subject and the predicate (a) express a complete thought and (b) can stand alone, you have an *independent clause.*

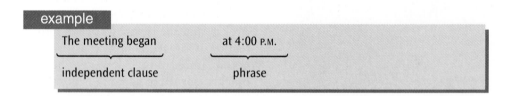

The meeting began	at 4:00 P.M.
independent clause	phrase

A *phrase* is a group of related words that does not contain a subject and a predicate and cannot stand alone or be punctuated as a sentence. The following are examples of phrases:

at the house

in the box

on the job

during the interview

If a clause is dependent, it cannot stand alone.

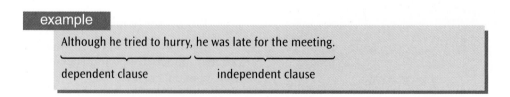

Although he tried to hurry,	he was late for the meeting.
dependent clause	independent clause

He was late for the meeting although he tried to hurry.

independent clause   dependent clause

NOTE: When a dependent clause begins a sentence, use a comma before the independent clause. However, when an independent clause begins a sentence, do not place a comma before the dependent clause.

Agreement Between Pronoun and Antecedent (Referent)

A pronoun has to agree in gender and number with its antecedent.

Susan went on *her* vacation yesterday.

The *people* who quit said that *they* deserved raises.

Problems often arise when a singular indefinite pronoun is the antecedent. The following antecedents require singular pronouns: *anybody, each, everybody, everyone, somebody,* and *someone*.

Anyone can pick up *their* applications at the job placement center.

Anyone can pick up *his* or *her* applications at the job placement center.

Problems also arise when the antecedent is separated from the pronoun by numerous words.

Even when the best *employee* is considered for a raise, *they* often do not receive it.

Even when the best *employee* is considered for a raise, *he or she* often does not receive it.

Agreement Between Subject and Verb

Writers sometimes create disagreement between subjects and verbs, especially if other words separate the subject from the verb. To ensure agreement, ignore the words that come between the subject and verb.

Her *boss* undoubtedly *think* that all the employees want promotions.

Her *boss* undoubtedly *thinks* that all the employees want promotions.

incorrect

> The *employees* who sell the most equipment *is* going to Hawaii for a week.

correct

> The *employees* who sell the most equipment *are* going to Hawaii for a week.

If a sentence contains two subjects (a compound subject) connected by *and,* use a plural verb.

incorrect

> Joe and Becky *was* both selected employee of the year.

correct

> Joe and Becky *were* both selected employee of the year.

incorrect

> The bench workers and their supervisor *is* going to work closely to complete this project.

correct

> The bench workers and their supervisor *are* going to work closely to complete this project.

Add a final *s* or *es* to create most plural subjects or singular verbs, as follows:

Plural Subjects	Singular Verbs
boss*es* hire	a boss hir*es*
employe*es* demand	an employee demand*s*
experiment*s* work	an experiment work*s*
attitud*es* change	the attitude chang*es*

If a sentence has two subjects connected by *either . . . or, neither . . . nor,* or *not only . . . but also,* the verb should agree with the closest subject. This also makes the sentence less awkward.

example

> Either the salespeople or the warehouse worker deserv*es* raises.
> Not only the warehouse worker but also the salespeople deserv*e* raises.
> Neither the salespeople nor the warehouse worker deserv*es* raises.

Singular verbs are used after most indefinite pronouns such as the following:

another	everything
anybody	neither
anyone	nobody
anything	no one

each	nothing
either	somebody
everybody	someone
everyone	something

Anyone who works here *is* guaranteed maternity leave.

Everybody wants the company to declare a profit this quarter.

Singular verbs often follow collective nouns such as the following:

class	organization
corporation	platoon
department	staff
group	team

The *staff is* sending the boss a bouquet of roses.

Comma Splice

A *comma splice* occurs when two independent clauses are joined by a comma rather than separated by a period or semicolon.

incorrect

Sue was an excellent employee, she got a promotion.

Several remedies will correct this error.
1. Separate the two independent clauses with a semicolon.

correct

Sue was an excellent employee; she got a promotion.

2. Separate the two independent clauses with a period.

correct

Sue was an excellent employee. She got a promotion.

3. Separate the two independent clauses with a comma and a *coordinating conjunction (and, but, or, for, so, yet).*

correct

Sue was an excellent employee, *so* she got a promotion.

4. Separate the two independent clauses with a semicolon (or a period), a conjunctive adverb, and a comma. *Conjunctive adverbs* include *also, additionally, consequently, furthermore, however, instead, moreover, nevertheless, therefore,* and *thus.*

Sue was an excellent employee; therefore, she got a promotion.

or

Sue was an excellent employee. *Therefore*, she got a promotion.

5. Use a *subordinating conjunction* to make one of the independent clauses into a dependent clause. Subordinating conjunctions include *after, although, as, because, before, even though, if, once, since, so that, though, unless, until, when, where,* and *whether.*

Because Sue was an excellent employee, she got a promotion.

Faulty or Vague Pronoun Reference

A pronoun must refer to a specific noun (its antecedent). Problems arise when (a) there is an excessive number of pronouns (causing vague pronoun reference) and (b) there is no specific noun as an antecedent. Notice that there seems to be an excessive number of pronouns in the following passage, and the antecedents are unclear.

Although Bob had been hired over two years ago, *he* found that *his* boss did not approve *his* raise. In fact, *he* was also passed over for *his* promotion. The boss appears to have concluded that *he* had not exhibited zeal in *his* endeavors for their business. Instead of being a highly valued employee, *he* was not viewed with pleasure by those in authority. Perhaps it would be best if *he* considered *his* options and moved to some other company where *he* might be considered in a new light.

The excessive and vague use of *he* and *his* causes problems for readers. Do these words refer to Bob or to his boss? You are never completely sure. To avoid this problem, limit pronoun usage, as in the following revision.

Although Bob had been hired over two years ago, he found that his boss, Joe, did not approve his raise. In fact, Bob was also passed over for promotion. Joe appears to have concluded that Bob had not exhibited zeal in his endeavors for their business. Instead of being a highly valued employee, Bob was not viewed with pleasure by those in authority. Perhaps it would be best if Bob considered his options and moved to some other company where he might be considered in a new light.

To make the preceding paragraph more precise, we have replaced vague pronouns (*he* and *his*) with exact names (*Bob* and *Joe*).

Fragments

A *fragment* occurs when a group of words is incorrectly used as an independent clause. Often the group of words begins with a capital letter and has end punctuation but is missing either a subject or a predicate.

Working with computers.
(lacks a predicate and does not express a complete thought)

The group of words may have a subject and a predicate but be a dependent clause.

Although he enjoyed working with computers.
(has a subject, *he*, and a predicate, *enjoyed*, but is a dependent clause because it is introduced by the subordinate conjunction *although*)

It is easy to remedy a fragment by doing one of the following:

- Add a subject.
- Add a predicate.
- Add both a subject and a predicate.
- Add an independent clause to a dependent clause.

Joe found that working with computers used his training.
(subject, *Joe*, and predicate, *found*, have been added)

Although he enjoyed working with computers, he could not find a job in a computer-related field.
(independent clause, *he could not find a job*, added to the dependent clause, *Although he enjoyed working with computers*)

Fused Sentence

A *fused sentence* occurs when two independent clauses are connected with no punctuation.

The company performed well last quarter its stock rose several points.

There are several ways to correct this error.

1. Write two sentences separated by a period.

The company performed well last quarter. Its stock rose several points.

2. Use a comma and a coordinating conjunction to separate the two independent clauses.

The company performed well last quarter, *so* its stock rose several points.

3. Use a subordinating conjunction to create a dependent clause.

correct

Because the company performed well last quarter, its stock rose several points.

4. Use a semicolon to separate the two independent clauses.

correct

The company performed well last quarter; its stock rose several points.

5. Separate the two independent clauses with a semicolon, a conjunctive adverb or a transitional word or phrase, and a comma.

correct

The company performed well last quarter; *therefore*, its stock rose several points.

correct

The company performed well last quarter; *for example*, its stock rose several points.

The following are transitional words and phrases, listed according to their use.

To Add

again	in addition
also	moreover
besides	next
first	second
furthermore	still

To Compare/Contrast

also	nevertheless
but	on the contrary
conversely	still
in contrast	

To Provide Examples

for example	of course
for instance	put another way
in fact	to illustrate

To Show Place

above	here
adjacent to	nearby
below	on the other side
elsewhere	there
further on	

To Reveal Time

afterward	second
first	shortly
meanwhile	subsequently
presently	thereafter

To Summarize

all in all	last
finally	on the whole
in conclusion	therefore
in summary	thus

Modification

A *modifier* is a word, phrase, or clause that explains or adds details about other words, phrases, or clauses.

Misplaced Modifiers

A *misplaced modifier* is one that is not placed next to the word it modifies.

incorrect

He had a heart attack *almost* every time he was reviewed by his supervisor.

correct

He almost had a heart attack every time he was reviewed by his supervisor.

incorrect

The worker had to *frequently* miss work.

correct

The worker frequently had to miss work.

Dangling Modifiers

A *dangling modifier* is a modifier that is not placed next to the word or phrase it modifies. To avoid confusing your readers, place modifiers next to the word(s) they refer to. Don't expect your readers to guess at your meaning.

incorrect

While working, tiredness overcame them.
(Who was working? Who was overcome by tiredness?)

correct

While working, the staff became tired.

After soldering for two hours, the equipment was ready for shipping. (Who had been soldering for two hours? Not the equipment!)

After soldering for two hours, the technicians prepared the equipment for shipping.

Parallelism

All items in a list should be parallel in grammatical form. Avoid mixing phrases and sentences (independent clauses).

We will discuss the following at the department meeting:
1. Entering mileage in logs (phrase)
2. All employees have to enroll in a training seminar. (sentence)
3. Purpose of quarterly reviews (phrase)
4. Some data processors will travel to job sites. (sentence)

We will discuss the following at the department meeting:
1. Entering mileage in logs
2. Enrolling in training seminars
3. Reviewing employee performance quarterly
4. Traveling to job sites

} phrases

At the department meeting, you will learn how to
1. Enter mileage in logs
2. Enroll in training seminars
3. Review employee performance quarterly
4. Travel to job sites

} phrase

PUNCTUATION

Apostrophe (')

Place an *apostrophe* before the final *s* in a singular word to indicate possession.

Jim's tool chest is next to the furnace.

Place the apostrophe after the final *s* if the word is plural.

example

The employees' reception will be held next week.

Don't use an apostrophe to make singular abbreviations plural.

incorrect

The EXT's will be shipped today.

correct

The EXTs will be shipped today.

Colon (:)

Use a *colon* after a salutation.

example

Dear Mr. Harken:

In addition, use a colon after an emphatic or cautionary word if explanations follow.

example

Note: Hand-tighten the nuts.
Caution: Wash thoroughly if any mixture touches your skin.

Finally, use a colon after an independent clause to precede a quotation, list, or example.

example

She said the following: "No comment."

These supplies for the experiment are on order: plastic hose, two batteries, and several chemicals.

The problem has two possible solutions: hire four more workers, or simply give everyone a raise.

NOTE: In the preceding examples, the colon follows an independent clause.

A common mistake is to place a colon after an incomplete sentence. Except for salutations and cautionary notes, whatever precedes a colon *must* be an independent clause.

incorrect

The two keys to success are: earning money and spending wisely.

Comma (,)

Writers often get in trouble with *commas* when they employ one of two common "words of wisdom."

- When in doubt, leave it out.
- Use a comma when there is a pause.

Both rules are inexact. Writers use the first rule to justify the complete avoidance of commas; they use the second rule to sprinkle commas randomly throughout their writing. On the contrary, commas have several specific conventions that determine usage.

1. Place a comma before a coordinating conjunction *(and, but, or, for, so, yet)* linking two independent clauses.

example

You are the best person for the job, *so* I will hire you.

We spent several hours discussing solutions to the problem, *but* we failed to decide on a course of action.

2. Use commas to set off introductory comments.

example

First, she soldered the components.

In business, people often have to work long hours.

To work well, you need to get along with your coworkers.

If you want to test equipment, do so by 5:25 P.M.

3. Use commas to set off sentence interrupters.

example

The company, started by my father, did not survive the last recession.

Mrs. Mittleman, the proprietor of the store, purchased a wide array of merchandise.

4. Set off parenthetical expressions with commas.

> A worker, it seems, should be willing to try new techniques.
>
> The highway, by the way, needs repairs.

5. Use commas after each item in a series of three or more.

> Fred, Helene, and Ron were chosen as employees of the year.
>
> We found the following problems: corrosion, excessive machinery breakdowns, and power failures.

6. Use commas to set off long numbers.

> She earns $100,000 before taxes.

NOTE: Very large numbers are often written as words.

> Our business netted over $2 million in 2005.

7. Use commas to set off the day and year when they are part of a sentence.

> The company hired her on September 7, 2005, to be its bookkeeper.

NOTE: If the year is used as an adjective, do not follow it with a comma.

> The 2003 corporate report came out today.

8. Use commas to set off the city from the state and the state from the rest of the sentence.

> The new warehouse in Austin, Texas, will promote increased revenues.

> **NOTE:** If you omit either the city or the state, you do not need commas.

> **example**
>
> The new warehouse in Austin will promote increased revenues.

Dash (—)

A *dash*, typed as two consecutive hyphens with no spaces before or after, is a versatile punctuation mark used in the following ways.

1. After a heading and before an explanation.

> **example**
>
> Forecasting—Joe and Joan will be in charge of researching fourth-quarter production quotas.

2. To indicate an emphatic pause.

> **example**
>
> You will be fired—unless you obey company rules.

3. To highlight a new idea.

> **example**
>
> Here's what we can do to improve production quality—provide on-the-job training, salary incentives, and quality controls.

4. Before and after an explanatory or appositive series.

> **example**
>
> Three people—Sue, Luci, and Tom—are essential to the smooth functioning of our office.

Ellipses (. . .)

Ellipses (three spaced periods) indicate omission of words within quoted materials.

> **example**
>
> "Six years ago, prior to incorporating, the company had to pay extremely high federal taxes."
>
> "Six years ago, . . . the company had to pay extremely high federal taxes."

Exclamation Point (!)

Use an *exclamation point* after strong statements, commands, or interjections.

> **example**
>
> You must work harder!
> Do not use the machine!
> Danger!

Hyphen (-)

A *hyphen* is used in the following ways.

1. To indicate the division of a word at the end of a typed line. Remember, this division must occur between syllables.
2. To create a compound adjective.

> **example**
>
> He is a well-known engineer.
> Until her death in 2005, she was a world-renowned chemist.
> Tom is a 24-hour-a-day student.

3. To join the numerator and denominator of fractions.

> **example**
>
> Four-fifths of the company want to initiate profit sharing.

4. To write out two-word numbers.

> **example**
>
> Twenty-six people attended the conference.

Parentheses ( )

Parentheses enclose abbreviations, numbers, words, or sentences for the following reasons.

1. To define a term or provide an abbreviation for later use.

> **example**
>
> We belong to the Society for Technical Communication (STC).

2. To clarify preceding information in a sentence.

> The people in attendance (all regional sales managers) were proud of their accomplishments.

3. To number items in a series.

> The company should initiate (1) new personnel practices, (2) a probationary review board, and (3) biannual raises.

Period (.)

A *period* must end a declarative sentence (independent clause).

> I found the business trip rewarding.

Periods are often used with abbreviations.

D.C.	Mrs.	A.M. or a.m.
e.g.	Ms.	P.M. or p.m.
Mr.		

It is incorrect to use periods with abbreviations for organizations and associations.

S.T.C. (Society for Technical Communication)

STC (Society for Technical Communication)

State abbreviations no longer use periods.

KS. (Kansas)
MO. (Missouri)
TX. (Texas)

KS
MO
TX

Question Mark (?)

Use a *question mark* after direct questions.

> **example**
>
> Do the lab results support your theory?
> Will you work at the main office or at the branch?

Quotation Marks (" ")

Quotation marks are used in the following ways.

1. When citing direct quotations.

> **example**
>
> He said, "Your division sold the most compressors last year."

NOTE: When you are citing a quotation within a quotation, use double quotation marks
(" ") and single quotation marks (' ').

> **example**
>
> Kim's supervisor, quoting the CEO, said the following to explain the new
> policy regarding raises: "'Only employees who deserve them will receive merit
> raises.'"

2. To note the title of an article or a subdivision of a report.

> **example**
>
> The article "Robotics in Industry Today" was an excellent choice as the basis of your
> speech.
> Section III, "Waste Water in District 9," is pertinent to our discussion.

When using quotation marks, abide by the following punctuation
conventions:

- Commas and periods always go *inside* quotation marks.

> **example**
>
> She said, "Our percentages are fixed."

- Colons and semicolons always go *outside* quotation marks.

> **example**
>
> He said, "The supervisor hasn't decided yet"; however, he added that the deci-
> sion would be made soon.

- Exclamation points and question marks go inside the quotation marks if the quoted material is either exclamatory or a question. However, if the quoted material is not exclamatory or a question, then these punctuation marks go outside the quotation marks.

> **example**
>
> John said, "Don't touch that liquid. It's boiling!"

(Although the sentence isn't exclamatory, the quotation is. Thus, the exclamation point goes inside the quotation marks.)

> **example**
>
> How could she say, "We haven't purchased the equipment yet"?

(Although the quotation isn't a question, the sentence is. Thus, the question mark goes outside the quotation marks.)

Semicolon (;)

Semicolons are used in the following instances.

1. Between two independent clauses *not* joined by a coordinating conjunction.

> **example**
>
> The light source was unusual; it emanated from a crack in the plastic surrounding the cathode.

2. To separate items in a series containing internal commas.

> **example**
>
> When the meeting was called to order, all members were present, including Susan Bailey, the president; Ruth Schneider, the vice president; Harold Holbert, the treasurer; and Linda Hamilton, the secretary.

MECHANICS

Abbreviations

Never use an abbreviation that your reader will not understand. A key to clear technical writing is to write on a level appropriate to your reader. You may use the following familiar abbreviations without explanation: *Mrs., Dr., Mr., Ms.,* and *Jr.*

A common mistake is to abbreviate inappropriately. For example, some writers abbreviate *and* as follows:

> **example**
>
> I quit my job & planned to retire young.

This is too colloquial for professional technical writing. Spell out *and* when you write.

The majority of abbreviation errors occur when writers incorrectly abbreviate states and technical terms. (We list many computer-related abbreviations in Chapter 13.)

States

Writers often abbreviate the names of states incorrectly. Use the U.S. Postal Service abbreviations in addresses.

Abbreviations for States

AL	Alabama	MT	Montana
AK	Alaska	NC	North Carolina
AZ	Arizona	ND	North Dakota
AR	Arkansas	NE	Nebraska
CA	California	NV	Nevada
CO	Colorado	NH	New Hampshire
CT	Connecticut	NJ	New Jersey
DE	Delaware	NM	New Mexico
FL	Florida	NY	New York
GA	Georgia	OH	Ohio
HI	Hawaii	OK	Oklahoma
IN	Indiana	OR	Oregon
IA	Iowa	PA	Pennsylvania
ID	Idaho	RI	Rhode Island
IL	Illinois	SC	South Carolina
KS	Kansas	SD	South Dakota
KY	Kentucky	TN	Tennessee
LA	Louisiana	TX	Texas
ME	Maine	UT	Utah
MD	Maryland	VT	Vermont
MA	Massachusetts	VA	Virginia
MI	Michigan	WA	Washington
MN	Minnesota	WV	West Virginia
MS	Mississippi	WI	Wisconsin
MO	Missouri	WY	Wyoming

Technical Terms

Units of measurement and scientific terms must be abbreviated accurately to ensure that they will be understood. Writers often use such abbreviations incorrectly. For example, "The unit measured 7.9 cent." is inaccurate. The correct abbreviation for centimeter is *cm*, not *cent*. Use the following abbreviation conventions.

Technical Abbreviations for Units of Measurement and Scientific Terms

absolute	abs	ampere	amp
alternating current	AC	ampere-hour	amp-hr
American wire gauge	AWG	amplitude modulation	AM

angstrom unit	Å	dram	dr
atmosphere	atm	electromagnetic force	emf
atomic weight	at wt	electron volt	eV
audio frequency	AF	elevation	el (or elev)
azimuth	az	equivalent	equiv
barometer	bar.	Fahrenheit	F
barrel, barrels	bbl	farad	F
billion electron volts	BeV	faraday	f
biochemical oxygen demand	BOD	feet, foot	ft
		feet per second	ft/sec
board foot	bd ft	fluid ounce	fl oz
Brinell hardness number	BHN	foot board measure	fbm
		foot-candle	ft-c
British thermal unit	Btu	foot-pound	ft lb
bushel	bu	frequency modulation	FM
calorie	cal	gallon	gal
candela	cd	gallons per day	GPD
Celsius	C	gallons per minute	GPM
center of gravity	cg	grain	gr
centimeter	cm	grams	g (or gm)
circumference	cir	gravitational acceleration	g
cologarithm	colog		
continuous wave	CW	hectare	ha
cosine	cos	hectoliter	hl
cotangent	cot	hectometer	hm
cubic centimeter	cc	henry	H
cubic foot	cu ft (or ft^2)	hertz	Hz
cubic feet per second	cfs	high frequency	HF
		horsepower	hp
cubic inch	cu in. (or in.^{3})	horsepower-hours	hp-hr
cubic meter	cu m (or m^3)	hour	hr
cubic yard	cu yd (or yd^3)	hundredweight	cwt
current (electric)	I	inch	in.
cycles per second	CPS	inch-pounds	in.-lb
decibel	dB	infrared	IR
decigram	dg	inner diameter or inside dimensions	ID
deciliter	dl		
decimeter	dm	intermediate frequency	IF
degree	deg		
dekagram	dkg	international unit	IU
dekaliter	dkl	joule	J
dekameter	dkm	Kelvin	K
dewpoint	DP	kilocalorie	kcal
diameter	dia	kilocycle	kc
direct current	DC	kilocycles per second	kc/sec
dozen	doz (or dz)	kilogram	kg

kilohertz	kHz	parts per million	ppm
kilojoule	kJ	pascal	pas
kiloliter	kl	positive	pos or +
kilometer	km	pound	lb
kilovolt	kV	pounds per square inch	psi
kilovolt-amperes	kVa	pounds per square inch absolute	psia
kilowatt-hours	kWH	pounds per square inch gauge	psig
lambert	L	quart	qt
latitude	lat	radio frequency	RF
length	l	radian	rad
linear	lin	radius	r
linear foot	lin ft	resistance	r
liter	l	revolution	rev
logarithm	log.	revolutions per minute	rpm
longitude	long.	second	s (or sec)
low frequency	LF	secant	sec
lumen	lm	specific gravity	sp gr (or SG)
lumen-hour	lm-hr	square foot	ft^2
maximum	max	square inch	$in.^2$
megacycle	mc	square meter	m^2
megahertz	MHz	square mile	mi^2
megawatt	MW	tablespoon	tbs (or tbsp)
meter	m	tangent	tan
microampere	μamp	teaspoon	tsp
microinch	μin.	temperature	t
microsecond	μsec	tensile strength	ts
microwatt	μw	thousand	m
miles per gallon	mpg	ton	t
milliampere	mA	ultra high frequency	UHF
millibar	mb	vacuum	vac
millifarad	mF	very high frequency	VHF
milligram	mg	volt-ampere	VA
milliliter	ml	volt	V
millimeter	mm	volts per meter	V/m
millivolt	mV	volume	vol
milliwatt	mW	watt-hour	whr
minute	min	watt	W
nautical mile	NM	wavelength	WL
negative	neg or −	weight	wt
number	no.	yards	y (or yd)
octance	oct	years	y (or yr)
ounce	oz		
outside diameter	OD		
parts per billion	ppb		

Capital Letters

Capitalize the following:

1. Proper nouns.

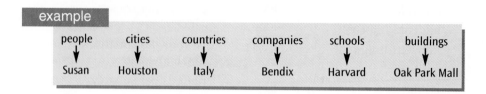

people	cities	countries	companies	schools	buildings
↓	↓	↓	↓	↓	↓
Susan	Houston	Italy	Bendix	Harvard	Oak Park Mall

2. People's titles (only when they precede the name).

> example
>
> Governor Sally Renfro
>
> *or*
>
> Sally Renfro, governor
>
> Technical Supervisor Todd Blackman
>
> *or*
>
> Wes Schneider, the technical supervisor

3. Titles of books, magazines, plays, movies, television programs, and CDs (excluding the prepositions and all articles after the first article in the title).

> example
>
> *Scream*
> *The Catcher in the Rye*
> *The New Yorker*
> *The Taming of the Shrew*
> *Friends*

4. Names of organizations.

> example
>
> Girl Scouts
> Kansas City Regional Home Care Association
> Kansas City Regional Council for Higher Education
> Programs for Technical and Scientific Communication
> American Civil Liberties Union
> Students for a Democratic Society

5. Days of the week, months, and holidays.

Monday

December

Thanksgiving

6. Races, religions, and nationalities.

American Indian

Jewish

Polish

7. Events or eras in history.

the Vietnam War

the Depression

8. North, South, East, and West (when used to indicate geographic locations).

They moved from the North.

People are moving to the Southwest.

NOTE: Don't capitalize these words when giving directions.

We were told to drive south three blocks and then to turn west.

9. The first word of a sentence.
10. Don't capitalize any of the following:

Seasons—spring, fall, summer, winter

Names of classes—sophomore, senior

General groups—middle management, infielders, surgeons

Numbers

Write out numbers one through nine. Use numerals for numbers 10 and above.

10	12
104	2,093
536	5,550,286

Although the preceding rules cover most situations, there are exceptions.

1. Use numerals for all percentages.

2 percent   18 percent   25 percent

2. Use numerals for addresses.

12 Elm   935 W. Harding

3. Use numerals for miles per hour.

5 mph   225 mph

4. Use numerals for time.

3:15 A.M.

5. Use numerals for dates.

May 31, 1948

6. Use numerals for monetary values.

$45 $.95 $2 million

7. Use numerals for units of measurement.

14'   6 3/4"   16 mm 10 V

8. Do not use numerals to begin sentences.

568 people were fired last August.

Five hundred sixty-eight people were fired last August.

9. Do not mix numerals and words when writing numbers. When two or more numbers appear in a sentence and one of them is 10 or more, figures are used.

We attended 4 meetings over a 16-day period.

10. Use numerals and words in a compound number adjective to avoid confusion.

The worker needed six 2-inch nails.

SPELLING

The following is a list of commonly misspelled or misused words. You can avoid many common spelling errors if you familiarize yourself with these words.

accept, except	cite, site, sight
addition, edition	council, counsel
access, excess	desert, dessert
advise, advice	disburse, disperse
affect, effect	fiscal, physical
all ready, already	forth, fourth
assistants, assistance	incite, insight
bare, bear	its, it's
brake, break	loose, lose
coarse, course	miner, minor

passed, past	stationery, stationary
patients, patience	their, there, they're
personal, personnel	to, too, two
principal, principle	whose, who's
quiet, quite	your, you're
rite, right, write	

PROOFREADER'S MARKS

Figure 19.1 illustrates proofreader's marks and how to use them.

Spelling

In the following sentences, circle the correctly spelled words within the parentheses.

1. Each of the employees attended the meeting (accept except) the line supervisor, who was out of town for job-related travel.

2. The (advise advice) he gave will help us all do a better job.

3. Management must (affect effect) a change in employees' attitudes toward absenteeism.

4. Let me (site cite sight) this most recent case as an example.

5. (Its It's) too early to tell if our personnel changes will help create a better office environment.

6. If we (lose loose) another good employee to our competitor, our production capabilities will suffer.

7. I'm not (quite quiet) sure what she meant by that comment.

8. (Their There They're) budget has gotten too large to ensure a successful profit margin.

9. We had wanted to attend the conference (to too two), but our tight schedule prevented us from doing so.

10. (You're Your) best chance for landing this contract is to manufacture a better product.

In the letter on page 660, correct the misspelled words.

Fragments and Comma Splices

In the following sentences, correct the fragments and comma splices by inserting the appropriate punctuation or adding any necessary words.

1. She kept her appointment with the salesperson, however, the rest of her staff came late.

2. When the CEO presented his fiscal year projections, he tried to motivate his employees, many were not excited about the proposed cuts.

3. Even though the company's sales were up 25 percent.

PROOFREADER'S MARKS

Symbol	Meaning	Mark on Copy	Revision
ℓ	delete	hire better peoples ℓ	hire better people
(tr) or ∩	transpose	computer systme	computer system
⌣	close space	we ne ed 40	we need 40
#	insert space	three#mistakes	three mistakes
¶	begin paragraph	¶The first year	The first year
RUN IN	no paragraph	financial. We earned twelve dollars.	financial. We earned twelve dollars.
⊏	move left	⌈ Next year	Next year
⊐	move right	⌉Next year	Next year
⊙	insert period	happy employee ⊙	happy employee.
⌃	insert comma	For two years	For two years,
⌃	insert semicolon	We need you come to work.	We need you; come to work.
⌃	insert colon	dogs:brown, black, and white.	dogs: brown, black, and white.
△	insert hyphen	first rate	first-rate
⌄	insert apostrophe	Marys car	Mary's car
⌄ / ⌄	insert quotation marks	Like a Rolling Stone	"Like a Rolling Stone"
SP	spell out	4 chips	four chips
cap or ≡	capitalize	the meeting	The meeting
lc	lower case	the Boss	the boss
∧	insert	the left margin	the left margin

Figure 19.1   Proofreader's Marks

4. The supervisor wanted the staff members to make suggestions for improving their work environment, the employees, however, felt that any grievances should be taken directly to their union representatives.

5. Which he decided was an excellent idea.

March 5, 2005

Joanna Freeman
Personel Director
United Teletype
1111 E. Street
Kansas City, MO 68114

Dear Ms. Freeman:

Your advertizmemt in the Febuary 18, 2005, Kansas City Star is just the opening I have been looking for. I would like to submit my quallifications.

As you will note in the inclosed resum, I recieved an Enginneering degree from the Missouri Institute of Technology in 2000 and have worked in the electronic enginneering department of General Accounts for three years. I have worked a great deal in design electronics for microprocesors, controll systems, ect.

Because your company has invented many extrordinary design projects, working at your company would give me more chances to use my knowlege aquired in school and through my expirences. If you are interested in my quallifications, I would be happy to discuss them futher with you. I look foreward to hearing from you.

Sincerly,

Bob Cottrell

Bob Cottrell

6. Because their machinery was prone to malfunctions and often caused hazards to the workers.

7. They needed the equipment to complete their job responsibilities, further delays would cause production slowdowns.

8. Their client who was a major distributor of high-tech machinery.

9. Robotics should help us maintain schedules, we'll need to avoid equipment malfunctions, though.

10. The company, careful not to make false promises, advertising their product in media releases.

In the letter on page 661, correct the fragments and comma splices.

Punctuation

In the following sentences, circle the correct punctuation marks. If no punctuation is needed, draw a slash mark through both options.

May 12, 2005

Maurene Pierce
Dean of Residence Life
Mann College
Mannsville, NY 10012

Subject: Report on Dormitory Damage Systems

Here is the report you authorized on April 5 for an analysis of the current dormitory damage system used in this college.

The purpose of the report was to determine the effectiveness of the system. And to offer any concrete recommendations for improvement. To do this, I analyzed in detail the damage cost figures for the past three years, I also did an extensive study of dormitory conditions. Although I had limited manpower. I gathered information on all seven dormitories, focusing specifically on the men's athletic dorm, located at 1201 Chester. In this dorm, bathroom facilities, carpeting, and air conditioning are most susceptible to damage. Along with closet doors.

Nonetheless, my immediate findings indicate that the system is functioning well, however, improvements in the physical characteristics of the dormitories, such as new carpeting and paint, would make the system even more efficient.

I have enjoyed conducting this study, I hope my findings help you make your final decision. Please contact me. If I can be of further assistance.

Rob Harken

Rob Harken

1. John took an hour for lunch (, ;) but Joan stayed at her desk to eat so she could complete the project.

2. Sally wrote the specifications (, ;) Randy was responsible for adding any needed graphics.

3. Manufacturing maintained a 93.5 percent production rating in July (, ;) therefore, the department earned the Golden Circle Award at the quarterly meeting.

4. In their year-end requests to management (, ;) supervisors asked for new office equipment (, ;) and a 10 percent budget increase for staffing.

5. The following employees attended the training session on stress management (, :) Steve Janasz, purchasing agent (, ;) Jeremy Kreisler, personnel director (, ;) and Karen Rochlin, staff supervisor.

6. Promotions were given to all sales personnel (, ;) secretaries, however, received only cost-of-living raises.

7. The technicians voted for better work benefits (, ;) as an incentive to improve morale.

8. Although the salespeople were happy with their salary increases (, ;) the technicians felt slighted.

9. First (, ;) let's remember that meeting schedules should be a priority (, ;) and not an afterthought.

10. The employee (, ;) who achieves the highest rating this month (, ;) will earn 10 bonus points (, ;) therefore (, ;) competition should be intense.

In the letter below, no punctuation has been added. Instead, there are blanks where punctuation might be inserted. First, decide whether any punctuation is needed (not every blank requires punctuation). Then, insert the correct punctuation—a comma, colon, period, semicolon, or question mark.

January 8_ 2005

Mr_ Ron Schaefer
1324 Homes
Carbondale_ IL_ 34198

Dear Mr_ Schaefer_

Yesterday_ my partners_ and I read about your invention in the Herald Tribune_ and we want to congratulate you on this new idea_ and ask you to work with us on a similar project_

We cannot wait to begin our project_ however_ before we can do so_ I would like you to answer the following questions_

• Has your invention been tested in salt water_
• What is the cost of replacement parts_
• What is your fee for consulting_

Once_ I receive your answers to these questions_ my partners and I will contact you regarding a schedule for operations_ We appreciate your design concept_ and know it will help our business tremendously_ We look forward to hearing from you_

Sincerely_

Elias Agamenyon

Elias Agamenyon

Agreement (Subject/Verb and Pronoun/Antecedent)

In the following sentences, circle the correct choice to achieve agreement between subject and verb or pronoun and antecedent.

1. The employees, though encouraged by the possibility of increased overtime, (was were) still dissatisfied with their current salaries.

2. The supervisor wants to manufacture better products, but (they he) doesn't know how to motivate the technicians to improve their work habits.

3. The staff (was were) happy when the new manager canceled the proposed meeting.

4. Anyone who wants (his or her their) vote recorded must attend the annual board meeting.

5. According to the printed work schedule, Susan and Tom (work works) today on the manufacturing line.

6. According to the printed work schedule, either Susan or Tom (work works) today on the manufacturing line.

7. Although Larry is responsible for distributing all monthly activity reports, (he they) failed to mail them.

8. Every one of the engineers asked if (he or she they) could be assigned to the project.

9. Either the supervisor or the technicians (is are) at fault.

10. The CEO, known for her generosity to employees and their families, (has have) been nominated for the humanitarian award.

In the following memo, find and correct the errors in agreement.

MEMO

DATE: October 30, 2005
TO: Tammy West
FROM: Susan Lisk
SUBJECT: REPORT ON AIR HANDLING UNIT

There has been several incidents involving the unit which has resulted in water damage to the computer systems located below the air handler.

The occurrences yesterday was caused when a valve was closed creating condensation to be forced through a humidifier element into the supply air duct. Water then leaked from the duct into the room below causing substantial damage to four disc-drive units.

To prevent recurrence of this type of damage, the following actions has been initiated by maintenance supervision:

- Each supervisor must ensure that their subordinates remove condensation valves to avoid unauthorized operation.
- Everyone must be made aware that they are responsible for closing condensation valves.
- The supply air duct, modified to carry away harmful sediments, are to be drained monthly.

Maintenance supervision recommend that air handlers not be installed above critical equipment. This will avoid the possibility of coil failure and water damage.

Capitalization

In the following memo, nothing has been capitalized. Capitalize those words requiring capitalization.

date: december 5, 2005
to: jordan cottrell
from: richard davis
subject: self-contained breathing apparatus (scba) and negative pressure respirator evaluation and fit-testing report

the evaluation and fit-testing have been accomplished. the attached list identifies the following:

- supervisors and electronic technicians who have used the scba successfully.
- the negative pressure respirators used for testing in an isoamyl acetate atmosphere.

fit-testing of waste management personnel will be accomplished annually, according to president chuck carlson. new supervisors and technicians will be fit-tested when hired.

all apex corporation personnel located in the new york district (12304 parkview lane) must submit a request form when requesting a respirator or scba for use. any waste management personnel in the north and south facilities not identified on the attached list will be fit-tested when use of scba is required. if you have any questions, contact chuck carlson or me (richard davis, district manager) at ext. 4036.

Grammar Quiz

The following sentences contain errors in spelling, punctuation, verb and pronoun agreement, sentence structure (fragments and run-ons), and modification. Circle the letter corresponding to the section of the sentence containing the error.

> **example**
>
> When you recieve the salary increase, your family will celebrate the occasion.
>   Ⓐ   B   Ⓒ

1. Each department manager should tell <u>his</u> subordinates to <u>advise</u> of any neg-
                            A   B
   ative <u>occurrences</u> regarding <u>in-house</u> training.
          C   D

2. The <u>lawyers'</u> new offices were similar to <u>their</u> former ones ____ the offices
      A   B   C
   were on a <u>quiet</u> street.
           D

3. New York City is <u>divided</u> into five <u>boroughs</u> <u>:</u> Manhattan, the Bronx,
               A   B   C
   Queens, Brooklyn, <u>and</u> Staten Island.
             D

4. Fashion consultants explain that <u>clothes</u> create a strong impression ____ so
                       A   B
   they advise executives ____ to <u>choose</u> wardrobes carefully.
           C   D

5. Everyone should make sure that <u>they are</u> well represented in union meetings<u>;</u>
                      A   B
   otherwise<u>,</u> management could become <u>too</u> powerful.
      C   D

6. The employment agency<u>,</u> <u>too</u> busy to return the telephone calls from
                A B
   prospective clients<u>,</u> <u>are</u> harming business opportunities.
            C D

7. The supervisory staff <u>are</u> making decisions based on scheduling <u>,</u> but all
              A   B
   <u>employees</u> want to <u>ensure</u> quality control.
     C   D

8. Because the price of cars has risen dramatically ____ most people keep <u>their</u>
                           A   B
   cars longer ____ to save <u>capital</u> expenditures.
      C   D

9. The <u>department's</u> manager heard that a merger was possible ____ but he de-
      A   B
   cided <u>to</u> keep the news <u>quiet</u>.
     C   D

10. The manager of the department believed that her <u>employee's</u> were excellent <u>,</u>
                                                                A   B

    but she decided that no raises could be given ____ because the price of stocks
                                                                C

    <u>was</u> falling.
    D

11. The detailed report from the audit ____ of the department gave good <u>advise</u>
                                              A   B

    about how to restructure <u>,</u> so we are <u>all ready</u> to do so.
                               C   D

12. When my boss and <u>I</u> looked at the books <u>,</u> we found these problems <u>;</u> lost in-
                       A   B   C

    voices, unpaid bills <u>,</u> and late payments.
                    D

13. The most dedicated staff members <u>are</u> <u>accepted</u> for the on-site training ses-
                                            A   B

    sions because <u>:</u> they are responsive to criticism, represent the <u>company's</u> fu-
                         C   D

    ture, and strive for improvements.

14. Despite her assurances to the contrary <u>,</u> there <u>is</u> still three unanswered ques-
                                              A   B

    tions <u>:</u> who will make the payments, when will these payments occur, and
          C

    why is <u>there</u> a delay?
               D

15. The reputation of many <u>companies</u> often <u>depend</u> on one employee <u>who</u>
                               A   B   C

    <u>represents</u> that company.
    D

16. Either my monthly activity report or my <u>year-end</u> report <u>are</u> due today <u>,</u> but
                                               A   B   C

    my computer is broken <u>;</u> therefore, I need to use yours.
                               D

17. Today's American <u>manpower</u> <u>,</u> according to many <u>foreign</u> governments,
                               A   B   C

    <u>suffers</u> from lack of discipline.
    D

18. Rates are increasing next year because <u>:</u> fuel, <u>maintenance,</u> and insurance <u>are</u>
                                          A   B   C

    all higher <u>than</u> last year.
               D

19. Many colleges have <u>long-standing</u> football rivalries <u>,</u> one of the most famous
                                   A   B

    ones <u>is</u> between KU and KSU <u>(two universities in Kansas)</u>.
          C   D

20. <u>Everyone</u> who <u>wants</u> to enroll in the business school should do so before
       A   B

    June if <u>they</u> can to <u>ensure</u> getting the best classes.
             C   D

21. Because John wanted high visibility for his two <u>businesses</u> , he paid top dol-
                                                    A   B
    lar ____ and spent long hours looking for appropriate <u>cites</u>.
        C   D

22. Many people apply for jobs at Apex , however , only a few <u>are</u> <u>accepted</u>.
                            A   B   C   D

23. Because most cars break down <u>occasionally</u> , all drivers should know how to
                             A   B
    change a flat tire , and how to signal for <u>assistance</u>.
              C   D

24. Arriving on time, working <u>diligently</u>, and closing the office securely—every-
                          A   B
    one <u>needs</u> to know that <u>they are</u> responsible for these job duties.
        C   D

25. Mark McGwire not only hit the ball <u>further</u> <u>than</u> other players, but also he
                                     A   B
    hit more dingers than other players <u>who</u> hit <u>fewer</u> homers.
                                C   D

ACTIVITIES

REFERENCES

CHAPTER 1

Bacon, S. "Want to Get Ahead? Learn to Get Along." *The Kansas City Star,* July 25, 1993: F16.

"Best Practice Database." Best Practices, LLC, February 26, 2003. http://www.bestpracticedatabase.com.

"Business Identity." Crane's, February 13, 2004. http://www.crane.com/business/businessidentity/.

"Deming's 14 Points." Deming Website—The Deming Philosophy, June 3, 1998. http://deming.ces.clemson.edu/pub/den/deming_philosophy.htm.

"Empowering Your Employees." Six Sigma Consultants, February 26, 2003. http://www.dosixsigma.com/sixsigmadefinition.htm.

Hughes, Michael A. "Managers: Move from Silos to Channels." *Intercom* (March 2003): 9–11.

"Individual's and Teams' Roles and Responsibilities." GOAL/QPC, February 24, 2003. http://www.goalqpc.com/index.htm.

McNeill, Angie E. "Self-Directed Work Teams." *Intercom* (September/October 2000): 15–17.

Miller, Carolyn R., et al. "Communication in the Workplace." *Center for Communication in Science, Technology, and Management.* October 1996. http://www.chass.ncsu.edu/ccstm/pubs/no2/index.html.

"Pitney Bowes Study Finds U.S. Workers Less Overwhelmed Despite Increased E-mail Volumes." ebizChronicle.com, August 21, 2000. http://www.ebizchronicle.com/spl_reports/august/pitneybowes_email2.htm.

"Planning Job Choices." NaceWeb. January 15, 2004. http://www.naceweb.org/press/display.asp?year=2004&prid=184.

"SAE's Approach to Total Quality Management," February 24, 2003. http://www.sae.org/about/quality.htm.

"SGA." Bloom Consultancy, February 26, 2003. http://www.blomconsultancy.nl/SGA-english.htm.

"Six Sigma." GCL Management Consultants, February 26, 2003. http://www.gclconsultancy.co.uk/six_sigma.htm.

CHAPTER 2

Albers, Michael J. "Single Sourcing and the Technical Communication Career Path." *Technical Communication* 50 (August 2003): 336–343.

Campbell, Kim S., et al. "Leader-Member Relations as a Function of Rapport Management." *Journal of Business Communication* 40 (2003): 170–194.

Cargile-Cook, Kelli. "Usability Testing in the Technical Communication Classroom." *Intercom* (September/October 1999): 36–37.

Carter, Locke. "The Implications of Single Sourcing for Writers and Writing." *Technical Communication* 50 (August 2003): 317–320.

Quinn, R. E., et al. "A Competing Values Framework for Analyzing Presentational Communication in Management Contexts." *Journal of Business Communication* 28 (1991): 213–232.

Chapter 3

Adams, Rae, et al. "Ethics and the Internet." *Proceedings: 42nd Annual Technical Communication Conference.* April 23–26, 1995: 328.

Anderson-Hancock, Shirley A. "Society Unveils New Ethical Guidelines." *Intercom* 42 (July/August 1995): 6–7.

Barker, Thomas, et al. "Coming into the Workplace: What Every Technical Communicator Should Know—Besides Writing." *Proceedings: 42nd Annual Technical Communication Conference.* April 23–26, 1995: 38–39.

Bowman, George, and Arthur E. Walzer. "Ethics and Technical Communication." *Proceedings: 34th Annual Technical Communication Conference.* May 10–13, 1987: MPD–93.

Bremer, Otto A., et al. "Ethics and Values in Management Thought." *Business Environment and Business Ethics: The Social, Moral, and Political Dimensions of Management.* Edited by Karen Paul. Cambridge, MA: Ballinger, 1987: 61–86.

"Code for Communicators." Society for Technical Communication, March 10, 2004. http://www.stermc.org/resources/resource_code.htm.

"Email Statistics." MailiWant.com, March 2001. http://www.mailiwant.com/stats.html.

"Ethical Principles for Technical Communicators." The Society for Technical Communication, December 9, 2003. http://www.stc.org/ethical.asp.

Gerson, Steven M., and Sharon J. Gerson. "A Survey of Technical Writing Practitioners and Professors: Are We on the Same Page." *Proceedings: 42nd Annual Technical Communication Conference.* April 23–26, 1995: 44–47.

Girill, T. R. "Technical Communications and Ethics." *Technical Communication* 34 (August 1987): 178–179.

Guy, Mary E. *Ethical Decision Making in Everyday Work Situations.* New York: Quorum Books, 1990.

Harkness, Holly E. "Sarbanes-Oxley & New Opportunities." *Intercom* (January 2004): 16–18.

Hartman, Diane B., and Karen S. Nantz. "Send the Right Messages About E-Mail." *Training & Development* (May 1995): 60–65.

Johnson, Dana R. "Copyright Issues on the Internet." *Intercom* (June 1999): 17.

LeVie, Donald S. "Internet Technology and Intellectual Property." *Intercom* (January 2000): 20–23.

Mollison, Andrew. "U.S. Loses Its Lead in High School and College Grads." Cox Newspapers, November 6, 2001: 1–3.

Moore, Linda E. "Serving the Electronic Reader." *Intercom* (April 2003): 17.

Perlin, Neil. "Technical Communication: The Next Wave." *Intercom* (January 2001): 4–8.

Schulzrinne, Henning. "Electronic Mail (Email), Fax and Postal Mail," November 19, 2002. http://www.cs.columbia.edu/~hgs/internet/email.html.

"Summary of Sarbanes-Oxley Act of 2002." American Institute of Certified Public Accountants, January 8, 2004. http://www.aicpa.org/index.htm.

Turner, John R. "Ethics Online: Looking Toward the Future." *Proceedings: 42nd Annual Technical Communication Conference.* April 23–26, 1995: 59–62.

Wilson, Catherine Mason. "Product Liability and User Manuals." *Proceedings: 34th International Technical Communication Conference.* May 10, 1987: WE-68–71.

Chapter 4

"Around the World," March 26, 2003. http://www2.coca-cola.com/ourcompany/aroundworld.html.

Cardarella, Toni. "Business." *The Kansas City Star.* September 30, 2003. http://www.kansascity.com/mld/kansascity/business/6877765.htm?lc.

Courtis, John K., and Salleh Hassan. "Reading Ease of Bilingual Annual Reports." *Journal of Business Communication* 39 (October 2002): 394–413.

Flint, Patricia, et al. "Helping Technical Communicators Help Translators." *Technical Communication* 46 (May 1999): 238–248.

General Motors, March 28, 2003. http://www.gm.com.

Grimes, Diane Susan, and Orlando C. Richard. "Could Communication Form Impact Organizations' Experience with Diversity?" *Journal of Business Communication* 40 (January 2003): 7–27.

Horton, William. "The Almost Universal Language: Graphics for International Documents." *Technical Communication* 40 (November 1993): 682–693.

Hussey, Tim, and Mark Homnack. "Foreign Language Software Localization." *Proceedings: 37th International Technical Communication Conference.* May 20–23, 1990: RT-44–47.

Jordan, Katrina. "Diversity Training: What Works, What Doesn't, and Why?" *Civil Right Journal* (Fall 1999) 1.

King, Janice M. "The Challenge of Communicating in the Global Marketplace." *Intercom* 42.5 (May/June 1995): 1, 28.

McInnes, R. "Workforce Diversity: Changing the Way You Do Business." Diversity World, February 12, 2003. http://www.diversityworld.com/workforce_diversity.htm.

"Multilingual Features in Windows XP Professional," August 24, 2001. http://www.microsoft.com/windowsxp/pro/techinfo/planning/multilingual/default.asp.

Nethery, Kent. "Let's Talk Business," February 16, 2003. http://www.cuspomona.edu/~cljones/powerpoints/chap02/sld001.htm.

Rains, Nancy E. "Prepare Your Documents for Better Translation." *Intercom* 41.5 (December 1994): 12.

Sanchez, Mary. "KC hospitals seek to overcome language barriers." *The Kansas City Star,* January 7, 2003: A1.

Scott, Julie S. "When English Isn't English." *Intercom* (May 2000): 20–21.

Skettino, Michele. "America in the 21st Century: A Marketer's Guide to the Nation." *Interep Research Division.* 13–15. August 24, 2004. http://site.interep.com/apps/Research/resweb.nsf/bf25ab0f47ba5dd785256499006b15a4/dd35caf6283fb50585256a4c0058a200/$FILE/A21C001.pdf.

Skettino, Michele. *America in the 21st Century: A Marketer's Guide to the Nation,* a report published by Interep Research Division, no date, 13–15.

St. Amant, Kirk R. "Communication in International Virtual Offices." *Intercom* (April 2003): 27–28.

St. Amant, Kirk R. "Designing Web Sites for International Audiences." *Intercom* (May 2003): 15–18.

St. Amant, Kirk R. "International Integers and Intercultural Expectations." *Intercom* (May 1999): 27–28.

St. Amant, Kirk. "International Integers and Intercultural Expectations." *Intercom* (May 1999): 25–27.

Swenson, Lynne V. "How to Make (American) English Documents Easy to Translate." *Proceedings: 34th International Technical Communication Conference.* May 10, 1987: WE-193–195.

"Swiss Fight Encroachment of English." *The Kansas City Star,* December 7, 2002: A16.

Walmer, Daphne. "One Company's Efforts to Improve Translation and Localization." *Technical Communication* 46 (May 1999): 230–237.

Weiss, Edmund H. "Twenty-five Tactics to 'Internationalize' Your English." *Intercom* (May 1998): 11–15.

"What Is the 'Business Case' For Diversity?" Society for Human Resource Management, February 12, 2003. http://www.shrm.org/diversity/businesscase.asp.

CHAPTER 5

Broughton, Philip Delves. "Boss's Angry Email Sends Shares Plunging." *London Daily Telegraph,* April 6, 2001.

Christopher, Glen. "eMail Etiquette." *Update*. September 2001: 1–3.

Cobbs, Chris. "r u using 2 much IM style in ur writing at school?" *The Kansas City Star,* November 6, 2002: F5.

"E-mail Etiquette Guidelines for *Connecticut State Government.*" July 29, 1999. http://www.state.ct.us/cmac/policies/netiqu.htm.

"E-mail Etiquette," 1997. http://srd.yahoo.com/drst/81419/ *http://www.iwillfollow.com/email.htm.

"E-mail Statistics." *MailiWant.com,* March 2001. http://www.mailiwant.com/ stats.html.

Hoffman, Jeff. "Instant Messaging in the Workplace." *Intercom* (February 2004): 16–17.

Jesdanun, Anick. "Buying 'Stamps' for E-mail May Slow Down Spammers." *The Kansas City Star,* March 8, 2004: B6.

Kim, H. Young. "Looking Toward the Electronic Future in the Classroom: Using Electronic Mail." *Proceedings: 42nd Annual International Technical Communication Conference.* (April 1995): 51–54.

MacIntyre, John. "E-mail Is Changing Business and Driving Storage Growth." EMC^2, November 13, 2002. http://www.emc.com/information_generation/ in_depth_archive/12112000_posthaste.jsp.

Miller, Carolyn, et al. "Communication in the Workplace." *Center for Communication in Science, Technology, and Management.* (October 1996).

Munter, Mary, et al. "Business E-mail: Guidelines for Users." *Business Communication Quarterly* 66 (March 2003): 29.

"Pitney Bowes Study Finds U.S. Workers Less Overwhelmed Despite Increased E-mail Volumes." ebizChronicle.com, August 21, 2000. http://www.ebizchronicle.com/ spl_reports/august/pitneybowes_email2.htm.

Rodenbough, Danielle. "Keeping an Eye and Ear on Employee Activity." *Small Business Monthly* (September 2003): 29.

Rose, Lance. *Netlaw: Your Rights in the Online World.* Berkeley, CA: McGraw-Hill, 1995.

Schwabach, Bob. "You've got e-mail, and how: Billions every day." *The Kansas City Star,* February 16, 1999: D13.

Shipman, John. "ASCII." January 1, 1995. http://www.nmt.edu/tcc/help/g/ascii.html.

"The Transition to General Management." Harvard Business School, 1998. http://www.hbs.edu/gm/index.html.

Wickman, Rosanne B. "Memo: Go easy on e-mail." *The Kansas City Star,* Steptember 10, 2000: D1.

Wong, Edward. "Forwarded E-mail Has Executive Under Fire." *New York Times,* April 9, 2001. http://www.siliconinvestor.com/stocktalk/msg.gsp?msgid=15635409.

CHAPTER 7

"Benefits of Being Online," Resumes Online, March 25, 1998. http://www.kbweb.com/ resume/.

Bloch, Janel M. "Online Job Searching: Clicking Your Way to Employment." *Intercom* (September/October 2003): 11–14.

Dikel, Margaret F. "The Online Job Application: Preparing Your Resume for the Internet." September 1999. http://www.dbm.com/jobguide/eresume.html.

Dixson, Kirsten. "Crafting an E-mail Resume." *BusinessWeek Online,* November 13, 2001. http://www.businessweek.com.

Drakeley, Caroline A. "Viral Networking: Tactics in Today's Job Market." *Intercom* (September/October 2003): 5–7.

Hansen, Randall S. "Scannable Resume Fundamentals: How to Write Text Resumes," December 9, 2003. http://www.quintcareers.com/scannable_resumes.html.

Hartman, Peter J. "You Got the Interview. Now Get the Job!" *Intercom* (September/October 2003): 23–25.

Kallick, Rob. "Research Pays Off During Interview." *The Kansas City Star*, March 23, 2003: D1.

Kendall, Pat. "Electronic Resumes," December 9, 2003. http://www.reslady.com.

LeVie, Donald S., Jr. "Resumes: You Can't Escape." *Intercom* (April 2000): 8–11.

Lorek, L. A. "Searching On-line." *The Kansas City Star*, August 23, 1998: D1.

McNair, Catherine. "New Technologies and Your Resume." *Intercom* (June 1997): 12–14.

Mulvaney, Mary Kay. "The Information Interview: Bridging College and Beyond." *Business Communication Quarterly* 66 (September 2003): 66–72.

"Polishing Your ASCII Text Resume," December 8, 2003. http://www.job-hunt.org.

"Powerful Resume as Easy as A-B-C." *The Kansas City Star*, April 13, 2003: D1.

Ralston, Steven M., et al. "Helping Interviewees Tell Their Stories." *Business Communication Quarterly* 66 (September 2003): 8–22.

Robart, Kay. "Submitting Resumes via E-Mail." *Intercom* (July/August 1998): 12–14.

"Scannable Resume," December 9, 2003. http://jobsearchtech.about.com.

Skarzenski, Emily. "Tips for Creating ASCII and HTML Resumes." *Intercom* (June/July 1996): 17–18.

Stafford, Diane. "Show Up Armed with Answers." *The Kansas City Star*, August 3, 2003: L1.

Stern, Linda. "New Rules of the Hunt." *Newsweek* (February 17, 2003): 67.

CHAPTER 8

Benson, Philippa J. "Writing Visually: Design Considerations in Technical Publications." *Technical Communication* 32 (Fourth Quarter 1985): 35–39.

Everson, Larry. "Recent Trends in Technical Writing." *Technical Communication* (Fourth Quarter 1990): 396–398.

Keyes, Elizabeth. "Typography, Color, and Information Structures." *Technical Communication* 40 (November 1993): 638–654.

Schriver, Karen A. "Quality in Document Design." *Technical Communication* (Second Quarter 1993): 250–251.

Southard, Sherry G. "Practical Considerations in Formatting Manuals." *Technical Communication* 35 (August 1988): 173–178.

Watzman, Suzanne. "The Approachable Page." *Proceedings: 34th International Technical Communication Conference*. May 10, 1987: VC-85–86.

Watzman, Suzanne. "Visual Literacy and Document Productivity." *Proceedings: 34th International Technical Communication Conference*. May 10, 1987: ATA-48–49.

Wise, Mary R. "Using Graphics in Software Documentation." *Technical Communication* 40 (November 1993): 677–681.

CHAPTER 9

Horton, William. "The Almost Universal Language: Graphics for International Documents." *Technical Communication* (November 1993): 682–693.

Reynolds, Michael, and Liz Marchetta. "Color for Technical Documents." *Intercom* (April 1998): 5–7.

CHAPTER 10

Dorazio, Pat. Interview. June 2000.

LeVie, Donald S., Jr. "Contracting: Flat Fee or Hourly Rates?" *Intercom* (September/October 2000): 18–21.

CHAPTER 12

Daughtery, Shannon. "The Usability Evaluation: A 'Discount' Approach to Usability Testing." *Intercom* (December 1997): 16–20.

Dorazio, Pat. Interview. June 2000.

Allen, Douglas. Executive Director of Information Services at Johnson County Community College. Interview. April 7, 1998.

Berners-Lee, Tim. "Realising the Full Potential of the Web." *Technical Communication* 46 (February 1999): 79–82.

Berst, Jesse. "Seven Deadly Web Site Sins." ZDNet, January 30, 1998. http://www.zdnet.com/anchordesk/story/story_1716.html.

Black & Veatch. April 15, 1998. http://www.bv.com.

Byrne, DiAnn. "Marketing on the World Wide Web: How to Get There and What to Do." *Intercom* (January 1996): 8–9, 43.

Dorazio, Pat. Interview. June 2000.

Eddings, Earl, Kim Buckley, Sharon Coleman Bock, and Nathaniel Williams. Software Documentation Specialists at PDA. Interview. January 12, 1998.

"E-Mail." *PC Webopedia*, April 24, 1997. http://www.pcwebopedia.com/e_mail.htm.

"Email: the ideal marketing tool." Nua Internet Surveys, December 24, 2000. http://www.qdata.net/fullnews.asp?newsid=583.

"E-mail Etiquette," 1997. http://www.iwillfollow.com/email.htm.

"E-Mail Etiquette Guidelines for Connecticut State Government," July 29, 1999. http://www.state.ct.us/cmac/policies/netiqu.htm.

"E-mail Statistics." MailiWant.com. http://www.mailiwant.com/stats.html. March 2001.

"Extranet." Whatis.com, January 13, 1998. http://whatis.com/extranet.htm.

"Firewall." *PC Webopedia*, January 23, 1998. http://www.pcwebopedia.com/firewall.htm.

Fisher, Sharon, et al. *Riding the Internet Highway: Deluxe Edition*. Indianapolis, IN: New Riders, 1994.

"FNC Resolution: Definition of 'Internet,'" October 30, 1995. http://www.fnc.gov/Internet_res.html.

Freeman, Lawrence H., and Terry R. Bacon. *Shipley and Associates Style Guide*. Bountiful, UT: Shipley Associates, 1990.

"From Tech Writer to Web Writer." Applied Communications Group, April 22, 1998. http://www.acgtech .com/content/write_to_web.htm.

Gateway 2000. February 4, 1998. http://www.gw2k.com.

Goldenbaum, Don. "Writing for Online—Do's and Don'ts." *Making Useful Software: A Digest of Software Development News and Trends* (Summer 1999): 1–2.

Goldenbaum, Don, and George Calvert. Applied Communications Group. Interview. January 10, 1998.

Hartman, Diane B., and Karen S. Nantz. "Send the Right Message About E-Mail." *Training & Development* (May 1995): 60–65.

Hemmi, Jane A. "Differentiating Online Help from Printed Documentation." *Intercom* (July/August 2002): 10–12.

Henselmann, Mary Anne. "Designing an Online Help System Before the Interface Is Ready." *Society for Technical Communication: 44th Annual Conference, 1997 Proceedings*. May 1997: 475–476.

Houten-Kemp, Mary. "E-Mail Tips: E-mail Help for the Newcomer," 1999. http://everythingemail.net/email_help_tips.html.

"Human-Computer Interaction." Rensselaer Polytechnic Institute: Academic Programs, April 21, 1998. http://www.llc.rpi.edu/acad/hcicert.htm.

"Internet." *PC Webopedia*, April 24, 1997. http://www.pcwebopedia.com/Internet.htm.

"Internet Traffic Growing Quickly." *The Kansas City Star*, April 16, 1998: B1, B9.

"Intranet." Whatis.com, January 8, 1998. http://whatis.com/intranet.htm.

"Intranet Applications." The Complete Intranet Resource, January 8, 1998. http://intrack.com/intranet/iapp.

Kim, H. Young. "Looking Toward the Electronic Future in the Classroom: Using Electronic Mail." *Proceedings: 42nd Annual International Technical Communication Conference.* April 1995: 51–54.

Krol, Ed. *The Whole Internet: User's Guide & Catalog.* Sebastopol, CA: O'Reilly & Associates, 1992.

Laquey, Tracy. *The Internet Companion: A Beginner's Guide to Global Networking.* New York: Editorial, 1993.

Legeros, Michael. "Etiquette and Email: Rules for Online Behavior." *Intercom* 42 (July/August 1995): 10–11.

Mazur, Beth. "Coming to Grips with WWW Color." *Intercom* (February 1997): 4–5.

McAlpine, Rachel. "Passing the Ten-Second Test." Wise-Women.com, April 22, 2002. http://www.wise-women.org/features/tenseconds/index.shtml.

McGowan, Kevin S. "The Leap Online." *Intercom* (September/October 2000): 22–25.

Moore, Linda E. "Serving the Electronic Reader." *Intercom* (April 2003): 16–17.

"Optin Email Advertising." *Submit Express, Newsletter #20,* March 31, 2000. http://srd.yahoo.com/goo/%22number+of+email+users%22/18/*http://submitexpress.com/newsletters/march_31_00.html.

Pratt, Jean A. "Where Is the Instruction in Online Help Systems." *Technical Communication* (February 1998): 33–37.

Rogers, David J., and Monica C. Peri. "E-mail and Tattoos: A Primer on Netiquette." *Intercom* (December 2000): 10–11.

Rose, Lance. *Netlaw: Your Rights in the Online World.* Berkeley, CA: McGraw-Hill, 1995.

Schwabach, Bob. "You've Got E-mail, and How: Billions Every Day." *The Kansas City Star,* February 16, 1999: D13.

Scott, Michon M. "Learning to Make Web Pages the Hard Way." *Intercom* (January 1998): 16–17.

Sherwood, Kaitlin Duck. "A Beginner's Guide to Effective Email," February 24, 2000. http://www.webfoot.com/advice/email.top.html?Yahoo.

Shipman, John. "ASCII," January 1, 1995. http://www.nmt.edu/tcc/help/g/ascii.html.

Shuler, Lynn McGuire. "Critiquing a Web Site." *Intercom* (November 1999): 21–22.

Spyridakis, Jan H. "Guidelines for Authoring Comprehensible Web Pages and Evaluating Their Success." *Technical Communication* 47 (August 2000): 359–382.

St. Amant, Kirk R. "Designing Web Sites for International Audiences." *Intercom* (May 2003): 15–18.

Stevens, Dawn M. "101 Standards for Online Communication." *Society for Technical Communication: 44th Annual Conference, 1997 Proceedings.* May 1997: 410–412.

Timpone, Donna. "Help! Six Fixes to Improve the Usability of Your Online Help." *Society for Technical Communication: 44th Annual Conference, 1997 Proceedings.* May 1997: 306.

Wagner, Carol A. "Using HCI Skills to Create Online Message Help." *Society for Technical Communication: 44th Annual Conference, 1997 Proceedings.* May 1997: 35.

"Web Sites that Work: Roger Black's Six Rules for Web Design." FastCompany.com., January 3, 2004. http://www.fastcompany.com/online/10/rogerblack.html.

Wickman, Rosanne B. "Memo: Go Easy on E-mail." *The Kansas City Star,* September 10, 2000: D1.

Wilkinson, Theresa. "How to Increase Performance on a Web Site." *Intercom* (January 2000): 38–39.

Wilkinson, Theresa A. "Web Site Planning." *Intercom* (December 1997): 14–15.

Williams, Thomas R. "Guidelines for Designing and Evaluating the Display of Information on the Web." *Technical Communication* 47 (August 2000): 383–396.

"World Wide Web." *PC Webopedia,* January 26, 1998. http://www.pcwebopedia.com/World_Wide_Web.htm.

Yeo, Sarah C. "Designing Web Pages That Bring Them Back." *Intercom* (March 1996): 12–14.

Zubak, Cheryl Lockett. "Developing Help for the Web: Designs, Trends, Strategies." *Intercom* (January 2001): 9–15.

Zubak, Chery L. "Choosing a Windows Help Authoring Tool." *Intercom* (January 1996): 10–11, 36.

CHAPTER 14

Hacker, Diane. *Research and Documentation in the Electronic Age.* Bedford Books, 1997. http://www.bedfordbooks.com/rd/index.html.

MLA: The Modern Language Association of America on the Web. SilverPlatter International, NV, 1997. http://www.mla.org.

CHAPTER 16

Robert, Henry M. "Robert's Rules of Order Revised." *Constitution Society.* September 15, 2004. http://www.constitution.org/rror/rror—00.htm.

CHAPTER 17

"Allocation of Technology Needs." Survey. May 1, 2005.

Cooper, Jack. AAA Computers. Interview. May 15, 2005.

"Hardware/Software Specifications and Costs." *PCWorld.* May 12, 2005. http://www.pcworld.com/.

Norton, Susan. *Biostaffing Annual Report.* January 15, 2005. http://www.Biostaff.com/Annual report.htm.

Zohorsky, Darrell. "The Winning Elements of Business Proposals." *About.com.* September 20, 2004. http://sbinformation.about.com/cs/bizlettersamples/a/proposal.htm.

CHAPTER 18

Buffet, Warren. "Preface." In *A Plain English Handbook: How to Create Clear SEC Disclosure Documents,* by the office of Investor Education and Assistance, U.S. Securities and Exchange Commission, 1–2. Washington, D.C.: U.S. Securities and Exchange Commission, August 1998.

"Communications: Oral Presentations." University of Surrey. September 18, 2004. http://www.surrey.ac.uk/Skills/pack/pres.html.

Keller, Julia. "Is PowerPoint the Devil?" *Chicago Tribune.* September 20, 2004. http://www.siliconvalley.com/mld/siliconvalley/5004120.htm.

O'Brien, Tim. "In Front of a Group." *The Kansas City Star,* July 15, 2003: C5.

Starr, Linda. Education World 2000, June 29, 2003. http://www.educationworld.com @tech/tech013.shtml.

"Tips and Techniques." AT&T, August 12, 2003. http://www.att.com/virtualmeetings/tips_audiobasics.html.

"Top Ten Tips for Using Voicemail Professionally and Effectively." Inserve.com, July 14, 2003. http://www.inserve.com/r_VmailTips.html.

Warfield, Anne. "Do You Speak Body Language?" *T+D Magazine—American Society for Traning & Development.* April 2001: 1.

"What Is Videoconferencing?" VideoBridging.com, July 14, 2003. http://www.videobridging.com/basics_of_videoconferencing.html.

INDEX